P9-AGA-605

The Biomarker Guide

Interpreting Molecular Fossils in Petroleum and Ancient Sediments

Kenneth E. Peters

Chevron Overseas Petroleum Inc.
San Ramon, California

J. Michael Moldowan

Chevron Oil Field Research Company
Richmond, California

Prentice Hall, Englewood Cliffs, New Jersey 07632

Library of Congress Cataloging-in-Publication Data

PETERS, KENNETH E.
The biomarker guide : interpreting molecular fossils in petroleum and ancient sediments / Kenneth E. Peters, J. Michael Moldowan.
p. cm.
Includes bibliographical references and index.
ISBN 0-13-086752-7
1. Petroleum—Prospecting. 2. Biogeochemical prospecting.
3. Biochemical markers. I. Moldowan, J. M. (J. Michael).
II. Title.
TN271.P4P463 1993
622'.1828—dc20 91-43815
CIP

Editorial/production supervision
and interior design: *Jean Lapidus*
Cover design: *Joseph DiDomenico*
Source: an adaptation of the drawing by
Annick Restlé courtesy of Pierre Albrecht
Copy editor: *Maria Caruso*
Pre-press buyer: *Mary Elizabeth McCartney*
Manufacturing buyer: *Susan Brunke*
Acquisition editor: *Betty Sun*
Editorial assistant: *Maureen Diana*

The publisher offers discounts on this book when ordered in bulk quantities. For more information, write: Special Sales/Professional Marketing, Prentice Hall, Professional & Technical Reference Division, Englewood Cliffs, NJ 07632.

Printed in the United States of America

10 9 8 7 6 5 4 3 2 1

ISBN 0-13-086752-7

PRENTICE-HALL INTERNATIONAL (UK) LIMITED, *London*
PRENTICE-HALL OF AUSTRALIA PTY. LIMITED, *Sydney*
PRENTICE-HALL CANADA, INC., *Toronto*
PRENTICE-HALL HISPANOAMERICANA, S.A., *Mexico*
PRENTICE-HALL OF INDIA PRIVATE LIMITED, *New Delhi*
PRENTICE-HALL OF JAPAN, INC., *Tokyo*
SIMON AND SCHUSTER ASIA PTE. LTD., *Singapore*
EDITORA PRENTICE-HALL DO BRASIL, LTDA., *Rio de Janeiro*

Dedicated to

Vanessa and Mary

Contents

EXECUTIVE SUMMARY ix

HOW TO USE THE "GUIDE" xi

ACKNOWLEDGMENTS xv

1 INTRODUCTION TO BIOLOGICAL MARKERS 1

1. Introduction 1

1.1 Diagenesis, Catagenesis, Metagenesis, and Biodegradation 5

1.2 Organic Chemistry 10

1.2.1 Alkanes—the sigma bond, 11

1.2.2 Alkenes—the pi bond, 12

1.2.3 Aromatics—benzene, 13

1.2.4 Structure notation, 14

1.2.5 Acyclic alkanes, 15

1.2.6 Cyclic alkanes, 15

1.2.7 The "isoprene rule," 17

1.2.8 Stereochemistry and nomenclature, 25

1.2.9 Stereoisomerization, 32

1.2.10 Stereochemistry of selected biomarkers, 34

1.3 Examples of Biomarkers, Their Abundance, and Precursors 40

1.3.1 Bacteriohopane, 42

1.3.2 Cholestane, 43

1.3.3 Porphyrins and acyclic diterpanes, 45

2 FUNDAMENTALS OF BIOMARKER SEPARATION AND ANALYSIS 48

2. Fundamentals 48

2.1 Supporting Geochemistry 48

2.2 Organization of the Chevron Biomarker Laboratory 49

2.3 Sample Availability, Quality, and Selection 50

2.3.1 Rock Samples, 51

2.3.2 Oil Samples, 52

2.4 Sample Cleanup and Separations 52

2.4.1 Column chromatography (cleanup), 53

2.4.2 High-performance liquid chromatography (separation), 54

2.4.3 Internal standards and preliminary analyses, 56

2.5 Gas Chromatography/Mass Spectrometry 57

2.5.1 Gas chromatography in GCMS, 59

2.5.2 Mass spectrometry in GCMS, 61

2.5.3 GCMS operating modes, 68

2.5.4 Mass spectra, compound identification, and quantitation, 89

2.5.5 Examples GCMS data problems, 97

2.5.6 Error analysis, 107

3 GUIDELINES TO INTERPRETATION 110

3. Guidelines 111

3.1 Correlation, Source, and Depositional Environment 111

3.1.1 Concepts, 111

3.1.2 Nonbiomarker parameters for correlation, source, and depositional environment, 115

3.1.3 Biomarker parameters for correlation, source, and depositional environment, 140

3.1.4 Geochemical characteristics of petroleum from carbonate versus shale source rocks, 208

3.1.5 Marine versus terrestrial organic matter input, 209

3.2 Maturation 210

3.2.1 Concepts, 210

3.2.2 Nonbiomarker maturity parameters, 213

3.2.3 Criteria for ideal biomarker maturity parameters, 220

3.2.4 Biomarker maturity parameters, 221

3.3 Biodegradation 252

3.3.1 Concepts, 252

3.3.2 Nonbiomarker biodegradation parameters, 253

3.3.3 Biomarker biodegradation parameters, 255

3.3.4 Effects of biodegradation on determination of maturity and correlation, 264

4 PROBLEM AREAS AND FURTHER WORK 266

4. Problem areas 266

4.1 Migration 266

4.2 Kinetics 269

4.3 Correlation 270

4.3.1 Test for indigenous bitumen, 270

4.3.2 Hydrous pyrolysis of immture source rocks, 271

4.4 Isotopic Signatures of Individual Biomarkers 271

4.5 Age Determinations from Biomarkers 273

4.6 Source Input and Depositional Environment 274

GLOSSARY **276**

REFERENCES **309**

INDEX **347**

Executive Summary

Biological markers (biomarkers) are complex "molecular fossils" derived from once living organisms. Because biological markers can be measured in both oils and source rock bitumens, they provide a method to relate the two (correlation) and can be used to interpret the characteristics of the petroleum source rock(s) when only oils are available. Biomarkers are also useful because they can provide information on the organic matter in the source rock (source), environmental conditions during its deposition and burial (diagenesis), the thermal maturity experienced by a rock or oil (catagenesis), the degree of biodegradation, some aspects of source rock mineralogy (lithology), and age.

Biomarker and nonbiomarker geochemical parameters are best used together to provide the most reliable geologic interpretations to help solve exploration problems. Prior to biomarker work, oil and rock samples are typically "screened" using nonbiomarker analyses. The strength of biomarker parameters is that they provide more detailed information needed to answer exploration questions on source, depositional environment, thermal maturity, and the biodegradation of oils, than by using nonbiomarker analyses alone.

Distributions of biomarkers can be used to correlate oils and bitumens. For example, C_{27}-C_{28}-C_{29} steranes or monoaromatic steroids distinguish oil-source families with high precision. Cutting-edge analytical techniques, such as linked-scan gas chromatography-mass spectrometry-mass spectrometry (GCMSMS), provide sensitive measurements for correlation of light oils and condensates, where biomarkers are typically in low concentration.

Different depositional environments are characterized by different assemblages of organisms and biomarkers. Commonly recognized classes of organisms include bacteria, algae, marine algae, and higher plants. For example, some rocks and related oils contain botryococcane, a biomarker produced by the lacustrine, colonial alga *Botryococcus braunii*. *Botryococcus* is an organism that thrives only in lacustrine environments. Marine, terrestrial, deltaic, and hypersaline environments also show characteristic differences in biomarker composition.

The distribution, quantity, and quality of organic matter (organic facies) are factors which help determine the hydrocarbon potential of a petroleum source rock. Optimal preservation of organic matter during and after sedimentation occurs in oxygen-depleted (anoxic) depositional environments, which commonly lead to rich, oil-prone petroleum source rocks. Various biomarker parameters, such as the C_{35}-homohopane index, can indicate the degree of oxicity under which marine sediments were deposited.

Biomarker parameters are an effective means to rank the relative maturity of petroleums throughout the entire oil-generative window. The rank of a petroleum can be correlated with regions within the oil window (e.g., early, peak, or late generation). This information can provide a clue to the quantity and quality of the oil that may have been generated and, coupled with quantitative petroleum conversion measurements (e.g., basin modeling programs), can help evaluate the timing of petroleum migration.

Biomarkers can be used to determine source and maturity, even for biodegraded oils. Ranking systems have been developed based on the loss of *n*-paraffins acyclic isoprenoids, steranes, terpanes, and aromatic steroids during biodegradation.

Biomarkers in oils provide information on the lithology of the source rock. For example, the absence of rearranged steranes can be used to indicate petroleums derived from clay-poor (usually carbonate) source rocks. Abundant gammacerane in some petroleums appears linked to hypersaline (evaporitic) conditions during deposition of the source rock.

Biomarkers can provide information on the age of the source rock for petroleum. For example: (1) oleanane is a biomarker characteristic of angiosperms (flowering plants) found only in Tertiary and Cretaceous rocks and oils, (2) dinosterane is a marker for marine dinoflagellates, possibly distinguishing Mesozoic and Tertiary from Paleozoic source input, and (3) unusual distributions of *n*-paraffins and cyclohexylalkanes are characteristic of *Gloeocapsamorpha prisca* found in early Paleozoic samples.

Continued growth in the geologic applications of biomarker technology is anticipated, particularly in the areas of age-specific biomarkers, the use of biomarkers to indicate source organic matter input and sedimentologic conditions, correlation of oils and rocks, and understanding the global cycle of carbon. New developments in analytical instrumentation and methods and the use of biomarkers in understanding petroleum migration and kinetics are likely. Finally, early investigations suggest that biomarkers may become important tools for understanding oil production and environmental problems.

How to Use the "Guide"

The Guide was written for a diverse audience which might include: (1) students and exploration geologists with little chemical background, but a desire for increased general knowledge of what biomarkers can do, (2) geochemical coordinators with both specific and general questions about which biomarker and/or non-biomarker parameters might best answer a regional exploration problem, (3) geochemists with extensive chemical backgrounds, who require detailed information on specific parameters or methodology, (4) managers, who require a concise explanation for a particular term used in a report, (5) refinery process chemists requiring a more detailed knowledge of petroleum, and (6) environmental scientists interested in a technology useful for characterizing petroleum in the environment.

As a consequence of the diverse audience for which it is intended, the guide has been designed to be used in two modes:

1. a "learning mode," involving reading of successive sections in order to build knowledge (primarily for those readers who want an integrated summary and overview of biomarker and related technologies),
2. a "selected access mode," involving use of the table of contents and the index to look up detailed information on specific parameters or procedures.

To satisfy the needs of the diverse audience, we have chosen to prepare the Guide in a logical sequence which progressed from basic to more complex discussion. An alternate approach we considered was to place discussion of organic chemistry and/or fundamentals of biomarker separation and analysis in an appendix. For this work, we decided that the logical sequence of discussion is a superior choice because an understanding of these introductory chapters is *essential* for proper use of the parameters. For example, if the reader does not understand the concepts of stereochemistry and nomenclature (Chapter 1), the discussion of thermal maturity based on sterane isomerization reactions (Chapter 3) will be unclear. Similarly, the explanation of mass analysis (Chapter 2) is a requisite for understanding the difference between mass spectra and mass chromatograms discussed later in the text (Chapter 3).

The text in each chapter is supplemented by numerous references to related sections in the Guide and to the literature. Various parts of the Guide, such as "notes" and portions of other sections, are presented in italics to alert the reader that detailed discussion follows.

The following is a brief overview of each chapter in the Guide.

1 INTRODUCTION TO BIOLOGICAL MARKERS

The discussion summarizes the origin and processes affecting the distribution, preservation, and alteration of biomarkers in sedimentary rocks. A brief overview of organic chemistry includes explanations of structural nomenclature and stereochemistry necessary for understanding biomarker parameters. The chapter concludes with examples of the structures and nomenclature for several biomarkers, their precursors in living organisms, and geologic alteration products. Readers who feel confident of their grasp of these vital fundamentals may wish to skip Chapter 1.

2 FUNDAMENTALS OF BIOMARKER SEPARATION AND ANALYSIS

This chapter describes the multidisciplinary and multiparameter approach to geochemical studies, the relationship between biomarker and nonbiomarker analyses, and how to select (screen) samples for analysis. The concepts of gas chromatography (GC; separation) and mass spectrometry (MS; analysis) are explained. Many of these fundamentals, such as the difference between a mass chromatogram and a mass spectrum, or between selected ion and linked-scan modes of analysis, are critical to understanding later discussion of biomarker parameters.

Several topics in italics (including analytical procedures, internal standards, examples of GCMS data problems, and error analyses) are necessary fundamentals for understanding later discussions of biomarker parameters in Chapters 3 and 4. If the reader is comfortable with the topics in italics, they can be skipped.

3 GUIDELINES TO INTERPRETATION

The chapter is divided into three parts, each beginning with a brief discussion of concepts and how biomarker and nonbiomarker analyses are used in the following areas of petroleum geochemistry:

1. Correlation, source, and depositional environment
2. Maturation
3. Biodegradation

In Sections 3.1 and 3.2, the biomarker parameters are arranged in groups of related compounds in the order: terpanes, steranes, aromatic steroids, and porphyrins. Within each group the parameters are roughly arranged according to the frequency of their use in biomarker studies, with the more commonly used parameters first. Critical information on the frequency of use of the parameter, its specificity for answering the particular problem, and the means for its measurement are highlighted in bold print above the discussion for each biomarker parameter. In Section 3.3, the parameters are arranged in groups showing increasing resistance to biodegradation.

Most readers will probably use Chapter 3 by finding selected parameters in the Index, Glossary, or Contents. For this reason, discussions of each biomarker parameter contain extensive references to the literature and text. The reader should remember that discussion of a parameter in one part of Chapter 3 is typically supplemented by additional discussion in the other two parts. For example, oleanane is discussed in Sections 3.1, 3.2, and 3.3 because it can provide information useful for: (1) correlation, source, depositional environment, (2) maturation, and (3) biodegradation.

4 PROBLEM AREAS AND FURTHER WORK

This chapter describes areas requiring further research, including the application of biomarkers to migration, the kinetics of petroleum generation, correlation, depositional environments, and age determination. Readers who feel they have samples or ideas that might contribute to ongoing research efforts in these areas are encouraged to contact the authors.

PURPOSE

This Guide provides a comprehensive discussion of the basic principles of biomarkers, their relationships with other parameters, and their applications to studies of maturation, correlation, source input, depositional environment, and biodegradation of the organic matter in petroleum source rocks and reservoirs. The Guide was prepared for a broad audience including students, company exploration geologists, geochemical coordinators, and geochemists for several reasons.

1. Biomarker geochemistry is a rapidly growing discipline with important worldwide applications to petroleum exploration and production.
2. Biomarker parameters are becoming increasingly prominent in exploration and production reports.
3. Different parameters are used within the industry, academia, service laboratories, and the literature.
4. The quality of biomarker data and interpretation can vary considerably depending on its source.

The objective of the Guide is to provide a single, concise source of information on the various biomarker parameters and to create general guidelines for the use of selected parameters. An important aim of this text is to clarify the relationships between biomarker and other geochemical parameters and to show how they can be used together to solve problems. The Guide is not intended to teach exploration geologists how to interpret raw biomarker data. This is a job for a biomarker specialist with years of training in instrumentation and organic chemistry. Such training cannot be provided by a "crash course" or "cookbook" approach without the consequence of serious interpretive errors and a tarnished view of the applicability of biomarkers in general.

A final objective of the Guide is to impart to each reader a feeling for the excitement and vigor of the new field of biomarker geochemistry. Expanding research efforts at geochemical laboratories worldwide have increased the rate of change and growth in our geochemical concepts. One rule of conduct appears appropriate: *A good geochemist is humble in his or her assertions*. Applications of many of the biomarker parameters in this book will undoubtedly improve with time and further research. We anticipate that more than a few readers of this book will be directly involved in making these improvements possible.

Acknowledgments

The examples of biomarker results from worldwide locations used to illustrate this book would not have been possible without the outstanding technical support of the Chevron biomarker team at Richmond. We thank F. Fago, P. Novotny, M. Pena, K. Smith, P. Lipton, and L. Wraxall for their continuing dedication to a challenging and sometimes daunting job. The quality of their work is a living tribute to the late W. K. Seifert and extends Chevron's reputation as the leader in biomarker technology.

We thank M. H. Koelmel, G. J. Demaison, N. H. Schultheis, E. L. Couch, and P. C. Henshaw for their vision and support in helping us to complete this major project.

The authors thank E. J. Gallegos (Chevron Reserach Company), C. Y. Lee, R. M. K. Carlson, P. Sundararaman, J. E. Dahl, and M. A. McCaffrey (Chevron Oil Field Research Company) for detailed discussions, technical assistance, and comments on the draft related primarily to biomarkers. Helpful reviews on these and other aspects of the draft were provided by R. P. Philp, M. Schoell, P. Albrecht, P. C. Henshaw, S. R. Jacobson, R. J. Hwang, and D. K. Baskin. We also acknowledge the excellent drafting, photographic, and stenographic support of A. Bycraft, D. Dumelle, D. K. Katayanagi, L. S. Rush, L. Steeley, J. J. Vesco, C. I. Fernandez, M. A. Kendall, M. T. Koziol, C. S. Murdock, A. Pacheco, S. M. Shields, L. A. Updike (Chevron Research and Technology Company), N. E. Breen, B. A. Maxon, and A. L. Whitlock (Chevron Oil Field Research Company), P. C. Garcia, V. W. Freeman, K. A. Morris, E. A. Lucero, and B. R. Barden (Chevron Overseas Petroleum Inc.).

ABOUT THE AUTHORS

Kenneth E. Peters attained B.A. and M.A. degrees in Geology from U.C. Santa Barbara and a Ph.D. in Geochemistry from UCLA in 1978. His experience includes over 14 years with Chevron Corporation; as a Research and Senior Research Geochemist at Chevron Oil Field Research Company in La Habra; Senior Research Geochemist with the Biomarker Group at Richmond; Geochemical Coordinator for Chevron U.S.A. Western Region, San Ramon; and currently Biomarker Coordinator for Chevron Overseas Petroleum, San Ramon, California. Dr. Peters teaches principles of petroleum geochemistry within Chevron and at universities, including California State University at Long Beach, the University of California at Berkeley, and Stanford University. He serves as Associate Editor for *Organic Geochemistry* and the *American Association of Petroleum Geologists Bulletin*. Dr. Peters and co-authors received the Organic Geochemistry Division of the Geochemical Society Best Paper Awards for publications in 1981 and 1989.

J. Michael Moldowan attained a B.S. degree in Chemistry from Wayne State University in Detroit, Michigan in 1968 and a Ph.D. in Chemistry from the University of Michigan in 1972. After a post-doctoral fellowship in marine natural products with Professor Carl Djerassi at Stanford University, he joined Chevron's Biomarker group in 1974, where he remains as Biomarker Section Supervisor. The Chevron Biomarker team, led by the late Dr. Wolfgang K. Seifert in the mid 1970s to the early 1980s, is largely credited with pioneering the application of biological marker technology to petroleum exploration. Dr. Moldowan served as Chairman-elect in 1985 and Chairman in 1986 of the Division of Geochemistry of the American Chemical Society. Dr. Moldowan and co-authors received the Organic Geochemistry Division of the Geochemical Society Best Paper Awards for publications in 1978 and 1989.

1

Introduction to Biological Markers

The discussion summarizes the origin and processes affecting the distribution, preservation, and alteration of biomarkers in sedimentary rocks. A brief overview of organic chemistry includes explanations of structural nomenclature and stereochemistry necessary for understanding biomarker parameters. The chapter concludes with examples of the structures and nomenclature for several biomarkers, their precursors in living organisms, and geologic alteration products. Readers who feel confident of their grasp of these vital fundamentals may wish to skip Chapter One.

1. INTRODUCTION

Biological markers or **biomarkers** (Eglinton et al., 1964; Eglinton and Calvin, 1967) are **molecular fossils,** meaning that these compounds are derived from formerly living organisms. Biomarkers are complex organic compounds composed of carbon, hydrogen, and other elements. They are found in rocks and sediments and show little or no change in structure from their parent organic molecules in living organisms.

Note: *Some of the oldest unequivocally indigenous biomarkers are found in 1.4 to 1.7 billion year-old mid-Proterozoic rocks from the McArthur Basin in northern Australia (Jackson et al., 1986). Subtle differences in the biomarkers between the marine and lacustrine rocks in this sequence indicate that depositional environments can be distin-*

guished using these compounds, even in the Proterozoic. Summons and Walter (1990) tabulate identified biomarkers in Proterozoic rocks and oils from worldwide localities and describe how these compounds can be used to infer the depositional environment, mineralogy, and type of organic matter input in their source rocks.

Biomarkers are useful because their complex structures reveal more information about their origins than other compounds. Unlike biomarkers, methane (CH_4) and graphite (pure carbon) for example, are comparatively less informative because virtually any organic compound will generate these products when heated sufficiently.

Note: *Although methane and other hydrocarbon gases are simple compounds compared to biomarkers, they still contain useful information on their origin and geologic history (e.g., Schoell, 1988).*

Distinguishing characteristics of a biomarker include the following:

1. The compound shows a structure indicating that it was, or could have been, a component in living organisms.
2. The parent compound is in high concentration in the organisms, which show widespread distribution.
3. The principal identifying structural characteristics of the compound are chemically stable during sedimentation and early burial.

Photosynthesis is the only major means by which new organic carbon is synthesized on earth (Fig. 1.1). It is the ultimate source for nearly all living organic matter, and accounts for most organic matter buried in sediments and rocks, including biomarkers. Two kinds of photosynthesis, bacterial and green plant, can be represented by the following generalized reaction:

$$2H_2A + CO_2 \xrightarrow{\text{light}} 2A + (CH_2O) + H_2O \qquad (1.1)$$

(CH_2O) stands for organic matter in the form of a carbohydrate such as the sugar glucose $(CH_2O)_6$. Polysaccharides (i.e., polymerized sugars) are the main form in which photosynthesized organic matter is stored in living cells. They represent the biosynthetic starting material for all reactions that yield other organic compounds. **Respiration** is the reverse of this reaction.

Because energy from light is required for photosynthesis, photosynthetic organisms are restricted to the euphotic zone in lakes and the ocean, and to the land. **Phototrophs** are organisms that derive energy for photosynthesis from light. Phototrophic microorganisms are the most important photosynthesizers in aquatic environments, while higher plants dominate the land. This fundamental difference affects the types of organic matter deposited in sediments from these environments (see Sec. 3.1.2.4.1).

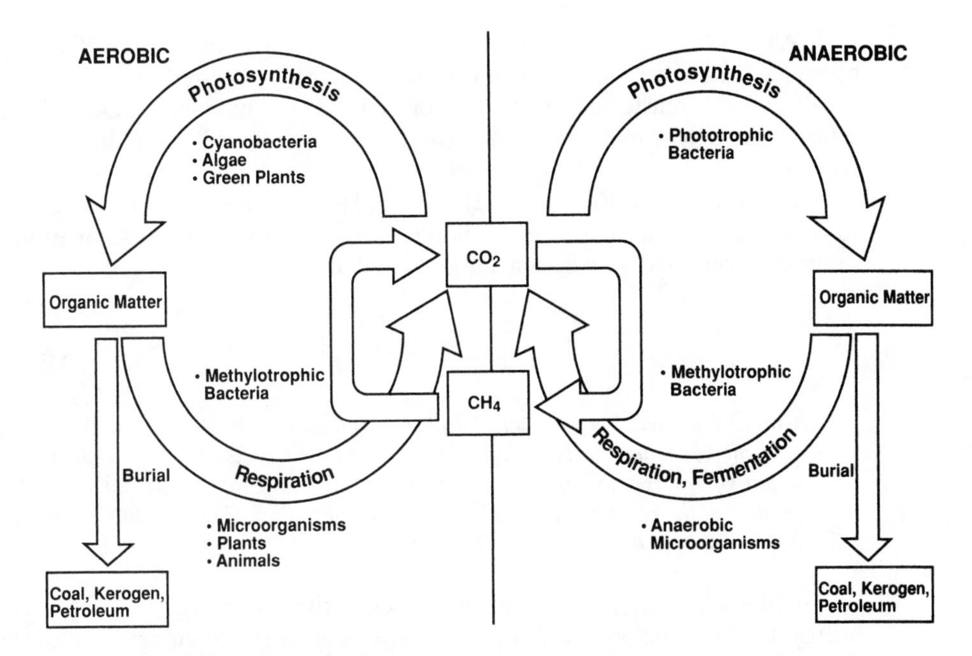

Figure 1.1 Generalized redox cycle for organic carbon. Production of new organic matter by photosynthetic fixation of carbon from CO_2 (primary productivity) can occur with (aerobic, left) or without oxygen (anaerobic, right) as a byproduct. Respiration and other processes result in the nearly complete oxidation of this organic matter back to CO_2. Burial of organic matter in sediments represents a small leak in the carbon cycle. As discussed in the text, it is estimated that less than 0.1 percent of primary productivity becomes preserved in sediments. Metamorphism, uplift, and erosion, or combustion of fossil fuels eventually return stored organic carbon to the cycle.

When organisms grow phototrophically, the rate of fixation of carbon from CO_2 into carbohydrate exceeds the rate of respiration. In fact, the whole carbon cycle on earth (Fig. 1.1) is based on a net positive balance in the rate of photosynthesis over the rate of respiration.

Green plant photosynthesis is a variation of Reaction Eq. 1.1.

$$2H_2O + CO_2 \xrightarrow{\text{light}} H_2O + (CH_2O) + O_2 \tag{1.2}$$

This process is an oxidation-reduction (redox) reaction. The oxidation reaction in Reaction Eq. 1.2 is

$$2H_2O \longrightarrow 4H + O_2 \tag{1.2.1}$$

Water is split to produce reducing power (hydrogen atoms), with O_2 as a byproduct. The reduction reaction is

$$4H + CO_2 \longrightarrow (CH_2O) + H_2O \tag{1.2.2}$$

As shown, the H_2O product is derived from reduction of CO_2. The necessary hydrogen atoms are derived from a donor molecule or "reductant" (in this case H_2O), and are transferred to an acceptor molecule or "oxidant," CO_2. After the reaction, the donor molecule is referred to as "oxidized," while the acceptor is "reduced."

In green plants (Reaction 1.2), H_2A is H_2O and 2A is O_2. In bacteria H_2A is some oxidizable substrate and 2A is the product of its oxidation. For example, in sulfur bacteria H_2A is hydrogen sulfide (H_2S) and A is sulfur

$$2H_2S + CO_2 \xrightarrow{\text{light}} H_2O + (CH_2O) + 2S \qquad (1.3)$$

Note: *The equation for bacterial photosynthesis clearly shows it to be an oxidation-reduction reaction. In bacteria, water is one of the products of the reduction of CO_2, and it differs from the oxidizable component, H_2S. In green plants, the oxidant and the product of the reduction cannot be readily distinguished without isotopic labeling experiments because both are water (Eq. 1.2).*

Chlorophyll plays a critical role in converting the energy in light to chemical energy stored in the products of photosynthesis. This stored energy is released when the products are brought together and undergo combustion or respiration (slow, enzyme catalyzed combustion). Forest fires and the muscles of an athlete both utilize this type of stored energy.

All photosynthetic organisms possess some type of chlorophyll. Chlorophyll-a (Fig. 1.27) is the principal chlorophyll in higher plants, most algae, and cyanobacteria. There are many other chlorophylls which absorb light in different ranges of wavelength. For example, purple and green bacteria contain bacteriochlorophylls. Accessory pigments, such as carotenoids, are at least indirectly involved with the capture of light, and are found with chlorophylls in many phototrophic organisms. Biomarkers derived from chlorophylls, bacteriochlorophylls, and carotenoids in living organisms are found in ancient rocks and petroleum.

Photosynthetically fixed carbon is eventually degraded by organisms to either methane or carbon dioxide (Fig. 1.1). For example, methane can be produced from organic matter in surficial, anaerobic sediments by methanogenic bacteria. Carbon dioxide is a product of various organisms which utilize fermentation, anaerobic respiration, or aerobic respiration to produce energy. As another example, when methane moves into aerobic environments, it can be oxidized to CO_2 by methylotrophic bacteria. Details on these various organisms can be found in any modern discussion of microbial metabolism (e.g., Brock and Madigan, 1991).

Fossil fuels represent organic matter that has been temporarily removed from the carbon cycle caused by burial in the lithosphere. One can view all fossil fuels, including finely disseminated organic matter, coal, and petroleum, as excess organic carbon that has escaped an otherwise remarkably efficient carbon cycle (Fig. 1.1, lower left and right).

Areas of concentrated petroleum occurrence can be predicted using mapping methods that integrate geology, geophysics, and geochemistry (Demaison, 1984). In the last 20 years, major technological breakthroughs have established biomarker geochemistry as a powerful component in this petroleum exploration approach.

Biomarkers are used to correlate oils with each other and with their source rocks, thus, improving understanding of reservoir relationships, petroleum migration pathways, and possible new exploration plays. Biomarkers can be used to evaluate thermal maturity and/or biodegradation, thus, providing vital information needed to properly evaluate the distribution and producibility of petroleum in basins. Biomarkers provide critical information on regional variations in the character of oils and source rocks as controlled by organic matter input and characteristics of the depositional environment. Finally, biomarkers offer the possibility of better understanding the kinetics of petroleum generation and basinal thermal history. Earlier works that review biomarkers, their origins, and applications include those by Connan (1981), Philp (1982), Mackenzie (1984), Petrov (1987), Tissot and Welte (1984), Johns (1986), and Waples and Machihara (1991).

The following material provides the necessary background for a basic understanding of the biological markers and their applications. Geologists and other scientists who study this material will be in a better position to make meaningful decisions on the usefulness of biomarkers for their particular projects. They will also have the background needed to critically evaluate the quality of biomarker data and interpretations obtained from contractors or other sources. However, because of the complexity of this material, the reader who is inexperienced in biomarker technology is cautioned against relying solely on his or her own interpretation of raw biomarker data.

1.1 Diagenesis, Catagenesis, Metagenesis and Biodegradation

Biomarkers exist because their basic structures remain intact through processes associated with sedimentation and diagenesis (Fig. 1.2). In this book, the term **diagenesis** refers to the biological, physical, and chemical alteration of organic matter in sediments prior to significant changes caused by heat.

In sedimentary rocks that have undergone diagenesis, the organic matter consists of kerogen, bitumen, and minor amounts of hydrocarbon gases. **Kerogen** is particulate organic debris which is insoluble in organic solvents and consists of mixtures of macerals and reconstituted degradation products of organic matter. **Macerals** (Sec. 3.1.2.5) are recognizable remains of different types of organic matter that can be differentiated under the microscope by their morphologies (Stach et al., 1982). They are analogous to minerals in a rock matrix, but differ in having less

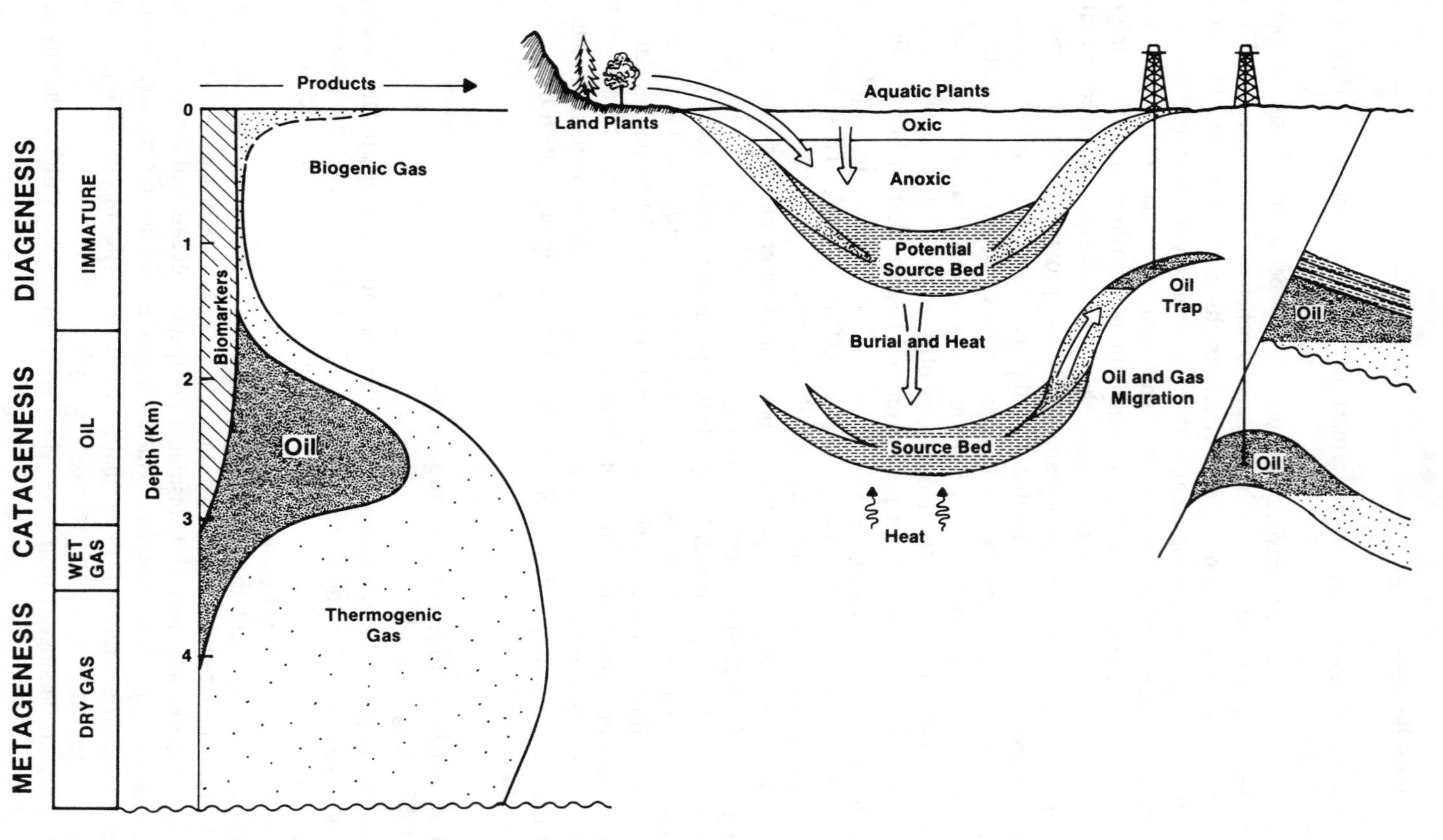

Figure 1.2 Generalized evolution of organic matter during and after sedimentation. The basic structural characteristics of many biomarkers survive diagenesis and much of catagenesis prior to their complete destruction during late catagenesis and metagenesis. Depth scale can vary depending on various factors, including geothermal gradient and type of organic matter.

well-defined chemical compositions. Part of the kerogen represents reconstituted low molecular-weight degradation products of biomass (e.g., Durand, 1980; Tissot and Welte, 1984). In addition, Tegelaar et al. (1989) list recognized biomacromolecules (e.g., cellulose, proteins, and tannins), and rank their relative potential for preservation by incorporation into kerogen through cross-linkage during diagenesis.

Bitumen consists of in situ hydrocarbons and other organic compounds dispersed within fine-grained sedimentary rocks that can be extracted using organic solvents. Biomarkers are found both free in the bitumen and bound to the kerogen in petroleum source rocks. They also are present in oils (crude oils) which migrate from the fine-grained source rocks to the reservoir rocks. As used in the text, **petroleum** is a general term which refers to solid, liquid, and gaseous materials composed dominantly of chemical compounds of carbon and hydrogen, and including bitumen, hydrocarbon gases, and oil.

Sedimentation and early burial typically result in the complete destruction of organic matter by oxidation and other processes. It has been estimated that during earth's history, less than 0.1 percent of the carbon produced as biomass by plants becomes preserved in sediments and available for petroleum-related processes (Tissot and Welte, 1984). Organic matter is preserved in sediments only under very special circumstances as described.

The quantity and quality of organic matter preserved during diagenesis of a sediment ultimately determines the petroleum-generative potential of the rock. Various factors play a role in the preservation of organic matter during sedimentation and burial, notably the oxygen content of the water column and sediments, water circulation, organic productivity, and sedimentation rate (Demaison and Moore, 1980; Emerson, 1985). The relative importance of these factors remains controversial and probably differs between depositional environments.

Under an **oxic** water column (over 2.0 ml oxygen/l water), aerobic bacteria and other organisms degrade organic matter settling from the euphotic zone. This respiratory process creates a demand for oxygen. If sufficient organic matter remains after exhausting all available oxygen, anaerobic organisms continue to oxidize the organic matter using nitrate or sulfate as primary oxidants. The boundary between aerobic and anaerobic metabolism (oxic vs. anoxic) can occur in the water column or in the bottom sediments.

Under an **anoxic** water column (less than about 0.1 ml oxygen/l water), aerobic degradation of organic matter is severely reduced because both metazoa (multicellular aerobic organisms) and aerobic bacteria generally require higher levels of oxygen. Below about 0.1 ml oxygen per l water, bioturbation of the bottom sediments is not observed because of the absence of metazoa, leaving only anaerobic bacteria to rework the organic matter. The lack of bioturbation allows the development of fine laminations recording depositional cycles, which are commonly observed in effective petroleum source rocks. **Euxinic** sediments are deposited under marine, anoxic conditions with free hydrogen sulfide (H_2S) produced by sulfate reducing bacteria (Raiswell and Berner, 1985).

Anaerobic degradation of organic matter is thermodynamically less efficient than aerobic degradation (Claypool and Kaplan, 1974). This observation supports the prevailing belief that anoxia is the main cause for apparent enhanced preservation of hydrogen- and lipid-rich organic matter in petroleum source rocks (Demaison and Moore, 1980). Where oxic conditions exist, the organic matter is largely destroyed during sedimentation and diagenesis, even when organic productivity is high. For example, polar regions in the modern oceans commonly show very high primary productivity, but the oxic bottom sediments are low in organic carbon.

Pederson and Calvert (1990) contend that organic productivity rather than anoxia is the major control on the accumulation of organic-rich, marine sediments. They cite references based primarily on laboratory incubation experiments that suggest the rates of destruction of organic matter under oxic and anoxic conditions are similar and cannot be used to invoke enhanced preservation of organic matter under anoxic conditions. In an example from the central Gulf of California, they observed no increased carbon content in alternating anoxic versus oxic sediments (Calvert, 1987). Data from anoxic sediments in the Black Sea suggest that organic carbon accumulation rates are not anomalously high compared to equivalent, oxygenated environments (Calvert et al., 1991).

We operate under the working assumption that anoxia is important in the development of organic-rich petroleum source rocks. Some arguments suggest that this is reasonable.

1. Even if equivalent rates of aerobic and anaerobic degradation occur in laboratory incubation experiments, anaerobic reactions in nature may slow or stop prior to completion because of lack of sufficient oxidants such as sulfate. Because of bioturbation, oxic sediments are better ventilated than anoxic sediments and depleted oxygen is rapidly replaced. Anoxic sediments approach closed systems.
2. Pedersen and Calvert (1990) discuss the effects of oxic versus anoxic deposition on the quantity rather than quality of preserved organic matter. Evidence suggests that anoxic conditions favor preservation of hydrogen-rich, oil-prone organic matter. For example, like Calvert (1987), Peters and Simoneit (1982) observed similar total organic carbon in alternating laminated (anoxic) and homogenous (oxic) diatomaceous oozes in the Gulf of California. However, the laminated zones in these sediments contain more hydrogen-rich organic matter as indicated by higher Rock-Eval hydrogen indices and lower oxygen indices than the homogenous zones.
3. If anoxia is unimportant in the preservation of organic matter, it is difficult to explain the general correspondence between oil-prone, organic-rich petroleum source rocks and faunal or sedimentologic features indicating anoxia.
4. Bitumens in petroleum source rocks show biomarker and supporting parameters indicating highly reducing to anoxic conditions (e.g., porphyrin and pristane/phytane ratios, homohopane indices).

Differences in the type of organic matter and the characteristics of the depositional environment result in lateral and vertical variations of organic facies within the same source rock. An **organic facies** is a mappable subdivision of a stratigraphic unit that can be distinguished on the basis of the organic components (Jones, 1987). Progress in our understanding of paleo-oceanographic controls on petroleum source rock deposition as discussed has improved regional mapping of organic facies and our ability to predict favorable areas for further exploration (e.g., Demaison et al., 1983). Biomarker compositions can be used to distinguish oils from different source rocks, but are also useful in showing regional variations in organic facies within the same source rock or within oils from the same source rock. These applications are possible because the biomarker patterns in oils are inherited from their respective source rocks.

Catagenesis is the process by which the organic matter in rocks is thermally altered by burial and heating at temperatures in the range of about 50 to 150°C under typical burial conditions requiring millions of years. During catagenesis, biomarkers undergo structural changes that can be used to gauge the extent of heating of their host sediments or oils migrated from these sediments. Furthermore, because the biomarkers in any rock represent a distinctive set of contributing organisms, their distribution in the bitumen remaining in an effective source rock is a "fingerprint" that can be used to relate it to a migrated oil which may be many kilometers away.

At temperatures in the range of about 150 to 200°C, prior to greenschist metamorphism, organic molecules are cracked to gas in the process called **metagenesis.** Biomarkers are severely reduced in concentration or absent because of their instability under these conditions.

Biomarkers are powerful geochemical tools, partly because many are highly resistant to **biodegradation** (Sec. 3.3). For example, biodegraded seep oils or asphalts commonly contain unaltered biomarkers that can be used for comparisons with nonbiodegraded oils.

Note: *Because of their resistance to weathering and biodegradation, biomarkers have been successfully applied in archaeology. For example, asphalt was used in the mummification process practiced by ancient Egyptian dynasties from at least 2600 B.C. Surface deposits of asphalt are rare in Egypt and the closest major asphalts are in the Dead Sea area. Rullkötter and Nissenbaum (1988) analyzed biomarkers in asphalts from four Egyptian mummies, including that of Cleopatra, dating from ca. 900 B.C. to the early second century A.D. The three younger mummy asphalts show sterane and triterpane biomarker distributions almost identical to those for a modern floating block of asphalt from the Dead Sea. These samples lack diasteranes and show high gammacerane, characteristic of petroleum from clay-poor carbonate or evaporitic source rocks (Sec. 3.1.3). Floating asphalt blocks from the Dead Sea are believed to originate from a Senonian limestone deposited under hypersaline conditions (Rullkötter et al., 1985). The oldest mummy asphalt shows biomarker distributions unlike the others, indicating a different origin, possibly because trade of Dead Sea asphalt into Egypt only began after 900 B.C.*

1.2 Organic Chemistry

Carbon, hydrogen, oxygen, and nitrogen are the principal elements in living organisms. With the exception of oxygen, these elements are rare components in the earth's crust compared to silicon and the light metals. Several unusual characteristics of carbon make it the vital element around which the chemistry of life, organic chemistry, has evolved.

The atomic structure of carbon allows it to form a greater diversity of compounds than any other element. Atomic orbital theory describes the approximate orientation of electron clouds around a single atom of carbon in the uncombined state (Fig. 1.3). The outer electron shell of carbon consists of four electrons (i.e., carbon shows a "valence" of four). Two electrons occupy the 2s orbital which is spherical in shape. The remaining two electrons each occupy a different dumbbell-shaped 2p orbital whose axis is at right angles to the other 2p orbitals. One of the 2p orbitals in carbon does not contain an electron. Each of the four electrons can be shared with other elements that are able to complete their electronic shells by sharing electrons to form covalent bonds.

A unique feature of carbon is that it can share electrons with other carbon atoms, resulting in large molecules dominated by carbon-carbon bonds. A few other elements contain four electrons in their outer shell and can form repetitive covalent

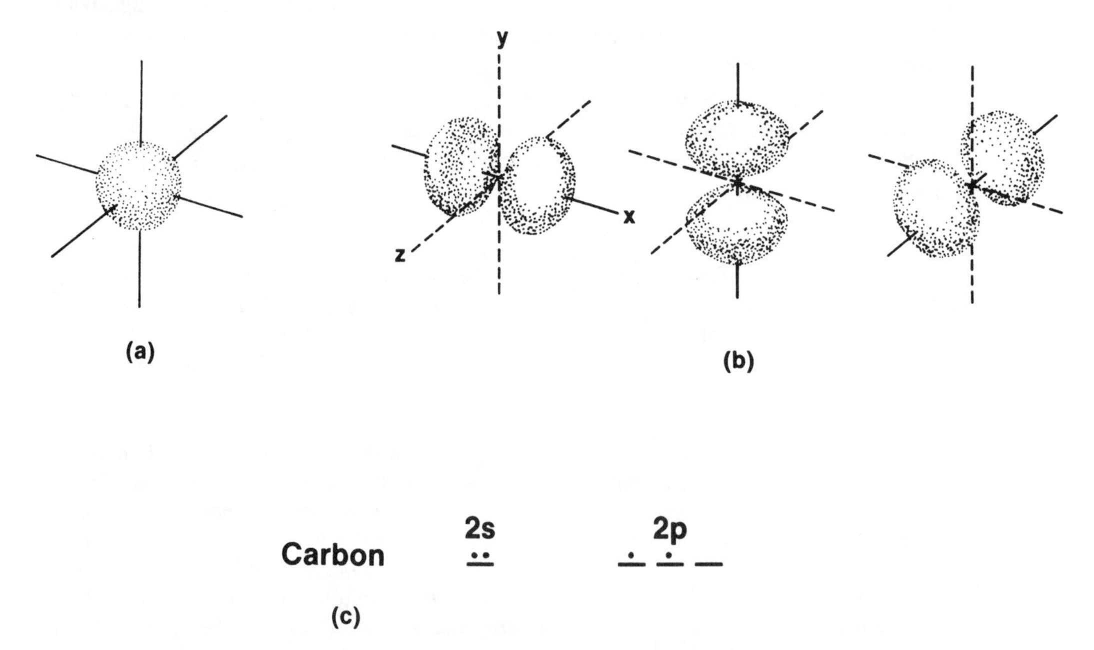

Figure 1.3 Atomic orbitals in the outer shell of carbon showing: (a) the spherical 2s orbital, (b) the three orthogonal, dumbell-shaped 2p orbitals, and (c) the distribution of electrons in the outer shell orbitals (after Morrison and Boyd, 1966).

bonds with atoms of the same element, but they are not stable under earth conditions. Thus, although silicon can covalently bond to other silicon atoms, silicon-silicon compounds are not stable in earth's atmosphere and readily oxidize to silica (SiO_2).

1.2.1 Alkanes—the sigma bond. When carbon combines with other atoms, the 2s and 2p electron orbitals hybridize into different orbital configurations. One 2s and three 2p orbitals can hybridize to form four equivalent sp3 orbitals (Fig. 1.4). The four sp^3 electron orbitals that can be formed by the valence shell of carbon are directed away from the central carbon atom at angles of 109.5° from each other [Fig. 1.4(b)]. By sharing an electron with each of four hydrogen atoms, a single carbon atom can satisfy the valence requirement of eight electrons. Thus, four hydrogen atoms bound to carbon result in the highly stable compound methane, where the hydrogens are located at the corners of a perfect tetrahedron with carbon in the center (Fig. 1.4(c)).

In compounds like methane, where carbon atoms are bound exclusively by sp^3 hybridized linkages, the carbon atoms are called "saturated" and the resulting strong

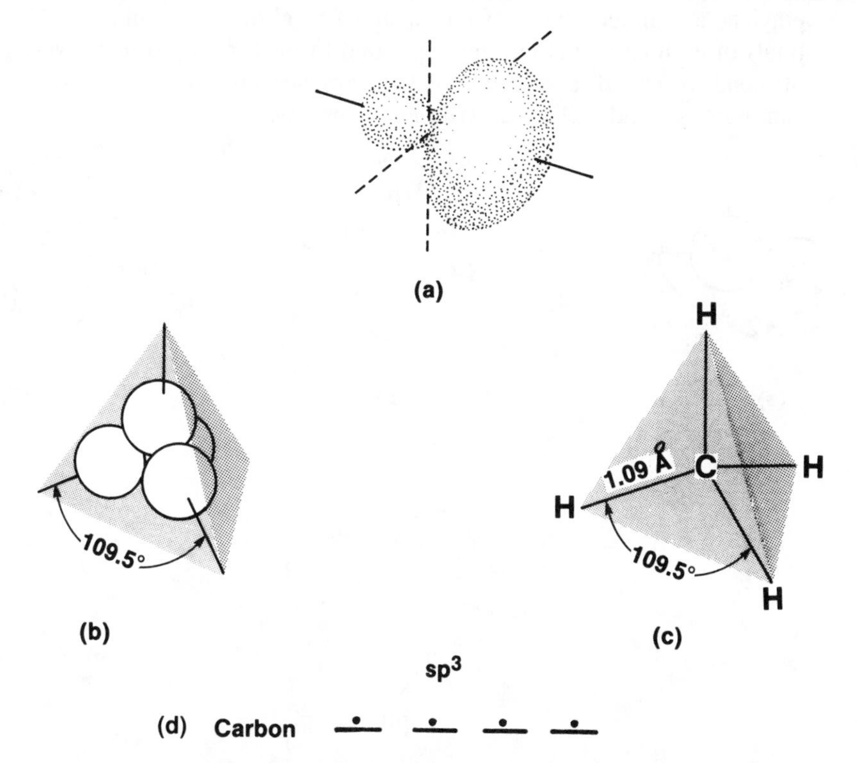

Figure 1.4 Hybridized sp^3 atomic orbitals of carbon: (a) approximate shape of a single orbital, (b) four orbitals with the axes directed toward the corners of a tetrahedron, (c) shape and dimensions of a methane molecule (CH_4) and (d) distribution of electrons in the outer shell orbitals (after Morrison and Boyd, 1966).

bonds are known as "sigma" or covalent bonds. When a sigma bond is the only link between two atoms in a molecule, it is called a single (versus double) bond. Stable molecules of carbon and hydrogen which contain only single bonds are called saturated hydrocarbons or **alkanes.**

1.2.2 Alkenes—the pi bond. The 2s and 2p electron orbitals of carbon can also hybridize into p and sp^2 orbitals. In this configuration the four valence electrons of carbon are divided between one p and three sp^2 orbitals. The three sp^2 orbitals, each containing one electron, are co-planar and are directed away from the central carbon atom at 120° from each other [Fig. 1.5(a)]. The p orbital contains the remaining electron and is oriented at 90° to the co-planar sp^2 orbitals. The p orbital for the methyl radical in the figure (b) contains one electron. Such hydrocarbon (alkyl-) radicals are unstable species, but can be intermediates in reactions of organic compounds.

When two p orbitals occur on adjacent atoms (e.g., C=C or C=0) a "pi" or double (versus single) bond results, as shown for ethylene in Fig. 1.5(c) and (d). Pi bonds typically are more reactive than sigma bonds. Thus, the carbon atoms in ethylene are linked by two bonds: a sigma bond formed from the sp^2 hybridized orbitals of each carbon atom, and a pi bond formed from p orbital overlap. This type of bond is called a double bond. Hydrocarbons containing double bonds are "unsaturated" and include **alkenes** and **aromatics.**

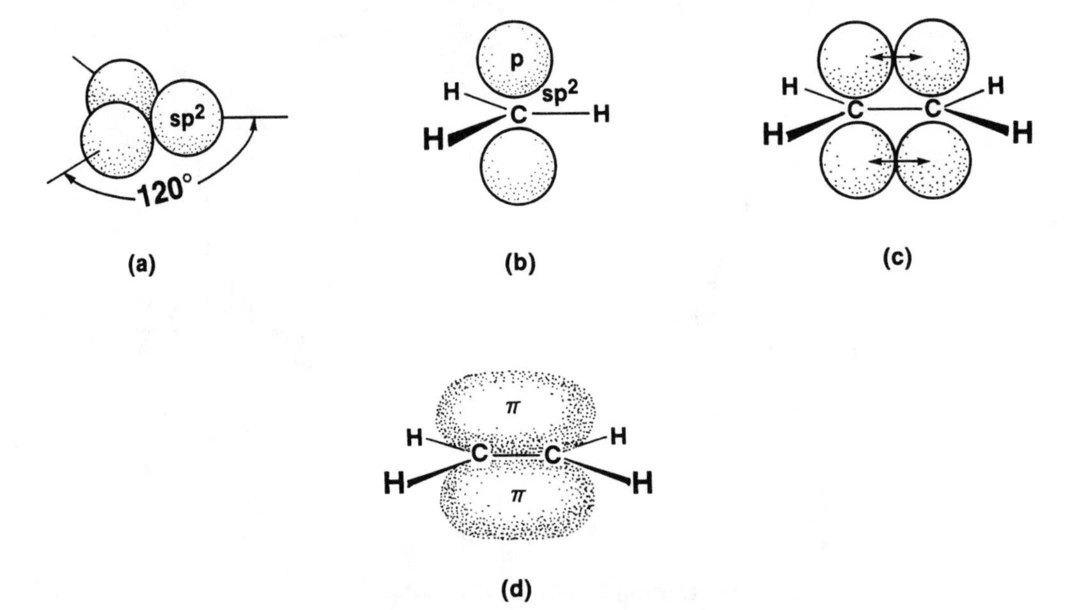

Figure 1.5. (a) Hybridized sp^2 atomic orbitals of carbon lie on axes directed toward the corners of an equilateral triangle. (b) A methyl radical contains a single p orbital occupied by one electron above and below the plane of the sigma bonds. (c) An ethylene molecule showing the interaction of adjacent p orbitals to give (d) a pi bond (cloud) above and below the plane of the sigma bonds (after Morrison and Boyd, 1966).

1.2.3 Aromatics—benzene. Hydrocarbons can be divided into two broad classes: (1) aliphatic hydrocarbons, including alkanes and alkenes as discussed, and (2) aromatic hydrocarbons. Although aromatics contain pi bonds, many are highly stable.

Benzene is the simplest aromatic hydrocarbon (Fig. 1.6). The six carbon atoms of benzene are sp^2 hybridized and connected in a flat hexagonal ring by sigma bonds. As in ethylene, each carbon atom has a single electron in the p orbital available for overlap in a pi bond with its neighbors. The pi bonds between each carbon atom in the benzene ring are equivalent. The six p electrons are shared equally or "delocalized" among the donating carbon atoms, forming a doughnut-shaped cloud of pi electrons above and below the ring. This arrangement of double bonds is more stable than an isolated double bond.

Note: *Various aromatic hydrocarbons are found in most petroleums, while alkenes are rare or absent. Unlike the more stable aromatics, alkenes are readily hydrogenated to alkanes under conditions of deep sediment burial. In general, molecules containing functional groups such as isolated double bonds, hydroxyl (—OH), carboxyl (—COOH), or thiol (—SH) substituents are typically more reactive and tend to be progressively eliminated during burial.*

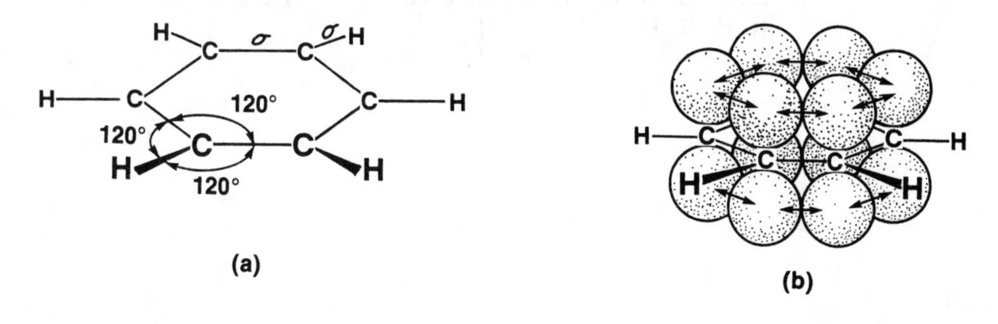

(c)

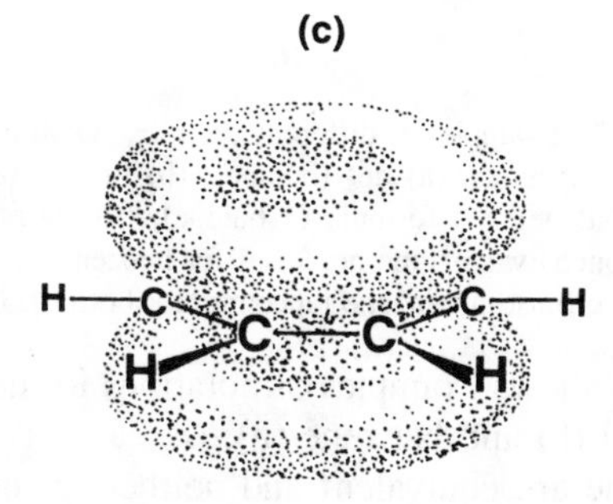

Figure 1.6 View of the benzene molecule (the simplest aromatic hydrocarbon) showing: (a) only the sigma bonds, (b) overlap of p orbitals to form pi bonds, and (c) clouds of "delocalized" pi electrons above and below the plane of the sigma bonds (after Morrison and Boyd, 1966).

1.2.4 Structure notation. Several different notations are used by chemists to describe the structures of organic compounds (Fig. 1.7). The most precise, but complex notation shows all outer shell electrons as a series of dots. Butane in Fig. 1.7(a) is drawn in dot notation. In the figure (a), two electrons are shared in each sigma bond between carbon atoms or between each carbon and hydrogen atom. These two bonding electrons can be replaced by lines (b), or omitted (c). When the bonds are not shown, the proper combinations of atoms to satisfy valence requirements (stoichiometry) are assumed. For example, each carbon atom in saturated hydrocarbons has four sigma bonds. The most abbreviated notation consists only of a zig-zag pattern of lines (d) which roughly depict a two-dimensional projection of the carbon skeleton of the molecule without the accompanying hydrogen atoms. Each angle and terminus of a line in this notation represents a single carbon atom with the appropriate number of hydrogens to satisfy valence requirements. Thus, in all structural formulas a line ending in nothing signifies a methyl group. The zig-zag notation is preferred for complex organic molecules including biomarkers, but is commonly used in combination with the other methods.

H H H H
H : C : C : C : C : H
H H H H

(a)

H H H H
H - C - C - C - C - H
H H H H

(b)

$CH_3CH_2CH_2CH_3$

(c)

(d)

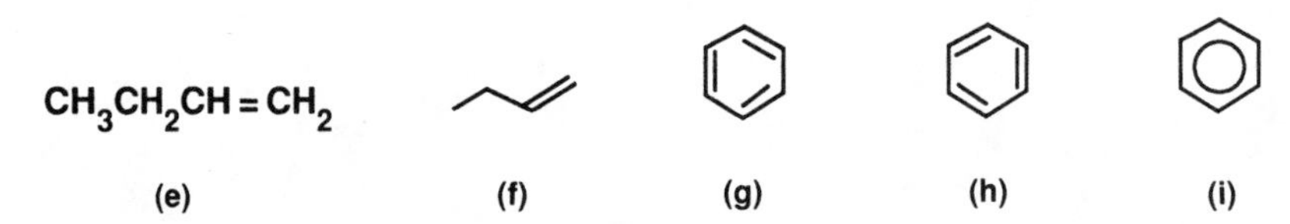

Figure 1.7 Examples of different chemical notations used to describe structures of organic compounds: (a) dot (butane), (b) line (butane), (c) formula (butane), (d) zig-zag (butane), (e) formula (1-butene), (f) zig-zag (1-butene), (g) and (h) two strictly nonequivalent line notations for benzene, (i) notation indicating delocalization of electrons on benzene [accounts for both (g) and (h)].

Figure 1.7 shows examples of notations for unsaturated hydrocarbons including 1-butene (e) and (f) and benzene (g), (h), and (i). As discussed, all carbon-carbon bonds in benzene are equivalent and neither (g) nor (h) are true representations of the delocalized cloud of electrons in this structure. Figure 1.7(i) is commonly used to denote the delocalized electron cloud in benzene, although notations (g), (h), and (i) are all acceptable.

Complex organic molecules typically require additional symbolic notation, particularly when stereochemical differences must be shown (see following).

1.2.5 Acyclic alkanes. Methane is the simplest compound in a series of saturated hydrocarbons (containing only hydrogen and carbon) with the formula C_nH_{2n+2}. These compounds are called **alkanes** or **paraffins.** By increasing n, a "homologous" series of compounds results, as shown in the table below. Compounds in the table with a linear arrangement of carbon atoms are called *normal* alkanes or *normal* paraffins. Compounds with a nonlinear arrangement of carbon atoms are called *iso*- or branched alkanes or paraffins.

Compounds with the same molecular formula, but different arrangements of their structural groups, are called **isomers.** Methane, ethane, and propane show only one isomer each. Butane, however, shows two compounds with the molecular formula C_4H_{10} as indicated in Fig. 1.8.

The number of isomers increases exponentially with increasing number of carbon atoms in each compound. For example, the Table 1.2.5 shows that there are over 366,000 different possible isomers of eicosane ($C_{20}H_{42}$).

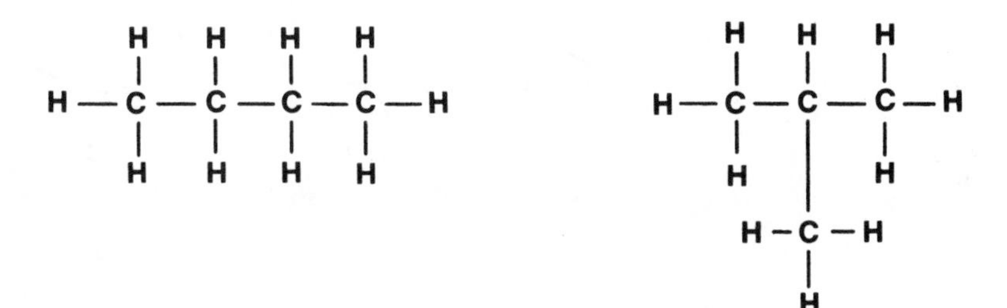

Figure 1.8 Butane (C_4H_{10}) is found as two structural isomers. *n*-Butane (nC_4) is one homolog in the series of *n*-paraffins (or *n*-alkanes).

1.2.6 Cyclic alkanes. Although cyclic alkanes containing almost any number of carbon atoms occur, only those containing combinations of five (cyclopentyl) or six (cyclohexyl) carbons are known to occur commonly in petroleum. Alkanes containing rings are sometimes called **naphthenes.** Cyclopentane and cyclohexane are simple cyclic alkanes that are commonly depicted in two dimensions [Fig. 1.9(a) and (b)]. Cyclohexane consists of a puckered six-carbon ring [(Fig. 1.9(c) and (d)]. Figure 1.9(c) is a more revealing projection of the three-dimensional, preferred "chair" shape or conformation of hexane. This is the same conformation of the hexane ring contained in the much more complex biomarker compounds. For example, the structure of the biomarker called **hopane** is shown in the common two-dimensional notation in Fig. 1.9(e). Hopane as shown in the projected three-dimensional

TABLE 1.2.5 Selected *n*-Alkane Homologs, Their Boiling Points, and Total Isomers with the Same Molecular Formula.

Number of carbons, n	Name	Formula	Boiling point, °C	Possible isomers
1	Methane	CH_4	−164	1
2	Ethane	C_2H_6	−89	1
3	Propane	C_3H_8	−42	1
4	Butane	C_4H_{10}	0	2
5	Pentane	C_5H_{12}	36	3
6	Hexane	C_6H_{14}	69	5
7	Heptane	C_7H_{16}	98	9
8	Octane	C_8H_{18}	126	18
9	Nonane	C_9H_{20}	151	35
10	Decane	$C_{10}H_{22}$	174	75
11	Undecane	$C_{11}H_{24}$	195	159
12	Dodecane	$C_{12}H_{26}$	216	355
20	Eicosane	$C_{20}H_{42}$	343	366,319
30	Triacontane	$C_{30}H_{62}$	450	4×10^9

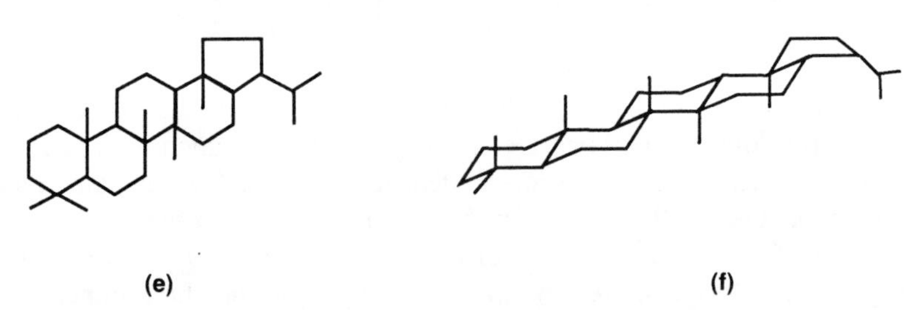

Figure 1.9 Some cyclic alkanes (naphthenes) include: (a) cyclopentane, (b) cyclohexane, (c) and (d) projected three-dimensional "chair" and "boat" conformations of hexane, respectively, (e) hopane in standard two-dimensional notation (a C_{30} pentacyclic triterpane), and (f) a three-dimensional projection of hopane.

notation in Fig. 1.9(f) consists of several fused cyclohexyl rings, each in the chair conformation, and a fused cyclopentyl ring.

> ***Note:*** *Because they are part of large multiring structures, the cyclohexyl rings in biomarkers are found mainly in the "chair" conformation. Cyclohexane, however, can show the "chair" [Fig. 1.9(c)], "boat" [(Fig. 1.9(d)], and "twisted" (or stretched) conformations. Both the chair and boat conformations are characterized by approximately tetrahedral carbon bond angles. While the chair form is quite rigid, the boat form is not, and it can flex into the twisted conformation without bond angle deformation. The various conformations of cyclohexane are inseparable using techniques such as distillation because they interconvert very rapidly at room temperature.*

1.2.7 The "isoprene rule." **Isoprene** (methylbutadiene) is the basic structural unit composed of five carbon atoms that is found in all biomarkers (Fig. 1.10). Compounds composed of isoprene subunits are called **terpenoids, isoprenoids,** or **isopentenoids** (Nes and McKean, 1977). All organisms, ranging from bacteria to man, biosynthesize or require these substances. The "isoprene rule" states that biosynthesis of these compounds occurs by polymerization of appropriately functionalized C_5-isoprene subunits. Unlike other biopolymers, such as proteins or polysaccharides, terpenoids are not readily depolymerized because they are joined together by covalent carbon-carbon bonds.

The saturated terpenoids (no double bonds) are divided into families based on the approximate number of isoprene subunits they contain. The various terpenoid families are composed of a wide variety of cyclic and acyclic structures (e.g., Devon and Scott, 1972; Simoneit, 1986), some of which are shown in Fig. 1.10. Hemiterpanes (C_5), monoterpanes (C_{10}), sesquiterpanes (C_{15}), diterpanes (C_{20}), and sesterterpanes (C_{25}) contain one, two, three, four, and five isoprene subunits, respectively. Triterpanes and steranes (C_{30}) differ in structure, but are derived from six isoprene subunits, while tetraterpanes (C_{40}) contain eight. Saturated terpenoids containing nine or more isoprene units (C_{40}^{+}) are called polyterpanes. Because their precursors are chemically reactive and highly volatile, the utility of hemiterpanes and monoterpanes as biomarkers is limited and they will not be discussed further.

> ***Note:*** *Natural rubber is a polyunsaturated polymer of isoprene with a molecular weight in the thousands. Although found in a variety of plants, most commercial rubber is obtained from Hevea brasiliensis, a tree belonging to the Euphorbiaceae family from the Amazon which is now cultivated in southeast Asia, Indonesia, and other localities. The Mayans from Central America were the first to collect and use rubber. They played a game whose objective was to bounce a rubber ball through a stone hoop high on a wall. Such ball courts are found at many Mayan archeological sites.*

The original compounds found to obey the isoprene rule were called isoprenoids. The discovery of additional chemical structures made it clear that the linkage between isoprene subunits can vary: (1) head-to-tail (regular) linkage, or (2) other (irregular) linkages. Farnesane (Fig. 1.10) is an example of a regular, acyclic

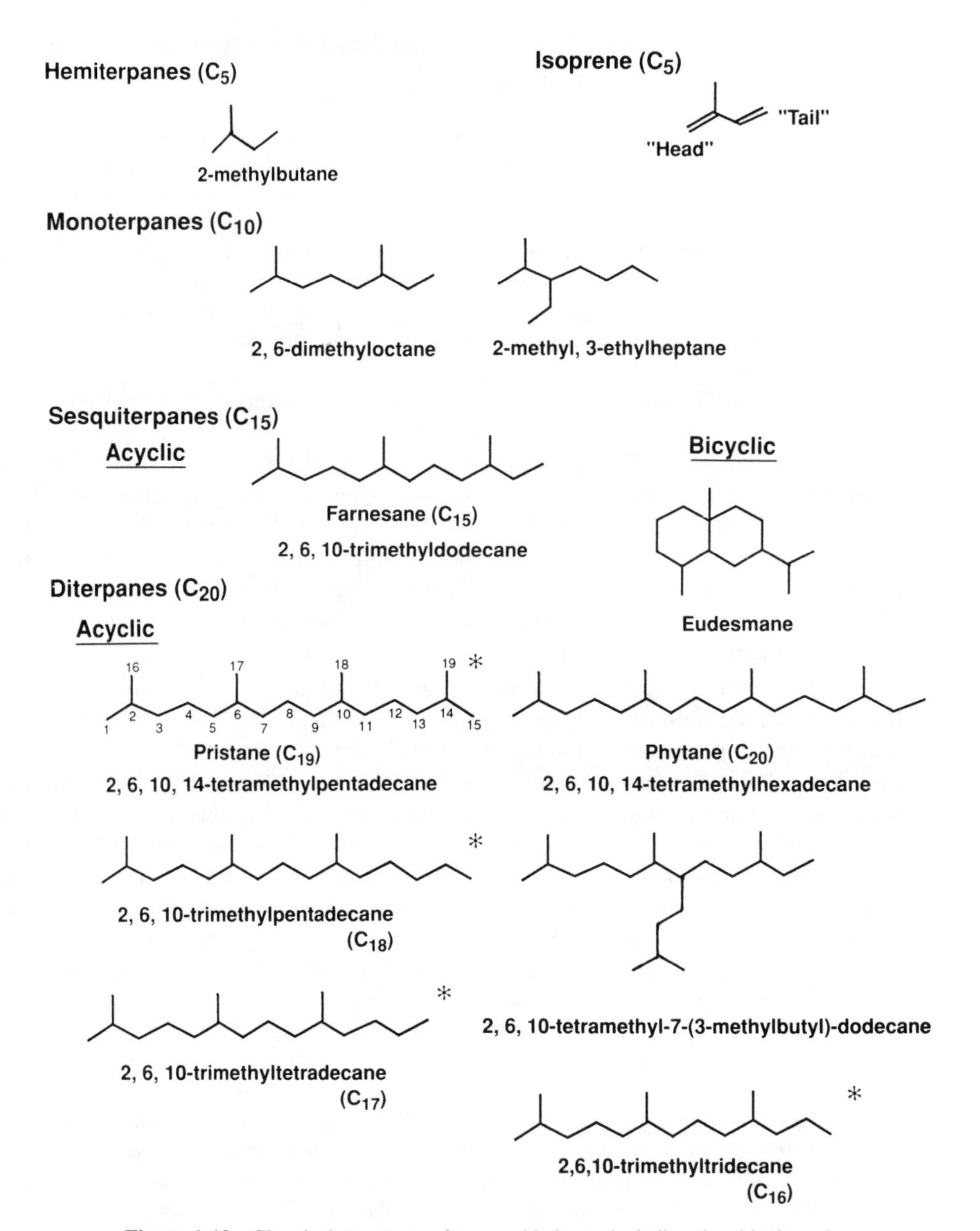

Figure 1.10 Chemical structures of terpenoid classes including the ubiquitous isoprene (C_5) subunit. Most of the examples shown are saturated compounds (e.g., monoterpanes versus monoterpenes) because these are typical of the biomarkers in petroleum. Pristane is used as an example of the numbering system for acyclic isoprenoids. Labeling systematics for cyclic biomarkers are in Figure 1.12. Asterisks mark compounds that contain fewer carbon atoms than their parent class of terpanes.

Bicyclic

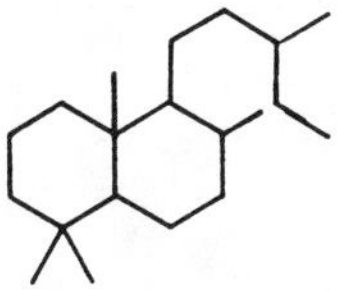

Tetracyclic

Labdane

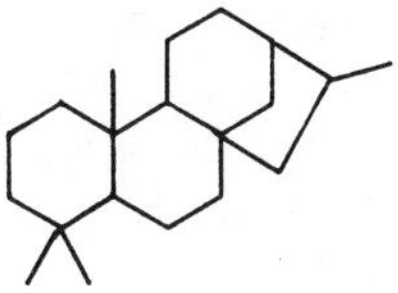

Kaurane

Tricyclic

*

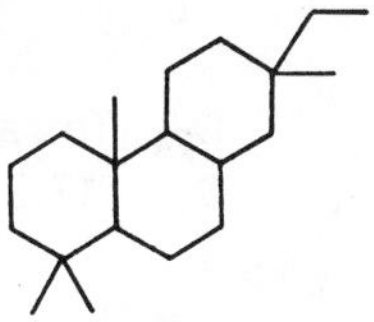

Pimarane

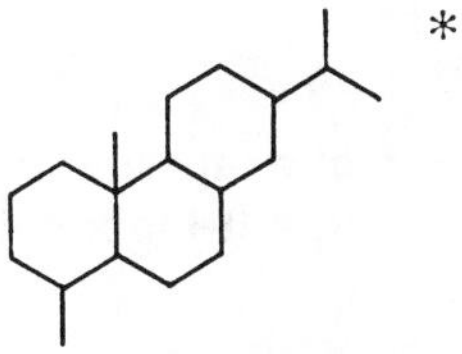

Fichtelite

Sesterterpanes (C_{25})

Tail-to-Tail Linkage

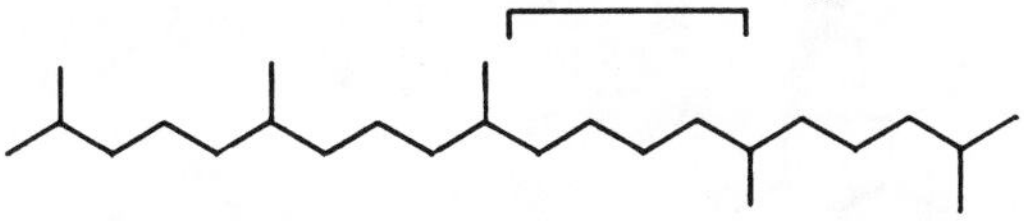

2,6,10,15,19-Pentamethyleicosane ($C_{25}H_{52}$)

Figure 1.10 *(cont.)*

Triterpanes (C_{30})

Acyclic

Tail-to-Tail Linkage

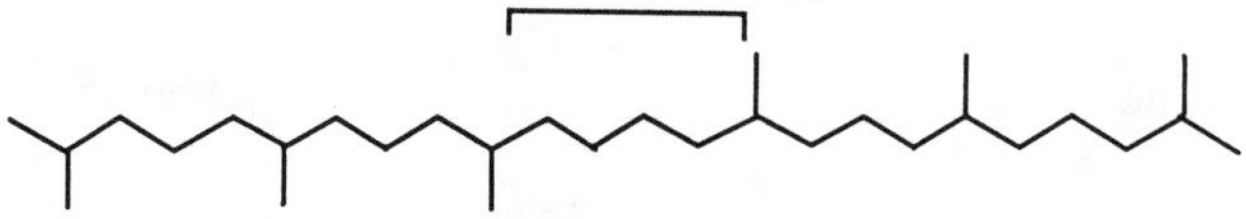

Squalane ($C_{30}H_{62}$)

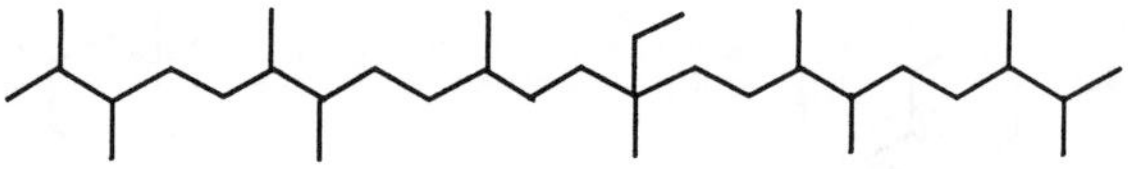

Botryococcane

Tricyclic

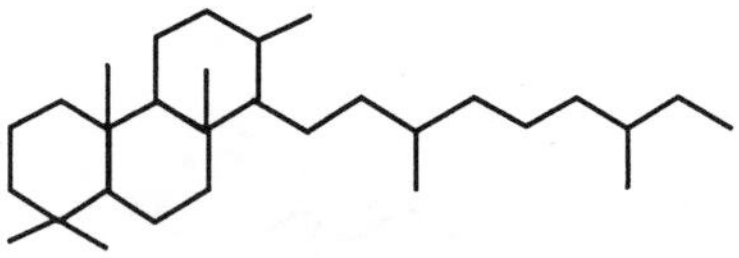

Tricyclohexaprenane
(C_{30} extended ent-Isocopalane)

Tetracyclic

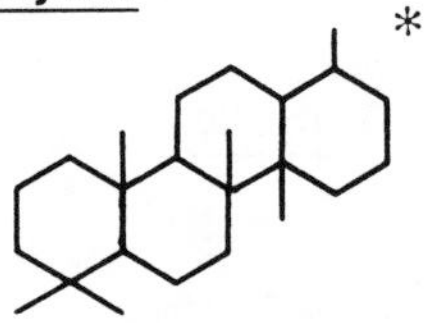

*

C_{24} Tetracyclic Terpane
(C_{24} 17,21-Secohopane)

Pentacyclic

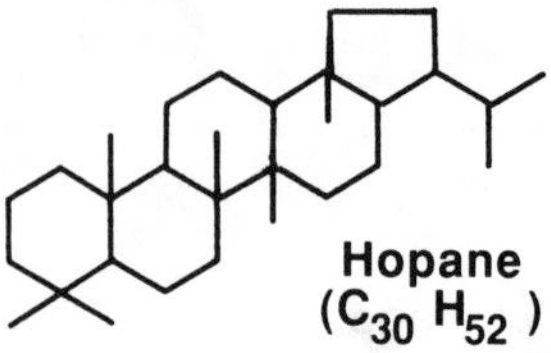

Hopane
($C_{30}H_{52}$)

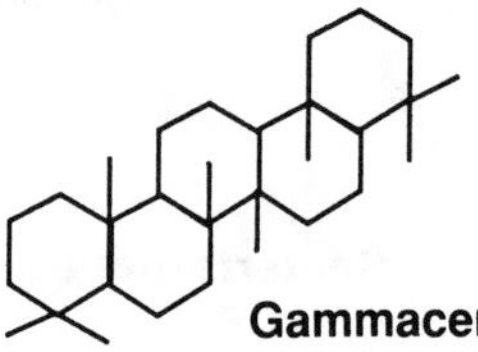

Gammacerane

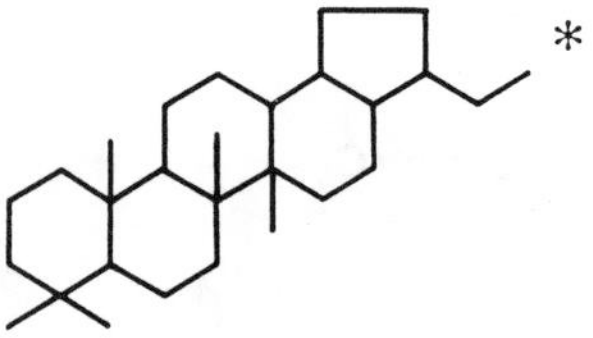

*

28,30-Bisnorhopane

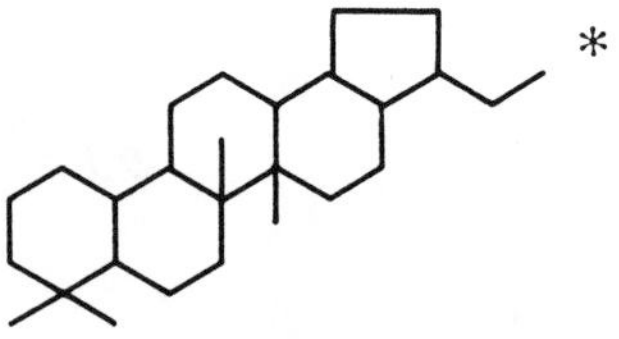

*

25,28,30-Trisnorhopane

Figure 1.10 (*cont.*)

Steranes (C_{30})

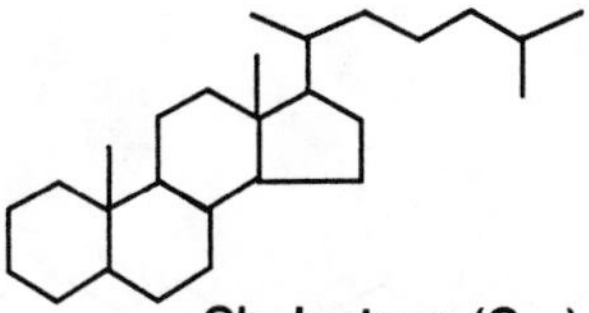

Cholestane (C_{27})

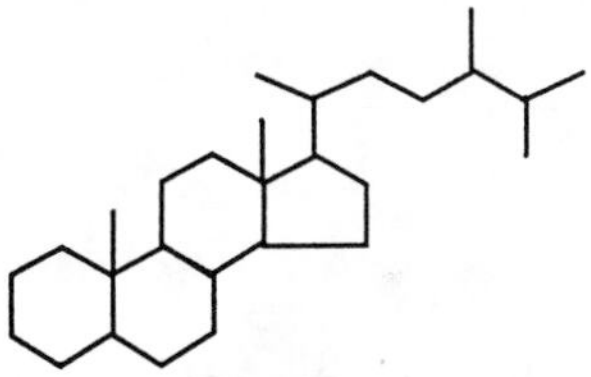

Ergostane (C_{28})

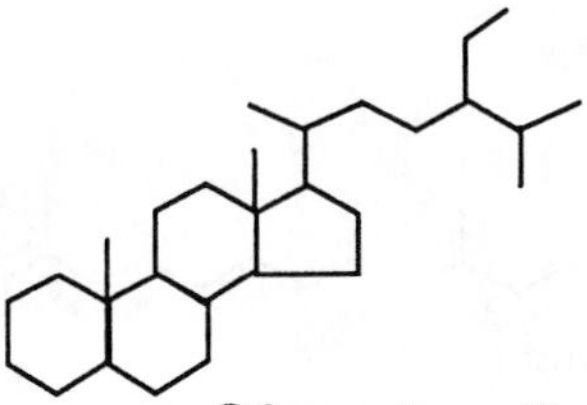

Stigmastane (C_{29})

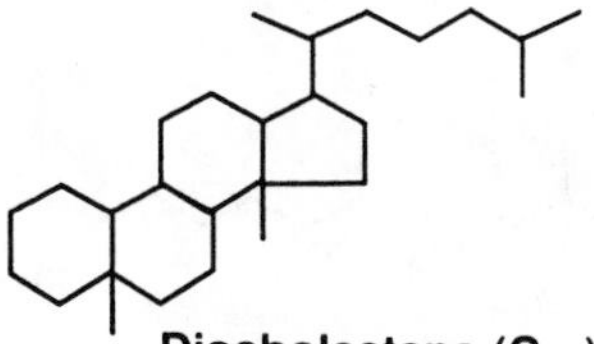

Diacholestane (C_{27})

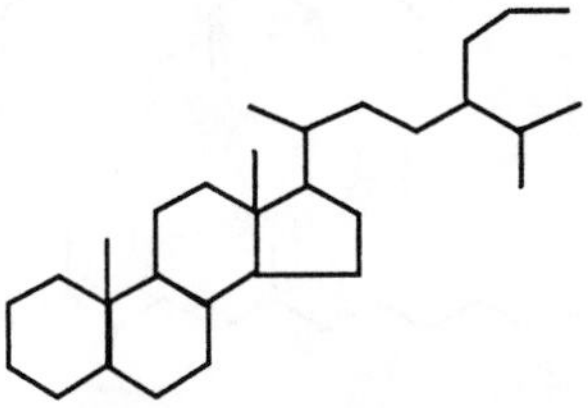

24-n-propylcholestane (C_{30})

Figure 1.10 (*cont.*)

Tetraterpanes (C_{40})

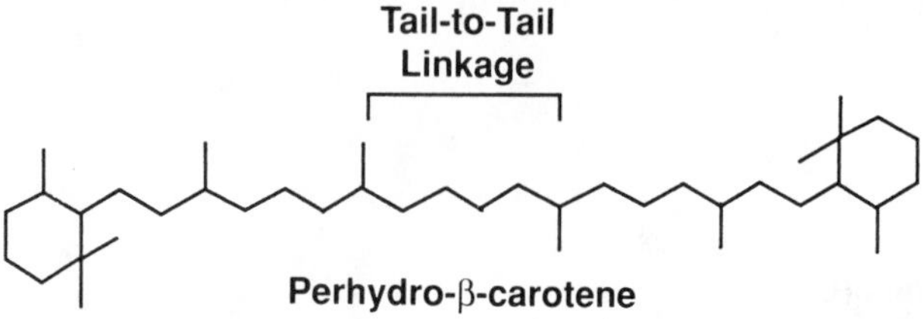

Tail-to-Tail
Linkage

Perhydro-β-carotene

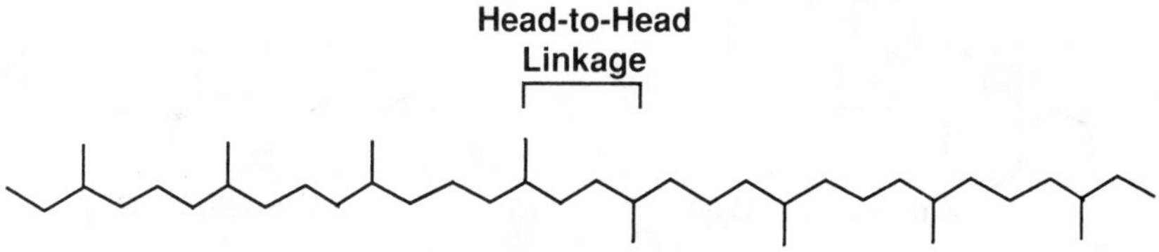

Head-to-Head
Linkage

Bis-Phytane ($C_{40}H_{82}$)

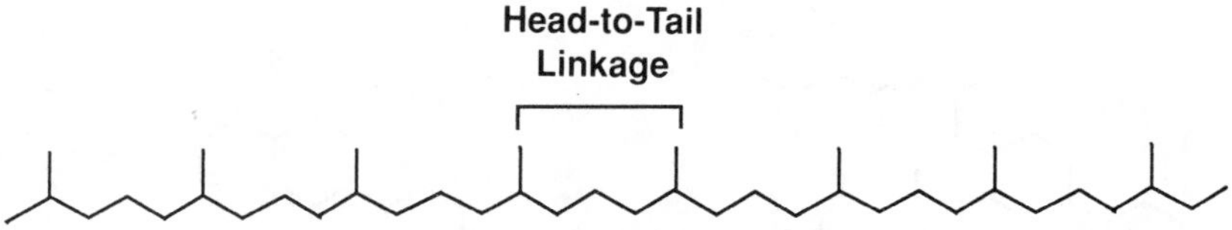

Head-to-Tail
Linkage

Regular C_{40} -Isoprenoid

Tail-to-Tail
Linkage

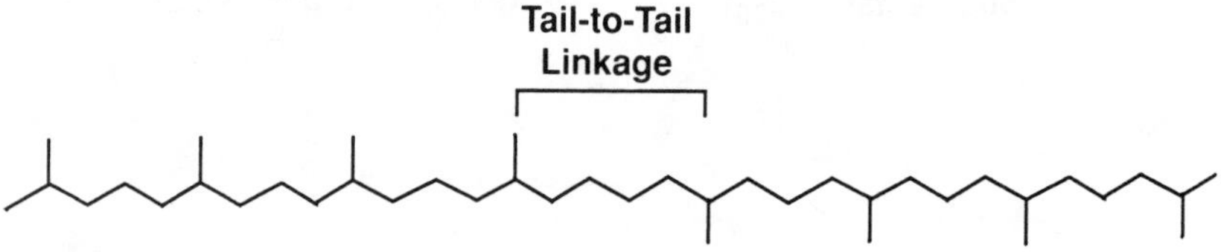

Lycopane ($C_{40}H_{82}$)

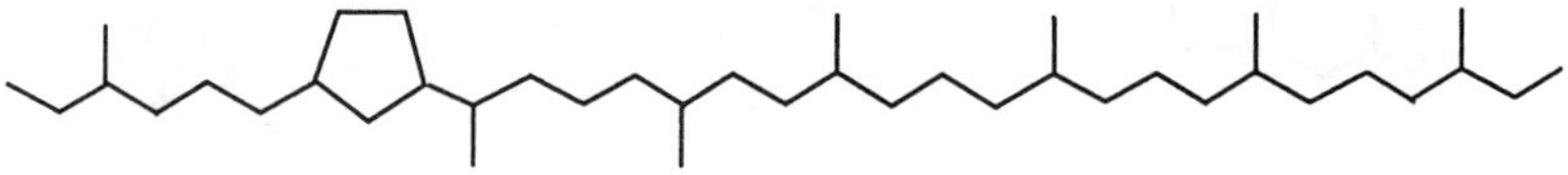

Irregular C_{40} -Isoprenoid With Cyclopentane Ring

Polyterpenes (C_{40}+)

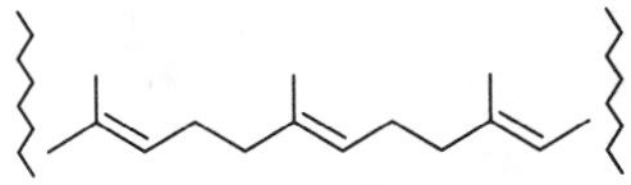

Trans-rubber

Figure 1.10 (*cont.*)

(not containing rings) isoprenoid consisting of three head-to-tail linked isoprene units. Squalane and bis-phytane (Fig. 1.10) are examples of irregular acyclic isoprenoids. Squalane contains six isoprene units with one tail-to-tail linkage, while bis-phytane contains a head-to-head linkage.

The geochemistry of acyclic isoprenoids is reviewed by Volkman and Maxwell (1986). Figure 1.10 shows other examples of head-to-tail or regular (Albaiges, 1980), head-to-head (Moldowan and Seifert, 1979; Petrov et al., 1990), tail-to-tail (Brassell et al., 1981), and other irregular isoprenoids, which differ in the order of attachment of the isoprene subunits. Included in the figure is an unusual case of three isomeric C_{40} compounds believed to be derived from Archaebacteria (Albaiges et al., 1985), which show head-to-head (bis-phytane), head-to-tail (regular C_{40}-isoprenoid), and tail-to-tail (lycopane) linkage.

Degraded, altered, or homologous structures may still be categorized in their parent terpenoid family. The precise number of carbon atoms in a given terpenoid family varies due to differences in source material, diagenesis, thermal maturity, and biodegradation. For example, the cyclic terpenoids **cholestane** (C_{27}), **ergostane** (C_{28}), and **stigmastane** (C_{29}) are three homologs in the sterane series (Fig. 1.10). These compounds only approximate the isoprene rule because they do not contain an integral number of isoprene units, $(C_5)n$. However, steranes show some terpenoid character. Terpenoids that do not strictly obey the isoprene rule can be explained by biochemical or other reactions that result in gain or loss of substituents.

Another example of a terpenoid that does not obey the isoprene rule is pristane. **Pristane** (C_{19}) is still classified as a diterpane although it contains one less methylene group ($-CH_2-$) than **phytane** (C_{20}), the next largest homolog in the regular, acyclic (linear), isoprenoid series. This pseudohomologous series of regular isoprenoids extends from farnesane (C_{15}) through C_{16}, C_{17}, C_{18}, and C_{19} to phytane (C_{20}) (Figure 1.10). The C_{16} to C_{19} compounds can be considered degraded diterpanes derived from the C_{20} parent by consecutive loss of methylene groups. Conversely, it is equally correct to consider the C_{16}, C_{17}, and C_{18} compounds as extended homologs of farnesane by addition of methylene groups to the linear portion of the hydrocarbon chain. Pristane, however, is not an extended homolog of the C_{18} acyclic isoprenoid. The nineteenth carbon atom in pristane (carbon atoms are numbered for pristane in Fig. 1.10) is a methyl branch (part of the fourth isoprene unit), not an additional carbon on the linear part of the chain.

> ***Note:*** *In Fig. 1.10 several of the compounds are degraded and contain fewer carbon atoms than their parent class of terpanes. These compounds are marked with asterisks. Also note botryococcane has four additional carbons compared to its parent class (triterpanes).*

Most regular isoprenoid alkanes containing 20 or less carbon atoms, including pristane, and phytane, appear to originate primarily from the phytol side chain of chlorophyll-a during diagenesis. For example, pristane can be derived by oxidation and decarboxylation of phytol, while phytane can be derived by dehydration and reduction. (See discussion on the use of Pristane/Phytane, Pr/Ph, to describe redox conditions in source rocks, Sec. 3.1.3.1.) Thus, pristane (C_{19}) and phytane (C_{20}) are usually the most abundant members of a whole series of regular head-to-tail linked isoprenoids.

> ***Note:*** *The regular C_{17} acyclic isoprenoid is low or absent in petroleum. Formation of a C_{17} homolog in the regular isoprenoid series is unfavorable because it would require cleavage of two carbon-carbon bonds in the C_{19} or C_{20} isoprenoid precursor rather than only one. For example, the C_{19} regular isoprenoid (pristane) has two methyl groups attached to the number 14 carbon. Cracking of one of these carbon-carbon bonds yields the C_{18} isoprenoid. Cracking of the carbon-carbon bond between C-13 and C-14 yields the C_{16} isoprenoid. Cracking of the carbon-carbon bond between C-2 and C-3 in phytane (Fig. 1.10) would result in a C_{17} compound, but this compound is no longer a member of the homologous isoprenoid series.*

Other sources for acyclic isoprenoids containing up to 20 carbon atoms include chlorophyll-b, bacteriochlorophyll-a, α- and β-tocopherols, carotenoid pigments, and the cell membranes of Archaebacteria (Goosens et al., 1984; Volkman and Maxwell, 1986). Archaebacteria have descended from primitive organisms existing during the early Proterozoic when today's "extreme" environments were probably more common. They include halophilic, thermoacidophilic, and methanogenic organisms with characteristics of both prokaryotes (bacteria and cyanobacteria) and eukaryotes (higher organisms) (Woese et al., 1978; Brock and Madigan, 1991).

Various irregular acyclic isoprenoids are important biomarkers. Botryoccocane is an example of an irregular acyclic isoprenoid (Fig. 1.10) that is a highly specific marker for lacustrine sedimentation (Sec. 3.1.3.1). Another unusual acyclic isoprenoid, 2,6,10-trimethyl-7-(3-methylbutyl)-dodecane (Fig. 1.10), has been isolated from Rozel Point crude oil, where it is the second most abundant alkane (Yon et al., 1982). This compound and related highly-branched isoprenoid structures appear derived from both marine and lacustrine phytoplanktonic algae or bacteria (Volkman and Maxwell, 1986) and may indicate hypersaline depositional conditions (Table 3.1.3–1).

> ***Note:*** *Smith (1968) accounts for normal paraffins in petroleum from lipids in living organisms, including naturally occurring n-paraffins, waxes, and fatty acids. The abundance of iso- (2-methyl-paraffins), and anteisoparaffins (3-methyl-paraffins) appears because of their biological origin. For example, 2-methyloctadecane is biosynthesized by Archaebacteria (Brassell et al., 1981). n-Paraffins, isoparaffins, and anteisoparaffins are examples of biomarkers that are not terpenoids.*

Although not biomarkers, many of the more volatile hydrocarbons in petroleum show evidence of an origin from biomarker precursors in living organisms. Further, many highly branched hydrocarbons, and those containing aromatic or cyclohexane rings, can be explained as originating from fragmentation of terpenoid precursors. Mair et al. (1966) show that in Ponca City crude, 2,6-dimethyloctane (0.50 vol. %) and 2-methyl-3-ethylheptane (0.64%) are more abundant than all other 49 possible structural isomers (0.44%) because they are derived from terpenoid precursors (two isoprene units, Fig. 1.11).

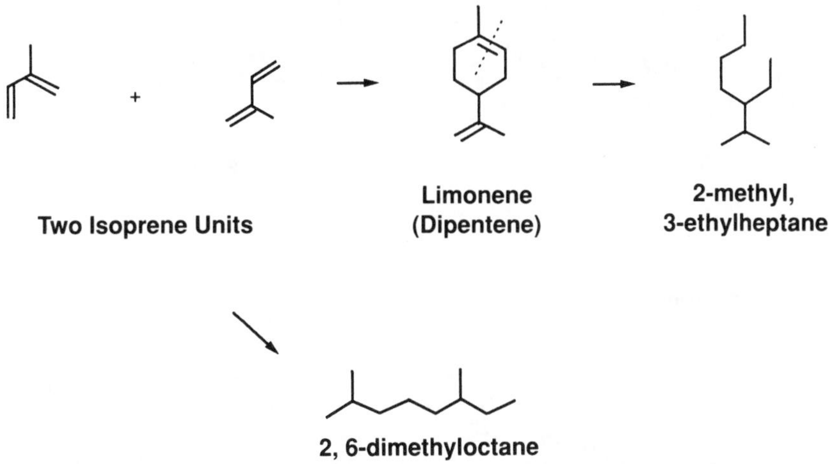

Two Isoprene Units

Limonene (Dipentene)

2-methyl, 3-ethylheptane

2, 6-dimethyloctane

Figure 1.11 Proposed origin of two monoterpanes, 2,6-dimethyloctane and 2-methyl,3-ethylheptane, each by linkage of two isoprene subunits (after Mair et al., 1966). These and many other low molecular-weight hydrocarbons common in petroleum obey the isoprene rule and can be explained as thermal breakdown products of terpenoids. Note the head-to-tail linkage of isoprene subunits in 2,6-dimethyloctane.

1.2.8 Stereochemistry and nomenclature. The spatial or three-dimensional relationship between atoms in molecules, called **stereochemistry,** is essential to understand the structures of biomarkers and how they are used in geochemical studies. Each carbon atom and the rings in biomarker molecules are labeled systematically. Fig. 1.10 shows the labeling for carbon atoms in the acyclic isoprenoid, pristane. The labeling system for steranes and triterpanes is shown in Fig. 1.12. This system appears haphazard because it is the result of the chronology of structural elucidation. In the text, a capital "C" followed immediately by a subscript number refers to the number of carbon atoms in a particular compound (e.g., the C_{27} sterane, cholestane, contains 27 carbon atoms). A capital "C" followed by a dash and number refers to a particular position within the compound (e.g., C-20 in the steranes is the carbon atom at position 20 in Fig. 1.12). The nomenclature "*n*C" followed by a number indicates the normal paraffin containing the specified number of carbon atoms (e.g., nC_{19} refers to the *n*-paraffin with 19 carbon atoms). Rings are also specified in succession from left to right as the A-ring, B-ring, C-ring, and so forth (e.g., Fig. 1.12).

Table 1.2.8 summarizes the more common nomenclature used to modify the structural specifications of biomarkers. For example, if a carbon is absent in a compound named after a parent (source) compound where it is present, the prefix "nor"

Steranes

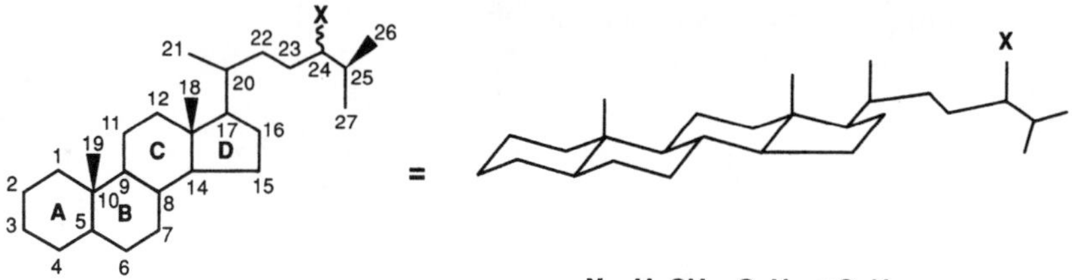

X = H, CH_3, C_2H_5, nC_3H_7

Triterpanes

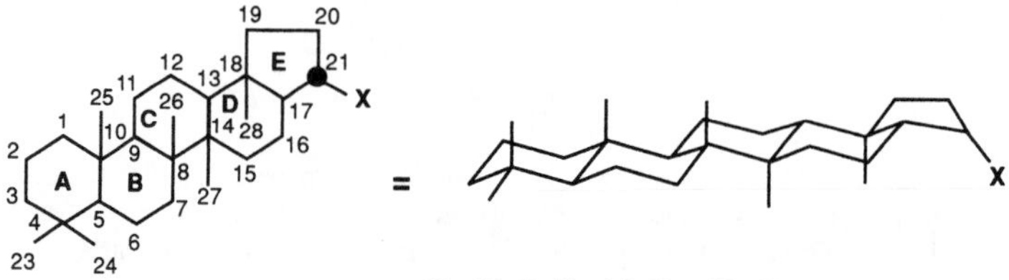

X = H, C_2H_5, iC_3H_7, iC_4H_9 . . . iC_8H_{17}

Figure 1.12 Labeling system for carbon atoms in steranes and triterpanes (cyclic terpenoids) with examples of projected three-dimensional structures.

TABLE I.2.8 Common Modifiers and Other Nomenclature Related to Biomarkers.

Modifier	Effect on biomarker
homo-	One additional carbon on structure
bis-, tris-, tetrakis- (also di-, tri-, tetra-)	Two, three, four additional carbons
pentakis-, hexakis- (also penta-, hexa-)	Five, six additional carbons
seco-	Cleaved C—C bond (specified)
nor-	One less carbon on structure
des-A (or de-A)	Loss of A-ring from structure
iso-	Methyl shifted on structure
neo-	Methyl shifted from C-18 to C-17 on hopanes
α	Asymmetric carbon in ring with functional group (usually H) down
β	Asymmetric carbon in ring with functional group (usually H) up
R	Asymmetric carbon in acyclic moiety obeying convention (Sec. 1.2.8.3) in a clockwise direction
S	Asymmetric carbon in acyclic moiety obeying above convention in a counterclockwise direction

is used, preceded by the number of the absent atom. Thus, "25-norhopanes" are identical to hopanes (pentacyclic triterpanes) except that a methyl group (C-25) has been removed from its point of attachment at the C-10 position (Fig. 1.12; Sec. 3.3.3). If two or three atoms are missing, the prefix "bisnor" or "trisnor," respectively, is used. Thus, 28,30-bisnorhopanes lack the C-28 and C-30 methyl groups found in hopanes. The 8,14-secohopanes consist of hopanes where the bond between the number 8 and 14 carbon atoms in the C-ring has been broken.

1.2.8.1 Chirality. Saturated carbon atoms are linked to their substituents by means of four, single covalent bonds which radiate outward toward the corners of a tetrahedron (Fig. 1.4). If all four substituents differ, the carbon atom at the center of the tetrahedron is asymmetric or "chiral" and two mirror-image structures, or **stereoisomers,** of the compound can be formed by transposing any two of the substituents. These stereoisomers show the same molecular formulae, but differ in the same manner as right- and left-handed gloves and are referred to as **enantiomers** (Fig. 1.13). In a molecule containing more than one asymmetric center, inversion of all the centers usually leads to the enantiomer (mirror image). Inversion of less than all the asymmetric centers yields a **diastereomer** or **epimer** as discussed in the following.

However, molecules that contain asymmetric carbon atoms, but are superimposable on their mirror images, are neither chiral nor asymmetric. For example, consider pristane (Fig. 1.14), which contains two asymmetric carbon atoms at the C-6 and C-10 positions. When the asymmetric carbon atoms are in the 6S,10R configuration or the 6R,10S configuration, the structures are identical and are called meso-pristane. The other pair of 6R,10R- and 6S,10S-pristane isomers are related as mirror images, and therefore, are chiral, and are enantiomers of each other. (For determining R versus S nomenclature, see following.)

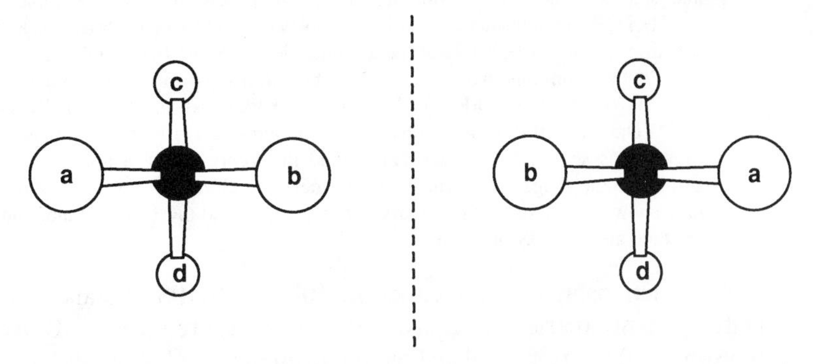

Figure 1.13 Asymmetric carbon atoms (black) are linked to four *different* substituents (symbolized as a, b, c, and d) by means of single covalent bonds which radiate outward toward the corners of a tetrahedron. Two mirror-image structures (similar to your right and left hands) are possible at each asymmetric center in a molecule. The vertical dashed line represents a mirror plane. The two structures shown are enantiomers because they are mirror images and cannot be superimposed without breaking bonds.

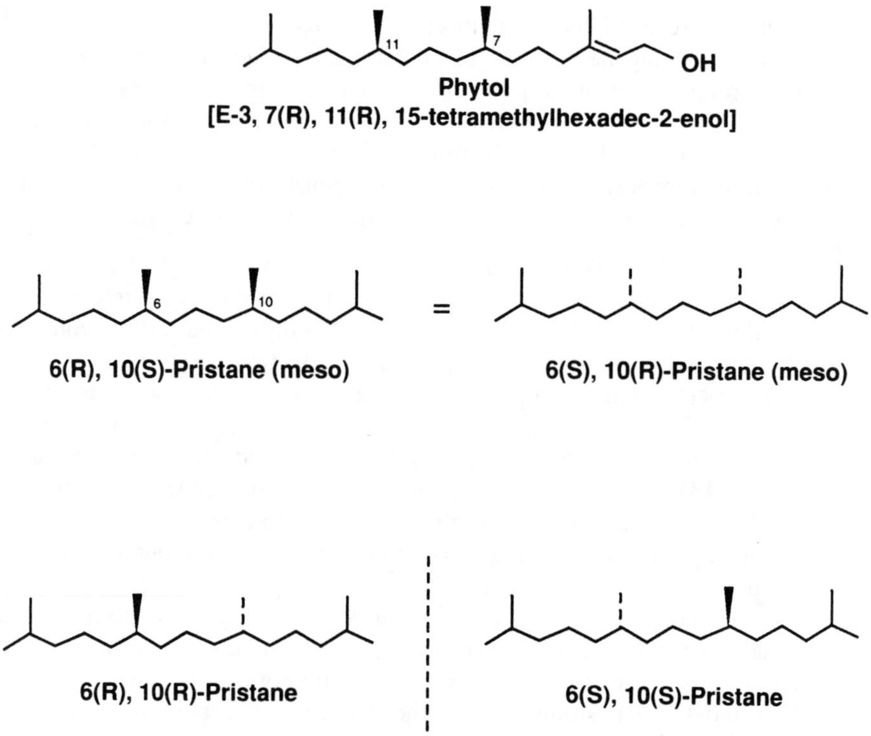

11 7 OH

Phytol

[E-3, 7(R), 11(R), 15-tetramethylhexadec-2-enol]

6 10 =

6(R), 10(S)-Pristane (meso) **6(S), 10(R)-Pristane (meso)**

6(R), 10(R)-Pristane **6(S), 10(S)-Pristane**

Figure 1.14 Phytol is a precursor of pristane and shows the same absolute stereochemistry as the biological (meso) configuration of pristane in immature sediments. Because of the hydroxyl group and double bond in phytol, application of the stereochemical rules of nomenclature discussed in the text result in a different numbering and stereochemical description for the asymmetric centers in the molecule [E-3,7(R),11(R),15-tetramethylhexadec-2-enol] compared to mesopristane. However, the absolute stereochemistry at the asymmetric centers in these molecules is identical. Pristane contains two asymmetric carbon atoms at positions C-6 and C-10. The 6(R),10(S)- and 6(S),10(R)-configurations are identical and are called meso-pristane. The 6(R),10(R)- and 6(S),10(S)-configurations are mirror images (enantiomers) as shown by the vertical dashed line representing a mirror plane. Like gammacerane (Fig. 1.15), these enantiomers are chiral but not asymmetric. Pristane shows two diasteromeric pairs; meso-pristane and 6R,10R-pristane, and meso-pristane and 6S,10S-pristane.

The relationship between either 6R,10R- or 6S,10S-pristane and meso-pristane is neither enantiomeric nor equivalent, but "diastereomeric." **Diastereomers** are stereoisomers that are not related as mirror images and show different physical and chemical properties (e.g., different melting and boiling points). Diastereomers are discussed further in Sec. 1.2.8.3.

In the strict sense, chirality is not equivalent to asymmetry. While all asymmetric molecules are chiral, certain chiral molecules are not asymmetric. For example, gammacerane (Fig. 1.15) is a biomarker that contains an axis of symmetry. If

Symmetry Axis

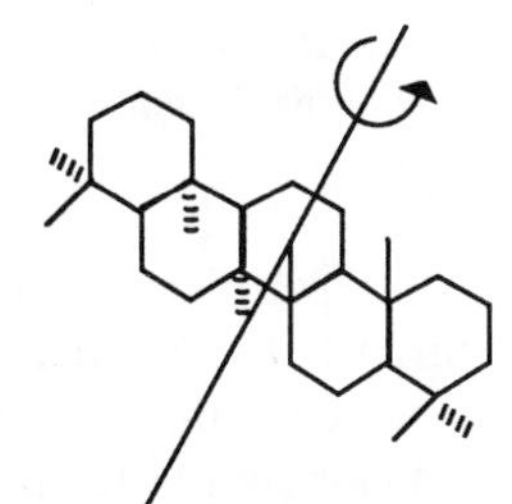

Gammacerane

Gammacerane Enantiomer

Figure 1.15 Gammacerane is an example of a chiral molecule which is not asymmetric. If rotated 180° about the symmetry axis, the resulting configuration is equivalent to that of the nonrotated molecule. However, gammacerane is not superimposable on its mirror image and these two compounds thus represent enantiomers.

rotated 180° about this axis, the resulting configuration is equivalent to that of the nonrotated molecule. Nevertheless, the mirror image of gammacerane is not superimposable and thus, represents an enantiomer. The 6R,10R- and 6S,10S-pristane isomers (Fig. 1.14) show the same level of symmetry as gammacerane (i.e., 180° axis of symmetry) and are therefore chiral, but not asymmetric. The reader is referred to texts on stereochemistry for more detailed discussions (Mislow, 1965; Cahn et al., 1966).

1.2.8.2 Optical Activity. Enantiomers show very similar chemical properties and cannot be separated using the achiral stationary phases in typical gas chromatographic columns (see following.) They can be separated using gas chromatography that employs chiral stationary phases, but this technique is generally not applied in geochemistry.

One property of enantiomers in solution is that the direction in which they rotate plane-polarized light differs. The direction and extent of rotation of plane-polarized light induced by an enantiomer is equal, but opposite to that induced by the other enantiomer. Nonbiologic (abiotic) synthesis of molecules containing asymmetric centers generally results in a racemic (50 : 50) mixture of right- and left-handed molecules. These mixtures are not optically active because the rotation of each enantiomer cancels that of the other.

However, many biologically formed compounds are optically active; that is, they rotate plane-polarized light clockwise (to the right or dextrorotatory) or counterclockwise (to the left or levorotatory). A unique feature of living organisms is that the enzymes responsible for biosynthesis of cellular materials are chiral and generate biomolecules that show only one configuration at certain asymmetric centers.

Note: *The preceding may indicate that either life originated only once, and that the original organisms contained only one configuration at each asymmetric center in their* *enzymes, or that the alternate "mirror-image organisms" became extinct. For example, enzymatic generation of lactic acid by an organism results in only the levorotatory compound, while abiotic synthesis results in a racemic mixture of the right- and left-handed lactic acid.*

Most petroleums show optical activity, evidence that they were formed, at least in part, by products from once-living organisms. Silverman (1971) provides classic evidence that biological lipids, dominantly steroids and triterpenoids, represent a major source for petroleum. Compared to other fractions, the 425 to 450°C distillation cut in a typical petroleum shows a low stable carbon isotope ratio and high optical rotation, implying a high concentration of relatively unaltered parent molecules. This distillation range corresponds to a molecular-weight range identical to that of many steroids and triterpenoids. These compounds are solids (melting points over 200°C) under conditions of deposition and are resistant to biodegradation and removal from sediments during diagenesis.

1.2.8.3 Stereochemical Nomenclature (R, S, α, and β). Stereochemical information is included in the names of biomarkers. For example, both 5α(H)- and 5β(H)-steranes are found in petroleum, but they differ in physical characteristics. The name "5α(H),14α(H),17α(H),20R cholestane" describes a single compound among millions that could be found in petroleum. A grasp of these stereochemical designations is critical for understanding the applications of biomarker geochemistry, especially those related to thermal maturation (Sec. 3.2.4).

Figure 1.16 is a conformational projection of cholestane (a C_{27} sterane) as it might appear in three dimensions. As an example, the figure shows the stereochemical designations for hydrogen atoms at positions 3 and 5 in the A-ring. (Normally, hydrogen atoms are not shown in these types of drawings.) The "α" hydrogens are below and the "β" hydrogens are above the plane of the molecule. In this particular

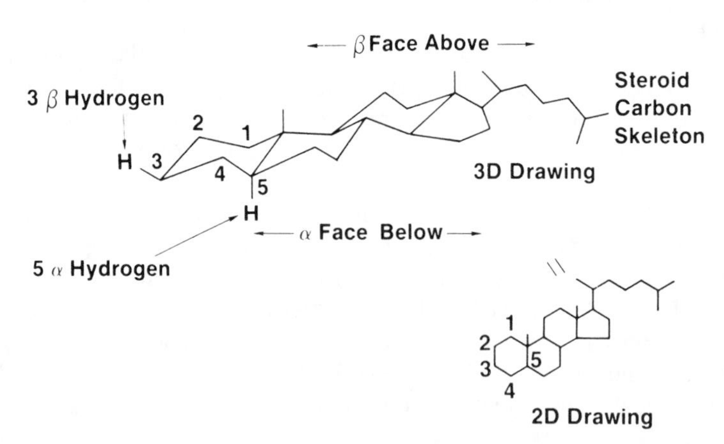

Figure 1.16 Procedure for designating α (below the plane) and β (above the plane) hydrogens or other groups for a polycyclic biomarker. The example is the C_{27} sterane, cholestane.

case, the 3β hydrogen is described as "equatorial" because the bond between the hydrogen and the position-3 carbon is oriented in the plane defined by the ring system. The 5α hydrogen is "axial" because the bond between the hydrogen and the position-5 carbon is oriented as an axis perpendicular to the plane defined by the ring system.

The two possible configurations of an asymmetric carbon atom are called "R" and "S" depending on a simple convention (Cahn et al., 1966), described as follows: *(Step 1.)* Orient the molecule so the smallest substituent, for example, a hydrogen atom (Fig. 1.17) points away from the viewer and the asymmetric carbon is closer. Imagine the bond between the asymmetric carbon atom and the smallest substituent represents the steering column in your car, while the remaining three substituents are the steering wheel. *(Step 2.)* Rank the three remaining substituents forming the steering wheel from largest to smallest based on atomic number; for isotopes priority is given to higher atomic weight; for different carbon substituents, the larger the group, the higher the priority. For example, an isopropyl group is larger than an *n*-butyl group, which is larger than an ethyl group. (If there are two carbon atoms attached to the atom next to the asymmetric carbon, the ranking is higher than if there is only one carbon atom. An isopropyl has a higher ranking than an *n*-propyl group or even an *n*-butyl group because the former has the carbon in question bound to two

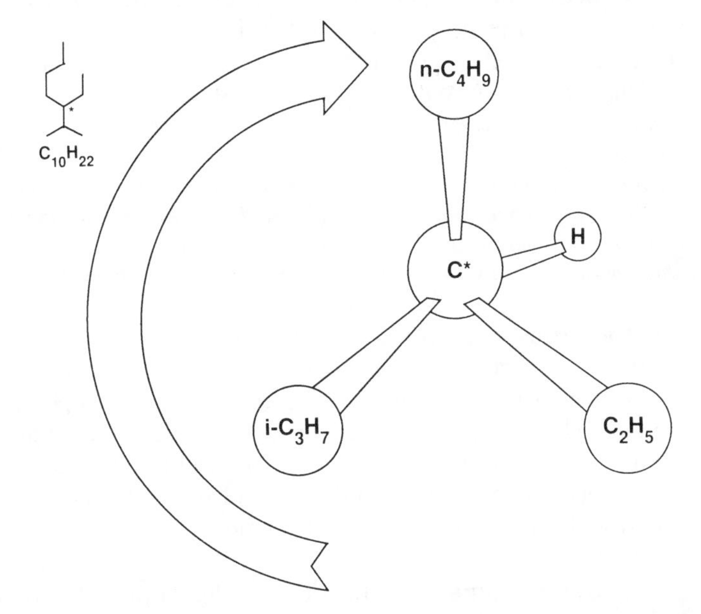

Figure 1.17 Application of the convention (text and Fig. 1.18) for describing stereochemistry of the asymmetric center in 2-methyl-3-ethylheptane ($C_{10}H_{22}$, a monoterpane). The two-dimensional stick notation on the left shows the location of the asymmetric center with an asterisk. The curved arrow shows the clockwise decrease in ranking priority for the three substituents nearest the observer. The three-dimensional example shows an "R" configuration at the asymmetric carbon atom.

carbon atoms compared to one for the latter, i.e., Fig. 1.17.) *(Step 3.)* Draw an imaginary arrow from the highest to the next highest ranked group on the steering wheel. The asymmetric carbon is in the R configuration (Latin "*rectus*," meaning right) if the arrow points clockwise on the wheel, or in the S configuration ("*sinister*," meaning left) if it points counterclockwise. The example, 2-methyl-3-ethylheptane, in Fig. 1.17 is in the R configuration according to these rules of nomenclature. This monoterpane was discussed in Sec. 1.2.7.

Figure 1.18 summarizes the three-step procedure for R and S stereochemical assignment using the C-20 position in cholestane as an example. The reader is encouraged to construct molecular models of cholestane and other biomarkers if a detailed understanding of stereochemistry is desired.

In the text, we restrict the use of R and S nomenclature to carbon atoms which are not part of a ring, and adopt the α (alpha = down) and β (beta = up) nomenclature common in natural product chemistry to describe asymmetric configurations at ring carbons. For example, cholestane can have the R or S configuration at C-20 or the α or β configuration at C-14 (Figs. 1.12 and 1.16).

Like pristane and cholestane, most biomarkers contain more than one asymmetric center. If at least one, but not all, of the asymmetric centers are the same, the resulting isomers that are not mirror images are called **diastereomers.** Pristane shows two diastereomeric pairs of compounds: (1) meso-pristane and 6R,10R pristane and (2) meso-pristane and 6S,10S pristane (Fig. 1.14). Unlike enantiomers, diastereomers typically show different chemical properties and are often separable by gas chromatography on normal, achiral stationary phases. Different diastereomers can show different mass spectra. (See following.)

> ***Note:*** *Stereoisomers are a special form of structural isomer. Most structural isomers, such as n-pentane and 2-methyl butane, show the same molecular formulae, but differ in the linkage between atoms. Stereoisomers show identical formulae and the same linkage between atoms; only the spatial arrangement of atoms differs.*

The R versus S and α versus β designations are a useful means of describing the relative configurations of compounds. On the other hand, the direction of rotation of plane-polarized light by a particular enantiomer is a physical property. There is no simple relationship between the absolute configuration of an enantiomer and its direction of rotation of plane-polarized light. The R versus S and α versus β designations are determined strictly on the basis of the convention as described (Cahn et al., 1966) without reference to optical rotation.

1.2.9 Stereoisomerization. The configurations at asymmetric centers imposed by enzymes in living organisms are not necessarily stable at the higher temperatures in buried sediments. Although the mechanism is unclear, "configurational isomerization" or **stereoisomerization** in saturated biomarkers occurs only at asymmetric carbon atoms where one of the four substituents is a hydrogen. Stereoisomerization could involve removal of a hydride ion (Ensminger, 1977) or a hydrogen radical (Seifert and Moldowan, 1980). The nearly planar carbocation (carbonium ion)

Step 1

Place the Lowest Priority Substituent Group (Smallest, Usually Hydrogen) so that it is <u>Behind</u> the Plane of the Paper and its Bond to the Asymmetric Carbon (C-20 of a Sterane in this Case) is Perpendicular to the Plane of the Paper

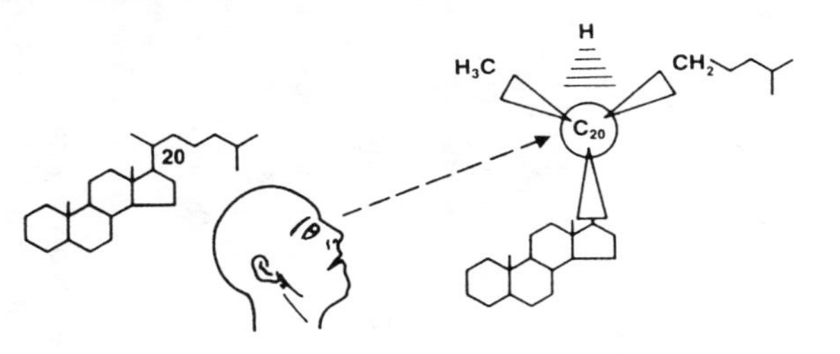

Step 2

Assign Priorities to the Substituent Groups that are in the Plane of the Paper. The Greater the Size, the Higher the Priority. (CH_3 < CH_2CH_3 < $CH(CH_3)_2$ etc.)

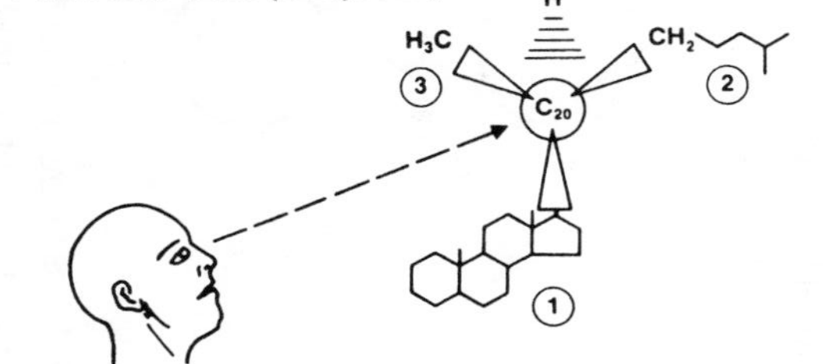

Step 3 (Final Step)

Make the Assignment by Noting in Which Direction (Clockwise or Counterclockwise) the Substituent Priorities <u>Decrease</u>

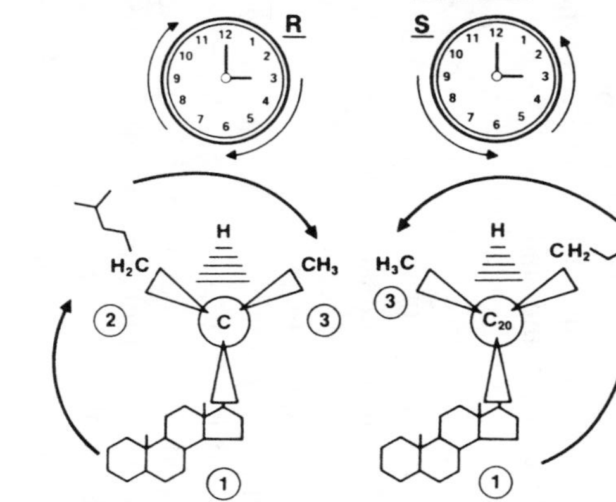

<u>Decreasing</u> Substituent Priority Size Counted in the Clockwise <u>Direction</u>

<u>Decreasing</u> Substituent Priority Size Counted in the <u>Counterclockwise</u> Direction

Figure 1.18 Three-step procedure for R versus S stereochemical assignments in noncyclic portions of biomarkers. *Concept for figure courtesy of R. M. K. Carlson.*

or radical [see sp^2 hybridization, Fig. 1.5(a)] can regain a hydrogen from the same side, resulting in the same configuration, or from the opposite side, resulting in the inverted configuration (Fig. 1.19). Depending on the kinetics of the reaction and the stability of the products, the resulting compound may show any proportion of the two possible configurations (R versus S when the asymmetric center is not part of a ring or α versus β when part of a ring). Configurational isomerization occurs only when cleavage and renewed formation of the bonds results in an inverted configuration compared to the starting asymmetric center. A complete discussion of organic chemical mechanisms is beyond the scope of this work. The reader may find such information in any organic chemistry textbook.

For acyclic (chain) carbon atoms that are asymmetric, the two possible isomer configurations generally show similar stabilities. For example, stereochemistry at the C-20 position in steranes proceeds during isomerization from nearly 100 percent 20R in shallow sediments to a nearly equal mixture of 20R and 20S diastereomers in deeply buried source rocks and petroleum. The "equilibrium" ratio for 20S/(20S + 20R) in C_{29} steranes is about 0.52 to 0.55, slightly favoring the 20S diastereomer (Sec. 3.2.4.2).

Because of steric forces imposed by a rigid cyclic structure, asymmetric centers that are part of a saturated ring system usually show two configurations with quite different thermal stabilities. Thus, the C-14 and C-17 positions in the C_{29} steranes each isomerize to the more stable β configuration during burial heating. The equilibrium ratio for $14\beta(H),17\beta(H)/[14\alpha(H),17\alpha(H) + 14\beta(H),17\beta(H)]$ is about 0.67 to 0.71 (Sec. 3.2.4.2).

Intramolecular (between different locations within the same molecule) rearrangements of hydrogen or alkyl groups can also occur and are important, for example, in the formation of diasterenes (rearranged sterenes) from sterenes during burial diagenesis. These reactions will be discussed (Sec. 3.2.4.2).

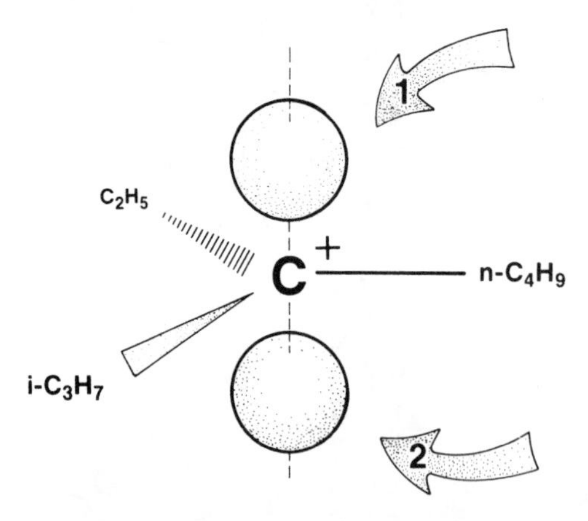

Figure 1.19 Example of nearly planar carbocation (carbonium ion) formed by removal of a hydride ion from the asymmetric carbon in 2-methyl-3-ethylheptane. The p orbital, indicated by the stippled electron clouds above and below the plane of the sigma bonds, does not contain an electron. Reattachment of a hydride ion from above (arrow 1) or below (arrow 2) the plane of the sigma bonds in the figure would result in the "R" or "S" configuration at the asymmetric carbon, respectively.

1.2.10 Stereochemistry of selected biomarkers. The stereochemistry of hydrogen and methyl groups, for example, at the 6 and 10 positions in pristane and phytane (Fig. 1.14), at the 5, 14, 17, and 20 positions in the steranes, and the 17 and 21 positions in the pentacyclic triterpanes (Fig. 1.12) determines the various isomers of these compounds and their chemical properties. The reader may wish to construct molecular models of some of the compounds that will be described to help visualize differences between the various stereoisomers.

1.2.10.1 Paraffins and Acyclic Isoprenoids. Because at least two hydrogens are bound to each carbon atom in *n*-paraffins, no asymmetric centers are possible. However, acyclic isoprenoids can show many asymmetric centers where methyl side branches are attached to the chain.

Phytol is a major precursor to the pristane and phytane in petroleum. Phytol shows a single stereoisomer (Fig. 1.14) as synthesized by enzymes in living organisms, where the asymmetric carbon atoms at positions 7 and 11 are both in the R configuration (Maxwell et al., 1973). Because phytol is an alcohol, carbon atoms are numbered consecutively from position 1 nearest the hydroxyl group. (See Fig. 1.14.) When the hydroxyl group and double bond in phytol are removed to produce pristane and phytane, the numbering of carbon atoms changes (Fig. 1.14). Positions 7 and 11 in phytol become positions 10 and 6 respectively in pristane and phytane. Loss of the hydroxyl group from phytol also changes the priority ranking of the substituents around the asymmetric carbon at C-7 (C-10 in pristane and phytane). Without changing the absolute stereochemistry of the groups around C-7, the rules as described require that the C-10 position in the resulting pristane or phytane be named as being in the S configuration.

Pristane in immature sediments shows a configuration dominated by 6R,10S (equivalent to 6S,10R; Fig. 1.14) stereochemistry. This stereochemistry is directly comparable to that at the 7 and 11 positions in phytol. (As previously described, loss of the hydroxyl group and double bond from phytol causes a change in the priority of substituents around position 11 resulting in a change from R to S.) During thermal maturation, isomerization at these positions results in an equilibrium mixture of 6R,10S; 6S,10S; and 6R,10R in the ratio 2 : 1 : 1 (Patience et al., 1980) (Fig. 1.14). Direct analysis of the eight possible isomers resulting from the three asymmetric centers in phytane is not possible using currently available gas chromatographic columns.

1.2.10.2 Terpanes. Hopanes with 30 carbon atoms or less show asymmetric centers at C-21 and at all ring junctures (C-5, C-8, C-9, C-10, C-13, C-14, C-17, and C-18). Hopanes with more than 30 carbon atoms are called **homohopanes,** the homo- prefix referring to additional methylene groups on the parent molecule, hopane. Common homohopanes show an extended side chain with an additional asymmetric center at C-22 (Fig. 1.20), which results in two peaks for each homolog (22R and 22S) on the mass chromatograms for these compounds (e.g., Fig. 2.12, peaks 22 to 35).

The hopanes are composed of three stereoisomeric series, namely 17α(H),21β(H)- and 17β(H),21β(H)-hopanes, and 17β(H),21α(H)-hopanes. Compounds in the $\beta\alpha$ series are called **moretanes** (Fig. 1.20). The α and β notations indicate whether the hydrogen atoms are below or above the plane of the rings,

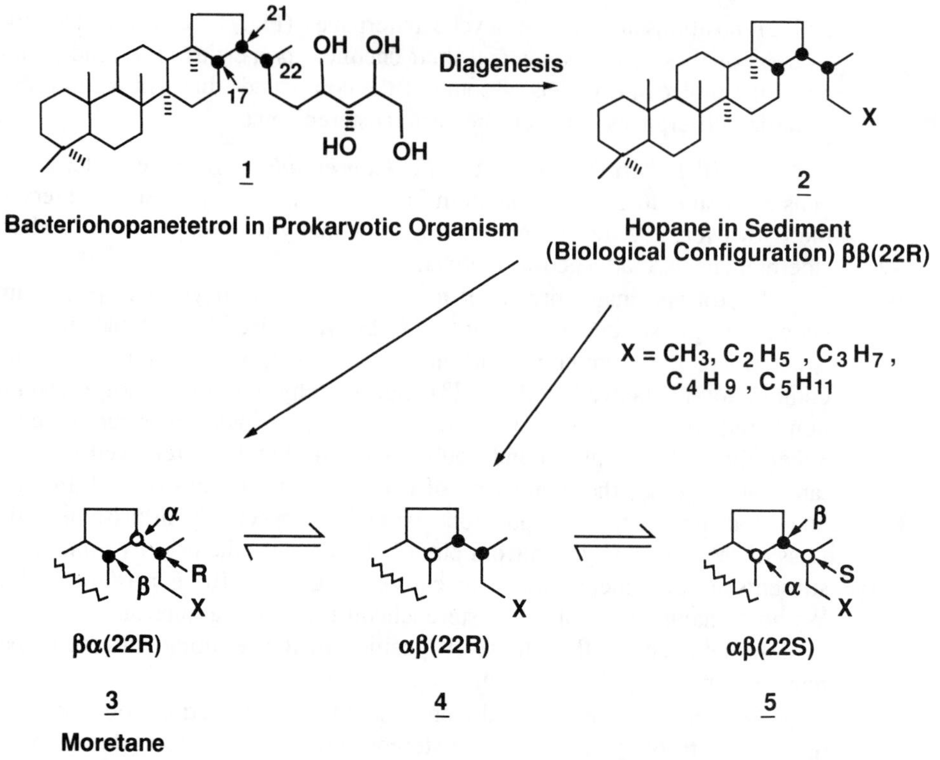

Bacteriohopanetetrol in Prokaryotic Organism

Hopane in Sediment (Biological Configuration) ββ(22R)

Diagenesis

X = CH_3, C_2H_5, C_3H_7, C_4H_9, C_5H_{11}

βα(22R) 3 Moretane

αβ(22R) 4

αβ(22S) 5

Figure 1.20 Origin of hopanes in petroleum from bacteriohopanetetrol (1) found in the lipid membranes of prokaryotic organisms (Peters and Moldowan, 1991). Stereochemistry is indicated by open (α) and solid (β) dots (hydrogen directed into and out of the page, respectively). The "biological" configuration [17β(H),21β(H),22R] imposed on bacteriohopanetetrol and its immediate saturated product (2) by enzymes in the living organism is unstable during catagenesis and undergoes isomerization to "geological" configurations (3,4,5) discussed in the text. The 17β,21α(H)-hydrocarbons (3) are called moretanes while all others are hopanes (2,4,5).

respectively, (Sec. 1.2.8.3). Hopanes with the 17α(H),21β(H) configuration ($\alpha\beta$) in the range C_{27} to C_{35} are characteristic of petroleum because of their greater thermodynamic stability compared to the other epimeric ($\beta\beta$ and $\beta\alpha$) series (Sec. 3.2.4). The $\beta\beta$ series is not generally found in petroleum (see following) because it is thermally unstable even during early catagenesis (Sec. 3.2.4.1). Hopanes of the $\alpha\alpha$ series are not natural products and it is unlikely that they occur above trace levels in petroleum (Bauer et al., 1983).

Note: *C_{28} $\alpha\beta$-hopanes are rare or absent in petroleum for reasons similar to those presented for the scarcity of regular C_{17} acyclic isoprenoids (Sec. 1.1.7; however, see Sec. 3.3—C_{28}-C_{34} 30-nor-17α(H)-hopanes). Formation of C_{28} hopanes would require cleavage of not one, but two carbon-carbon bonds attached to position 22 in the C_{35}*

hopanoid precursor (Fig. 1.20). Cleavage of the single carbon-carbon bond between positions 21 and 22 (resulting in C_{27} hopane) or cleavage of either of the other two carbon-carbon bonds attached to C-22 (resulting in C_{29} hopane) is far more likely than sequential cleavage of two carbon-carbon bonds. A C_{28} compound formed by processes involving loss of a methyl group from the hopanoid ring system would represent another compound class. For example, loss of the C-25 methyl group from a hopane results in a 25-norhopane (Sec. 3.3.3).

The major precursors for the hopanes in living organisms, bacteriohopanetetrol and related bacteriohopanes (Fig. 1.20), show 17β(H),21β(H)- or "biological" stereochemistry. The biological configuration is nearly flat, although puckering of the carbon-carbon bonds in the rings results in a three-dimensional shape. Like the sterols (as will be discussed), bacteriohopanetetrol is **amphipathic** because it contains both polar and nonpolar ends (Fig. 1.21). The flat configuration and amphipathic character appear necessary for bacteriohopanetetrol to fit into the lipid membrane structure (Rohmer, 1987). Because this stereochemical arrangement is thermodynamically unstable, diagenesis and catagenesis of bacteriohopanetetrol result in a transformation of the 17β(H),21β(H) precursors to the 17α(H),21β(H)-hopanes and 17β(H),21α(H)-moretanes. Similarly, the biological 22R configura-

Bacteriohopanetetrol

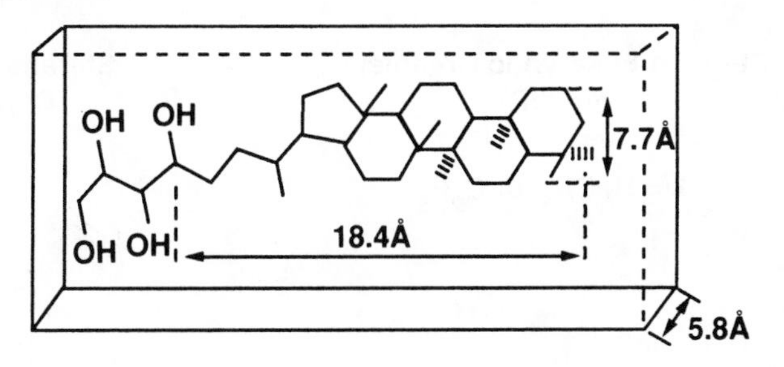

Polar End

Nonpolar End

Cholesterol

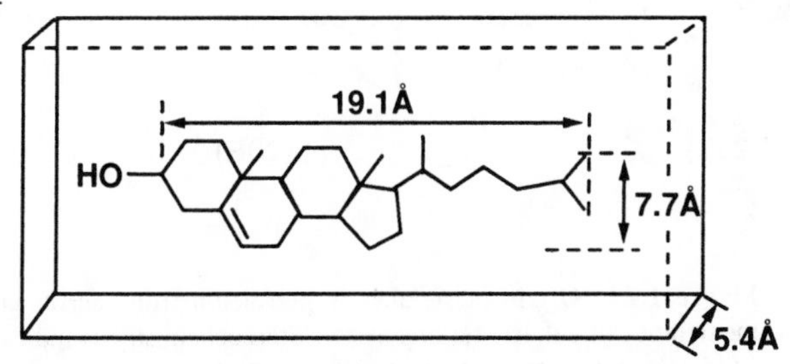

Figure 1.21 Bacteriohopanetetrol (a hopanoid in prokaryotes) and cholesterol (a steroid in eukaryotes) are similar in size and amphipathic character and both appear essential to lipid membranes in living organisms.

tion found in bacteriohopanetetrol converts to an equilibrium mixture of 22S and 22R $\alpha\beta$-homohopanes, showing a 22S/(22S + 22R) ratio for the C_{31} homolog of about C.57 to 0.62 in most crude oils (Seifert and Moldowan, 1980).

> ***Note:*** *It has recently been shown that bacteriohopanetetrol is synthesized by bacterial addition of a C_5 sugar (a D-ribose derivative) to a C_{30} hopanoid [possibly hop-22(29)-ene] (Flesch and Rohmer, 1988). In addition to bacteriohopanetetrol, other polyfunctional bacteriohopanoids (C_{35}) with additional functionalities on the side chain are known (Rohmer, 1987). The C-35 hydroxy group (—OH) is sometimes replaced by an amino group (—NH_2).*

1.2.10.3 Steranes. The sterols in eukaryotic organisms are precursors to the steranes in sediments and petroleum (Figs. 1.22 and 1.23) (Mackenzie et al.,

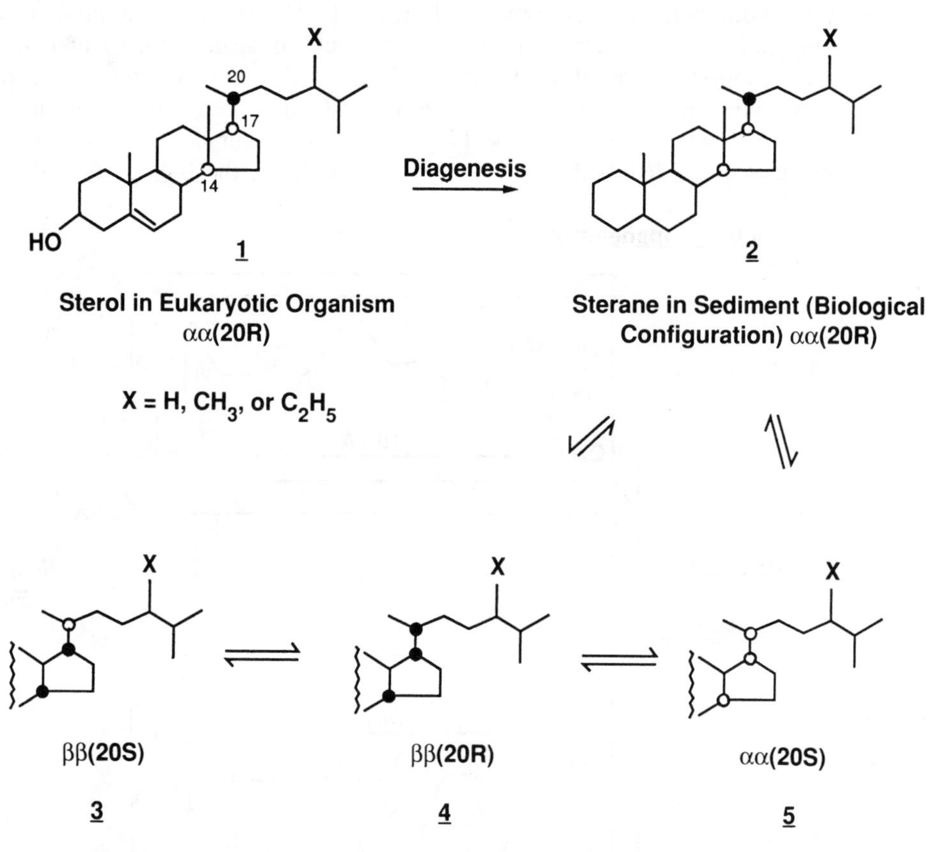

Figure 1.22 Origin of steranes in petroleum from sterols (1) found in the lipid membranes of eukaryotic organisms. Stereochemistry is indicated by open (α) and solid (β) dots (hydrogen directed into and out of the page, respectively). The "biological" configuration [14α,17α(H),20R] imposed on the sterol precursor and its immediate saturated product (2) by enzymes in the living organism is unstable during catagenesis and undergoes isomerization to "geological" configurations (3,4,5) discussed in the text.

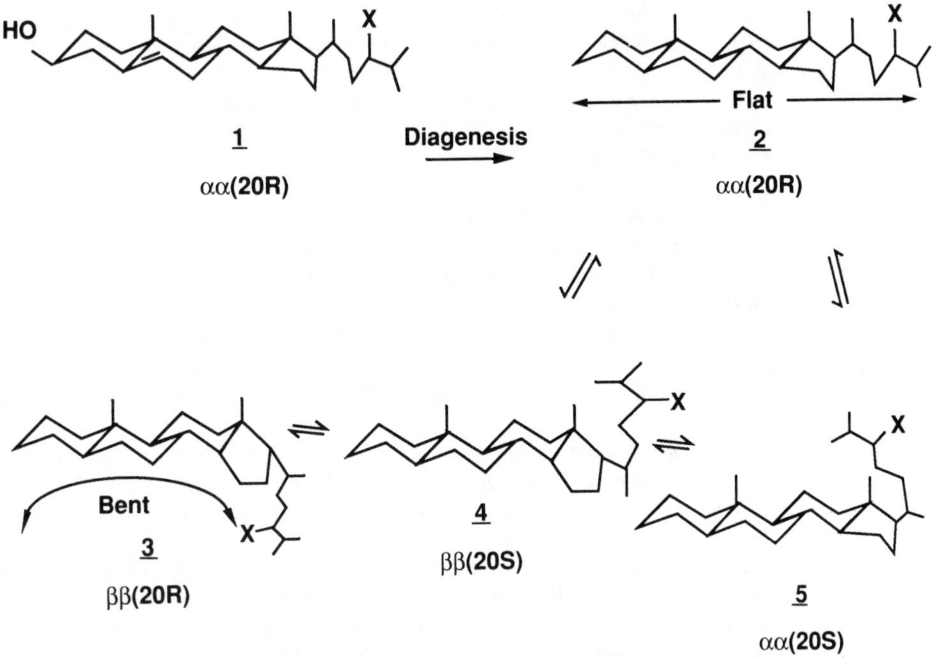

Figure 1.23 Origin of steranes from sterols as in Fig. 1.22, but projecting structures from three dimensions.

1982a; de Leeuw et al., 1989). Because of the large number of asymmetric centers in sterols, very complex mixtures of stereoisomers are possible. For example, cholesterol has eight asymmetric centers and thus, might be expected to show as many as 2^8 or 256 stereoisomers. However, because of the highly specific biosynthesis of cholesterol, only one stereoisomer of cholesterol exists in living organisms.

Sterols are amphipathic and show molecular dimensions similar to bacteriohopanetetrol. Like bacteriohopanetetrol in prokaryotes, the "flat" configuration allows sterols to fit into and increase the rigidity of cell membranes in eukaryotes (Fig. 1.25). The "flat" sterols show three-dimensional character (Fig. 1.23) and are unlike benzene, where all 12 carbon and hydrogen atoms lie in one plane (Fig. 1.6).

Sterols in living organisms show the following configuration: $8\beta(H),9\alpha(H),10\beta(CH_3),13\beta(CH_3),14\alpha(H),17\alpha(H)$,20R. The note that follows explains why some of these asymmetric positions, in particular those at C-14, C-17, and C-20, are of greater importance to the geochemist than others.

Note: *During diagenesis and catagenesis, the configurations at C-10 and C-13 cannot be changed because stereoisomerization mechanisms require that a hydrogen atom be attached to the asymmetric carbon atom. Furthermore, stereoisomerization at C-8 and C-9 does not occur because the biological configuration at these positions is energetically highly favorable. Thus, the biological configurations at these positions are not changed during diagenesis and catagenesis and they are of little use in characterizing these processes.*

Both the C-5 and C-24 positions in sterols from organisms consist of mixtures of α(H)- and β(H)-configurations. However, all sterols in living organisms appear to show only the 20R configuration, as will be discussed. The C-24 position in sterols can be R or S, but is usually either one or the other in a given organism. In sediments and petroleum however, sterols consist of mixed 24R and 24S ergostane (C_{28}) and stigmastane (C_{29}). The 24R and 24S mixture may also derive in part by diagenetic hydrogenation of a double bond at C-24 found in some sterols.

Tsuda et al. (1958) report the occurrence of the 20S epimer of fucosterol (sargasterol) in living organisms. Verification of this compound by additional work has been attempted without success and there are theoretical reasons for it not existing (Nes and McKean, 1977 and references therein).

Because most sterols carry a double bond at C-5, the stereochemistry at the C-5 position is determined largely by the reduction (hydrogenation) of this double bond during early diagenesis. This hydrogenation gives 5α- and 5β-mixtures favoring the 5α-epimer by 2 : 1 to 10 : 1. Stanols and some sterols that have saturated C-5 positions occur in minor concentrations in some organisms. These usually carry 5α(H)-stereochemistry and may become 5α(H)-steranes during early diagenesis. Bile acids (steroids secreted by the liver), however, carry 5β(H)-stereochemistry, but probably contribute little to sediments. Equilibration of stereochemistry at the C-5 position in steranes greatly favors the 5α(H)- over 5β(H)-stereochemistry. In thermally mature petroleum, the 5β(H)-compounds exist only as trace components, requiring special analytical techniques for detection. The 5β(H)-compounds are, therefore, generally disregarded in biomarker applications. However, when abundant, they can be useful for indicating low thermal maturity.

The important asymmetric centers during catagenesis of steranes are at C-14, C-17, and C-20 (Fig. 1.22). Partly because C-20 is in the sterol side chain, relatively free from steric effects imposed by the cyclic system, the biologically derived 20R isomer is converted to a near-equal mixture of 20R and 20S [at equilibrium 20S/(20S + 20R) = 0.52 to 0.55 for the C_{29} homologs]. Furthermore, the flat configuration imposed by the 14α(H),17α(H) stereochemistry in the sterol is lost in favor of the thermodynamically more stable 14β(H),17β(H) form (Fig. 1.23).

Isomerization of the 5α(H),14α(H),17α(H),20R configuration (αααR) inherited from living organisms results in increasing amounts of the other possible stereoisomers until the equilibrium ratio for αααR, αααS, αββR, and αββS is about 1 : 1 : 3 : 3 (Figs. 1.22 and 1.23). These distributions have been duplicated in: (1) laboratory experiments where αααR steranes were heated using platinum-carbon catalyst (Petrov et al., 1976; Seifert and Moldowan, 1979), and (2) theoretical calculations for several cholestane (C_{27}) isomers (van Graas et al., 1982; Pustil'nikova et al., 1980). Section 3.2.4.2 describes the use of sterane isomerization ratios as thermal maturity indicators for petroleum.

1.3 Examples of Biomarkers, Their Abundance, and Precursors

Although biomarkers are "trace" components in petroleum, they commonly represent the most abundant compounds whose structures are defined. For example, Yallourn lignite in Australia contains several hundred parts per million of a single C_{32} hopanoid acid (Ourisson et al., 1984). Although this amount may seem insignificant, a 1-meter cube of the rock, weighing about 2 tons contains approximately 1 kg of the acid, thus, making it easily the most abundant defined organic substance in the coal.

Figure 1.24 shows the quantitative distributions of various compound fractions and biomarker classes in an oil from Wyoming. Like many others, the Wyoming oil contains individual biomarkers in concentrations ranging from tens to hundreds of parts per million.

Bacteriohopanetetrol (a hopanoid), cholesterol (a steroid), and chlorophyll (a tetrapyrrole pigment) are three examples of complex compounds in living organisms that can be traced to biomarkers in petroleum (Figs. 1.20, 1.22, and 1.27). Among

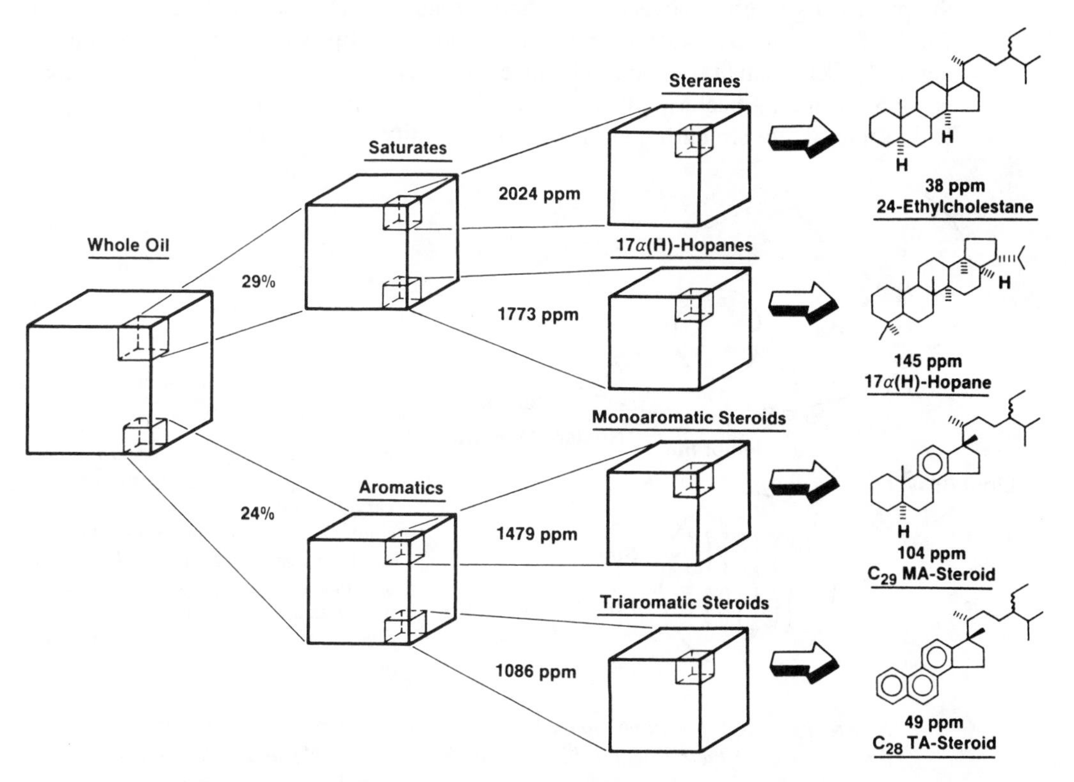

Figure 1.24 Biomarkers are "trace" components measured in parts per million (ppm) in most petroleums (e.g., 2024 ppm = 0.2024% steranes in oil). The figure shows the relative abundances of various compound fractions (Sec. 2.4) and individual biomarkers in a biodegraded oil (17° API, 3.2. wt % sulfur) from Hamilton Dome, Wyoming (GCMS No 81-3). Despite their low abundance compared to fractions such as the aromatics, biomarkers commonly represent some of the most abundant structurally-defined compounds in petroleum. Biomarkers used as examples in the figure (far right) include: C_{29} $5\alpha,14\alpha,17\alpha$(H),24S- and 24R-ethylcholestane, C_{30} $17\alpha,21\beta$(H)-hopane, C_{29} 5α(H),20R,24R- and 24S-monoaromatic steroid, and C_{28} 20S,24R and 24S-triaromatic steroid.

other compounds, bacteriohopanetetrol is converted to hopanes, cholesterol to cholestane, and chlorophyll to porphyrins, pristane, and phytane.

1.3.1 Bacteriohopane. Hopanes are abundant in sediments and petroleum compared to other compounds whose structures are known, because their hopanoid precursors are important membrane components in living cells and are resistant to degradation during diagenesis. Cell membranes are typically composed of two layers of **amphipathic** lipids (Fig. 1.25), which consist of polar and nonpolar ends (Fig. 1.21). The nonpolar end of an amphipathic lipid is hydrophobic, and thus, does not readily associate with water or other polar solvents. The polar end of an amphipathic lipid however, is hydrophilic and is readily soluble in water. In a cell membrane, the hydrophobic ends of the lipids face inward toward the center of the lipid bilayer and the polar ends face the aqueous environment surrounding the cell or the cytoplasm within the cell (mostly water). Figure 1.25 shows examples of lipid bilayers in a living cell. Cell membranes are dominated by acyclic lipids, but polycyclic triterpenoids, including bacteriohopanetetrol (Fig. 1.20), are also present. These apparently serve to add strength and increase the rigidity of the membrane.

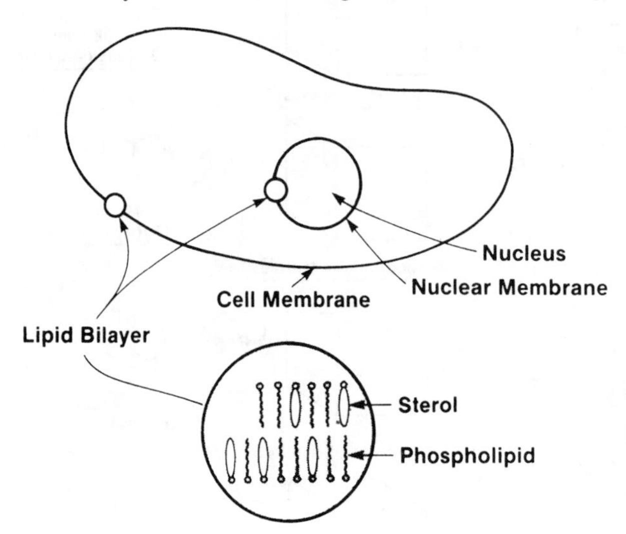

Figure 1.25 Bilayers composed of amphipathic lipids enclose the cell (prokaryotes) and internal organelles such as chloroplasts, mitochondria (not shown), and the nucleus (eukaryotes). Polar and nonpolar ends of the lipids are indicated by circles and ovals/wavy lines, respectively. *Concept for figure courtesy of R. M. K. Carlson.*

Note: *Extended homohopanoids, such as bacteriohopanetetrol in bacteria, were found only after their original discovery as hopanes in rocks and sediments. Unlike the amphipathic hopanoids, hopanes show low polarity and are readily extracted from sediments using nonpolar solvents like hexane. Consequently, hopanes were discovered early. On the other hand, hopanoids such as bacteriohopanetetrol contain side chains loaded with polar hydroxyl groups, while the rest of the molecule is comparatively nonpolar. These amphipathic compounds will only dissolve in a mixture of polar and nonpolar solvents. Bacteriohopanetetrol was only discovered after the development of natural product chemistry, when extraction with mixtures of polar and nonpolar solvents became more common.*

Bacteriohopanetetrol serves as a constituent of the cell membrane in prokaryotes (Ourisson et al., 1984). **Prokaryotes** such as bacteria and cyanobacteria are primitive organisms without nuclei or other organelles. Dehydration and reduction of bacteriohopanetetrol during diagenesis result in bacteriohopane, which belongs to the triterpane class of biomarkers (Fig. 1.20). Excluding **kerogen**, which is insoluble in organic solvents and represents about 90 percent of the organic carbon in sediments, it is calculated that hopanoids account for 5 to 10 percent of the remaining soluble organic carbon (Ourisson et al., 1984). Evidence shows that bacteriohopanetetrol can also be incorporated into kerogen and later released as hopane hydrocarbons during catagenesis (Mycke et al., 1987; Sinninghe Damsté and de Leeuw, 1990). In addition, significant amounts of the preserved organic matter in source rocks represent the remains of bacterial biomass itself (Ourisson et al., 1984).

1.3.2 Cholestane. Cholesterol is a member of the class of four-ringed alcohols called sterols (Fig. 1.21). It is found in the membranes of all eukaryotic cells, but very few, if any prokaryotes (Rohmer, 1987). **Eukaryotes** contain nucleated cells with organelles, such as mitochondria and chloroplasts (Fig. 1.25). All higher organisms are eukaryotes. In eukaryotic membranes some of the straight-chain lipids are replaced by sterols, whose tetracyclic structures apparently act to increase rigidity and strengthen the membrane (Ourisson et al., 1984). Most eukaryotic organisms contain substantial amounts of sterols, while investigations of several species of prokaryotes failed to reveal any above 0.0001 percent of dry cell weight (Schubert et al., 1968). However, a few reports suggest that sterol synthesis occurs in some prokaryotic species. Bouvier et al. (1976) identified 4,4-dimethyl- and 4α-methyl-5α-cholest-8(14)-en-3β-ol and their 8(14),24-diene analogs in abundance in the archaebacterium *Methylococcus capsulatus*. In more recent work Kohl et al. (1983), show substantial concentrations of cholest-8(9)-en-3β-ol in *Nannocystis exedens* (several percent). *Thioploca* and several other bacteria also contain 4,4-dimethyl sterols (McCaffrey et al., 1989).

Note: *Evidence suggests that the sterols in eukaryotes are structurally equivalent to the bacteriohopanetetrols in prokaryotes (Fig. 1.21) and that the biosynthetic pathway to sterols has been only slightly modified from that required to synthesize bacteriohopanetetrols (Rohmer et al., 1979). Both sterols and bacteriohopanetetrols appear derived from squalene, a terpenoid, because it is composed of isoprene subunits (Fig. 1.26). Because oxygen is required during biosynthesis of sterols, but is not needed for bacteriohopanetetrol, the biosynthetic pathway to sterols probably evolved during or after aerobic conditions were established on the primitive earth.*

Your body manufactures cholesterol, but it can also be ingested in food. The average person contains 300 to 600 mg of sterol/100 g of wet weight. A 150-pound (68 kg) individual contains about 306 g of cholesterol (Nes and McKean, 1977). A diet of fatty foods contributes to high levels of blood cholesterol commonly found in people suffering from arteriosclerosis. Arteriosclerosis occurs when cholesterol and other fatty substances become embedded in the walls of arteries, forming what are called plaques that gradually block arteries.

Cholesterol is transported in the blood by two classes of lipoproteins. Low density lipoproteins (LDL) carry cholesterol to the organs and tissues where it is used by cells. Excess LDL is responsible for deposits of cholesterol that result in plaque. For this reason, LDL is sometimes called "bad cholesterol." High density lipoproteins (HDL or "good cholesterol") are believed to carry excess cholesterol from cells to the liver, where it is removed from the system.

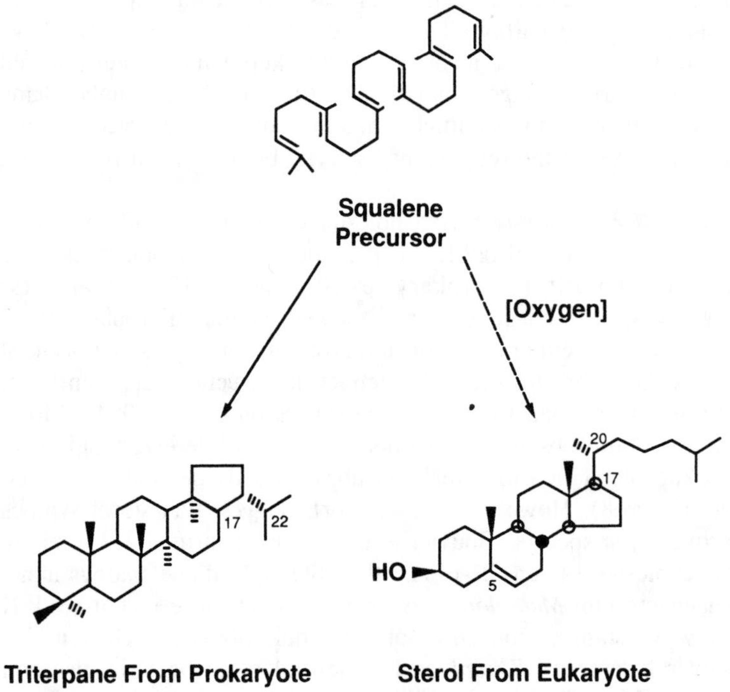

Squalene Precursor

[Oxygen]

Triterpane From Prokaryote

Sterol From Eukaryote

Figure 1.26 Hopanoids in prokaryotes (left) and steroids in eukaryotes (right) appear to be derived from the same precursor, squalene, by only slightly different enzymatic pathways. Because oxygen is required during biosynthesis of sterols, but is not needed for bacteriohopanetetrol, it appears that the biosynthetic pathway to sterols evolved after aerobic conditions were established on the primitive earth. The arrows in the figure do not imply that squalene is the direct precursor of the triterpane and sterol shown. Rather, there are several biosynthetic steps involved in these generalized reactions.

Cholesterol is converted by dehydration and reduction to cholestane during diagenesis (Fig. 1.22). Thus, like the hopanes, steranes are abundant in sediments, rocks, and petroleums compared to other compounds whose structures are known, because their precursors (sterols) are so common in living organisms.

Note: *Living organisms can be divided into three kingdoms: (1) archaebacteria, (2) eubacteria, and (3) eukaryotes. Unlike the archaebacteria and eubacteria, which are* *prokaryotes, the eukaryotes contain a membrane-bound nucleus and complex organelles for carrying out important functions. Two of the most common organelles include mitochondria, used in energy generation, and chloroplasts, used in photosynthesis. The morphologic differences between these kingdoms reflect fundamental differences in biochemistry.*

1.3.3 Porphyrins and acyclic diterpanes. Chlorophylls are tetrapyrrole pigments required for the photosynthetic fixation of carbon by plants and certain bacteria. They are by far the most abundant tetrapyrroles in the biosphere. Other tetrapyrroles include heme in the blood of many animals. The ratio of chlorophyll-to-heme type tetrapyrroles in living organisms is estimated at about 100,000 : 1 (Baker and Louda, 1986 and references therein).

Treibs (1936 and references therein) showed the link between chlorophyll-a in living photosynthetic organisms and porphyrins in petroleum, thus, providing the first strong evidence for an organic origin of petroleum. During diagenesis and catagenesis, a wide variety of chlorophyll-related pigments (e.g., Bidigare et al., 1990; Keely et al., 1990) are converted to various products, including three common constituents in petroleum: porphyrins and the acyclic diterpanes pristane and phytane (Fig. 1.27). Sedimentary porphyrins have been linked to chlorophylls in eukaryotes which yield a C_{32} carbon skeleton and to chlorophylls in prokaryotic bacteria having extended side chains. These transformations involve a complex series of reactions, similar, but not identical to those proposed by Treibs (1936). In simple terms, the reactions summarized by Baker and Louda (1983) and Filby and Berkel (1987) proceed as follows. During early diagenesis, magnesium is removed from the chlorophylls (demetallation) and they are defunctionalized to phorbides, chlorins, and purpurins. During middle diagenesis, these intermediates are aromatized in noncoaly sediments to "free base" porphyrins which chelate with metal ions (Lewan, 1984) to form immature metalloporphyrins (geoporphyrins) during late diagenesis. Subsequently, the immature metalloporphyrins undergo alteration during catagenesis and destruction during metagenesis. While most porphyrins appear derived from various chlorophylls, recent evidence from stable carbon isotope ratios of individual porphyrins suggests that at least one C_{32}-etioporphyrin originates from hemes (Boreham et al., 1989; Ocampo et al., 1989).

Apparently aromatization of chlorins need not always occur prior to metal incorporation. The major chlorin in Messel oil shale is a nickel complex of mesopyrophaeophorbide-a (Prowse et al., 1990). The structure of Ni-mesopyrophaeophorbide-a clearly indicates an origin from chlorophyll. However, this chlorin has been metallated without prior aromatization to form a porphyrin.

Porphyrins are an extremely complex group of metallated tetrapyrrolic compounds (Sundararaman, 1985). The structures of many sedimentary porphyrins have been fully or partly characterized (Chicarelli et al., 1987; Callot et al., 1990). Deoxophylloerythroetioporphyrins (DPEP) and etioporphyrins (etio) are the two major series of porphyrins in petroleum. They are most commonly complexed to vanadyl (VO^{2+}) or nickel (Ni^{2+}) ions in source rock bitumens and petroleum, although higher

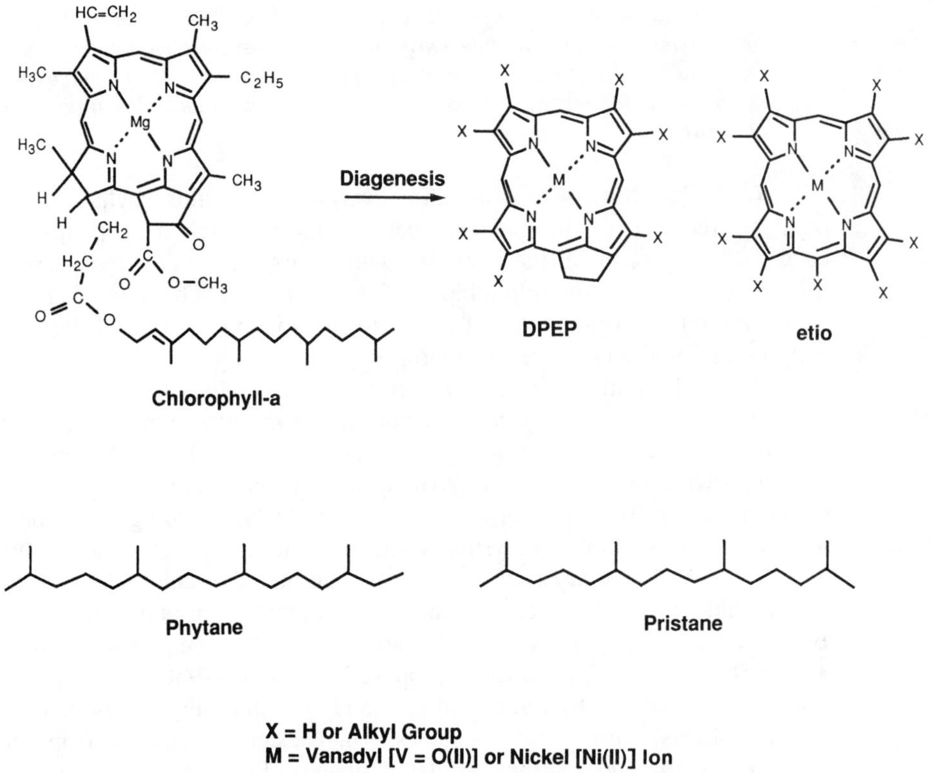

Diagenesis

Chlorophyll-a

DPEP

etio

Phytane

Pristane

X = H or Alkyl Group
M = Vanadyl [V = O(II)] or Nickel [Ni(II)] Ion

Figure 1.27 During diagenesis and catagenesis, chlorophylls are converted to several biomarkers commonly found in petroleum, including deoxophylloerythroetioporphyrins (DPEP), etioporphyrins (etio), pristane, and phytane. Note the tetrapyrrole nucleus and the phytyl side chain in chlorophyll-a. M = nickel (Ni^{2+}), vanadyl (VO^{2+}), or other anion.

plant coals are characterized by iron (Fe^{3+}), gallium (Ga^{3+}), and manganese (Mn^{3+}) porphyrins (Filby and Berkel, 1987). Two minor series are also known: rhodo-DPEP and rhodo-etio. Barwise and Whitehead (1980) also describe di-DPEP porphyrins.

Note: *"DPEP" and "etio" are commonly used as generic terms based on the number of exocyclic rings (one or none, respectively), in the porphyrin structure. Precise structures of a number of DPEP-type porphyrins in sediments having 5- to 7-membered rings at various exocyclic positions have been determined. Likewise, etio-type porphyrins used in this context may have acyclic alkyl substituents at any position on the porphyrin nucleus. Rhodo-porphyrins have an exocyclic fused benzene ring.*

"Abelsonite" is a unique solid bitumen composed of a crystalline nickel (II) DPEP porphyrin derived from one of the chlorophylls. It is found only in the Parachute Creek Member of the Green River Formation, Utah (Mason et al., 1990).

Most routine biomarker studies incorporate only a few porphyrin parameters (Sec. 3), because unlike many other biomarkers, porphyrins require different separation and analysis techniques (Sec. 3.1.3.4). For these reasons the discussion of porphyrins in this text is limited. Reviews by Baker and Louda (1986), Louda and Baker (1986), and Filby and Branthaver (1987) describe porphyrin geochemistry in more detail.

2

Fundamentals of Biomarker Separation and Analysis

This chapter describes the multidisciplinary and multiparameter approach to geochemical studies, the relationship between biomarker and nonbiomarker analyses, and how to select (screen) samples for analysis. The concepts of gas chromatography (GC; separation) and mass spectrometry (MS; analysis) are explained. Many of these fundamentals such as the difference between a mass chromatogram and a mass spectrum, or between selected ion and linked-scan modes of analysis, are critical to understanding later discussions of biomarker parameters.

Several topics in italics (including analytical procedures, internal standards, and examples of GCMS data problems) are necessary fundamentals for understanding the biomarker parameters in Chapters 3 and 4. If the reader is comfortable with the topics in italics, they can be skipped.

2. FUNDAMENTALS

2.1 Supporting Geochemistry

Many geochemical projects are complex because of large numbers of samples. Biomarker work should never be initiated on these projects without a preliminary appraisal of samples using less costly and more rapid geochemical methods. Large numbers of oils and/or rocks can be "screened" using simple geochemical tools such as total organic carbon (TOC), Rock-Eval pyrolysis, vitrinite reflectance (R_o), petrographic analysis of maceral composition (MOA), gas chromatography (GC), kerogen atomic H/C, and stable carbon isotope ratios. Using this practical and less expensive approach allows nonsource rocks to be eliminated and oils and source rocks to be grouped into families that can be submitted for further work using the more powerful biomarker parameters. After the analyses are completed on the selected samples, both biomarker and supporting data are combined to establish the most reliable interpretations.

Reliable biomarker interpretations are usually based on a dozen or more biomarker parameters. Individual biomarker parameters only become meaningful when they agree with:

1. other biomarker parameters,
2. supporting geochemical parameters,
3. reasonable geologic scenarios.

Only rarely does the initial biomarker assessment in a large study answer all of the possible questions. In many cases, additional questions are raised by the data requiring more samples and/or analyses. For example, anomalous results are sometimes encountered in the first biomarker data from a large sample suite. Reexamination of the data may reveal interfering peaks in the gas chromatography-mass spectrometry (GCMS) analysis, GC resolution problems resulting from lack of optimization of instrumental operating parameters, poor signal to noise ratios caused by low biomarker concentrations, instrument drift, or other problems. Although analytical problems will not be discussed in detail, several examples are provided in the text (Sec. 2.5.5). Recognition of these problems generally requires a trained geochemist, mass spectroscopist, or technician with extensive experience.

2.2 Organization of the Chevron Biomarker Laboratory

A well-balanced biomarker laboratory which properly addresses operating company exploration problems requires input and support from various disciplines. The following description of the Chevron Biomarker Laboratory may not represent the only good way to organize this type of group, but our experience shows that it works well.

The most critical stage in any project is the input required: (1) operating company personnel define the exploration problem and (2) the operating company Geochemical Coordinator functions as a vital liaison between the geologists and the biomarker specialists. The Coordinator evaluates the suitability of geochemical methods for answering the problem, assembles appropriate samples based on a clear understanding of regional geology, and prepares a detailed list of objectives for the biomarker specialist. The Coordinator generally screens samples by requesting the appropriate routine geochemical analyses (such as Rock-Eval pyrolysis, total organic carbon, vitrinite reflectance, and others) from an in-house laboratory or service company. After approvals are obtained from both operating company and laboratory managements for the biomarker work to proceed, the Geochemical Coordinator forwards samples to the Organic Geochemistry Section Supervisor for distribution.

The biomarker team consists of specialists with expertise covering several topics.

1. Chromatographic separation methods, including column chromatography and high performance liquid chromatography,
2. Natural product chemistry required to evaluate precursor-product relationships,
3. Biomarker and GCMS interpretive skills obtained by experience and knowledge of the literature,
4. Geology, required to evaluate the condition and quality of samples and to place the biomarker results in perspective for the exploration personnel,
5. Familiarity with conventional geochemical parameters and their applications.

Supporting technologies that are essential for smooth operation of the biomarker group include:

1. advanced mass spectral and electronics technology,
2. synthetic organic chemistry, nuclear magnetic resonance (NMR) spectroscopy, and X-ray diffraction crystallography for identification and structural elucidation of unknown compounds,
3. computer science, required for data collection and processing.

Additional support includes access to relevant literature on each exploration problem through computerized library searches, and supporting geochemical analyses conducted within the organization or elsewhere. Finally, constructive and timely peer-group review is essential to ensure a uniform high quality of released technical memoranda.

2.3 Sample Availability, Quality, and Selection

Sample availability and quality are complicating factors that can determine the success or failure of a biomarker study. The Geochemical Coordinator and regional geologists typically play a critical role in acquiring the proper samples with adequate and accurate sample information. Samples of potential source rocks may not be readily available. Source rocks may lie far beneath the reservoir and may not have been drilled or sampled. Produced oils may have migrated great distances from their source rocks. Critical samples may be owned by competitors, contaminated by drilling additives, lost because of improper storage, or depleted because of other work.

Critical information that should be included with rock and oil samples includes well name, operator, location, formation, age (rock or reservoir), depth (or production interval), and type of sample (core, sidewall, outcrop, cuttings, oil). Maps and lithology logs assist the investigators in evaluation of sample quality and geologic relationships. If man-made additives or migrated oil could represent contaminants, samples of these materials are helpful.

2.3.1 Rock samples. Rock sample quality generally decreases in the order: whole core, sidewall core, cuttings, outcrops. Outcrop samples are commonly weathered, resulting in alteration of organic matter. Biomarker analyses of cuttings are not recommended because they are most readily affected by flushing with drilling fluids. These fluids can alter or contaminate the indigenous biomarker distributions, thus, nullifying or limiting useful geochemical interpretations. When only outcrop or cuttings samples are available, critical geochemical questions can still be answered using biomarkers. However, these samples should be used only as a last resort, and they should be discussed on an individual basis with the biomarker personnel. Screening analyses such as Rock-Eval pyrolysis or gas chromatography can be used to eliminate samples showing unfavorable characteristics indicating weathering or contamination.

Except for research purposes, we do *not* recommend that biomarkers be applied to organic-lean rocks (< 1 wt % TOC) because they are unlikely sources for commercial accumulations of petroleum. Information on the quantity, quality, and thermal maturity of organic-lean rocks is best obtained using rapid, inexpensive screening techniques such as Rock-Eval pyrolysis. Furthermore, biomarker analyses of bitumens from organic-lean rocks can be misleading. In some organic-lean rocks, biomarkers that are characteristic of the organisms contributing to the depositional environment are overwhelmed by those associated with recycled organic matter. For example, Farrimond et al. (1989) noted that unlike organic-rich rocks, adjacent organic-lean Toarcian age rocks contained bitumens showing anomalously high maturities based on hopane and sterane isomerization ratios. They concluded that recycled organic matter in the organic-lean samples had experienced high thermal maturity prior to deposition of the sediments.

Rock sample size for biomarker analysis varies depending on organic richness. Typically about 50 g of rock is advisable, although successful analyses can be completed on some samples as small as 1 to 5 g. Core, sidewall, and outcrop samples can be brushed or scraped to remove mudcake and residues from marking pens. Weathered surfaces should be removed prior to sampling for geochemistry. Because of contamination, cores enclosed in wax-based sealing agents should be avoided.

Cuttings are normally washed with fresh or salt water to remove mudcake and are air-dried at low temperature (below about 50°C) on paper towels at the drill site. Cuttings washed with organic solvents must be avoided because these solvents extract biomarkers. Prior to analysis, cuttings should be examined using a binocular microscope and obvious contaminants such as walnut hulls or woodchips removed. We call this **negative picking.** We do not recommend **positive picking,** where a single lithology is selected for analysis from a mixture of lithologies in a given cuttings sample. Although informative about isolated lithologies, positive picking results in nonrepresentative samples.

2.3.2 Oil samples. Oils are best collected in glass containers. Metal containers can gradually react with emulsified water and leak. Oil can leach contaminants, such as phthalates, from plastic containers or rubber-lined caps. Teflon liners are recommended for bottled oils. Usually 250 cc of oil is sufficient for all geochemical analyses with some remaining for storage. However, certain samples as small as a few milligrams have been successfully analyzed for biomarkers because of high concentrations of these compounds. Each sample container should be carefully labeled using waterproof ink.

Care must be taken in collection of production oils because they may represent mixtures from various zones or fields. Bottled oils do not appear to be affected by biodegradation over periods of years under typical storage conditions. However, oils heavily biodegraded in the reservoir present special problems in analysis and interpretation of residual biomarkers compared to unaltered oils. Finally, condensates and high-gravity oils ($> 40°$ API) typically show few, if any biomarkers, because of extensive thermal degradation.

2.4 Sample Cleanup and Separations

Biomarkers are commonly trace components in petroleum and normally must be concentrated prior to analysis. Bitumens are extracted from ground sedimentary rocks using ultrapure solvents to avoid contamination. Because the oils and bitumens that contain biomarkers are complex mixtures, glass column and high-performance liquid chromatography are used to clean and separate them into fractions prior to analysis by GCMS. For example, the **saturate fraction** represents the non-aromatic organic compounds in petroleum, including normal and branched alkanes and cycloalkanes. The **aromatic fraction** contains organic compounds with one or more unsaturated rings such as monoaromatics (C_nH_{2n-6}) and polycyclic aromatic hydrocarbons as well as some compounds that contain sulfur, nitrogen, or oxygen. Various sophisticated mass spectral techniques such as GCMSMS multistage (i.e., triple quadrupole) mass spectrometry, and GC-high resolution-MS, offer the possibility of direct analysis of whole oils for biomarkers (Sec. 2.5.3.4).

Solvent extractions and separations are based on the "like-dissolves-like" principle. Solvents most readily dissolve solutes of approximately the same polarity. A polar molecule contains regions of opposite electrical charge. Methanol (CH_3OH), for example, shows positive and negative charge regions on the methyl and hydroxyl groups, respectively. Normal hexane (*n*-hexane) is nonpolar compared to methanol. In column chromatography, the oil or bitumen sample is poured onto alumina or other suitable adsorbent in a glass column and fractions of increasing polarity are obtained by pouring increasingly polar solvents through the column.

Details of typical sample cleanup and separation procedures are described in the following. Figure 2.1 shows a generalized flow chart of procedures.

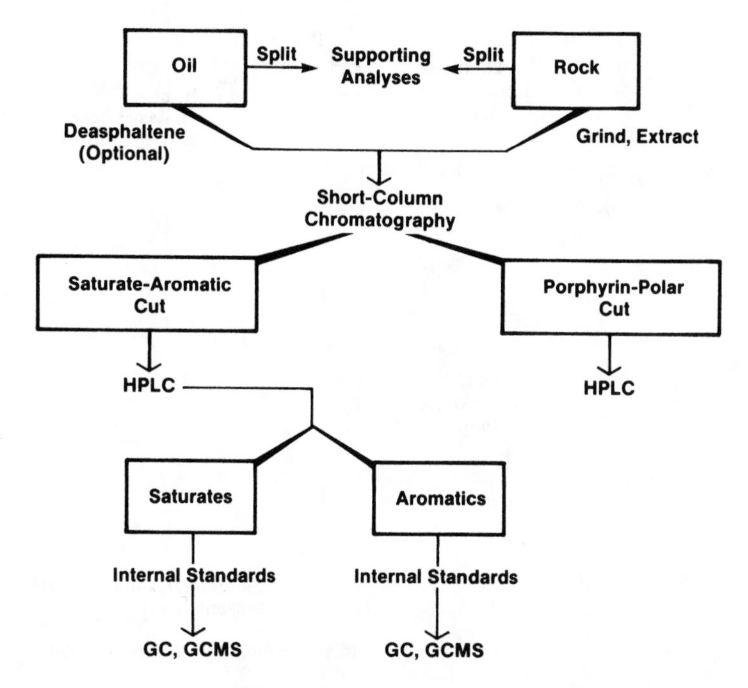

Figure 2.1 Flow chart showing procedures for separation of oils and bitumens into fractions for analysis. The "short-column chromatography" apparatus is shown in Fig. 2.2.

2.4.1 Column chromatography (cleanup)

2.4.1.1 Adsorption of Sample on Alumina. *The oil or bitumen sample is dissolved in methylene chloride (HPLC grade), mixed with 15 times its weight of alumina (Baker alumina oxide acid powder for chromatography, deactivated to Brockman activity II by adding water to 2.9 wt %) and distributed evenly at the bottom of a round bottom flask. The solvent is removed by rotoevaporation at low speed to avoid bumping and generation of fine alumina particles. Final rotoevaporation is done at full house vacuum (about 170–200 mBar) with the flask in a 40°C waterbath.*

2.4.1.2 Preparation and Loading of Column. *The chromatography column (2.5 × 30 cm with Teflon stopcock, medium grade glass frit, 29/42 outer ground glass joint) is half filled with diethyl ether : hexane (10 : 90 vol : vol) to which alumina (50 times the weight of the sample) is added (Fig. 2.2). The alumina with the adsorbed sample (see previous text) is loaded into the column and rinsed from the column walls onto the top of the clean alumina bed using a few milliliters of ether:hexane.*

Saturates and aromatics are eluted together from the column using 10 : 90 ether : hexane (HPLC grade; 10 times the weight of the clean alumina). The column can be extended for this step by adding another section of glass column having an inner 29/42 ground glass joint above the original column. Clean nitrogen or air (us-

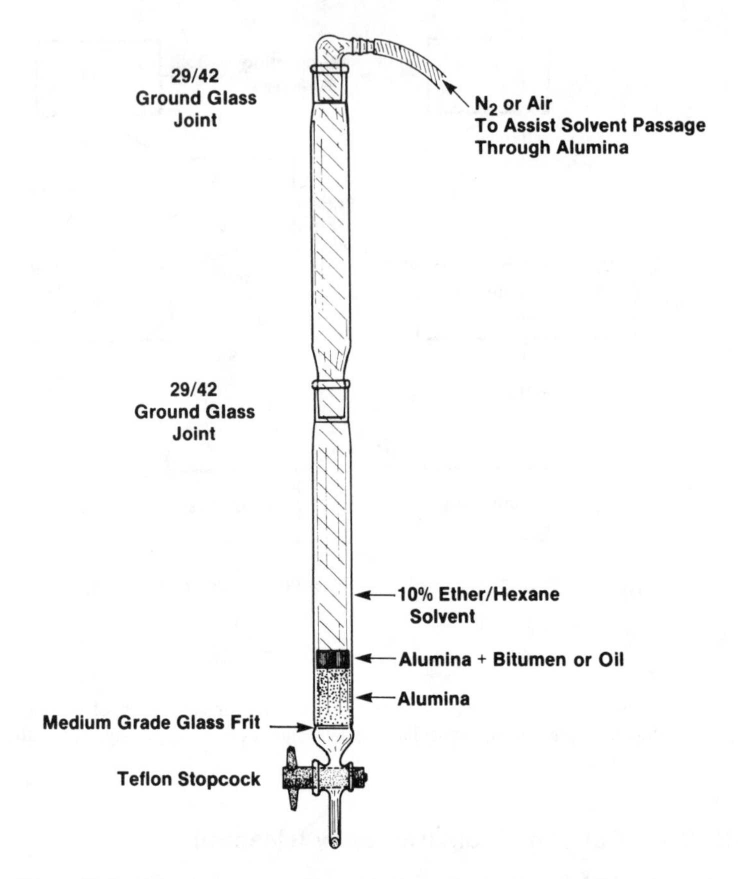

Figure 2.2 Alumina column for separation of saturate-aromatic and porphyrin-polar fractions from petroleum samples (short-column chromatography in Fig. 2.1). In this figure, the column containing the alumina and sample (bottom) has been extended by another section of column (top) to accommodate the ether:hexane mixture used to elute the fractions.

ing molecular sieve traps) can be used to assist passage of the solvent through the column.

Porphyrins and polar compounds are eluted together using 100 percent chloroform (HPLC grade) until a dark brown band comes off the column. The eluent is checked with a UV-visible spectrophotometer (350–600 nm) for quantitative isolation.

After rotoevaporation and weighing, the saturate-aromatic and porphyrin-polar fractions are ready for further separations using high-performance liquid chromatography (HPLC).

2.4.2 High-performance liquid chromatography (separation). *The cleaned up saturate-aromatic fraction of the oil or bitumen is separated using a Waters Model 590 HPLC pump equipped with a Whatman Partisil 10 Silica column*

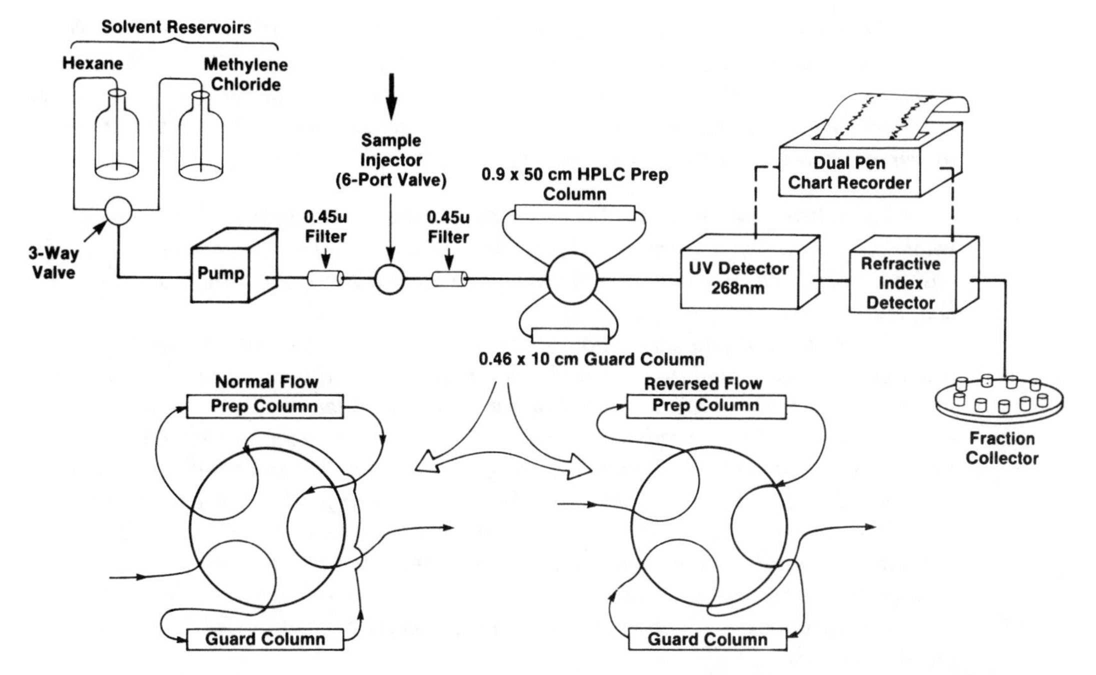

Figure 2.3 Configuration of the high performance liquid chromatograph (HPLC) for automated separation of saturate, aromatic, and polar fractions from petroleum samples. Except for the sample inlet valve, all operations are automated using a microprocessor in the pump. *Concept for figure courtesy of F. J. Fago.*

(9.4 mm I.D. × 50 cm) and a Brownlee Spheri-10 silica guard column (Fig. 2.3). The guard column protects the main column from irreversibly adsorbed compounds.

The eluent is divided into three cuts (saturates; mono-, di-, and triaromatics; and polar compounds) using an ISCO Foxy programmable fraction collector and 400 ml glass bottles. Under "normal flow" conditions (Fig. 2.3, lower left), the saturates (2 ml/minute for 15.4 minutes) and aromatics (9 ml/minute for 26.5 minutes) are eluted with hexane while methylene chloride is used for the polar compounds (7 ml/minute for 20 minutes). Before the polars are eluted from the column with methylene chloride, the backflush valve is switched, resulting in passage of the eluent from the injection port to the main column and back through the guard column. In this "reversed flow" valve position (Fig. 2.3, lower right), the direction of flow through the main column remains the same, but flow through the guard column is reversed (backflushed). The columns are partially reequilibrated with hexane (7 ml/minute for 12 minutes) before the next separation.

A Perkin Elmer LC 55 UV detector (254 nm; HPLC flow cell) and a Waters R 401 RI detector monitor the fractions using a Soltec Model 1242 two pen recorder. The cut point between saturates and aromatics is determined based on retention times of cholestane and monoaromatic steroid standards. (See following.) The cut between triaromatics and polars is made immediately after elution of dimethyl phenanthrene.

Fractions are rotoevaporated under house vacuum, transferred to tared vials, and rotoevaporated again using a 45°C water bath for 10 minutes and a vacuum of 35 mm Hg. Unusually low concentrations of biomarkers, as in certain condensates, may require further treatment of the saturates by urea adduction or molecular sieves to remove normal paraffins (Michalczyk, 1985) (e.g., Fig. 3.15).

2.4.3 Internal standards and preliminary analyses. *Subsamples of whole oil or bitumen and separated fractions are always taken for auxiliary geochemical analyses such as gas chromatography, stable carbon isotope ratios, weight percent sulfur, and API gravity.*

After cleanup and separation of the bitumen or oil into saturate and aromatic fractions, internal standards are added to facilitate quantitation of peaks. For example, a small aliquot of the saturate fraction is prepared for injection into a gas chromatograph (GC) for a preliminary evaluation of its behavior during GCMS. This sample is spiked with four branched paraffins (3-methylheptadecane, 3-methylnonadecane, 2-methyldocosane, and 3-methyltricosane) to facilitate peak quantitations and aid in determining whether further treatment is necessary prior to GCMS analysis. This GC analysis commonly shows that certain bitumens or heavy oils require dilution with a solvent such as hexane. Extremely waxy oils may require removal of n-paraffins by urea adduction to allow adequate concentrations of biomarkers for GCMS analysis.

Absolute quantities of biomarkers in oils and bitumens may be determined using steroid internal standards. These steroids do not occur in nature and offer the advantage of fragmenting to give the same principal ion as the steroids being measured. If absolute quantities of biomarkers are not determined, differences in detector response between the various compounds being analyzed are ignored. This approach is not valid when comparing compounds with different mass spectral characteristics.

We add 5β(H)-cholane to the saturate fraction as an internal standard for sterane and terpane quantitations prior to all GCMS analyses (Seifert and Moldowan, 1979). 5β(H)-cholane is not found in significant abundance in natural oils, does not interfere with the indigenous compounds, and fragments to give the same principal ion (mass/charge = m/z 217) by the same mechanism as other steranes (Fig. 2.4). Figure 2.12 (upper left) shows the m/z 217 mass chromatogram for an oil from Hamilton Dome, Wyoming, used as a standard (GCMS No. 81-3) including the internal standard peak identified as 5β(H)-cholane (peak 1 in the figure).

Two internal standards are added to the aromatic fraction (Fig. 2.4). For monoaromatic (MA) steroids, the standard is a synthetic C_{30} MA-steroid mixture of four epimers [5β(20S), 5α(20S), 5β(20R), and 5α(20R)] which fragment to give the same principal ion (m/z 253) as natural MA-steroids. Figure 2.13 shows the MA-steroid mass chromatogram for an oil standard from Carneros, California, including the four MA-steroid epimers used as the internal standards (peaks 17-20). For triaromatic (TA) steroids, the internal standard is a synthetic C_{30}-TA steroid mixture of two epimers (20S and 20R) which fragment to give the same principal ion (m/z 231) as natural TA-steroids. Figure 2.14 shows the TA-steroid mass chromatograms for

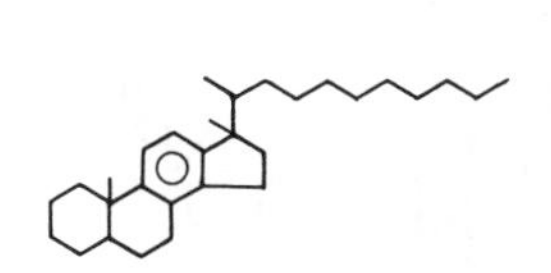

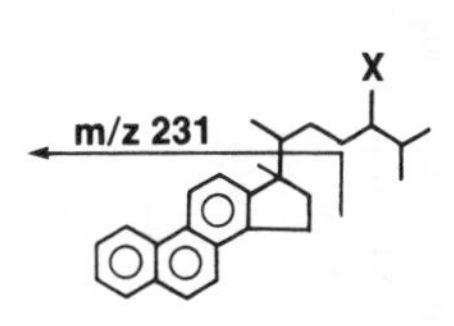

Figure 2.4 The steroids used as internal standards for the saturate [5β(H)-cholane] and aromatic (C_{30} mono- and triaromatic steroids) fractions are not found in petroleum in significant concentrations. However, these steroids fragment to give the same principal ions in the source of the mass spectrometer as the steroids being measured. This allows absolute quantities of biomarkers in the petroleum fractions to be determined. Peaks corresponding to the above internal standards are identified on mass chromatograms in Figs. 2.12, 2.13, and 2.14. C_{30} mono- and triaromatic steroid standards were prepared in collaboration with D. S. Watt and co-workers. Alternative approaches using various compounds as standards have been described (Rullkötter et al., 1984b; Abbott et al., 1984; Dahl et al., 1985; Mackenzie et al., 1985).

the standard oil from Wyoming, including peaks representing the above internal standards.

Some deuterated compounds are commercially available (e.g., Chiron Laboratories, Norway) for use as internal biomarker quantitation standards for GCMS. In deuterated standards, one or more deuterium atoms (a heavy isotope of hydrogen; atomic weight = 2) are synthetically substituted for hydrogen atoms (atomic weight = 1). The result is a molecule that shows very similar physical and chemical properties compared to the nondeuterated analog, but which is one unit heavier for each hydrogen atom that has been replaced by deuterium. Thus, the advantage of these deuterated standards is that they behave nearly the same as the compounds being analyzed under mass spectrometric conditions. A disadvantage is that the standards and compounds to be analyzed must be recorded on chromatograms with different mass/charge ratios.

2.5 Gas Chromatography/Mass Spectrometry

Computerized gas chromatography/mass spectrometry (GCMS; McFadden, 1973; Watson, 1985) is the principal method used to evaluate biomarkers (Fig. 2.5). A typical GCMS system performs six functions indicated in the figure.

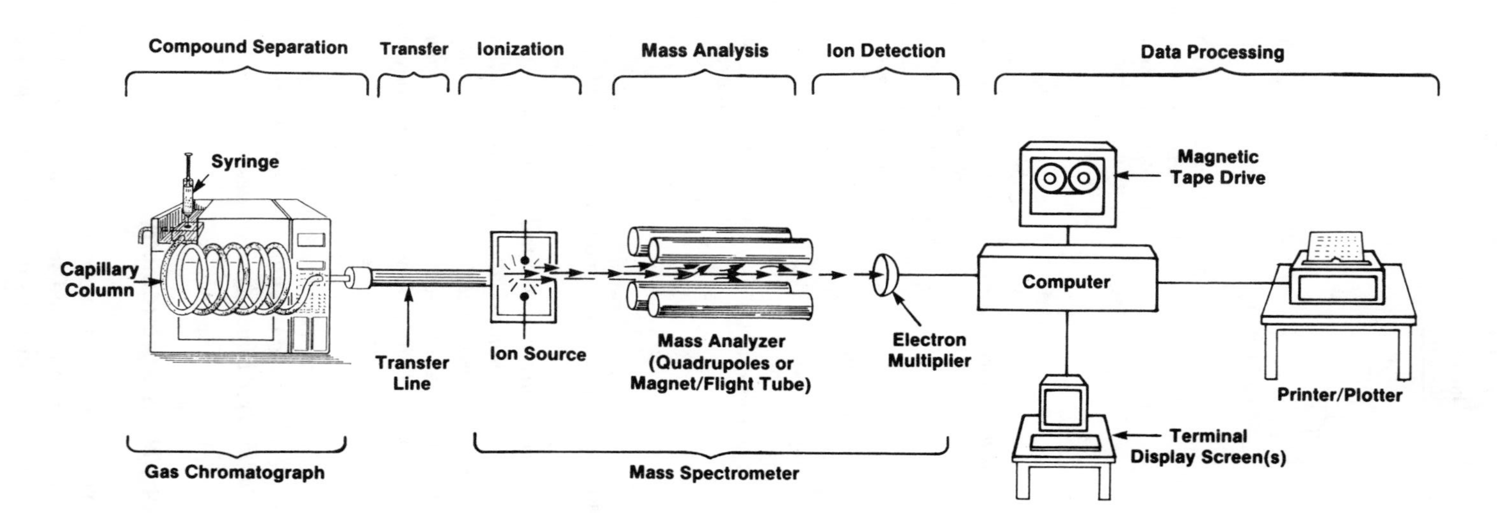

Figure 2.5 A typical gas chromatograph/mass spectrometer performs six functions (from left to right): (1) compound separation by gas chromatography, (2) transfer of separated compounds to the ionizing chamber of the mass spectrometer, (3) ionization and acceleration of the compounds down the flight tube, (4) mass analysis of the ions, (5) detection of the focussed ions by the electron multiplier, and (6) acquisition, processing, and display of the data by computer.

1. Compound separation by gas chromatography,
2. Transfer of separated compounds to the ionizing chamber of the mass spectrometer,
3. Ionization,
4. Mass analysis,
5. Detection of the ions by the electron multiplier,
6. Acquisition, processing, and display of the data by computer.

The GCMS can be used to detect and provisionally identify compounds using their relative GC retention times, elution patterns, and the mass spectral fragmentation patterns characteristic of their structures.

Stringent criteria are applied to GCMS procedures (Seifert and Moldowan, 1986) to ensure meaningful interpretations. For example, GCMS data are obtained using high-resolution capillary columns (generally 50 m long or more), high signal-to-noise output from a finely tuned mass spectrometer, and rapid scanning. (See following.)

2.5.1 Gas chromatography in GCMS

2.5.1.1 Compound Separation. The theory and practice of gas chromatography (sometimes called gas/liquid chromatography) is described extensively in the literature (e.g., Poole and Schuette, 1984).

A syringe is used to inject a known amount (typically $< 0.1\ \mu l$) of the saturate or aromatic fraction, which may or may not be dissolved in solvent (usually toluene), into the gas chromatograph (Fig. 2.6, top). The larger molecules are retained by the stationary phase (see following) at the head of the GC column in a process called "cold trapping." The temperature of the column is gradually raised using a temperature-programmed oven, causing the cold-trapped compounds to move. In gas chromatography, each injected sample is vaporized and mixed with an inert **carrier gas,** typically helium or hydrogen.

The gas (**mobile phase**) and sample mixture moves through a long, thin capillary column (typically 0.25 mm I.D., 60 m long) whose inner surface is coated with a film (about 0.25 μ thick) of nonvolatile liquid (**stationary phase**). Components are separated as they are repeatedly retained by the stationary phase and released into the mobile phase depending on their volatility and affinity for each phase (Fig. 2.6, bottom).

Flexible fused silica capillary columns have largely replaced the older glass or stainless steel capillary columns used in GCMS because of several advantages: (1) flexibility eliminates many difficulties related to installation, and (2) lower activity reduces peak tailing and loss of sample by adsorption on the stationary phase. Most published data have been obtained using methyl-polysiloxane (e.g., OV-101, DB-1, etc.) so that GCMS results can be readily compared. For best standardization of analytical conditions, an oil standard (e.g., at Chevron, GCMS No. 81-3, Hamilton Dome Field, Wyoming) is run with each set of samples to allow evaluation of

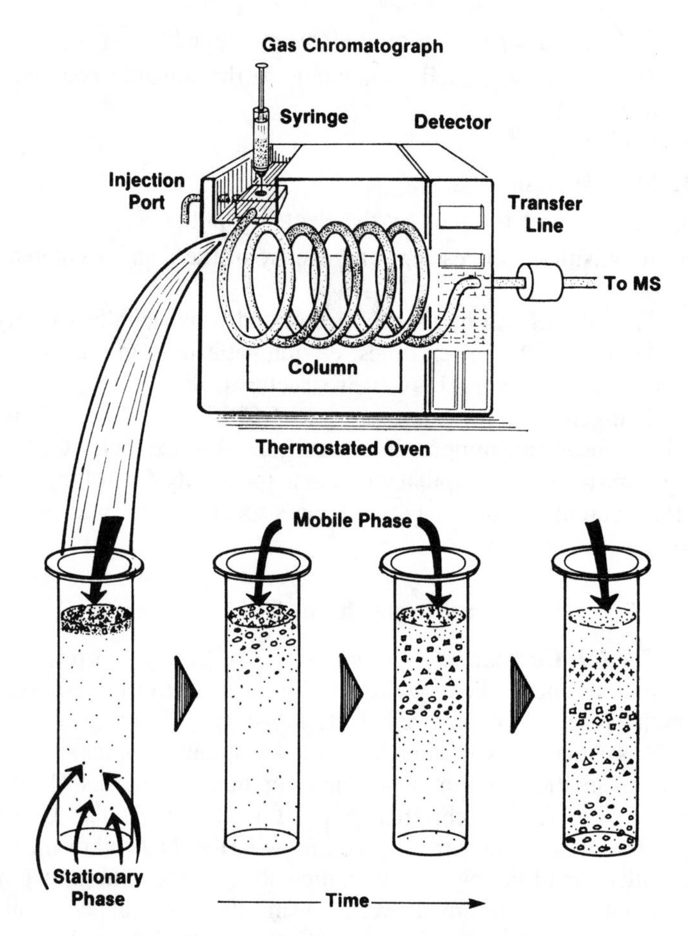

Figure 2.6 Detailed view of a typical gas chromatograph used to separate mixtures of compounds. The blowup (bottom) shows the separation of compounds during movement down the chromatographic column which results from their repeated partitioning between the mobile and stationary phases.

column performance in separating various isomers and to facilitate calibrations of sensitivity (response) for peak quantitation. Most biomarkers used for interpretation elute from the GC in the range nC_{24} to nC_{36} (between *n*-paraffins containing 24 and 36 carbon atoms) and are generally in much lower abundance than *n*-paraffins (Fig. 2.7). Exceptions include pristane and phytane, which elute prior to this range, and porphyrins, which do not elute under normal gas chromatographic conditions because of their high molecular weights and low volatilities.

2.5.1.2 Transfer. The interface between the gas chromatograph and mass spectrometer serves to transfer the separated compounds to the ion source (Fig. 2.5). In the past, various types of separators were used to concentrate the effluent from the gas chromatograph by removing much of the carrier gas. Low-carrier flow rates

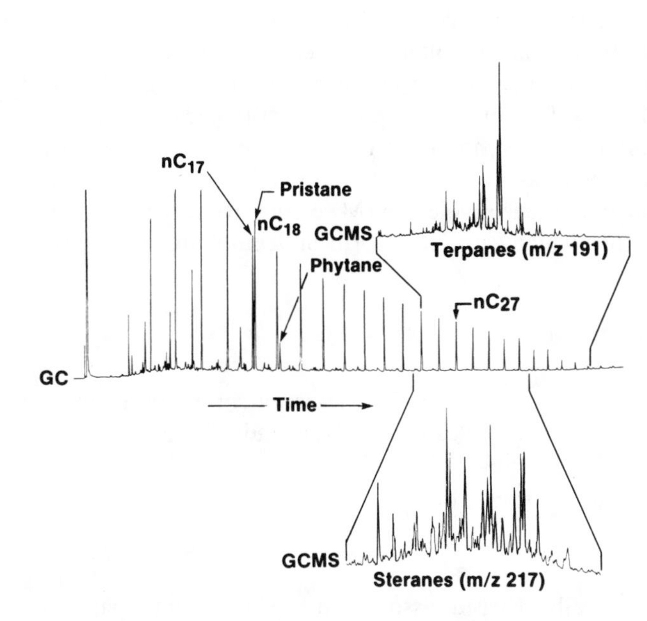

Figure 2.7 The gas chromatogram of an oil obtained by routine gas chromatography using a flame ionization detector (center) is dominated by the homologous series of *n*-paraffins. Mass chromatograms of the steranes and terpanes (bottom and top, respectively), in the same oil are obtained by gas chromatography using a mass spectrometer as detector (GCMS). Steranes and terpanes are monitored using their principal fragment ions at m/z 217 and m/z 191, respectively. Note that the sensitivity of this analysis allows detection of sterane and terpane peaks too small to see on the gas chromatogram. Many biomarkers, including the steranes and terpanes, elute from the gas chromatograph between the *n*-paraffins with 24 and 36 carbon atoms (nC_{24}-nC_{36}). Pristane and phytane are acyclic isoprenoid biomarkers which elute prior to this range of retention time.

for capillary columns and high-capacity diffusion pumps on newer GCMS systems have largely eliminated the need for these separators. Most GCMS systems now in use transfer all effluent directly to the ion source without using a separator. Flexible capillary columns can be introduced directly into the ion source, thereby eliminating loss of chromatographic resolution which previously arose because of the separator.

2.5.2 Mass spectrometry in GCMS. Detailed descriptions of mass spectrometry are available to supplement the following discussion (e.g., Burlingame et al., 1980; Watson, 1985).

2.5.2.1 Ionization. Electron impact (EI) is the usual mode of ionization in GCMS. Other modes of ionization include chemical ionization (CI) and field ionization (FI). These modes of ionization are discussed next.

2.5.2.1.1 Electron impact ionization. After the separated compounds elute from the high-resolution capillary column, they are analyzed by the mass spectrometer (Fig. 2.5). In **electron impact (EI) ionization,** eluting compounds pass directly from the column to the ionizing chamber (source) of a mass spectrometer where they are ionized by an electron beam. An electron impact source consists of a filament, an electron trap, a repeller, and appropriate focussing plates. The beam is generated by passing a current (< 1 mamp) through a 10 μ rhenium or tungsten wire filament. Resistance heats the filament and causes electrons to escape. The electrons are then accelerated using about 70 electron volt (eV) toward the trap. Typically, a mass range of 50 to 600 atomic mass units (amu) is scanned in three seconds or less. Pressures within the mass spectrometer are maintained at low levels ($< 10^{-5}$ Torr).

Most MS systems ionize the eluting compounds in the electron impact mode using 70 eV. The choice of 70 eV ionizing voltage is based on the empirical observation that molecules are most efficiently ionized in the range 50 to 90 eV. Below 50 eV, the electrons do not impart sufficient energy to the target molecules to cause the most efficient ionization. Above 90 eV, the electrons are so energetic that they do not react with the target molecules.

In electron impact ionization, each molecule (M) eluting from the GC is bombarded with energetic electrons which cause it to form **molecular ions** ($M^{+\cdot}$) as follows:

$$M + e^- \longrightarrow M^{+\cdot} + 2e^-$$

The molecular ion can undergo further fragmentation or rearrangement to form other ions ($F^{+\cdot}$, $F1^{+\cdot}$), neutral molecules (N1, N2), or radical ions.

$$M^{+\cdot} \longrightarrow F^{+\cdot} + N1$$

$$F^{+\cdot} \longrightarrow F1^{+\cdot} + N2$$

Fragment ions are electrically charged dissociation products from a parent ion. Fragment ions may dissociate further to form other electrically charged molecular or atomic moieties of successively lower formula weight. The fragment ions generated in the ion source chamber are accelerated toward the detector through the mass analyzer by a high-differential voltage (Fig. 2.5).

Ions formed in the source of the mass spectrometer are subsequently analyzed according to their mass to charge ratio using a magnetic or quadrupole mass spectrometer. In all cases, the positive ions are detected using an electron multiplier. The result is a characteristic fragmentation pattern or **mass spectrum** of molecule M (discussed in the following).

The mass of the molecular ion is useful in identifying each compound analyzed by GCMS. Using 70 eV ionizing energy, however, often reduces the molecular ion to very low levels relative to fragment ions. By lowering the ionizing energy to about 20 eV or less and injecting more sample, the operator can reduce the efficiency of ionization and thus, increase the relative abundance of the M+. ion compared to the fragment ions. This choice results in loss of absolute intensity.

Electron impact is the most commonly used ionization technique because it usually provides all the necessary spectral information required to identify an organic compound. This is particularly true for routinely analyzed compounds whose structures and retention times are already known from previous investigations. For routine work, compound identification is a necessary first step, but the principal objective is to quantify the compounds for use in biomarker parameters. For less routine work, where structural elucidation of unknown compounds is the principal objective, other ionization techniques are frequently used in conjunction with EI, as will be described.

2.5.2.1.2 Chemical ionization. Techniques other than electron impact such as chemical ionization (CI) GCMS or field ionization (FI) GCMS have been used to provide molecular weight information. Chemical ionization involves ionization of components by ion-molecule reactions rather than by electron impact or other forms of ionization. Ionization of a reagent gas, R (present in large excess) is typically initiated by electron impact, followed by ion-molecule reactions involving neutral molecules of interest (M) and the reagent gas ions (R^+). Molecules ionized in this way (M^+) can undergo further reactions, forming additional fragment ions (F^+, $F1^+$, etc.) and neutral species (N, N1, N2, etc.). The principal reactions are as follows:

$$R + e- \longrightarrow R^+ + 2e-$$
$$(R^+) + M \longrightarrow (M^+) + R$$
$$(M^+) \longrightarrow (F1^+) + N1$$
$$F^+ \longrightarrow (F2^+) + N2$$

Thus, in chemical ionization relatively high pressures (~1 mm Hg) of a reagent gas, typically a low molecular-weight hydrocarbon or ammonia, are introduced into the ion source of the mass spectrometer. This gas is ionized by the filament to give primary ions and secondary ions formed by fragmentation of the primary ions and/or recombination of the fragments. The compound to be analyzed (**analyte**) is introduced through the GC and reacts with the secondary ions in the ion source. The analyte is present in low concentration, usually less than 1 percent of the reagent gas. The ions are generally formed by proton transfer or hydride abstraction and the principal ion is usually $(M + 1)^+$ or $(M - 1)^+$. Further fragmentation is minor compared to EI spectra.

2.5.2.1.3 Field ionization. The same as CI mass spectrometry, field ionization (FI) is another "soft" ionization technique that can provide molecular weight information. In FI, a high voltage (~10^8 V/cm) is applied to an electrode system consisting of an anode or emitter (usually a 10 μ tungsten wire on which carbon "dendrites," that is, hair-like filaments, have been grown) and a cathode or extractor plate with a narrow slit. The compounds to be analyzed are introduced from the GC into the field of this electrode system where they acquire a positive charge by a quantum mechanical tunneling mechanism. FI spectra typically show only the molecular ion with few if any fragments. The use of FI or GC-FIMS for biomarkers has not been reported. However, the method shows potential applications and is being tested at various laboratories.

2.5.2.2 Mass Analysis. The ions are mass analyzed into a concentrated beam in the mass analyzer so that only positive ions of a given **mass/charge ratio** (m/z) impinge on the detector at any moment. The two principal methods used to analyze the ion beam use either magnets or quadrupole rods as described.

1. Magnetic mass spectrometers consist of single- or double-focusing systems. A double-focusing instrument (e.g., VG 7070H) uses an electrostatic sector to energy focus the ion beam prior to entering or after leaving the magnetic sector to achieve high-mass resolution of 2000 or more. For example, an electrostatic lens can be used to energy focus the beam into a magnetic field surrounding the curved flight tube (Fig. 2.8). Heavier ions are deflected less than light ions of the same charge in the magnetic field. **Scanning** the magnetic field by varying the field strength focuses ions of a given m/z onto the detector. Adjustable slits are used at the source and detector to improve the mass resolution of the system. The curved-flight-tube ion focusing arrangement typical of magnetic instruments is not shown in Fig. 2.5, which instead shows a mass analyzer composed of quadrupole rods (see following).

   Single focusing magnetic systems consist only of the magnet, do not have electrostatic focusing, and thus, achieve only low resolution (see following) of about 1000 or less. Both single- and double-focusing GCMS systems scan the magnetic field to obtain a mass spectrum (Sec. 2.5.2.3).

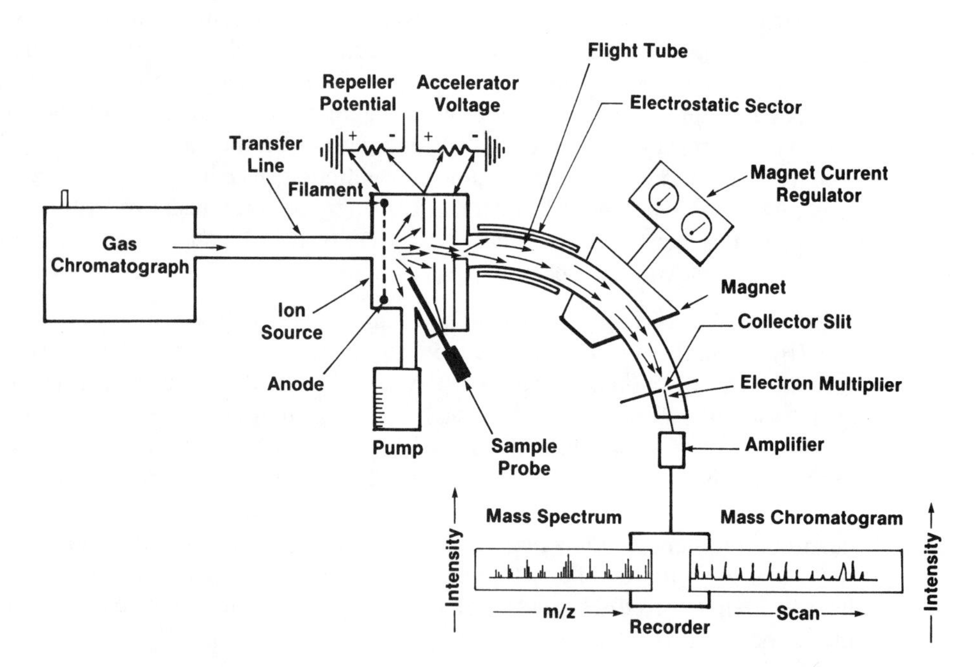

Figure 2.8 An "electrostatic mass analyzer" is an example of one type of mass spectrometer. Mass analysis can be accomplished using an electrostatic lens which focuses the ion beam into a magnetic field surrounding the curved flight tube. The flight paths of ions are changed (scanned) by varying the magnetic field strength. A narrow collector slit admits only ions of a given mass to the detector at any moment during the scan.

2. Single (e.g., Hewlett-Packard GC-MSD, VG Trio-1, or Finnigan 4000) or triple quadrupole (e.g., Finnigan TSQ-70) instruments (Fig. 2.9) and the three-dimensional "ion trap" system all make use of quadrupole RF (radio frequency) per electrical field to select ions of a given mass to charge ratio. "Quadrupole mass filters" (e.g., Finnigan 4000) use no focusing at all, but achieve mass selection because, for a given set of conditions, only ions with a narrow range of mass to charge ratios are able to remain within the instrument during mass analysis. The three-dimensional ion trap uses no focusing. In this case, the ions are trapped in three dimensions by RF electrical fields. Thus, only those ions of interest are trapped and detected.

   For example, a two-by-two arrangement of four parallel quadrupole rods can be used to scan the ion beam by varying a combination of RF and DC (direct current) currents within the rods. This arrangement is used as the example of the mass analyzer portion of Fig. 2.5 and is shown in greater detail in Fig. 2.9.

***Note:* *Resolution* *in mass spectrometry is defined as the ratio of mass (amu) to the difference in mass between two adjacent masses that the instrument is just able to separate completely (M/ΔM). Low resolution is considered to be about 1000 (no units), whereas high resolution is over about 2000. Most quadrupole instruments are low-resolution mass filters. High-resolution mass spectrometers can be used to separate ions of the same nominal mass but having different exact mass (e.g., see Sec. 2.5.5.8).*

2.5.2.3 Detection and the Principle of Scan Analysis. During a typical "scan" analysis, the detector measures ions over the mass range m/z 50 to m/z 600 every three seconds (i.e., over 500 ions in three seconds). Each peak (which represents one or more compounds) eluting from the GC yields a distribution of fragment ion masses. Thus, a "time-slice" of the ions generated by each peak is detected every three seconds. The principle of a scan analysis is demonstrated by a three-di-

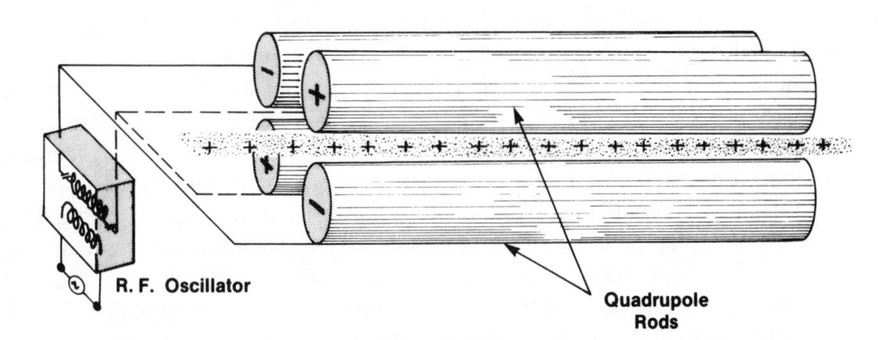

Figure 2.9 The quadrupole mass analyzer is an important component in quadrupole mass spectrometers. Mass analysis can be accomplished using four parallel quadrupole rods. By varying a combination of RF and DC currents within the rods, a beam of ions can be scanned, thus allowing only ions of given mass to reach the detector at any moment during the scan.

mensional diagram (Fig. 2.10). The x-, y-, and z-axes on this diagram represent the time or scan number (i.e., total number of scans at three seconds/scan; related to the GC retention time), the mass to charge ratio of the ions (m/z), and the detector response, respectively.

In Fig. 2.10, a "mass spectrum" is obtained by plotting m/z versus response at constant scan number (related to time). A "mass chromatogram" is obtained by plotting scan number versus response at a constant m/z. The difference between a mass chromatogram and a mass spectrum is critical for understanding the sections that follow.

Detection of the ions separated by the mass analyzer is achieved using a multistage electron multiplier whose output is either recorded directly onto an analog recorder or digitized and acquired by a data system. Virtually all MS units rely on electron multiplier (discrete dynode) detection where the electron beam is amplified by a cascade of collisions with special metal surfaces (Watson, 1985).

Each compound eluting from the GC yields a particular distribution of fragment ion masses called a mass spectrum (Figs. 2.10 and 2.18). Mass spectra of biomarkers are useful because they usually show the molecular mass (some of the molecules ionize, but do not fragment further) and characteristic fragmentation patterns that can be used to infer structures. Ideally, each GC peak represents a separate compound resulting in a unique mass spectrum that can be used for identification. In reality, most peaks are unresolved mixtures of two or more compounds, thus, complicating interpretation.

The magnitude of the total ion current for all mass spectra in each sample can be plotted versus retention time on a **reconstructed ion chromatogram** (TIC or RIC) to show a series of peaks that represent the relative amounts of eluting compounds. RIC and GC traces of petroleum are essentially identical, except an RIC requires MS detection while GC uses the more conventional flame ionization detector (FID).

2.5.2.4 Data Processing and Calibration. A computer is necessary to store and process the large amounts of data generated during GCMS operations. During a typical "scan run" requiring about 90 minutes, where the mass spectrometer scans the mass range of interest every three seconds, 1800 mass spectra are generated for each sample [(90 minutes × 60 seconds/minute)/(3 seconds/scan)]. Most GCMS data systems such as the Finnigan MAT INCOS or VG Opus systems consist of a central computer with one or more display terminals and various peripheral devices including magnetic tape drives for long-term storage of data and printer/plotters (Fig. 2.5). The computer contains a library of thousands of electron impact spectra for known compounds. As the computer acquires data from each GCMS analysis, unknown compounds in the injected sample can be provisionally identified by automated comparison of their spectra with the library spectra. The computer typically provides two or more best choices for the compound to be identified with accompanying purity and fit information. Purity and fit are each ranked on a scale of 0 to 1000, where 1000 represents a perfect correlation between the known and unknown spectra.

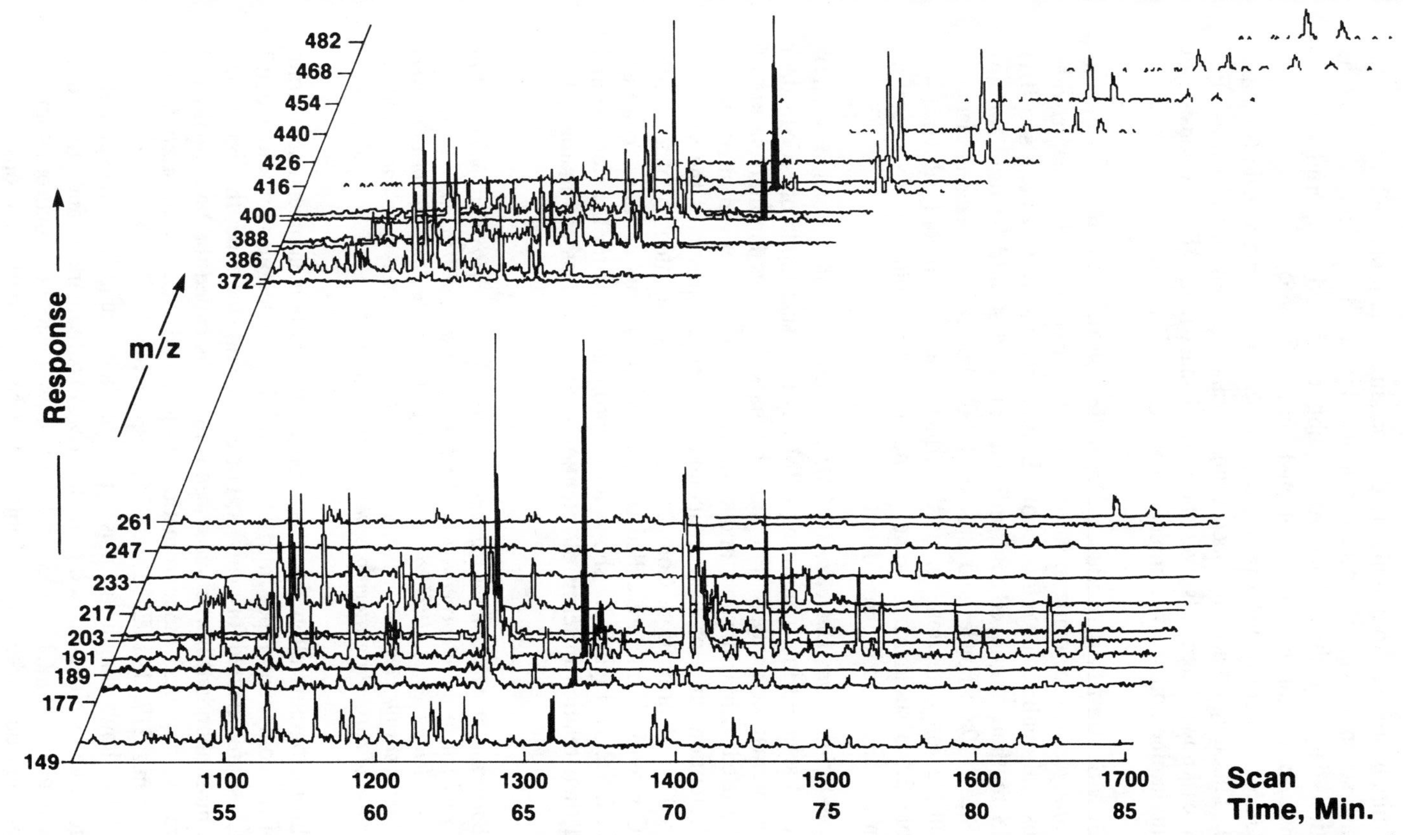

Figure 2.10 A three-dimensional plot demonstrates the principle of scan analysis in mass spectrometry. The x- (horizontal), y- (vertical), and z- (into page) axes represent the scan number or retention time, detector response, and mass to charge ratio (m/z), respectively. Scan and GC retention time are related, as indicated on the x-axis. We typically scan the mass range m/z 50 to 600 every three seconds. For example, the 1200th scan is complete after 60 minutes (3600 seconds) into an analysis. In this example, only selected m/z ratios from the scan analysis have been plotted to reduce the complexity of the display.

A mass spectrum is obtained by holding scan number constant and plotting m/z versus response. A mass chromatogram is obtained by holding m/z constant and plotting scan versus response. Each mass spectrum consists of a series of fragment ions that can be used to help elucidate the structure of a single compound. A mass chromatogram can be used to monitor a series of compounds of various molecular weights that all fragment to form a particular ion (e.g., m/z 217 is a fragment ion common to the pseudohomologous series of steranes).

Black peaks on the plot are the response for a single compound: C_{30} 17α,21β(H)-hopane (peak 20 in Fig. 2.12). The Wyoming oil (GCMS No. 81-3) we use as a standard contains 145 ppm of this compound (Fig. 1.24).

The terminals are used to manipulate data acquired and processed by the computer. Software accompanying the data system usually includes programs for instrument calibration, data acquisition, background subtraction, and monitoring the variation in intensity of one or more ions (selected ion monitoring).

Every GCMS system requires periodic calibration of the mass scale (m/z axis in Fig. 2.10). This may be accomplished using a standard compound such as "FC-43" which gives known fragments over the desired mass range. FC-43 consists of perfluorotributylamine and shows a molecular weight of 671 daltons.

2.5.3 GCMS operating modes. Depending on the available instrumentation, GCMS analyses can be made in various operating modes. Each mode provides different types and qualities of information. **Biomarker parameters are sensitive to the GCMS method employed and only one method should be used for any study** (e.g., Steen, 1986; Fowler and Brooks, 1990). For example, sterane data obtained in multiple ion detection versus GCMSMS mode commonly differ (Sec. 2.5.3.3). Table 2.5.3 summarizes some of the more important GCMS operating modes, which are discussed in more detail.

2.5.3.1 Multiple Ion Detection GCMS. Multiple ion detection (MID), sometimes called selected ion monitoring (SIM), is the usual mode of GCMS data acquisition for biomarker analysis. For most biomarker studies familiar classes of compounds are used such as hopanes and steranes. One ion of a given m/z together with the GC retention time for each compound is often diagnostic of the structure and can be used for identification. Computer-assisted plots of intensity of a specific ion versus GC retention time are called **mass chromatograms** or **mass fragmentograms** (Sec. 2.5.2.3). Figure 2.7 shows mass chromatograms for the steranes and terpanes, and their relation in terms of retention time on the gas chromatogram of an oil.

> ***Note:*** *Early use of mass chromatograms in organic geochemistry was pioneered at Chevron and led to a stereochemical understanding of steroid and other carboxylic acids in petroleum (Seifert, 1975) and the first practical method of fingerprinting steranes and terpanes for correlation purposes (Seifert, 1977). Since these early applications, the use of mass chromatography has grown rapidly.*

Mass chromatograms allow identification of the carbon number and isomer distribution of the compound type. In MID, several diagnostic ions for various compounds of interest are selected for analysis. For each compound type, the ion chosen is usually the most abundant in the mass spectrum and is called the **base peak.** For example, steranes, hopanes, monoaromatic steroids, and triaromatic steroids are monitored using m/z 217, 191, 253, and 231, respectively.

Figure 2.11 shows diagnostic mass spectrometric fragmentations (and their mass to charge ratios) which are used to monitor various biomarkers and other compounds in petroleum. Figs. 2.12, 2.13, and 2.14 show selected mass chromatograms for several common biomarker classes obtained in MID mode for the oils used as

TABLE 2.5.3. Modes of GCMS Analysis for Biomarkers.

MODE: SIM GCMS (MID GCMS) (Selected ion monitoring or multiple ion detection; Secs. 2.5.3.1 and 2.5.3.3)

METHOD: Scan only selected ions (e.g., m/z 217, 191, 253, etc.), using a dwell time of ~ 100 ms/ion. About 1 picomole is required for a reliable response when only two or three selected ions are monitored.

RESULT: Chromatograms of selected ions can be used as "fingerprints" for the selected compound type (e.g., steranes, hopanes, or monoaromatic steroids).

ADVANTAGES/DISADVANTAGES: Longer dwell time per ion results in better sensitivity compared to full-scan data. Requires knowledge of the retention times and fragmentation characteristics of the molecules to be studied. Forfeits full spectral data that are sometimes required to identify unknown compounds.

MODE: Full-scan GCMS (Sec. 2.5.3.2)

METHOD: Scan over range m/z 50 to m/z 600 every three seconds (i.e., over 180 ions/second). About 1 nanomole of each compound is required for a reliable spectrum, comparable to that required to produce a discernable peak by FID (flame ionization detector) analysis using GC.

RESULT: Provides both mass spectra for structural elucidation and chromatograms of all ions.

ADVANTAGES/DISADVANTAGES: Mass spectra allow tentative structural elucidation. Unlike MID mode, no data are discarded. Uses more computer disk space and shows lower sensitivity because of shorter dwell times than MID mode.

MODE: Parent mode GCMSMS (Sec. 2.5.3.4)

METHOD: In the parent ion mode of a triple quadrupole or other tandem instrument, the third quadrupole (Q3) monitors daughter ions (e.g., m/z 217) and the first quadrupole (Q1) scans for all possible precursor (parent) ions. Decomposition of the precursors is induced by collision (CAD) with neutral molecules in the second quadrupole (Q2) (Fig. 2.17).

ADVANTAGES/DISADVANTAGES: Improved selectivity and signal-to-noise ratio compared to other modes. More expensive than most GCMS instrumentation.

MODE: Parent mode MRM GCMS (Metastable reaction monitoring; Sec. 2.5.3.4)

METHOD: Monitor a selected daughter ion (e.g., m/z 217) formed from molecular ions that decompose in the first field-free region of a double focusing mass spectrometer. This is a commonly used form of GCMSMS.

ADVANTAGES/DISADVANTAGES: Similar advantages in selectivity and signal-to-noise ratio over other modes as in parent mode GCMSMS on a tandem instrument, except lower resolution of metastable ions can result in lower selectivity. Uses double focusing magnetic instrument which may be more readily available than a tandem instrument.

standards at Chevron (GCMS No. 81-3 and 257 from Hamilton Dome, Wyoming, and Carneros, California, respectively).

2.5.3.2 Full Scan GCMS. In "full scan" GCMS the magnet or quadrupoles scan nearly the entire spectrum of ions (i.e., 50–600 amu) generated at the source. Unlike MID, no data are lost, but extensive computer storage is required. Complete mass spectra for qualitative identification of compounds can be generated from full

COMPOUND CLASS	m/z	Diagnostic Fragmentations[1]
SATURATED HYDROCARBONS		
Alkylcyclohexanes	83	
Methylalkylcyclohexanes	97	
Terpanes[2]	123	(Drimane)
Tetracyclic Terpanes	191	
5α(H)-Steranes [5β(H)-Steranes]	149 (151)	
14α(H)-Steranes [14β(H)-Steranes]	217 (218)	
Steranes	261 + X	
17α(H)-Steranes [17β(H)-Steranes]	257 (259)	

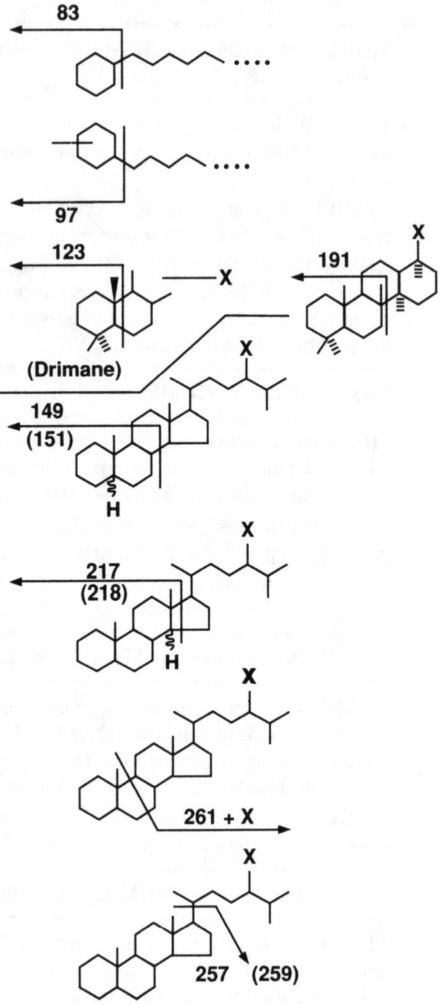

Figure 2.11 Diagnostic mass spectrometric fragmentations (and their mass to charge ratios) used to monitor various biomarkers and other compounds.

[1] Fragmentation arrows indicate carbon-carbon bond cleavage. Many fragmentations also include hydrogen transfers and rearrangements, which are not shown.

[2] Includes most terpanes with a drimane structure or substructure.

[3] Includes most terpanes with a cheilanthane structure or substructure such as degraded and extended hopanes, gammacerane, and oleanane.

[4] The ring D + E fragment is the same as that for the hopanes.

[5] The 25,28,30-trisnorhopanes have the m/z 163 fragment plus a m/z 177 fragment for ring A + B.

Compound	Ions (m/z)
14α(H)-Methylsteranes [14β(H)-Methylsteranes]	231 (232) 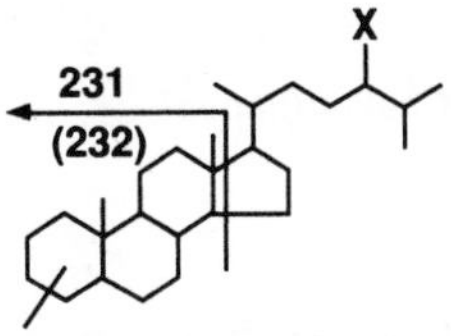
Dinosteranes	98 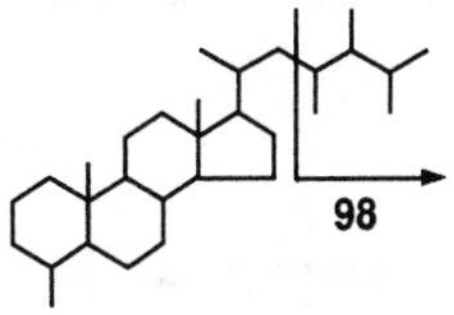
Diasteranes	259 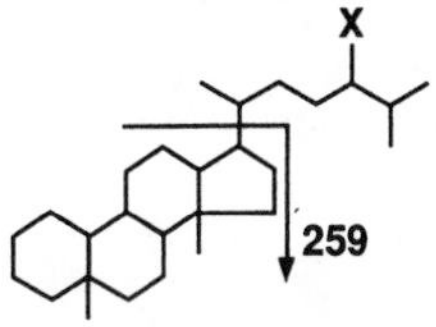
13α,17β(H)-Diasteranes	232 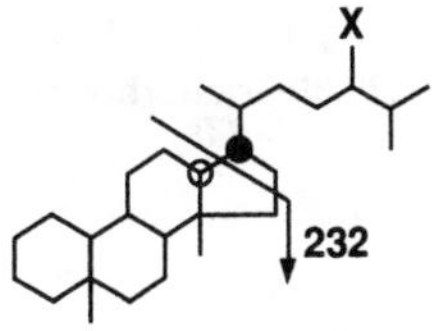
Gammacerane $C_{30}H_{52}$	191, 412 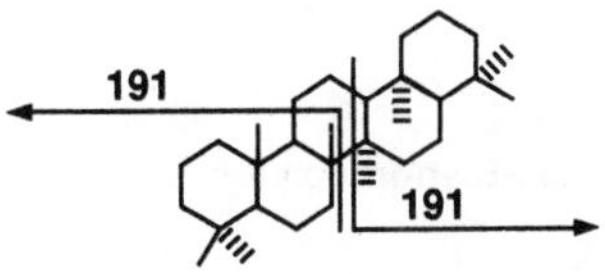
Oleanane $C_{30}H_{52}$	191, 412 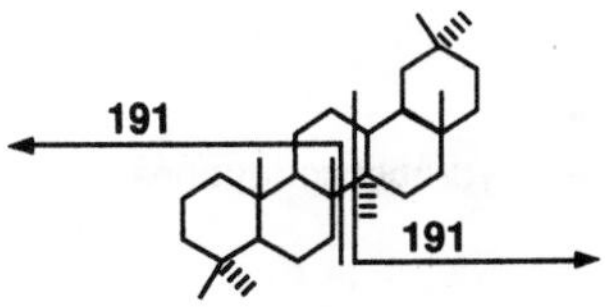

Figure 2.11 (*cont.*)

Cheilanthanes[3]	**191**
Hopanes (Ring A+B)[3]	**191**
Hopanes (Ring D+E) (C_{27}-C_{35})	**148+X (149, 163, 177, 191, 205, 219, 233, 247, 261)**
Hopanes	**369**
22,29,30-Trisnorhopane-II (Ts)	**149, 191**
25-Norhopanes (Ring A+B)	**177**
28,30-Bisnorhopanes[5]	**163, 191**
Hexahydrobenzohopanes	**191, 216+X, 94+X**

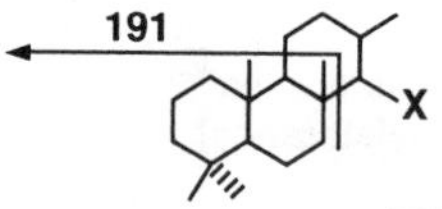

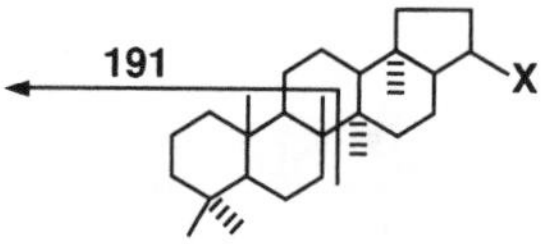

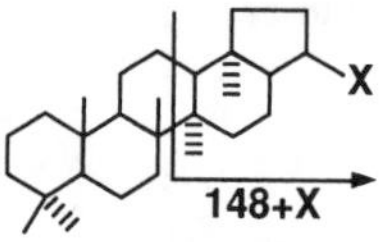

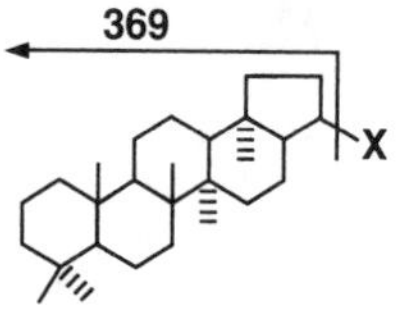

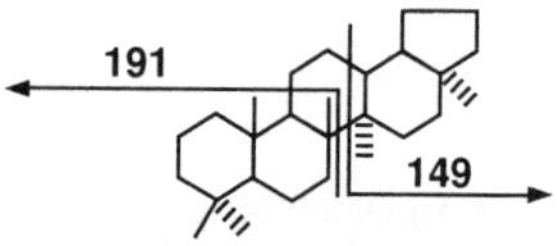

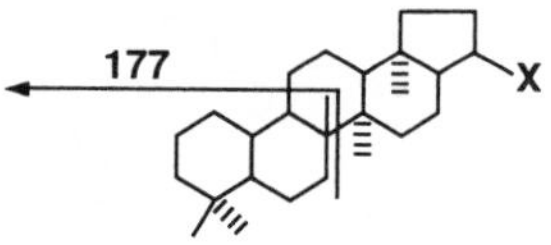

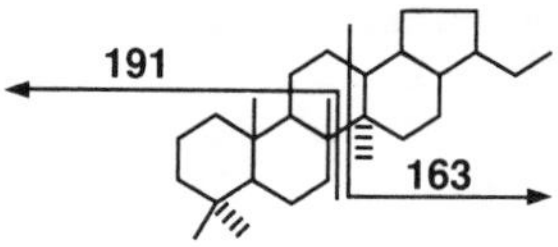

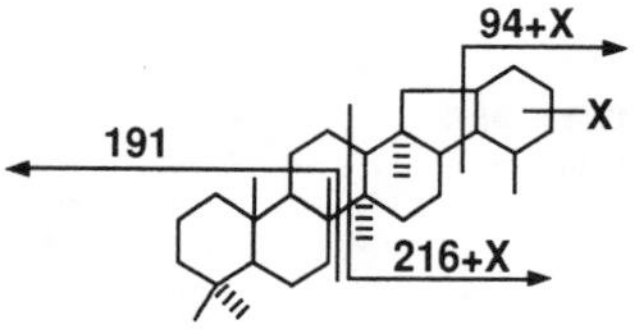

Figure 2.11 *(cont.)*

Botryococcane $C_{34}H_{70}$	238, 239, 294, 295	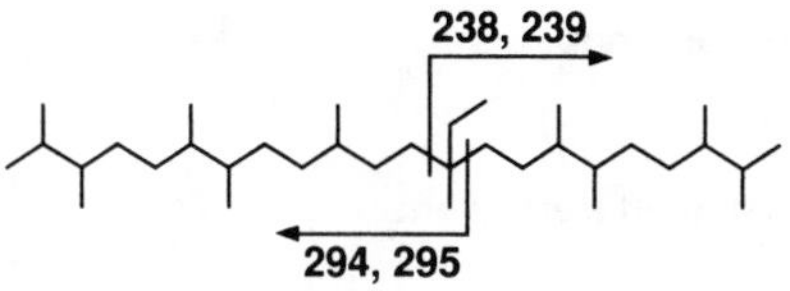 238, 239; 294, 295
β-Carotane $C_{40}H_{78}$	125, 558	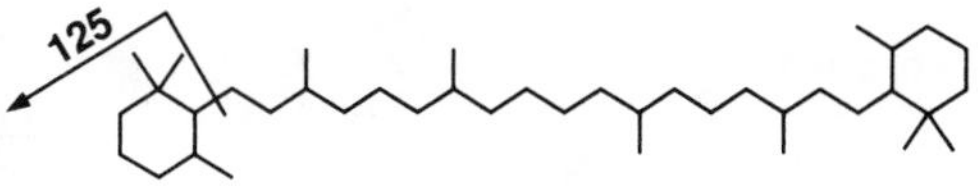 125
Regular Isoprenoids	113+70n	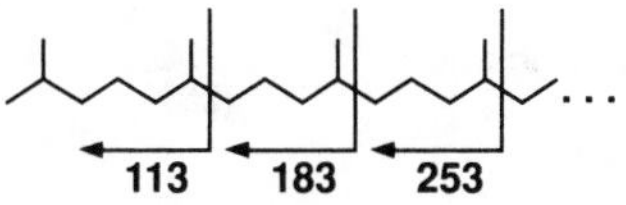 113; 183; 253
Head-to-Head Isoprenoids [1,1'- Bis(phytane)]	(323)	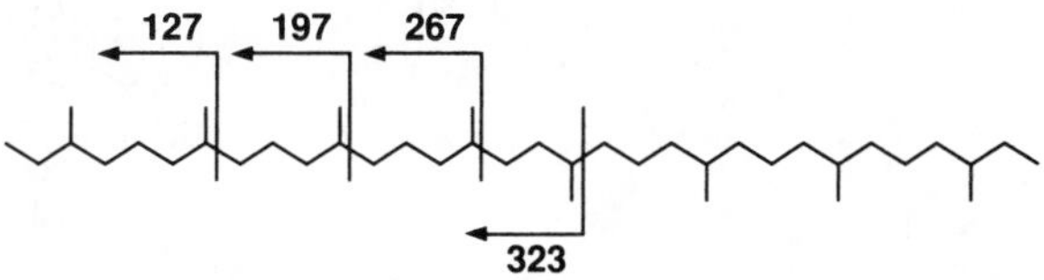 127; 197; 267; 323
Squalane $C_{30}H_{62}$	183, 422	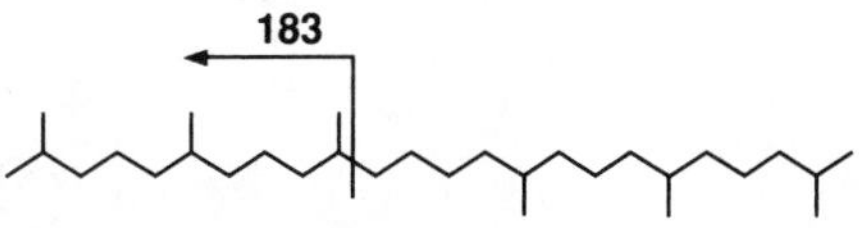 183

Figure 2.11 (*cont.*)

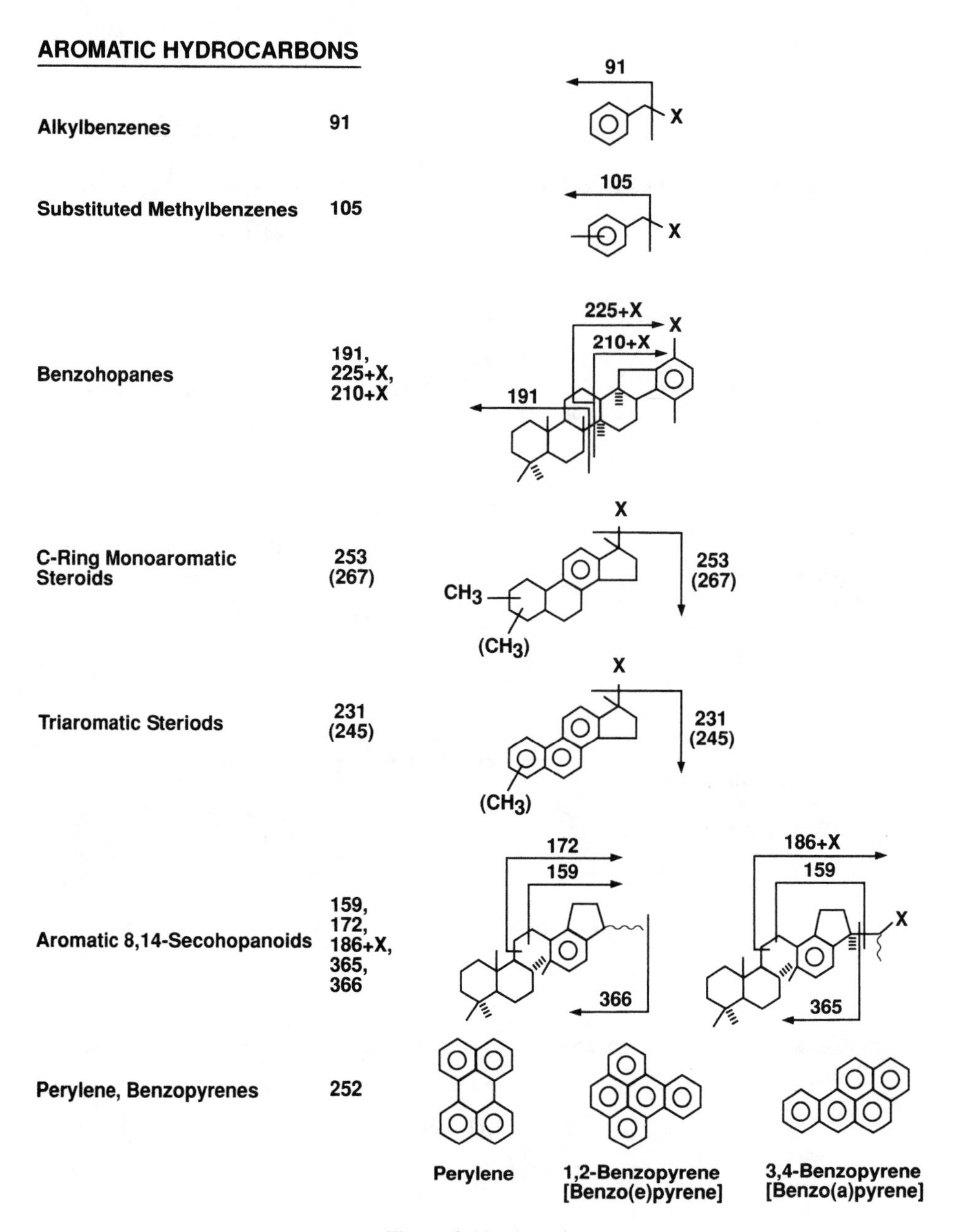

Figure 2.11 (*cont.*)

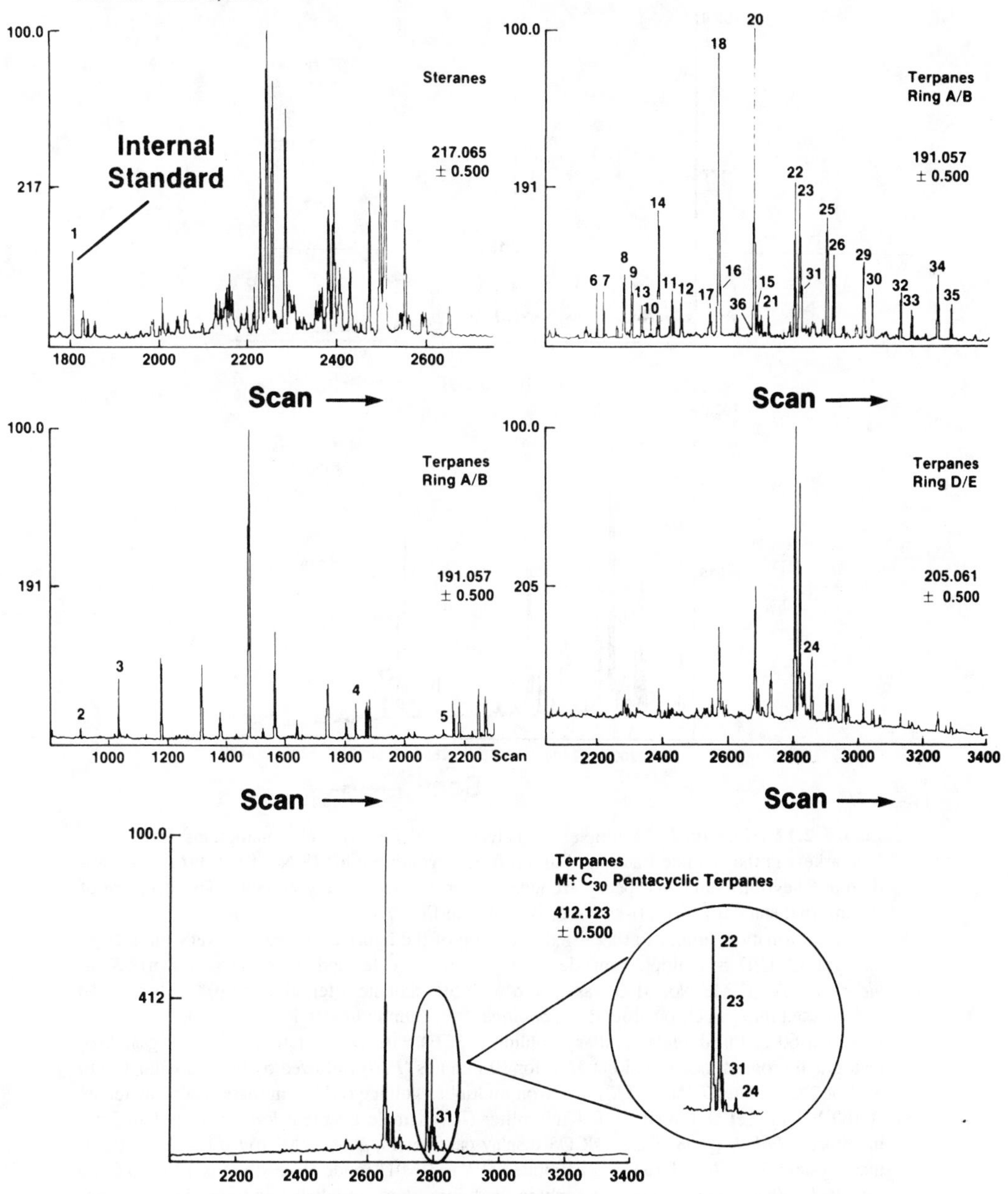
MID Mass Chromatogram
10/30/87 10:46:00
Sample: GCMS #81-3 GCL STD
.009768%CHOL
Conds: 117391 60M DB1 FSCP, P10, T150-325/2,
EMV160 1x10-6,.1UL
Range: G 1,3600 Label: N 0, 4.0
Data: VM7169 #1 Scans 1751 to 2750
Cali: VK7169 #1
Quan: A 0, 1.0 J 0
Base: U 20, 3
Internal Standard
Steranes
217.065
± 0.500
Terpanes
Ring A/B
191.057
± 0.500
Terpanes
Ring D/E
205.061
± 0.500
Terpanes
M+ C30 Pentacyclic Terpanes
412.123
± 0.500
Scan

Figure 2.12

MID Mass Chromatogram
10/30/87 10:46:00
Sample: GCMS #81-3 GCL STD .009768%CHOL
Conds: 117391 60M DB1 FSCP, P10, T150-325/2, EMV160 1x10-6,.1UL

Range: G 1,3600 Label: N 0, 4.0
Data: VM7169 #1 Scans 1751 to 2750
Cali: VK7169 #1
Quan: A 0, 1.0 J 0 Base: U 20, 3

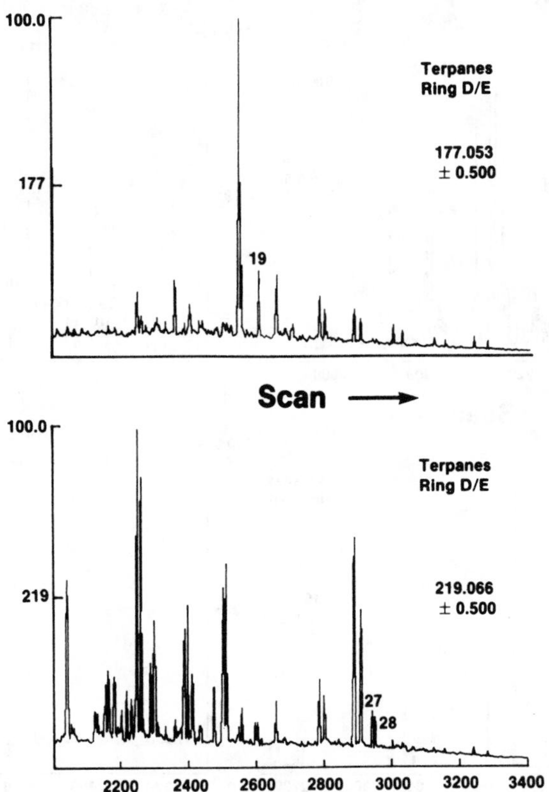

Figure 2.12 ***(continued)*** Multiple ion detection (MID) mass chromatograms for various biomarkers in the saturate fraction of an oil from Wyoming (GCMS No. 81-3) used as a standard at Chevron. Numbered peaks are identified in the accompanying table. The structure of the internal standard, 5β (H)-cholane, is shown in Fig. 2.4.

Information similar to that listed at the top of the figure accompanies every mass chromatogram. *MID* = multiple ion detection (analysis date and time also shown); *Sample* = sample GCMS No. 81-3 was spiked with 5β-cholane internal stand (0.009768 wt %) and injected into gas chromatograph; *Conditions* = column number 117391 (in log book) consists of a 60 m fused silica capillary column (*FSCP*) with *DB-1* stationary phase; gas chromatograph programmed to hold at 150° for 10 minutes (*P10*) followed by heating from 150 to 325° at 2°C/minute (*T150-325/2*), electron multiplier voltage (*EMV*) in mass spectrometer set at 160 V, gain set at 1×10^{-6}, 0.1 microliter (μL) sample injected; *Range*: scans 1 to 3600 monitored; *Label, Quan, Base*: INCOS display parameters; *Data*: VM7169 #1 = VG Micromass system 7169 No. 1 run and calibrated (*VK*) in MID mode; *Scans*: scans 1751 to 2750 recorded. When other scan ranges apply in the figure, they are labeled on the x-axis of the appropriate mass chromatogram.

As indicated in Table 2.5.3, m/z 217 = steranes, m/z 191 = ring A/B fragment in terpanes, m/z 177, 205, 219 = ring D/E fragment in terpanes, and m/z 412 = molecular ion of C_{30}-pentacyclic terpanes. All peaks are normalized to the fragment with the greatest mass spectrometric response (100% on y-axis).

Peak Number	Name	Carbon Number
1	5β-Cholane	24
2	Tricyclic (Cheilanthane) (C_{19})	19
3	Tricyclic (Cheilanthane) (C_{20})	20
4	Tetracyclic (C_{24})	24
5	Tetracyclic (C_{25})	25
6	Tricyclic (Cheilanthane) (C_{28})	28
7	Tricyclic (Cheilanthane) (C_{28})	28
8	Tricyclic (Cheilanthane) (C_{29})	29
9	Tricyclic (Cheilanthane) (C_{29})	29
10	Tetracyclic (C_{26})	26
11	Tricyclic (Cheilanthane) (C_{30})	30
12	Tricyclic (Cheilanthane) (C_{30})	30
13	22,29,30-Trisnorhopane-II (Ts)	27
14	22,29,30-Trisnorhopane (Tm)	27
15	17α(H)-30-nor-29-homohopane	30
16	18α(H)-30-norneohopane (C_{29}Ts)	29
17	Tricyclic (Cheilanthane) (C_{31})	31
18	17α,21β(H)-30-norhopane	29
19	17β,21α(H)-30-norhopane (normoretane)	29
20	17α,21β(H)-hopane	30
21	17β,21α(H)-hopane (moretane)	30
22	17α,21β(H)-29-homohopane 22S	31
23	17α,21β(H)-29-homohopane 22R	31
24	17β,21α(H)-29-homohopane 22S + 22R	31
25	17α,21β(H)-29-bishomohopane 22S	32
26	17α,21β(H)-29-bishomohopane 22R	32
27	17β,21α(H)-29-bishomohopane (22?)	32
28	17β,21α(H)-29-bishomohopane (22?)	32
29	17α,21β(H)-29-trishomohopane 22S	33
30	17α,21β(H)-29-trishomohopane 22R	33
31	Gammacerane	30
32	17α,21β(H)-29-tetrakishomohopane 22S	34
33	17α,21β(H)-29-tetrakishomohopane 22R	34
34	17α,21β(H)-29-pentakishomohopane 22S	35
35	17α,21β(H)-29-pentakishomohopane 22R	35
36	Oleanane	30

Figure 2.12 *(cont.)*

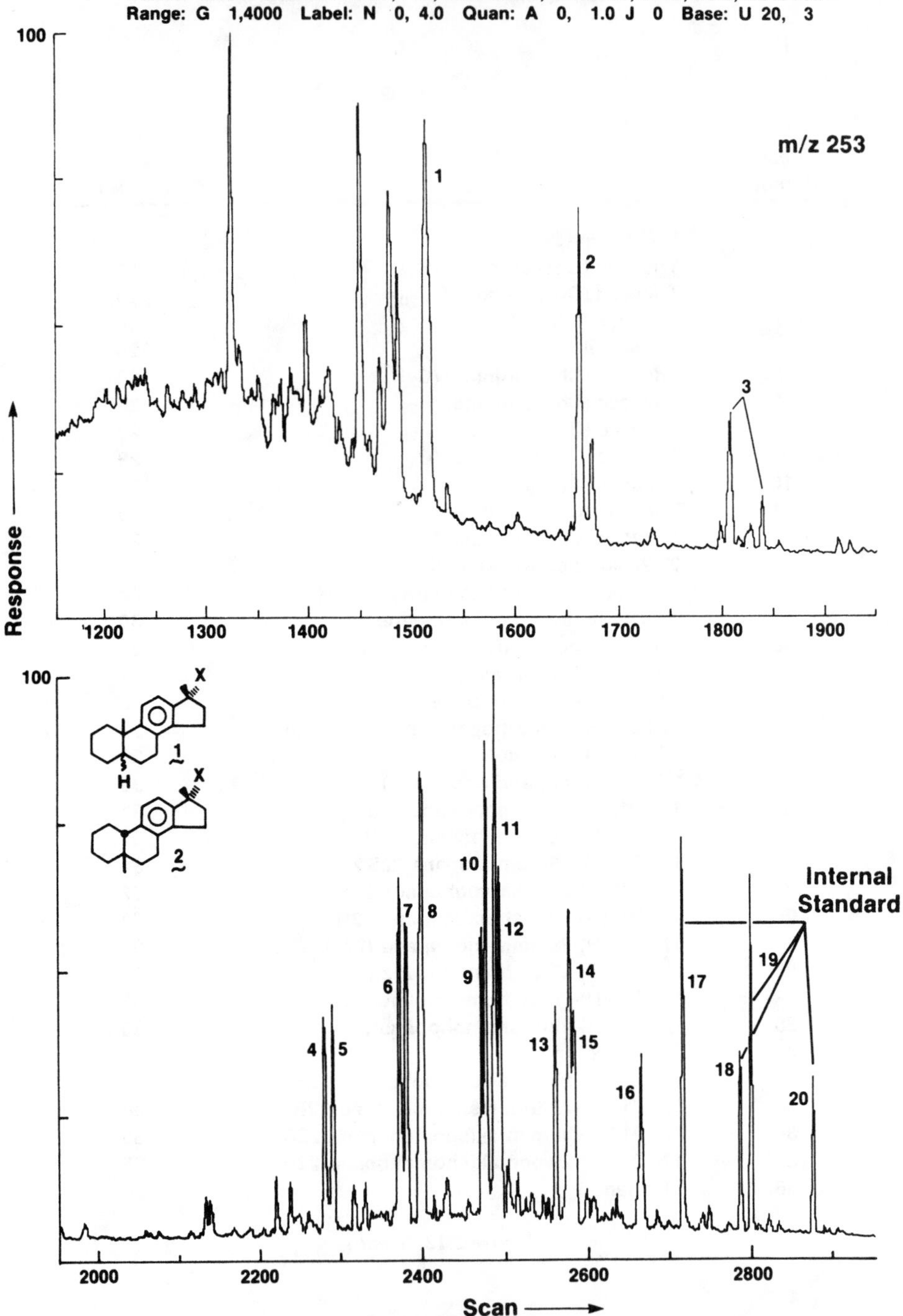

Figure 2.13 MID mass chromatograms for the monoaromatic (MA-) steroids in the aromatic fraction an oil standard from Carneros, California (GCMS No. 257) used at Chevron. This standard is more suitable than the Wyoming oil standard for MA-steroids (GCMS No. 81-3) because it contains all compounds used in routine studies. Numbered compounds are identified in the following table. Figure 2.4 shows the structures of the C_{30} MA-steroid standards.

Peak Number	Structure	Identification*	Carbon Number
1	?	Pregnane (X = Ethyl)	21
2	?	20-Methylpregnane (X = 2-Propyl)	22
3	?	20-Ethylpregnane (X = 2-Butyl)	23
4	1	5β-Cholestane 20S	27
5	2	Diacholestane 20S	27
6	1 2	5β-Cholestane 20R Plus Diacholestane 20R	27 27
7	1	5α-Cholestane 20S	27
8	1 2	5β-Ergostane 20S Plus Diaergostane 20S	28 28
9	1	5α-Cholestane 20R	27
10	1	5α-Ergostane 20S	28
11	1 2	5β-Ergostane 20R Diaergostane 20R	28 28
12	1 2	5β-Stigmastane 20S Diastigmastane 20S	29 29
13	1	5α-Stigmastane 20S	29
14	1	5α-Ergostane 20R	28
15	1 2	5β-Stigmastane 20R Diastigmastane 20R	29 29
16	1	5α-Stigmastane 20R	29
17	1	5β-n-Nonylpregnane 20S (X = 2-Undecyl)	30
18	1	5α-n-Nonylpregnane 20S	30
19	1	5β-n-Nonylpregnane 20R	30
20	1	5α-n-Nonylpregnane 20R	30

*Name of Sterane (Structure 1) or Diasterane (Structure 2) with the Same Side Chain (X) Structure

Figure 2.13 *(cont.)*

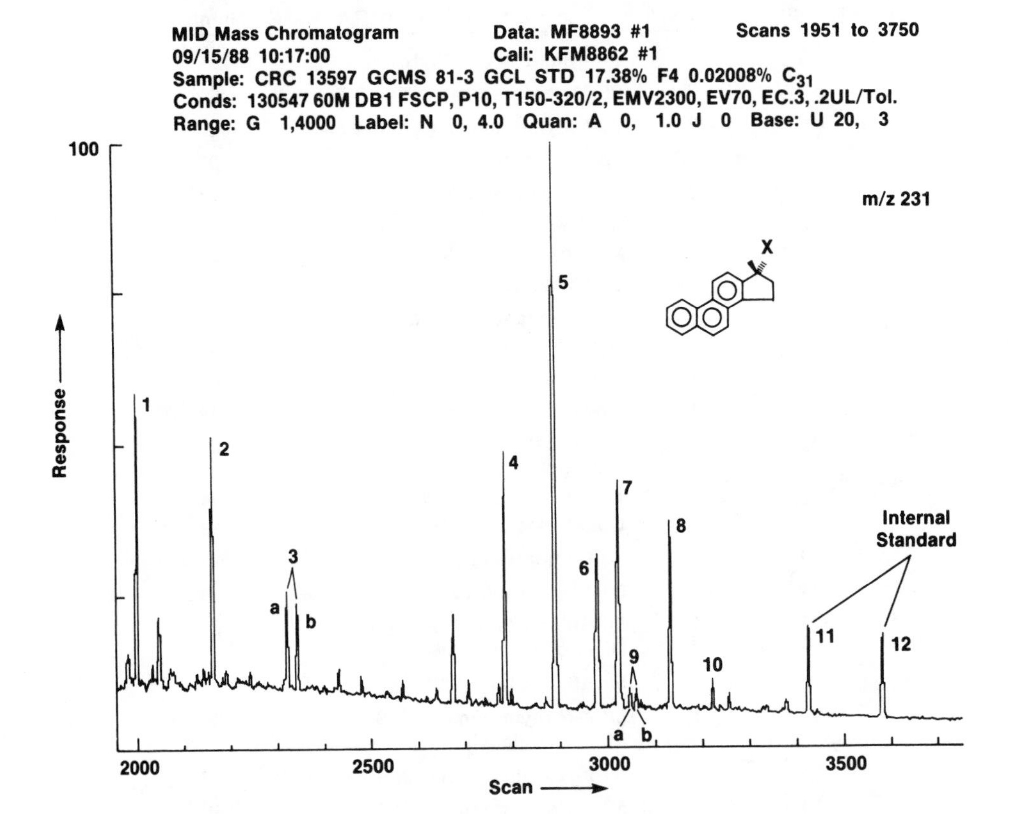

Figure 2.14 MID mass chromatogram for the triaromatic (TA-) steroids in the aromatic fraction of an oil from Wyoming (GCMS No. 81-3) used as a standard at Chevron. Numbered compounds are identified in the following table. Figure 2.4 shows the structures of the TA-steroid internal standards.

Peak Number	Identification*	Carbon Number
1	Pregnane (X = Ethyl)	20
2	20-Methylpregnane (X = 2-Propyl)	21
3	20-Ethylpregnanes (X = 2-Butyl) (a and b are Epimeric at C_{20})	22
4	Cholestane 20S	26
5	Cholestane 20R Plus Ergostane 20S (Coelution)	26, 27
6	Stigmastane 20S (24-Ethylcholestane 20S)	28
7	Ergostane 20R (24-Methylcholestane 20R)	27
8	Stigmastane 20R	28
9	24-n-Propylcholestane 20S (a and b are Epimeric at C_{24})	29
10	24-n-Propylcholestane 20R	29
11	20-n-Decylpregnane 20S (X = 2-Dodecyl)	30
12	20-n-Decylpregnane 20R	30

*Name of Sterane with the Same Side Chain (X) Structure

Figure 2.14 *(cont.)*

scan analysis. Full scan mode records several hundred ions per scan (~ 3 seconds), resulting in < 0.0075 second dwell time per mass. By comparison, MID acquisitions record at about 25 ions/second, resulting in a longer dwell time (~ 0.04 second/ion) and an order of magnitude better sensitivity and signal-to-noise ratio. Thus, MID is preferred over full scan data acquisition for quantitative biomarker analysis, even though the same kind of mass chromatograms can be displayed from a full scan run.

2.5.3.3 Benchtop Quadrupole GCMS. Benchtop quadrupole GCMS systems offer many of the capabilities of larger, more expensive floor models. For example, the Hewlett Packard Model 5970 GC-MSD (mass selective detector) and VG Instruments Trio-1 are benchtop GCMS systems designed primarily for routine and inexpensive biomarker analysis in the multiple ion detection mode (MID). Hwang et al. (1989) have shown that sterane and triterpane distributions obtained using the GC-MSD are in qualitative and quantitative agreement with those produced using the more versatile floor-model VG Micromass 7070H GCMS in MID mode. Screening prior to detailed biomarker analysis (Sec. 3.1.1.1.1) can be provided by benchtop quadrupole GCMS analysis. The GC-MSD can be used to provide mass chromatograms for about one dozen ions in a single analysis.

Benchtop systems are useful for many applications such as screening samples for more detailed biomarker analysis or rapid correlation of groups of oils. However, benchtop MID and GCMSMS (see following text) data are not interchangeable. Benchtop quadrupole systems are currently incapable of GCMSMS analyses of parent-daughter ion relationships. Certain maturity ratios such as the C_{29} 20S/(20S + 20R) (Sec. 3.2.4.2) are best determined by GCMSMS because of possible interfering peaks on MID analyses.

2.5.3.4 GCMSMS. **GCMSMS** (gas chromatography/mass spectrometry/mass spectrometry) is a technique based on the fact that complex organic molecules (parents) that are ionized in the ion source of a mass spectrometer break down to smaller charged fragments (daughters). Many of these daughter ions are characteristic of their parent molecules (e.g., m/z 217 is a daughter of most steranes). GCMSMS allows the operator to determine the parents of any selected daughter, as described in the following.

GCMSMS represents a powerful technique that allows determination of specific parent-daughter relationships with little interference from other reactions and their related ions. The analytical specificity of these methods eliminates most interference by coeluting GC peaks. This specificity may boost signal-to-noise (i.e., chemical noise) ratios to levels that are orders of magnitude better than those obtained using conventional GCMS in MID mode.

MID GCMS analysis of an oil for the m/z 217 fragment ion typically reveals a complex mixture of structural and stereoisomeric C_{27}, C_{28}, C_{29} and other steranes and may also include some nonsterane compounds. The bottom chromatogram in Fig. 2.15 is a simplified, hypothetical example of a mixture of three single sterane epimers at C_{27}, C_{28}, and C_{29}. Each epimer yields a base peak at m/z 217 on this trace. In one MSMS method, for example, the electrostatic and magnetic analyzers in a double focusing MS are linked so as to preserve a predetermined field strength between the two analyzers. GCMSMS analysis of the sterane parent ion transitions corresponding to m/z 372 to 217, m/z 386 to 217, and m/z 400 to 217, for example, allows the operator to generate separate mass chromatograms for the C_{27}-, C_{28}-, and C_{29}-steranes, respectively (Fig. 2.15).

GCMSMS analyses can be used to determine marine input to oils (C_{30} steranes), and for correlations using triangular diagrams of C_{27}-C_{28}-C_{29}- steranes, triaromatic steroids, and other compound types. Figure 2.12 includes an example of a m/z 217 trace for the standard oil from Wyoming, which is much more complex than the simplified, hypothetical mixture in Fig. 2.15. GCMSMS of the same standard oil (Fig. 2.16) allows differentiation of the steranes using mass chromatograms by carbon number.

GCMSMS includes linked and de-linked double focusing mass spectrometry and tandem (e.g., triple sector quadrupole) mass spectrometry. **Metastable reaction monitoring** (MRM) GCMS refers to monitoring those decompositions that occur in the first field-free region of a double focusing mass spectrometer that has been configured in the linked (Haddon, 1979) or de-linked (Gallegos, 1976; Warburton and Zumberge, 1982) mode. These methods are also called "selected metastable ion

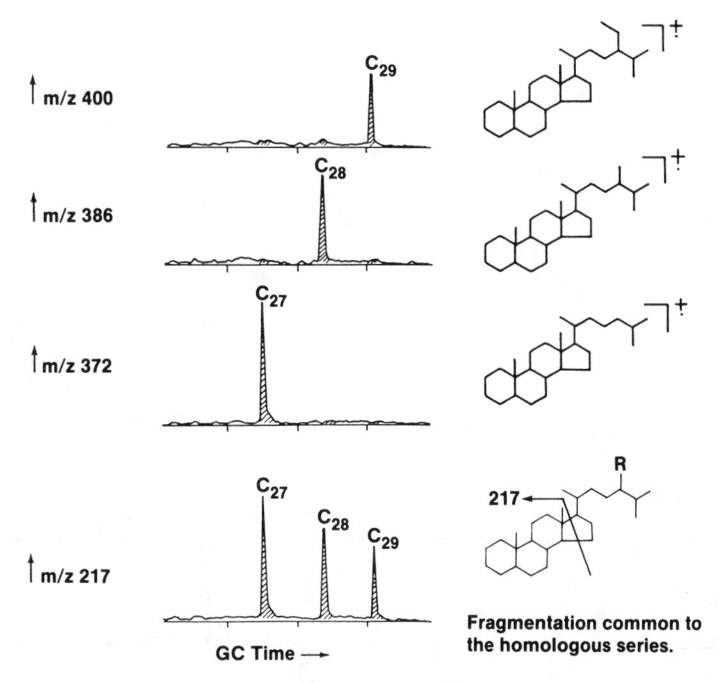

Figure 2.15 GCMSMS reduces interference by allowing measurement of specific parent/daughter relationships between ions. The bottom mass chromatogram shows a hypothetical mixture of three steranes, containing 27, 28, and 29 carbon atoms, respectively. The steranes were monitored using m/z 217, a fragment ion common to most steranes. In this very simple case, only one epimer of each sterane is present. A typical m/z 217 mass chromatogram for petroleum contains a much larger number of overlapping peaks representing the various epimers of several sterane homologs (e.g., Fig. 2.16). By monitoring specific transitions, for example, from a parent m/z of 400 (parent ion for C_{29} steranes) to a daughter m/z of 217, GCMSMS allows detection of each individual compound without interference.

monitoring" (SMIM, e.g., Steen, 1986). A triple sector instrument uses a neutral gas in the middle sector to induce decomposition and achieve GCMSMS. The following discussion describes how double focusing and triple quadrupole instruments are configured to accomplish this type of analysis.

1. Double focusing mass spectrometers can be configured in two ways for GCMSMS. In the first method, the electrostatic voltage is de-linked from the accelerating voltage. The magnet is set to "see" only the daughter fragments of interest and the accelerating voltage is scanned to measure the metastable transitions that occur in the first field-free region of the mass spectrometer. Only those parents that decompose to give the daughter of interest are detected. In the second method, called linked-scan, the magnet and electrostatic sectors are linked. For example, a linked-scan experiment could involve scanning the

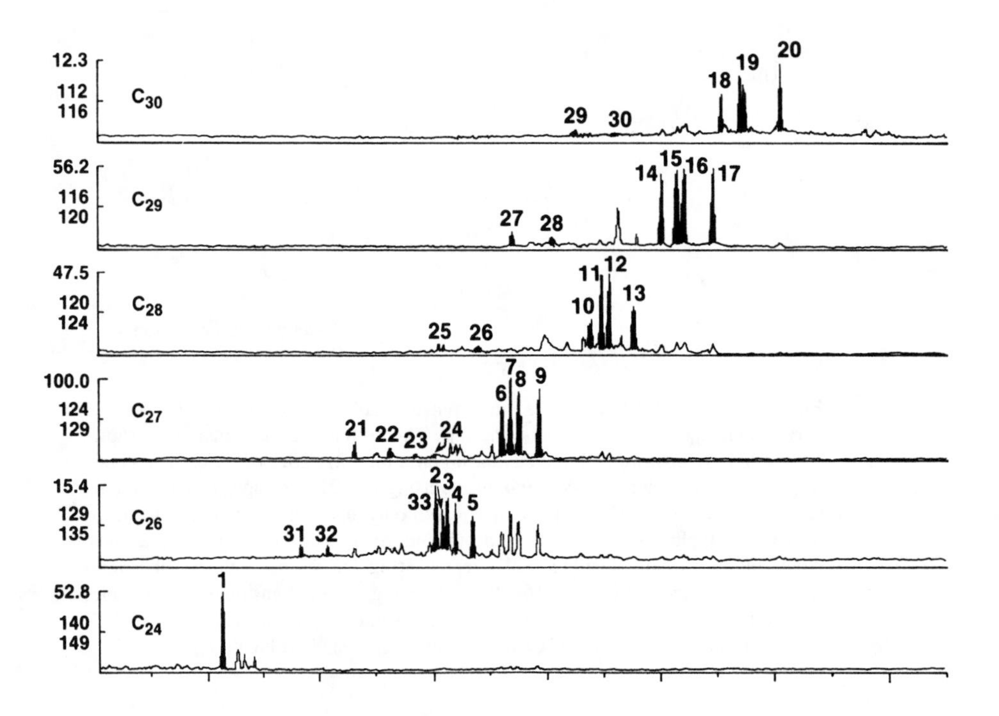

Figure 2.16 MRM GCMS analysis ($M^+ \rightarrow m/z$ 217) of the same oil (GCMS No. 81-3) shown in Fig. 2.12 (steranes) allows differentiation of sterane epimers by carbon number. Note the complex epimer distributions at each carbon number compared to the hypothetical example in Fig. 2.15. Also note that peaks corresponding to 6, 7, and 8 on the C_{27} chromatogram also appear on the C_{26} chromatogram. This is an interference which is caused by the inherent low resolution (~100) of the MRM method. Similar experiments using GCMSMS on a tandem mass spectrometer are typically performed at higher resolution (~1000) to eliminate this type of interference.

Peak Number	Name	Carbon Number
1	5β-Cholane	24
2	5α,14α,17α,27-norcholestane 20S	26
3	5α,14β,17β,27-norcholestane 20R	26
4	5α,14β,17β,27-norcholestane 20S	26
5	5α,14α,17α,27-norcholestane 20R	26
6	5α,14α,17α-cholestane 20S	27
7	5α,14β,17β-cholestane 20R	27
8	5α,14α,17β-cholestane 20S	27
9	5α,14α,17α-cholestane 20R	27
10	5α,14α,17α-ergostane 20S	28
11	5α,14β,17β-ergostane 20R	28
12	5α,14β,17β-ergostane 20S	28
13	5α,14α,17α-ergostane 20R	28
14	5α,14α,17α-stigmastane 20S	29
15	5α,14β,17β-stigmastane 20R	29
16	5α,14β,17β-stigmastane 20S	29
17	5α,14α,17α-stigmastane 20R	29
18	5α,14α,17α,24-*n*-propylcholestane 20S	30
19	5α,14β,17β,24-*n*-propylcholestane 20R+20S	30
20	5α,14α,17α,24-*n*-propylcholestane 20R	30
21	13β,17α-diacholestane 20S	27
22	13β,17α-diacholestane 20R	27
23	13α,17β-diacholestane 20S	27
24	13α,17β-diacholestane 20R	27
25	13β,17α-diaergostane 20S (24S+24R)	28
26	13β,17α-diaergostane 20R (24S+24R)	28
27	13β,17α-diastigmastane 20S	29
28	13β,17α-diastigmastane 20R	29
29	13β,17α-dia-24-*n*-propylcholestane 20S	30
30	13β,17α-dia-24-*n*-propylcholestane 20R	30
31	13β,17α-dia-27-norcholestane 20S	26
32	13β,17α-dia-27-norcholestane 20R	26
33	5α,14α,17α + 5α,14β,17β,21-norcholestanes	26

Figure 2.16 (*cont.*)

magnetic sector field strength (B) and the electric sector field strength (E) simultaneously, holding the accelerating voltage constant so as to maintain a constant ratio B to E. This constant value is determined by the ratio of the two field strengths which transmit the desired ions of a predetermined mass to charge ratio. For parent ion experiments, the sectors are linked according to the constant B^2/E. For daughter ion experiments, the sectors are linked according to the constant B/E.

2. Triple quadrupole and tandem magnetic instruments can be used for parent, daughter, and neutral loss experiments. They typically use a collision gas to cause decomposition of the compounds to be analyzed. These systems consist of three mass analyzers as discussed in Sec. 2.5.3.5.

The powerful MRM GCMS technique, together with triple stage quadrupole MS (Sec. 2.5.3.5), continue to expand in applications. While not truly GCMSMS as performed on a tandem mass spectrometer, MRM GCMS yields similar results. Thus, further references to GCMSMS in the text also include MRM GCMS unless otherwise stated. Because of the greater sensitivity of these techniques, they are likely eventually to replace conventional GCMS in biomarker applications.

2.5.3.5 Tandem or Triple Stage Quadrupole Mass Spectrometry. Tandem mass spectrometers show a major advantage over conventional GCMS systems because they can resolve individual compounds or compound families from complex petroleum mixtures using linked-scanning techniques, generally called GCMSMS (Table 2.5.3). Triple quadrupole mass spectrometers, such as the Finnigan TSQ-70, are the most common type of tandem mass spectrometer. The TSQ-70 consists of three quadrupoles connected in series (Fig. 2.17): (1) the first or parent quadrupole (Q1), (2) the middle or collision cell quadrupole (Q2), and (3) the third or daughter quadrupole (Q3). In the collision cell quadrupole, all ions formed in the ion source and selected by or passed through Q1 undergo collision-activated decomposition (CAD) with argon or another inert gas. The fragment ions formed in the collision process are separated or selectively monitored by the daughter quadrupole and recorded using an electron multiplier.

Because of the selectivity offered by the use of three quadrupoles, triple quadrupole mass spectrometry offers the possibility of direct analysis of whole oils without preparatory separations of fractions.

Triple quadrupole mass spectrometers can be operated in three GCMSMS modes.

- Parent
- Daughter
- Neutral loss

Philp et al. (1988) describe applications of the triple quadrupole mass spectrometer for determining parent/daughter and daughter/parent ion relationships. Philp and Oung (1992) show that the additional analytical power of MSMS over

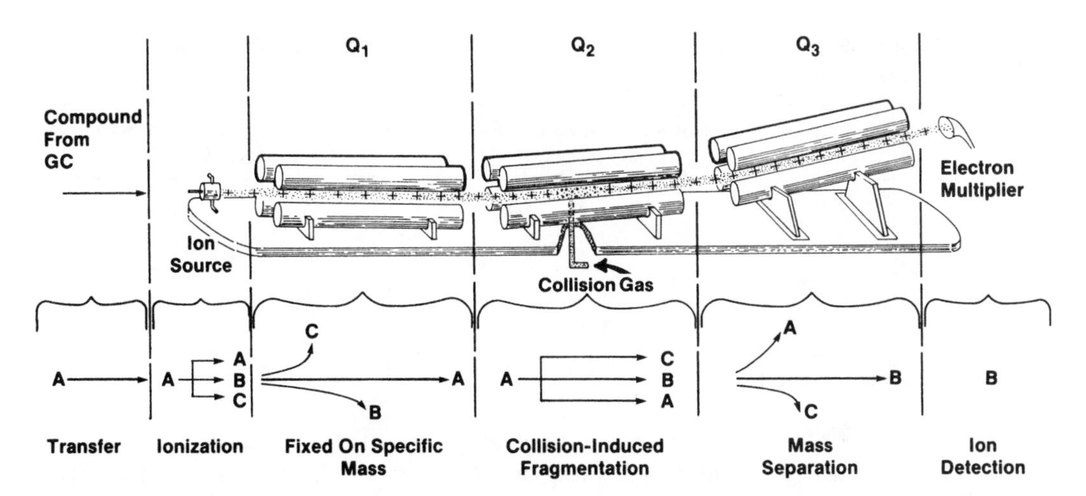

Figure 2.17 Schematic of a triple quadrupole mass spectrometer for GCMSMS analysis. Combined use of the three quadrupoles (Q_1, Q_2, and Q_3) allows highly selective compound detection (e.g., *only* daughter ion B from parent compound A). See text.

GCMS allows its use for rapid biomarker screening methods such as short column GCMSMS or direct insertion probe MSMS. These methods are faster than GCMS because little or no time is used for gas chromatographic separation of compounds. However, both short column and direct insertion techniques sacrifice stereochemical and structural information because epimers are not resolved.

Other tandem mass spectrometers combine various series of magnetic and electrostatic sectors and collision cells. A **hybrid mass spectrometer** combines a magnetic sector instrument with a collision cell quadrupole and a mass filtering quadrupole. These other types of tandem mass spectrometers offer some potential advantages over triple quadrupole instruments. For example, hybrid mass spectrometers with a high resolution magnet in place of Q1 allow the selection of parent ions at high resolution.

2.5.3.5.1 Parent mode GCMSMS. In the parent mode, ions formed in the ion source of the mass spectrometer enter the collision cell quadrupole (Q2 in Fig. 2.17) and react with the inert gas to form various daughter ions. One or more daughter ions for each biomarker family to be monitored are selected by the third quadrupole (Q3) to focus on the electron multiplier and be analyzed.

For example, each pseudohomolog in the series of C_{26} to C_{30} steranes differs by one methylene group (—CH_2—, i.e., mass 14 amu) from the adjacent pseudohomolog. GCMSMS can be used to differentiate between these pseudohomologs. Molecular ions (parents) of the C_{26} to C_{30} steranes consist of m/z 358, 372, 386, 400, and 414, respectively (Table 2.5.3.5.1). Each of these parent ions produces a major daughter ion at m/z 217 (Table 2.5.3.5.1) following collision with the inert gas in the collision cell (Q2, Fig. 2.17) that can be monitored using the daughter quadrupole (Q3). Both parents and daughters can be selectively monitored to improve signal-to-noise. Thus, by setting the parent quadrupole (Q1) to pass only ions

TABLE 2.5.3.5.1 Example of the Principle of Parent Mode GCMSMS Using C_{26} to C_{30} Steranes.

Carbon atoms in sterane	Parent ion separated by first quadrupole, Q1[1]	Daughter ion generated in second quadrupole, Q2[1]
26	358	217
27	372	217
28	386	217
29	400	217
30	414	217

[1] Refer to Figure 2.17.

having m/z 372 and by setting the daughter quadrupole to monitor m/z 217, we can obtain a mass chromatogram showing only C_{27} steranes. A signal is obtained only if the selected molecular ion produces a corresponding daughter ion.

The GCMSMS approach to identification of the C_{27} steranes described is based on the relationship between the parent (m/z 372) and daughter (m/z 217) fragments. This approach is superior to direct monitoring of m/z 372 in MID mode because MID is affected by interference. For example, C_{28} steranes can lose a methyl group (15 amu) during ionization, resulting in a fragment at m/z 371. However, some m/z 372 will also result from the C_{28} steranes because heavy isotopes of carbon (^{13}C) or hydrogen (D) are also present (Sec. 3.1.2.2). Compounds other than steranes may also fragment under MID GCMS conditions.

GCMSMS (including MRM GCMS) represents a significant refinement over routine GCMS. For example, monitoring m/z 217 by routine GCMS in MID mode provides a single mass chromatogram containing all steranes, many of which coelute. However, GCMSMS provides nearly complete separation of individual steranes by carbon number (Fig. 2.16) and mass chromatograms for each are obtained. These GCMSMS data are critical in constructing C_{27}, C_{28}, C_{29} sterane ternary diagrams for oil-oil and oil-source rock correlations. Routine MID GCMS, as offered by many service companies, cannot be used for constructing these diagrams as accurately as GCMSMS because of interference from compounds containing different numbers of carbon atoms. During any single GCMSMS analysis, several parent/daughter relationships can be monitored simultaneously, allowing resolution of several carbon numbers of the same family (e.g., C_{27} to C_{30} steranes) or of several different families of biomarkers (e.g., steranes, tricyclic terpanes, pentacylcic terpanes, etc.). GCMSMS analyses usually involve numerous parent/daughter relationships that must be deconvoluted by computer.

CAD GCMSMS is superior to other GCMS modes for deconvoluting biomarker compositions. For example, Fowler and Brooks (1990) compared sterane distributions and maturity parameters for oils and bitumens from the Egret Member of the Rankin Formation, offshore eastern Canada using four methods. Abundant 4-methylsteranes in these samples interfere with regular (4-desmethyl) steranes on m/z 217 mass chromatograms because both groups of steranes yield m/z 217 fragments. As expected, neither low- nor high-resolution MID GCMS systems were effective in removing the interference caused by the 4-methylsteranes because the fragment ions of both groups of steranes have the same atomic composition. MRM GCMS results for the same samples were more reliable than those from the two MID GCMS instruments, because specific parent-daughter transitions could be monitored. However, even more reliable results were obtained using CAD GCMSMS because of the greater resolution of both parent and daughter ions achieved using this method.

2.5.3.5.2 Daughter mode GCMSMS. In the daughter mode, a selected parent ion enters the collision cell where it undergoes collision-activated decomposition and a complete daughter mass spectrum is collected by setting the daughter quadrupole on full scan. This approach allows identification of components with specific molecular weights from their daughter ion spectra, which carry information similar to EI mass spectra (Philp and Oung, 1992; Gallegos and Moldowan, 1992).

2.5.3.5.3 Neutral loss mode GCMSMS. In the neutral loss mode, the parent and daughter quadrupoles are both set to scan over the desired mass range. The daughter quadrupole is set to scan at a given mass below the parent quadrupole based on hypothetical loss of that mass from the parent compounds as a neutral fragment. Thus, to monitor the distribution of sulfur-containing compounds that undergo a neutral loss of mass 32, the daughter quadrupole is set 32 mass units below the parent quadrupole. Only the compounds which lose the m/z 32 fragment will be detected.

2.5.4 Mass spectra, compound identification, and quantitation

2.5.4.1 Mass Spectra and Compound Identification. Mass spectra are an important tool for interpreting the structures of unknown compounds (McLafferty, 1980). A **mass spectrum** shows the mass of a given molecule and the masses of fragments from that molecule. One mass spectrum is generated approximately every three seconds as the detector scans through the desired mass range (usually 50–600 amu). Thus, each mass spectrum plots the mass/charge (m/z) ratio for ions impinging on the detector during each scan versus response (Fig. 2.10). Figure 2.18 shows mass spectra for several common biomarkers. Philp (1985) has compiled the mass spectra for a large number of biomarkers.

A combination of several analyses, which may include mass spectroscopy, nuclear magnetic resonance (NMR) spectroscopy, X-ray diffraction crystallography, and others, are necessary to prove the chemical structures of individual components eluting from the gas chromatograph. The structure of an unknown component is proven only when these analyses are identical for both the unknown and an authentic standard compound synthesized in the laboratory or when X-ray diffraction results are conclusive. Rigorous identification of compounds using mass spectra (McLafferty, 1980) combined with other methods such as two-dimensional NMR spectroscopy (Croasmun and Carlson, 1987) is beyond the scope of this book and will not be discussed further. The reader is referred to the following two examples of the use of NMR and X-ray crystallography in solving structural relationships in biomarkers. The structures of 18α(H)-oleanane and a spirotriterpane in a Nigerian crude oil were determined using X-ray crystallography by Smith et al. (1970). The

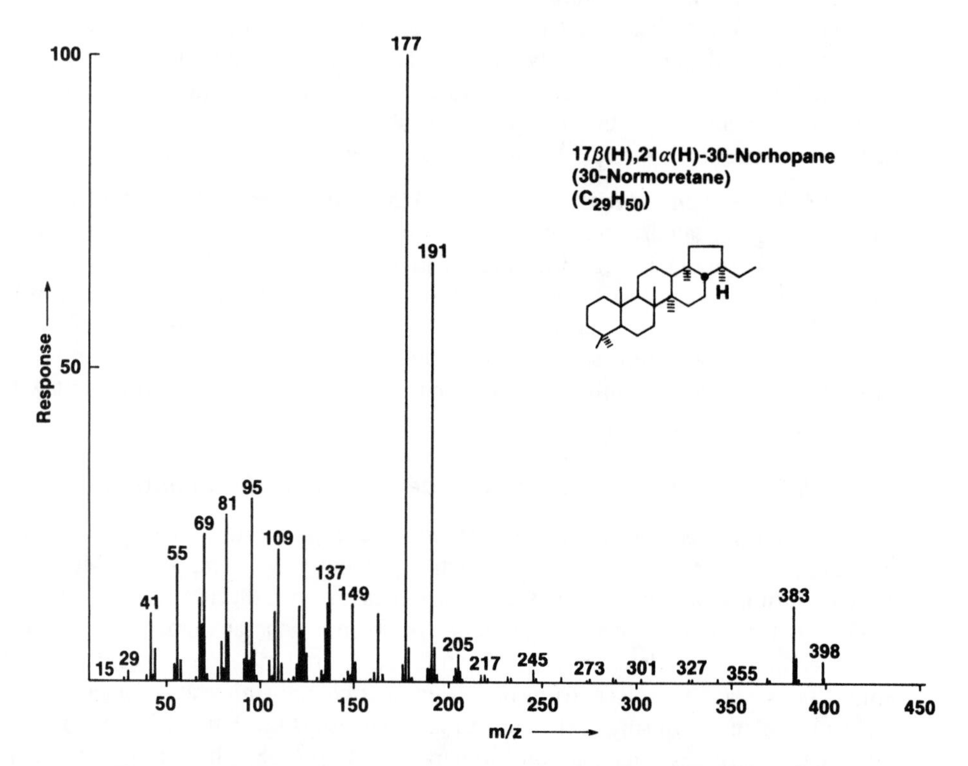

Figure 2.18 Mass spectra of several common biomarkers. The most intense peak for each spectrum (base peak) is defined as showing a response of 100 units on the y-axis. Philp (1985) shows mass spectra for a large number of biomarkers. Mass spectra of mixtures are difficult to interpret. For example, the mass spectrum for the mixture of β-carotane and β-carotene shows a base peak at m/z 123 while that for β-carotane is actually at m/z 125 (Murphy et al., 1967).

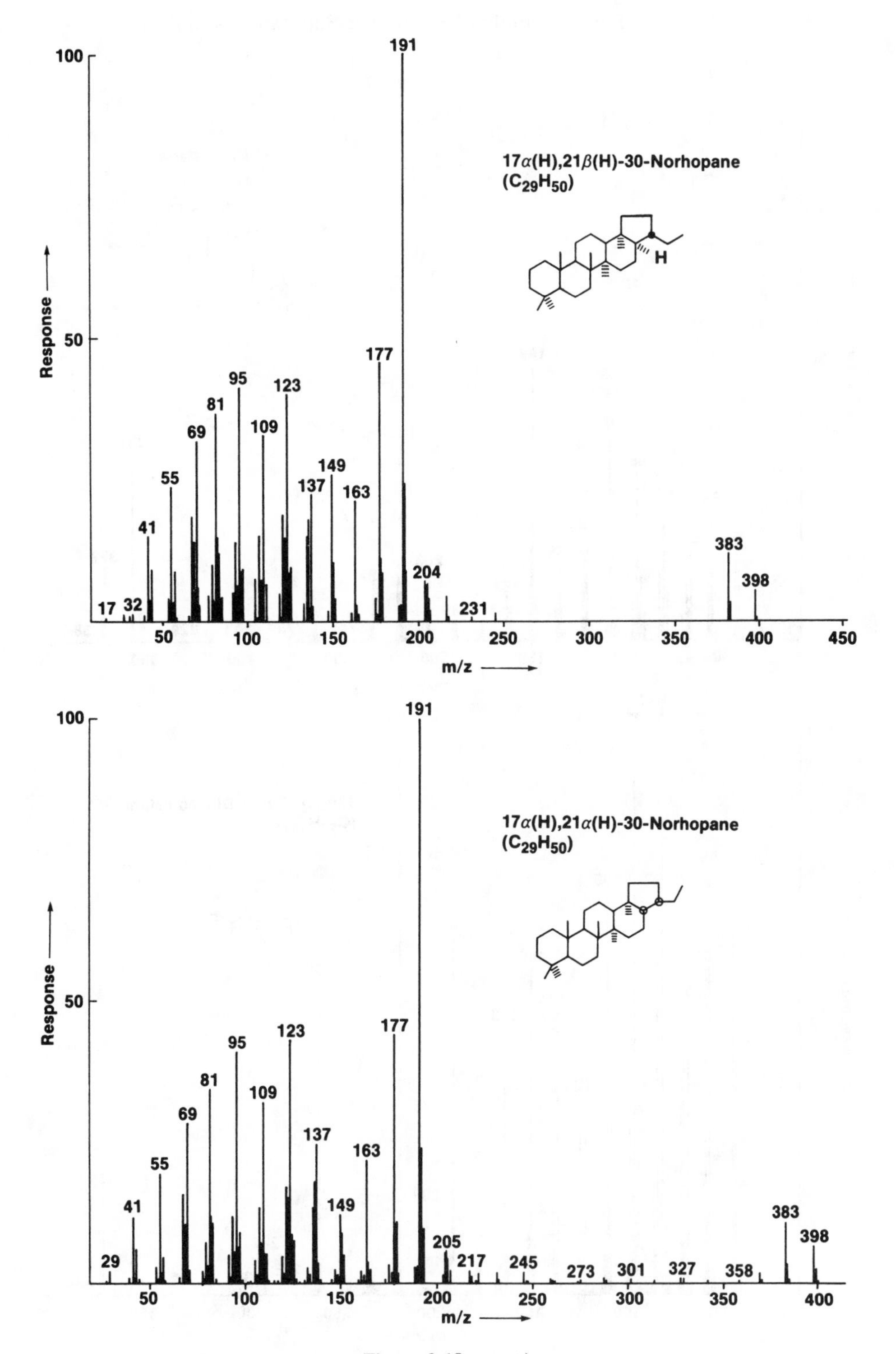

Figure 2.18 *cont.*)

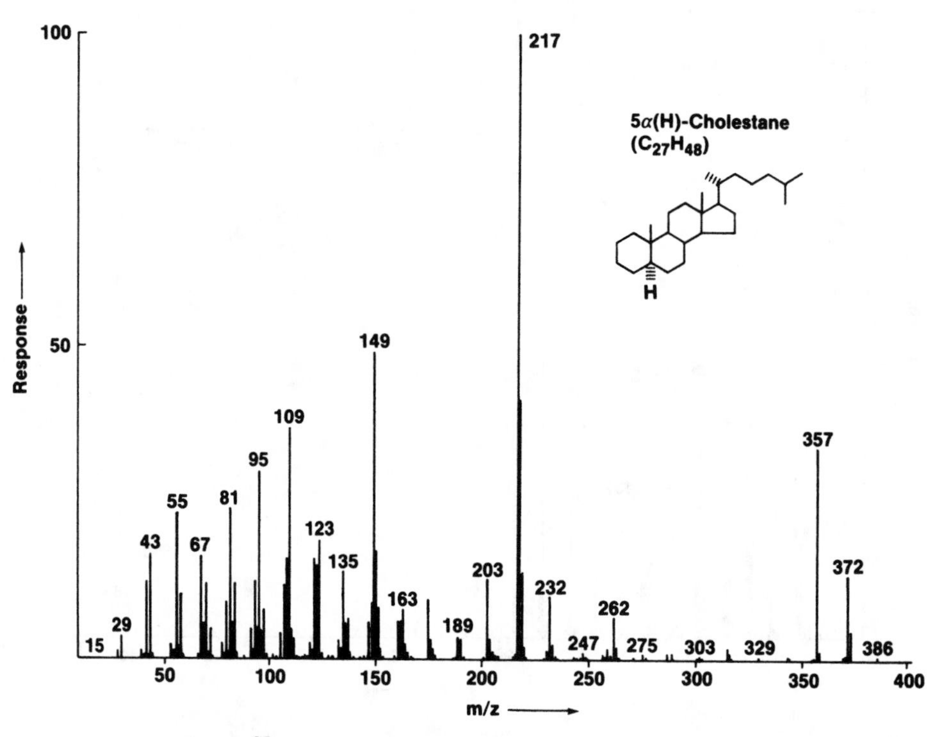

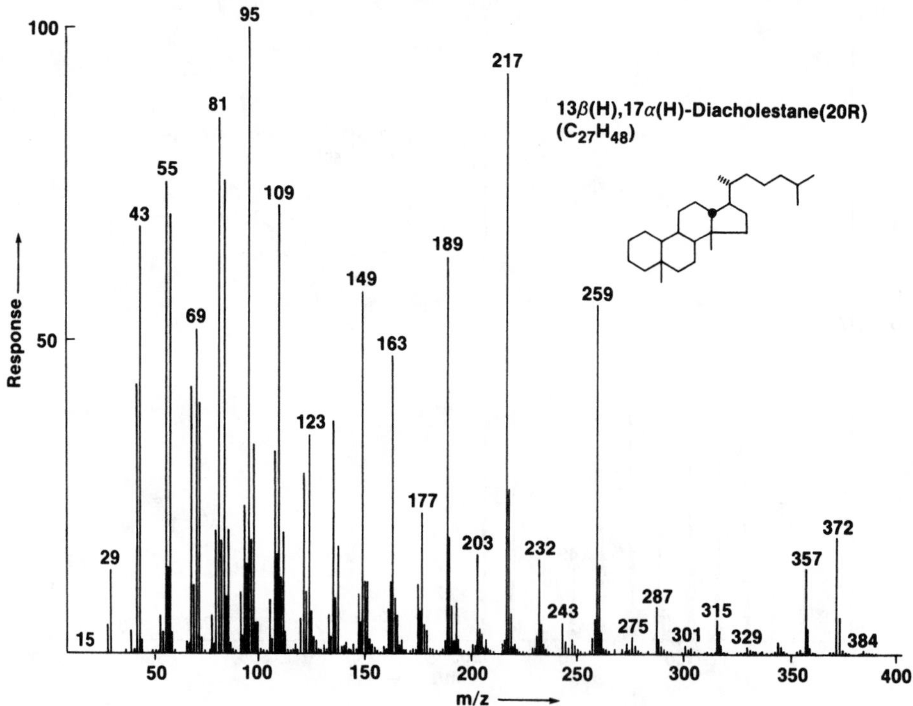

Figure 2.18 (*cont.*)

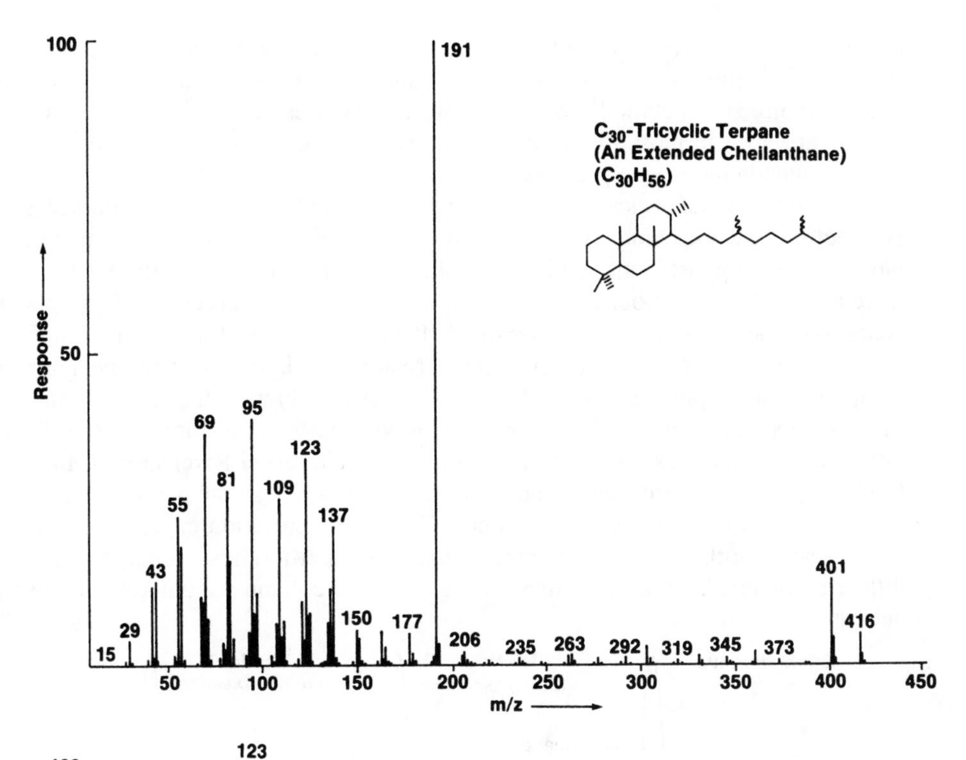

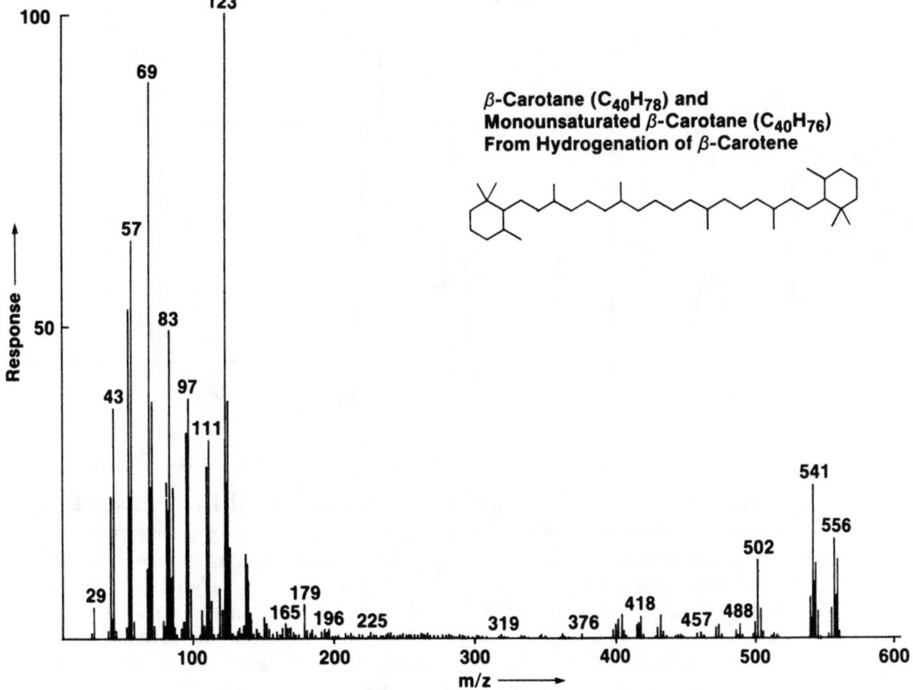

Figure 2.18 (*cont.*)

structure of 17α(H)-hopane in Green River oil shale was determined using ^{13}C NMR by Balogh et al. (1973). Fortunately, most routine geochemical investigations do not require such detailed proof of chemical structures because the same key compounds are used in most studies and the principal fragments and retention times of the compounds have been previously determined.

Provisional identification of unknown compounds is usually accomplished by **coinjection** and matching of mass spectra. Coinjection is a chromatographic technique used to support identifications of unknown compounds. A synthesized or commercial standard compound is mixed with the sample (a process called **spiking**) containing the compound to be identified. If the standard and unknown compounds coelute from the gas chromatograph, the relative peak intensity of the unknown compound on chromatograms of the mixture will be higher than that for the neat (unspiked) sample. Figure 2.19 shows the provisional identification of 18α(H)- and 18β(H)-oleanane peaks in a bitumen from the offshore Eel River area, California. Coelution of a standard compound with the unknown peak supports, but does not prove, that the compounds are identical. Commonly coelution experiments are repeated using another chromatographic column with a different stationary phase. Two different compounds that fortuitously coelute on one column are unlikely to do so on another.

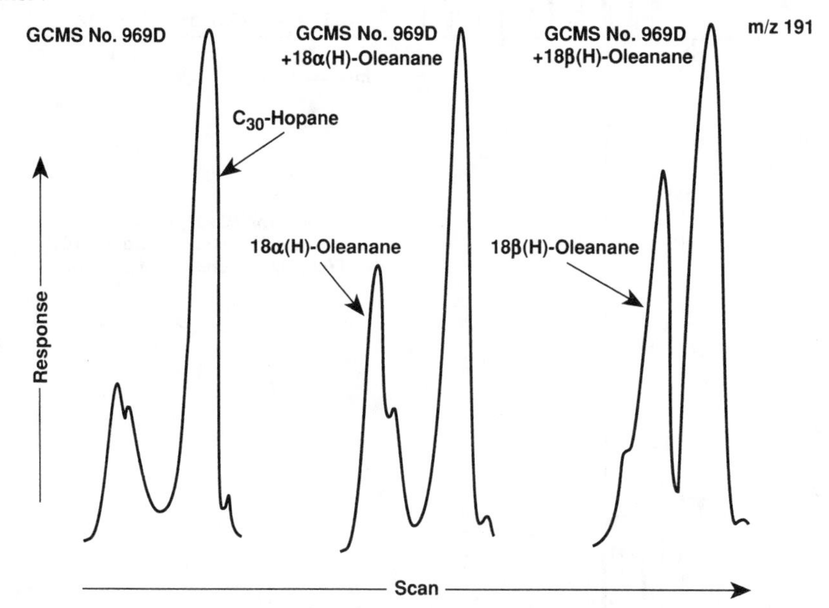

Figure 2.19 Provisional indentification of 18α(H)- and 18β(H)-oleanane in a bitumen from offshore Eel River Basin, California, by coinjection of authentic standards. The terpane mass chromatogram (m/z 191) for the neat (unspiked) bitumen (left) shows two unknown peaks which elute prior to the C_{30} 17α(H),21β(H)-hopane. Coinjection of synthetic 18α(H)-oleanane with the bitumen results in coelution with the first-eluting unknown peak (center). Note that the peak height for 18α(H)-oleanane relative to the C_{30} hopane has increased from the neat sample (left) to the coinjected sample (center). Coinjection of synthetic 18β(H)-oleanane with the bitumen results in coelution with the second-eluting unknown peak (right).

After coelution is established, matching of mass spectra for the unknown and the standard can now be used to infer that the two are identical. Figure 2.20 compares our mass spectrum for an unknown peak in an oil from the North Sea with a published mass spectrum for 25-nor-17α(H)-hopane (25-norhopanes as indicators of heavy biodegradation are discussed in Sec. 3.3.3). The similarity of the spectra, combined with coelution of the standard and the unknown peak, represents provisional identification of the compound. As indicated, rigorous proof of structure might also include coelution of the unknown and standard on various chromatograhic columns, and NMR or X-ray structural verification.

Crossplots of scan numbers for precursors versus proposed products can be used to assist in structural elucidation of homologs. For example, 25-norhopanes are believed to originate from the series of hopane homologs by loss of a methyl group at C-25 (Sec. 3.3.3). Thus, scan numbers for hopane homologs Ts and Tm (C_{27}), bisnorhopane (C_{28}), norhopane (C_{29}), hopane (C_{30}), and the C_{31} to C_{35} homohopanes (m/z 191) should show a linear relationship versus scan numbers for demethylated 25-norhopanes (m/z 177).

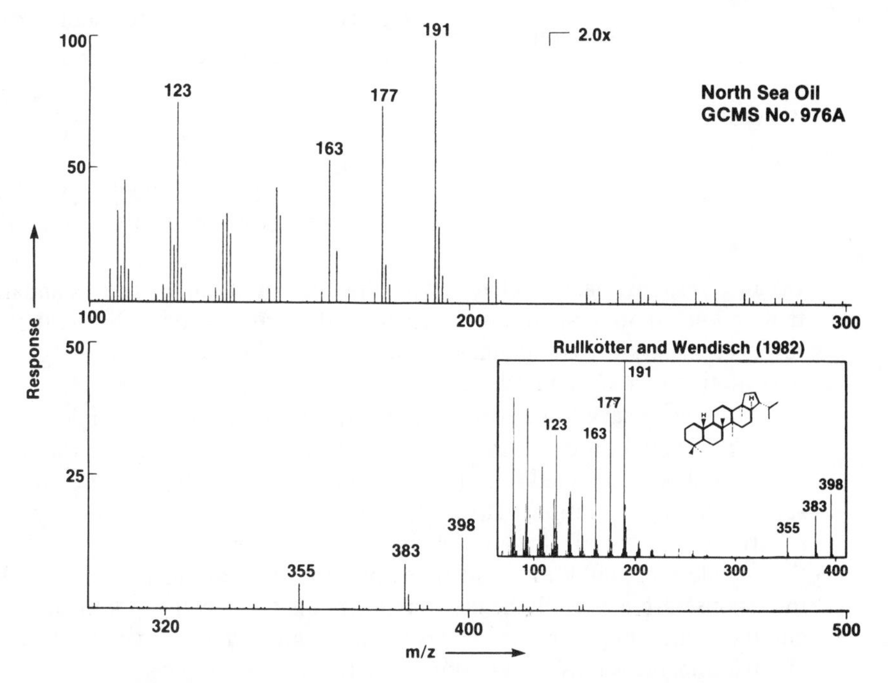

Figure 2.20 Mass spectrum of an unknown compound identified as a peak on the m/z 177 mass chromatogram of an oil from the North Sea. The similarity of the mass spectrum of the unknown compound with that of the published spectrum (box) assisted in the provisional indentification of the compound as 25-nor-17α(H)-hopane (structure in box). The mass spectrum of the compound shows major fragment ions at m/z 191 and m/z 177. Figure 3.64 shows m/z 191 and m/z 177 mass chromatograms for the same oil.

2.5.4.2 Biomarker Quantitation. Quantitation of individual compounds using mass chromatograms is handled at Chevron using automated programs such as those on the Finnigan INCOS computer system. The program is used to identify peaks in a GCMS analysis based on: (1) GC retention time relative to an internal standard (Sec. 2.4.3) coinjected with the sample, and (2) comparison of mass spectra of unknown compounds with a library of standard spectra.

Internal standards are described in detail in Sec. 2.4.3. For saturates we use 5β-cholane as the internal standard (Figs. 2.12 and 2.16) because it is not present in significant amounts in crude oils, its fragmentation in the mass spectrometer is similar to other steranes (Seifert and Moldowan, 1979), and it does not coelute with other steranes. Synthetic aromatic steroids that are not found in crude oils and do not coelute with natural aromatic steroids are used for quantitation of the monoaromatic and triaromatic steroids (Figs. 2.13 and 2.14).

The INCOS program quantitates the area of each identified peak on a given mass chromatogram and the internal standard, which has been added in a known amount from a standard solution. The amount of each compound in parts per million (ppm) of the sample is calculated using the following formulas:

$$\text{Amount compound} = \frac{\text{(Area compound peak)(Area standard)}}{\text{(Area standard)(Response factor)}}$$

where,

$$\text{Response factor} = \frac{\text{(Area compound)(Peak area/unit amount of compound)}}{\text{(Area standard)/(Unit amount of standard)}}$$

Because response factors change with instrument conditions, a "standard" oil containing known amounts of all the compounds is run periodically. At these times, the response factors for compounds are adjusted in the quantitation program, based on quantitation of the standard oil.

It is necessary to perform absolute quantitation for use of many biomarker ratios. For example, area ratios for two compounds showing similar fragmentation patterns could be substituted for ratios of amounts, if the samples were all run under the same operating conditions. However, different instrument conditions can change the relative responses of various compounds in a biomarker ratio, particularly when the samples are run at different times or with different instruments. Such variations in response are especially deleterious for ratios that involve compounds with different fragmentation characteristics such as regular steranes versus 17α(H)-hopanes and triaromatic versus monoaromatic steroids.

Similar approaches have been used by others for GCMS absolute biomarker quantitation (Rullkötter et al., 1984b; Mackenzie et al., 1985; Eglinton and Douglas, 1988; Requejo, 1992). These approaches also rely on internal standards that do not coelute with the natural biomarkers to adjust response factors for quantitation.

2.5.5 Examples of GCMS data problems. Some of the factors affecting the validity of GCMS data include: scan rate, sampling frequency, electronic zero level, background noise, sampling thresholds, amplifier saturation, availability of disk space, and sample size (GC column overload). Some of the more common GCMS data problems are described as follows.

2.5.5.1 Poor GC Resolution. *Chromatographic resolution of biomarker compounds is necessary for the most accurate interpretations. Older papers commonly show poorly resolved biomarkers compared to more recent work because of innovations in chromatographic column technology. Improper column maintenance or prolonged use without replacement can also result in poor resolution of peaks and peak tailing. Degrading column performance is obvious when chromatograms for a familiar standard mixture are compared through time.*

2.5.5.2 Poor Signal-to-Noise. *Poor signal-to-noise ratios can result from several different causes. Mass chromatograms of mixtures containing very low concentrations of biomarkers may show low signal-to-noise. Very low biomarker concentrations are typical of highly mature rock extracts, oils, or condensates, where nearly all biomarkers have been destroyed (Fig. 2.21). Some samples are inherently low in certain biomarker classes. For example, some crude oils from lacustrine source rocks are very low in steranes.*

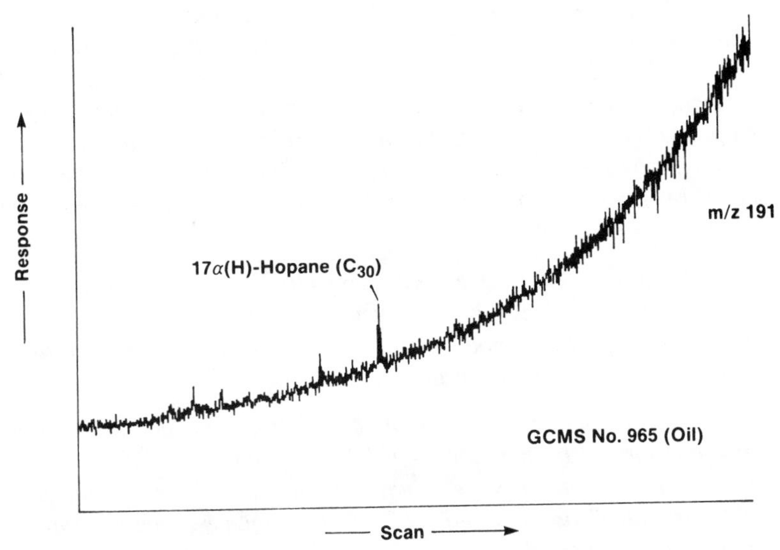

Figure 2.21 Terpane mass chromatogram or "fingerprint" (m/z 191) showing very low signal-to-noise ratios for a highly mature condensate (49°API) from the Eel River basin, California. The very low concentrations of biomarkers in this condensate preclude detailed biomarker analysis. Compare this fingerprint to those for various other oils in Fig. 3.11. The rising baseline is attributed to chromatographic bleed from the DB-1 column (see Fig. 2.23).

Mass spectrometer problems frequently cause low signal-to-noise. Dirty lenses, gain settings too high or too low, dirty source or quadrupole rods, weak multiplier, poor calibration or insufficient dwell time for the mass being analyzed, can all cause signal-to-noise problems. Mass spectrometer sensitivity and stability also vary according to manufacturer and model. A sample with low biomarker concentrations can show poor signal-to-noise on one instrument, while another will give acceptable results.

2.5.5.3 Ion Sampling Frequency. *Ion sampling frequency is an important parameter affecting the precision and accuracy of mass chromatography. The sampling frequency is usually expressed as the number of seconds per scan over a given ion. Each scan is recorded by the data system as a data point having a height above the baseline or intensity which is proportional to the number of ions arriving at the detector of the mass spectrometer. A typical biomarker peak representing a* C_{30} *compound might have an elution width spanning 12 to 15 seconds. Figure 2.22 compares sampling rates of 3 and 1.5 seconds and how these rates affect definition of the chromatographic peak. The peak sampled at a 1.5-second scan rate is better defined and shows a smoother profile than that at 3 seconds. This results in a more accurate and reproducible quantitation on repetitive runs using the 1.5 second scan rate. We generally complete quantitations of peaks using a sampling rate which results in at least 10 scans per peak or about 1.5 second per scan. A 3 second scan rate is usually suitable for mass spectra.*

2.5.5.4 Column Bleed. *The mass spectrum of DB-1 stationary phase column bleed at 325°C (maximum program temperature for most analyses) is shown in Fig. 2.23. Although column bleed can be minimized by programming low chromatograph temperatures, the effects of column bleed can be background subtracted. Excessive column bleed results in rising chromatographic baselines and is usually corrected by replacement of the column and/or reduction of the maximum oven temperature. Even with background subtraction, column bleed ions sometimes appear in mass spectra, complicating structural assessment.*

2.5.5.5 Column Overload. *Injection of too much sample, or injection of sample overly enriched in a particular compound class, can result in "negative" peaks, tailing peaks, poor resolution, and altered retention times. Examples of column overload are shown in Fig. 2.24.*

2.5.5.6 Data System Overload. *All data systems, including the INCOS system, are limited by the rate at which data can be acquired. When this rate is exceeded because of excessive signal from the mass spectral analyzer, the data system is "overloaded" or saturated. The effect on a mass chromatographic peak is a height limitation which results in flattening of the top of the peak and spurious quantitation (Fig. 2.25). However, the change in appearance of the overloaded peaks may be subtle and not readily detected by the interpreter. A software feature on the INCOS data system can be used to check scan points for computer saturation.*

2.5.5.7 Incorrect Mass Range Monitored. *Figure 2.26 (top) shows an example of overlap of m/z 218 on the m/z 217 trace for the Wyoming oil standard 81-3.*

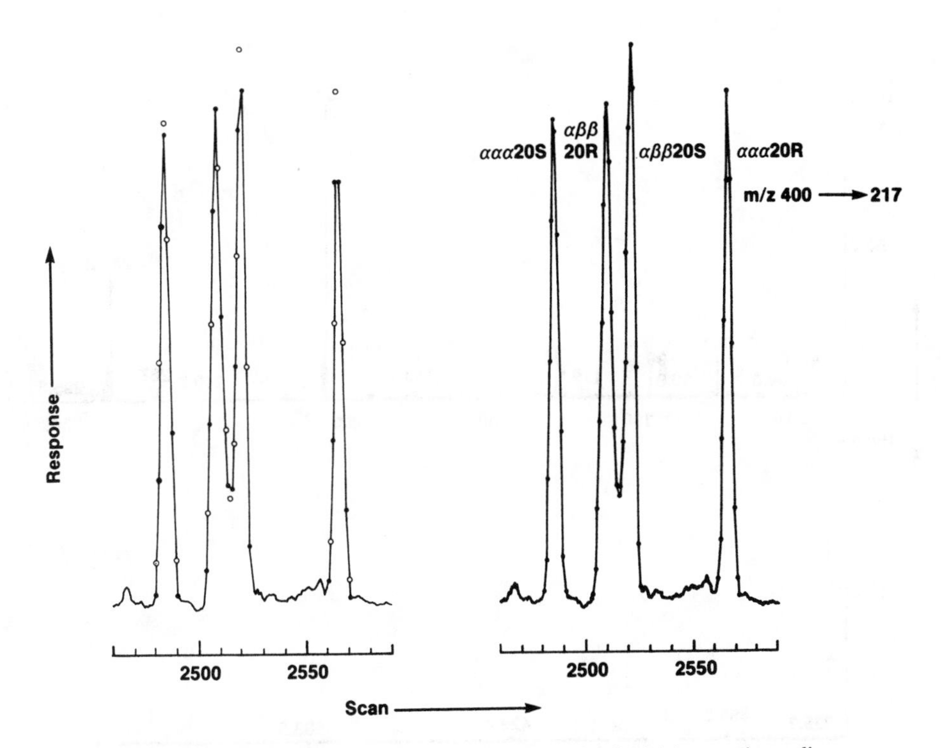

Figure 2.22 Comparison of the effects of 3 (left) and 1.5 (right) second sampling (scan) rates on definition of chromatographic peaks. The figure shows a portion of a mass chromatogram for steranes obtained by GCMSMS (M^+, m/z 400 → m/z 217). Note that the 3 second scan rate missed the top of the $\alpha\beta\beta$ 20S and $\alpha\alpha\alpha$ 20R peaks, causing a spurious increase in the 20S/(20S + 20R) ratio compared to that obtained using the 1.5 second scan rate. A 3 second scan results in only half the sampled points of a 1.5 second scan rate. For comparison, the chromatogram obtained using the 3-second scan rate shows points (open circles) that would have been available using a 1.5 second scan rate.

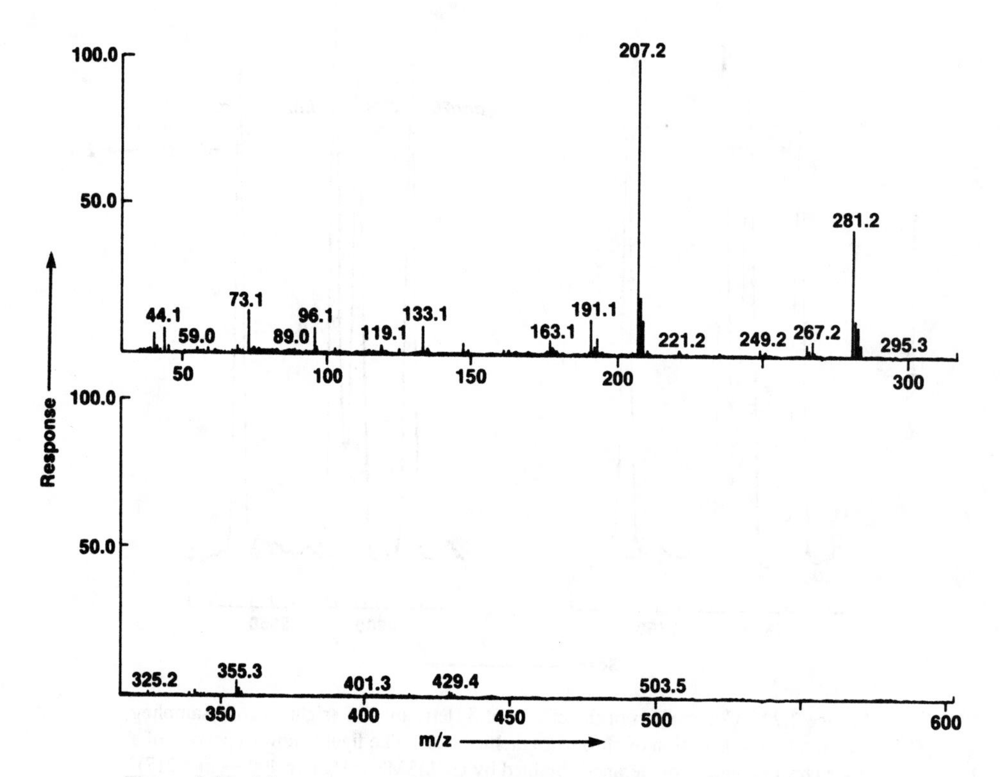

Figure 2.23 Mass spectrum of chromatographic column "bleed" from DB-1 stationary phase. Note that DB-1 column bleed shows a significant m/z 191 peak, which can account for a rising baseline in the m/z 191 chromatogram used for terpane analysis (e.g., Fig. 2.21).

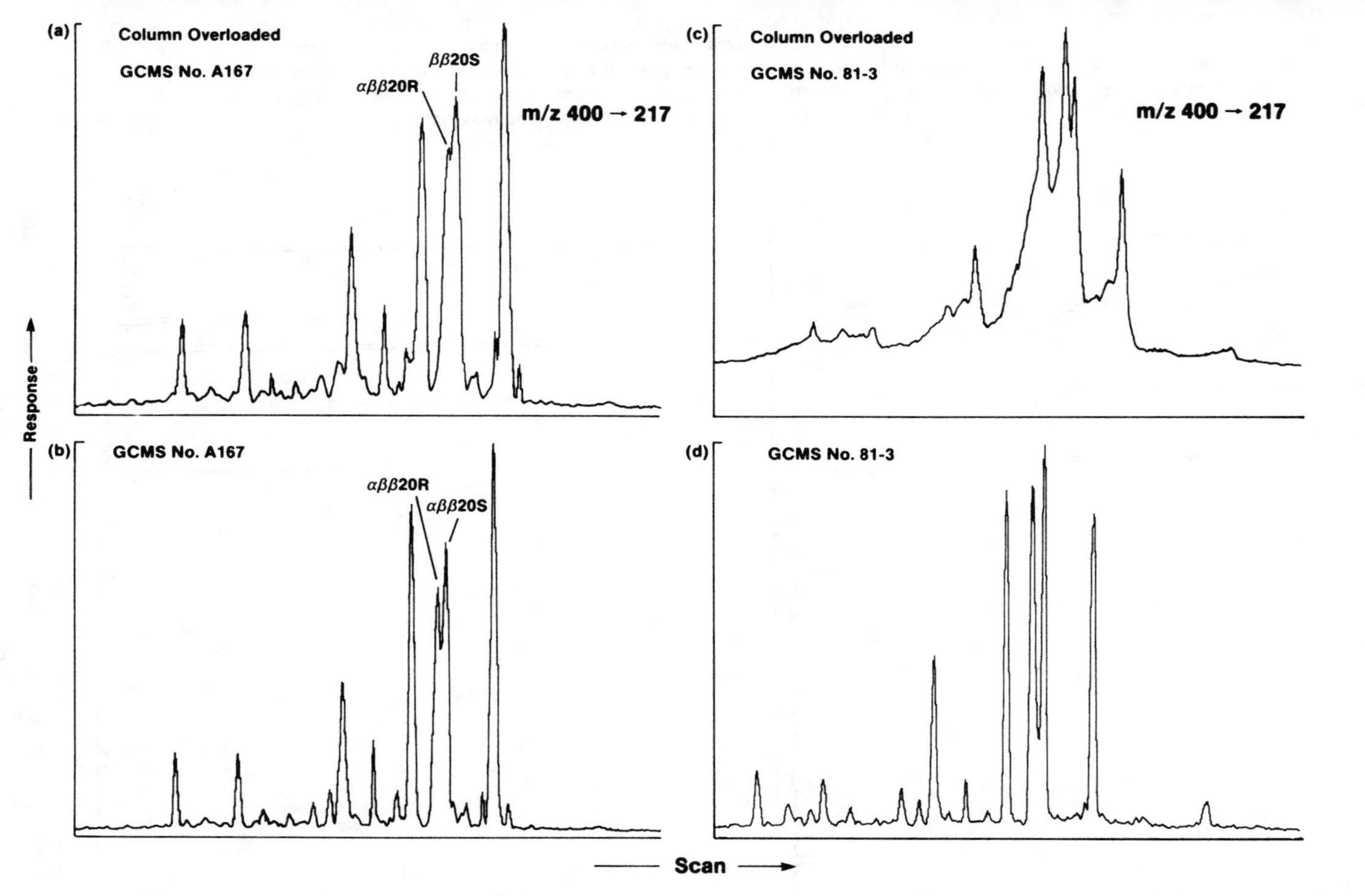

Figure 2.24 Examples of column overload on peak resolution in metastable reaction monitoring (MRM) GCMS mass chromatograms of C_{29} steranes (m/z 400 → 217): (a) Injection of 0.2 μl of a toluene solution of a sterane-rich seep oil from Turkey (GCMS No. A167) results in characteristics typical of an overloaded column, including poor resolution and peak broadening compared to (b). Poor resolution is particularly evident for the $\alpha\beta\beta$ 20S peaks; (b) injection of only 0.1 μl of the same solution as in (a) results in narrower peaks and acceptable resolution; (c) severe overloading of the column with a 0.2 μl toluene solution of the standard oil 81-3 results in very broad, poorly resolved peaks; (d) an analysis of a 0.1 μl toluene solution of the same oil as in (c) showing acceptable peak resolution.

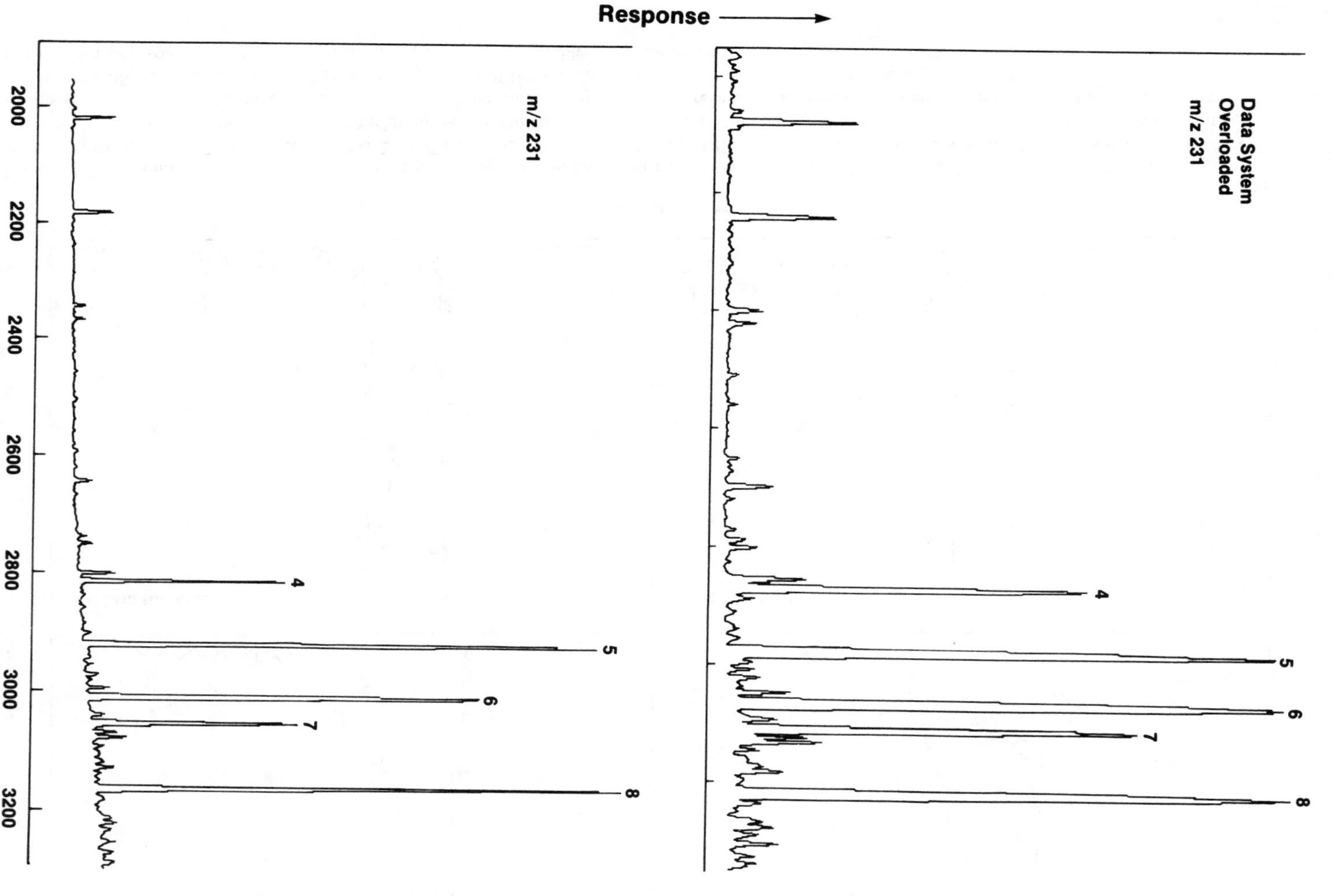

Figure 2.25 Example of the effect of a data system overload on the character of a m/z 231 (triaromatic steroids) mass chromatogram obtained by multiple ion detection (MID) GCMS of the aromatic fraction of a seep oil from Turkey (GCMS No. A169) (top). The five tallest peaks on the mass chromatogram have saturated the data system, resulting in overloading of the computer caused by too high a rate of data acquisition. Consequently, the size of these peaks is underestimated compared to the same peaks on the normal mass chromatogram (bottom). The top and bottom mass chromatograms were obtained by injecting 0.2 and 0.02 μl of "neat" (without solvent) sample, respectively. Peak numbers correspond to

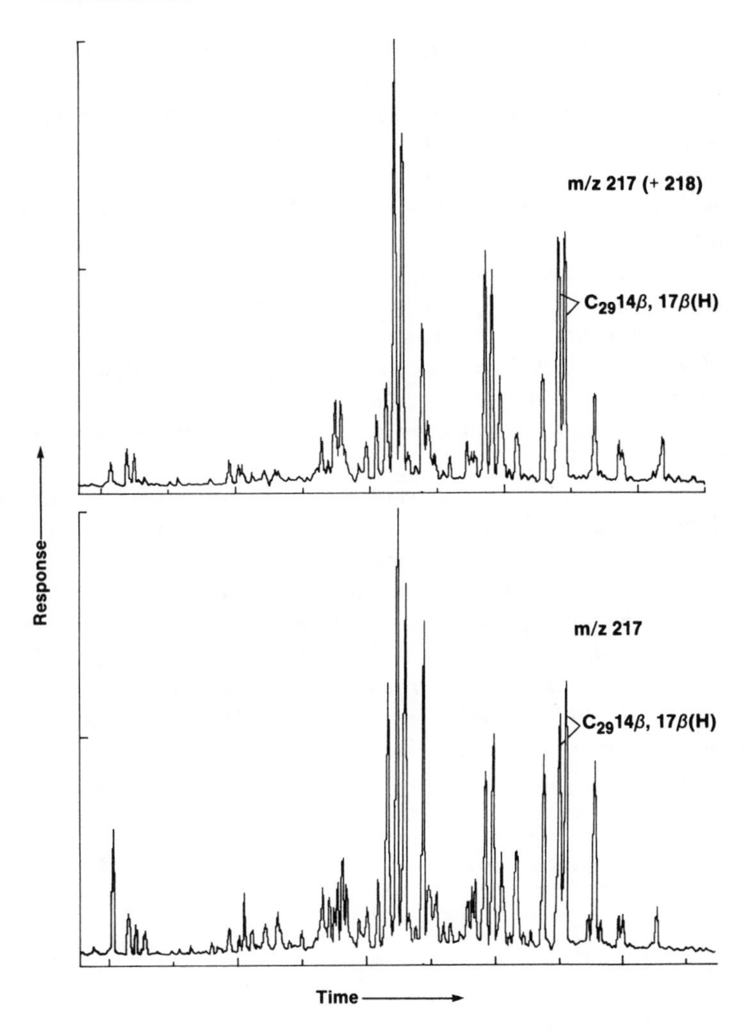

Figure 2.26 Effect of monitoring a slightly incorrect mass range on the sterane distribution (m/z 217) for oil GCMS No. 81-3. The top mass chromatogram is an example of overlap of m/z 218 on the m/z 217 trace. Note the unusually high C_{29} 14β,17β sterane peaks on this trace compared to the normal mass chromatogram at the bottom.

Note the unusually high C_{29} 14β,17β(H)-sterane peaks compared to the same peaks in an acceptable analysis of the same sample (bottom) where no overlap with m/z 218 occurs. The m/z 218 and m/z 217 overlap resulted from an incorrect mass calibration which set the m/z 217 "window" in the mass spectrometer at a position to collect data also from m/z 218. Less extreme cases of monitoring the incorrect mass range can result in more subtle but still incorrect results.

2.5.5.8 Interfering Peaks. *Several classes of compounds commonly interfere with the m/z 217 mass fragmentogram used for sterane analysis. Some pentacyclic triterpanes (notably 28,30-bisnorhopane which elutes from the GC among the C_{29} steranes; Moldowan et al., 1984) have significant m/z 217 fragments in their spectra. Mass spectra of 4-methyl steranes show a small m/z 217 fragment. In some samples 4-methyl steranes predominate over the 4-desmethyl steranes, interfering with their analysis on m/z 217.*

Another example of interfering peaks occurs when saturates and aromatics are analyzed together. Analysis of C-ring monoaromatic steroids requires selected ion monitoring of m/z 253, but many acyclic saturated hydrocarbons show the same major fragment, which interferes with the MA-steroid analysis. These compounds can be separated, however, by using a higher resolution (exact mass) mass spectrometer set at m/z 253.20 for monoaromatics or m/z 253.29 for saturates (Mackenzie et al., 1983a).

A severe example of interference is a case where the isotope peaks for two abundant compounds with molecular weights of 252 registered as major peaks on the m/z 253 mass chromatogram. Several analyses of the sample showed that the m/z 253 trace for each was not reproducible, primarily because of variations in the intensity of these two peaks. Figure 2.27 shows these two peaks on the m/z 253 trace for Beatrice oil from Well 11/30-2 in the Inner Moray Firth, U.K. This oil is a mixture of hydrocarbons from lacustrine Devonian and marine M. Jurassic source rocks (Peters et al., 1989). The Beatrice oil and M. Jurassic bitumen contain these unusual aromatic markers which are absent in other samples in the area. We have detected compounds with identical retention times on m/z 253 chromatograms in petroleum from onshore Eel River, California.

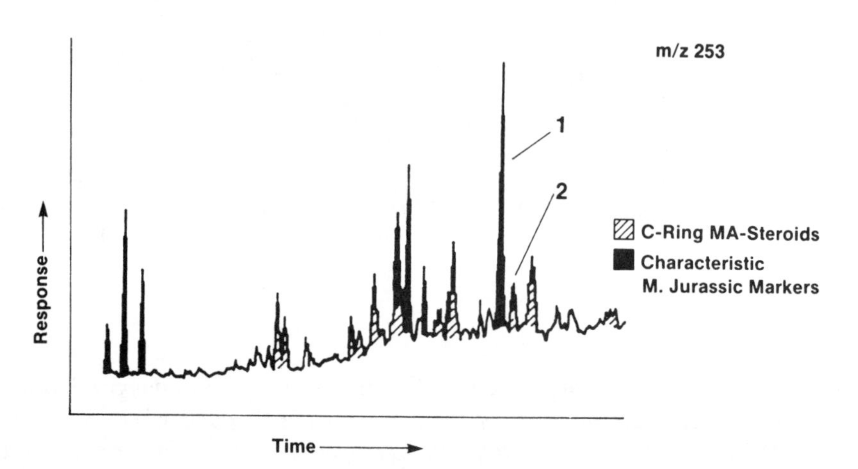

Figure 2.27 Mass chromatogram of m/z 253 for the aromatic fraction of Beatrice oil, Inner Moray Firth, U.K. Note the interference of polynuclear aromatics (black peaks) with analysis of the monoaromatic steroids (hachured peaks). Numbers refer to benzopyrenes described in text.

To provide structural information, mass spectra were obtained on the two most prominent aromatic peaks in the M. Jurassic bitumen (GCMS 997) and an Eel River bitumen (GCMS 969B). The spectra for each peak are essentially identical between the samples, showing an intense base peak at m/z 252 with a subordinate peak at m/z 126. Many polynuclear aromatic hydrocarbons generate prominent molecular ions at m/z 252, including 1,2- and 3,4-benzopyrene (Fig. 2.11). The ion at m/z 126 appears to represent a doubly charged species of the molecular ion. The mass spectra of these peaks and coinjection of authentic standards show that the two compounds represent 1,2- and 3,4-benzopyrene. Apparently the quantities of these compounds in the Beatrice oil are so large that an isotope peak, representing benzopyrenes containing one deuterium or ^{13}C *atom (Sec. 3.1.2.2), registers on the m/z 253 trace. Differences in the slit size used to monitor m/z 253 or in mass calibration between repeated analyses appear responsible for the variability of peak intensity for the isotope peaks of these compounds.*

2.5.5.9 Cold Spots in the Transfer Line. *Occasionally there may be a problem with heat applied to the section of tubing which interfaces the GC to the MS (transfer line). In modern instruments, the GC column is threaded through the transfer line so that the effluent from the column elutes directly at the ion source. A cold spot in the transfer line acts as a barrier to the transmission of compounds with boiling points above the temperature of the cold spot. The resulting chromatograms show loss of resolution on peaks for the higher boiling compounds while lower boiling compounds show normal resolution (Fig. 2.28).*

2.5.5.10 Dirty Ion Source. *A dirty ion source can cause increased background and lower signal-to-noise ratios. A dirty source prevents proper instrument calibration.*

2.5.5.11 Defocusing by n-Paraffins. *Elution of n-paraffins from the GC into the ion source can cause negative peaks on mass chromatograms of other compounds, simply by overwhelming (defocusing) the ion source (Fig. 2.29). Defocusing is particularly troublesome when analyzing paraffinic saturate fractions on GCMS systems where the MS source has a low-ion volume. Low-ion volumes are typical of certain early quadrupole instruments and mass selective detectors.*

2.5.5.12 Maximum Column Temperature Too Low. *If the maximum column temperature is set too low, the peaks for the higher boiling analytes will progressively broaden as the isothermal (constant temperature) portion of the temperature program is reached. The resulting data resemble those obtained from systems with a cold spot in the transfer line (see Sec. 2.5.5.9), except that the onset of peak broadening is less abrupt.*

2.5.5.13 Baseline Threshold Set Too High. *If the baseline threshold is set too high, the data system sets up a "default" baseline above the actual baseline. Using this default baseline as the peak base for quantitation causes changes in relative peak intensities which favor the tallest peaks on the chromatogram. The smallest peaks may be completely lost if they are lower in intensity than the threshold.*

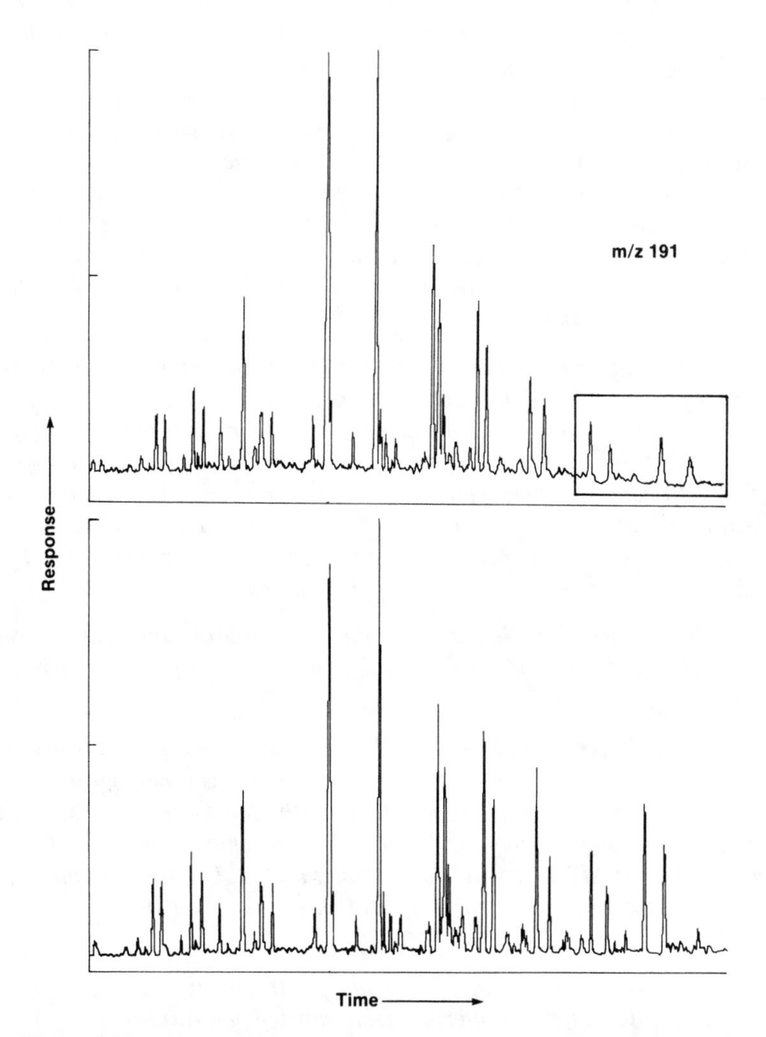

Figure 2.28 Example of the effects of a "cold spot" on the m/z 191 mass chromatogram for oil GCMS No. 81-3. The m/z 191 mass chromatogram at top shows peak broadening (boxed area) due to a cold spot in the transfer line, while the lower chromatogram shows acceptable resolution of peaks in this area.

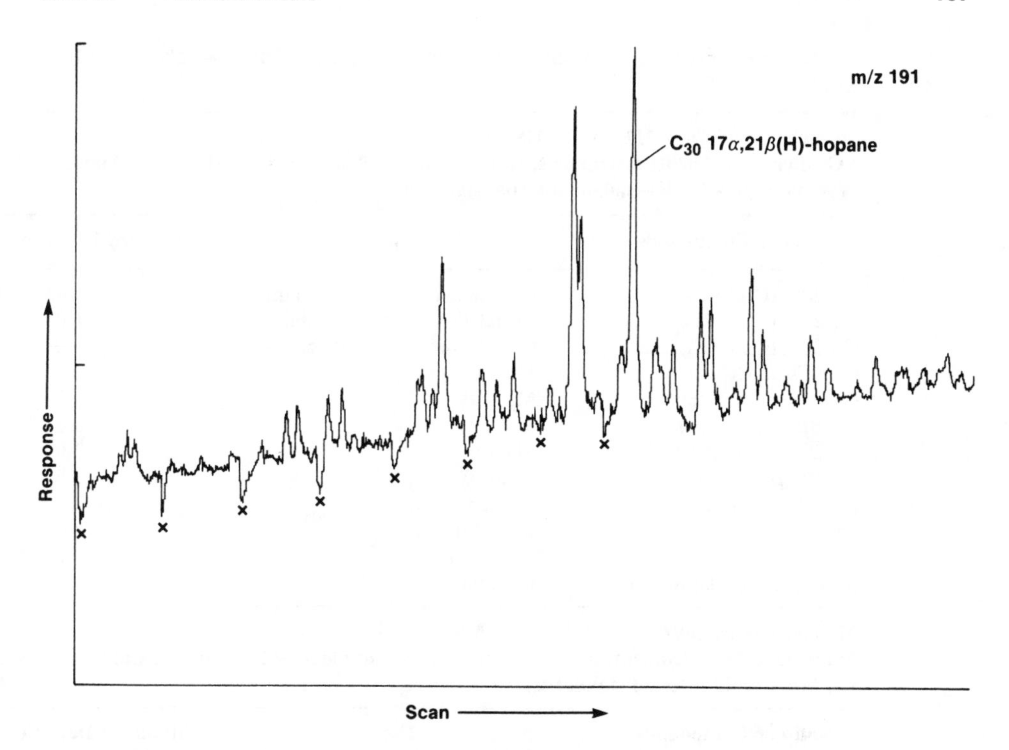

Figure 2.29 Terpane mass chromatogram (m/z 191) of a highly mature, paraffinic oil from the Nugget Sandstone, Pineview Field, Wyoming. The example shows "defocusing" caused by high concentrations of *n*-paraffins compared to biomarkers. The negative peaks marked by an "×" are at scan times corresponding to the *n*-paraffins. The character of the mass chromatogram can be improved by removal of the *n*-paraffins using molecular sieves or urea adduction.

2.5.6 Error analysis. *Table 2.5.6 is a compilation of error analyses for many of the parameters described in Chap. 3. These error analyses can be used with caution as a general guide to the significance of specific results, provided it is understood that standard deviations about the mean value for any parameter can differ between instruments and samples. Therefore, any experimental data that the reader wishes to evaluate with respect to tabulated error data must be obtained using: (1) optimum operating conditions, (2) the same instrument with identical settings, and (3) the same operating mode as specified in the table. Although less practical, the ideal approach to error evaluation would be for the analyst to include replicate analyses of standards with every sample suite.*

TABLE 2.5.6 Precision Calculations from Ten Consecutive Analyses of the Same Standard.[1]

Steranes ($M^+ \rightarrow m/z$ 217) by GCMSMS
VG Micromass 7070H instrument, saturate cut of oil sample 81-3, Hamilton Dome, Wyoming, plus 5β(H)-cholane internal standard.

Ratio or Compounds	Use	Standard Deviation (%)
C_{27} 20R/(C_{27} to C_{29})	Correlation (Ternary diagram)	1.4
C_{28} 20R/(C_{27} to C_{29})	Correlation (Ternary diagram)	1.0
C_{29} 20R/(C_{27} to C_{29})	Correlation (Ternary diagram)	1.0
C_{30}/(C_{27} to C_{30})	Marine Input	8.8
C_{29} 20S/(20S + 20R)	Maturity	2.0[2]
C_{29} $\beta\beta/(\beta\beta + \alpha\alpha)$	Maturity	5.6[2]
C_{27} 20S/(20S+20R)	Maturity	2.0
C_{27} $\beta\beta(\beta\beta + \alpha\alpha)$	Maturity	3.6
ppm C_{27} (in crude)	Concentration	9.9
ppm C_{28}	Concentration	8.0
ppm C_{29}	Concentration	10.2
ppm Total Regular Steranes	Concentration	9.4

Monoaromatics (m/z 253) by (MID)GCMS
Finnigan 4000 instrument, aromatic cut from oil sample 31042, Carneros, California, plus C_{30} MA-steroid internal standards.

Ratio or Compounds	Use	Standard Deviation (%)
C_{27}/(C_{27} to C_{29})	Correlation (Ternary diagram)	2.3
C_{28}/(C_{27} to C_{29})	Correlation (Ternary diagram)	1.3
C_{29}/(C_{27} to C_{29})	Correlation (Ternary diagram)	2.3
MA(I)/MA(I + II)	Maturity	6.6
ppm C_{27} (in crude)	Concentration	3.3
ppm C_{28}	Concentration	2.0
ppm C_{29}	Concentration	2.9
ppm Group I	Concentration	8.9
ppm Group II	Concentration	1.9
ppm Total MA-Steroids	Concentration	2.8

Triaromatics (m/z 231) by (MID)GCMS
Finnigan 4000 instrument, aromatic fraction of oil sample 81-3, Hamilton Dome, Wyoming, plus C_{30} TA-steroid internal standards.

Ratio or Compounds	Use	Standard Deviation (%)
TA(I)/TA(I + II)	Maturity	10.8
ppm Group I (in crude)	Concentration	13.9
ppm Group II	Concentration	3.5
ppm Total TA-Steroids (Group I + II)	Concentration	6.3

[1] For details of operating conditions see Peters et al. (1990).

[2] Based on ten consecutive analyses of the same sample, Steen (1986) notes standard deviations of 2.5% and 1.6% for the C_{29} 20S/(20S + 20R) and C_{29} $\beta\beta/(\beta\beta + \alpha\alpha)$ ratios, respectively.

TABLE 2.5.6 *(cont.)*

Others		
Ratio or Compounds	**Use**	**Standard Deviation (%)**
TA/(MA + TA) (MID GCMS)	Maturity	3.2
22S/(22S + 22R) (MRM GCMS)	Maturity	2.2
Diasteranes/Steranes	Maturity, depositional environment	~8
PMP (Porphyrins)	Maturity	~2

3

Guidelines to Interpretation

The chapter is divided into three parts, each beginning with a brief discussion of concepts and how biomarker and nonbiomarker analyses are used in the following areas of petroleum geochemistry:

1. Correlation, source, and depositional environment,
2. Maturation,
3. Biodegradation.

In Secs. 3.1 and 3.2, the biomarker parameters are arranged in groups of related compounds in the order: terpanes, steranes, aromatic steroids, and porphyrins. Within each group the parameters are arranged according to the frequency of their use in biomarker studies, with the more commonly used parameters first. Critical information on the frequency of use of the parameter, its specificity for answering the particular problem, and the means for its measurement are highlighted in bold print above the discussion for each biomarker parameter. In Sec. 3.3, the parameters are arranged in groups showing increasing resistance to biodegradation.

Most readers will probably use Chap. 3 by finding selected parameters in the Index, Glossary, or Contents. For this reason, discussions of each biomarker parameter contain extensive references to the literature and text. The reader should remember that discussion of a parameter in one part of Chap. 3 is typically supplemented by additional discussion in the other two parts. For example, oleanane is discussed in Parts 1, 2, and 3 because it can provide information useful for correlation, source, depositional environment, maturation, and biodegradation.

3 GUIDELINES

3.1 Correlation, Source, and Depositional Environment

3.1.1 Concepts. Genetic "correlations" of petroleums are based on the principle that the composition of organic components in a source rock is transmitted to the oil. This "similarity through heritage" can range from bulk properties, such as stable carbon isotope composition, to individual compound ratios, such as pristane/phytane. One advantage of biomarker correlations compared to those using only bulk parameters is that a variety of specific compounds are used for correlation.

Regardless of the parameters used, a basic rule applies to all correlations: **A positive correlation is not necessarily "proof" that samples are related.** For example, source rocks can show similar characteristics. A bitumen in one rock may appear similar to an oil, until further analyses show that other bitumens from different strata are even more similar to the oil. On the other hand, **a negative correlation is strong evidence for lack of a relationship between samples.**

Correlations between samples become more reliable when more parameters are used. This "multiparameter approach" was initiated early to answer correlation and other geochemical problems (Seifert and Moldowan, 1978). In this approach, independent methods such as biomarker and stable isotope analyses are always applied to support correlations. As will be discussed, certain geochemical parameters based on ratios of homologous biomarkers represent excellent source parameters, while some of the more useful maturity parameters are based on ratios of stereoisomers.

3.1.1.1 Correlation.

3.1.1.1.1 Oil and source rock screening. Prior to any oil-oil or oil-source rock correlation, prospective oils and source rocks must be selected or "screened." Oils can be rapidly and inexpensively screened using various techniques, including stable carbon isotopes (Sec. 3.1.2.2.2), gas chromatography (Sec. 3.1.2.1), and/or benchtop quadrupole GCMS (Sec. 2.5.3.3).

An effective petroleum source rock must satisfy requirements as to the quantity, quality, and thermal maturity of the organic matter. Tables 3.1.2.3(a) and 3.1.2.3(b) show the generally accepted criteria for describing the quantity and quality of organic matter in source rocks, respectively. Criteria for describing thermal maturity are in Table 3.2.2.1.1. For samples from wells, appropriate potential source rock intervals are selected for sampling based on evaluation of extensive Rock-Eval pyrolysis and TOC screening data in geochemical log format (Peters, 1986). Extensive sampling is recommended because significant changes in organic facies can occur, even laterally or vertically within the same source rock (e.g., Grantham et al., 1980; Espitalié et al., 1987; Burwood et al., 1990; Peters and Cassa, 1992). These initial screening results are followed by more detailed geochemical analyses of critical samples.

3.1.1.1.2 Oil-source rock correlation. Oil-source rock correlations are based on the concept that certain compositional parameters of a migrated oil do not differ significantly from those of the bitumen remaining in its source rock. Detailed oil-source rock correlations provide important information on the origin and possible paths of migration of oils that can lead to additional exploration plays. Ratios of adjacent homologs or compounds with similar structures such as the source-dependent biomarker ratios, as will be described, do not change from bitumen to migrated oil. For example, the ratio $C_{27}/(C_{27}$ to $C_{29})$ steranes used in C_{27}-C_{28}-C_{29} ternary diagrams (Sec. 3.1.3.2) does not differ significantly between bitumens and related oils throughout the oil-generative window.

Exploration is complicated by the fact that many oils migrate from their fine-grained, organic-rich source rocks to coarser grained reservoir rocks. Because both short- and long-distance migration of petroleum can occur and several potential source rocks are generally available in a given basin, the source for many oils remains problematic.

> ***Note:*** *Oils from the Monterey Formation in California are commonly found within or stratigraphically near the source rock. Oils in the Devonian pinnacle reefs of Western Canada appear to have migrated from nearby sources, while those at Athabasca have probably migrated up to 100 km or more from source rock to reservoir.*

The reader should be aware that compositional relationships between source rock bitumens and related, migrated oils *might* be obscured for several reasons. These potential problems are listed below in decreasing order of importance. Items (4) and (5) appear to show only very slight potential as problems in correlation studies based on limited available data.

1. *Bitumen from proposed source rock samples might contain undetected migrated oil or contaminants that are not representative of the indigenous hydrocarbons. Tests for indigenous bitumen are listed in Sec. 4.3.*
2. *Most oil-source rock correlations require equivalent or at least similar levels of thermal maturity for the bitumen and oil samples to be compared. In many cases, however, the available source rock candidates are either more or less mature than the oil. Artificial (laboratory) maturation, using methods such as hydrous and closed-tube pyrolysis, provides one approach to the solution of this problem (Sec. 4.3.2).*
3. *Oil-source rock studies are typically limited to a few selected potential source rocks. In nature, oils may represent a composite of migrated materials generated from thick sections of source rock. If the samples of potential source rocks in the study are not representative of the composite section which generated the oil, incorrect conclusions might result.*
4. *Expulsion/migration affects the distribution of compounds showing radically different molecular weights, polarities, or adsorptivities. For example, migrated oils are typically enriched in saturates and aromatics and depleted in*

NSO-compounds (nitrogen, sulfur, oxygen) and asphaltenes compared to related bitumens (Hunt, 1979; Tissot and Welte, 1984; Peters et al., 1990). Biomarkers showing only minor differences in polarity or configuration can show some changes in relative distributions due to migration through clay, as demonstrated in laboratory studies (Carlson and Chamberlain, 1986). These effects may reflect primary migration (expulsion) out of the fine-grained sediments of the source rock rather than secondary migration through coarse-grained carrier beds to the reservoir.

5. *Because migration may continue throughout much of the petroleum generation process, and because different compounds are generated at different times during this process, the composition of a reservoired petroleum may not exactly coincide with that being generated at a given time in the source rock. For example, Mackenzie et al. (1985) suggest that oil accumulations could be averaged mixtures of organic fluids representing a range of maturities. According to these authors, catagenesis results in decreases in concentrations of biomarkers that could exaggerate the contributions of less mature sources to a given reservoir.*

Despite these potential problems, when adequate source rock samples are obtained, successful oil-source rock correlations are the rule rather than the exception. However, certain major problems, such as origin of the Athabasca oils in Canada (Brooks et al., 1988) or the oils in Tertiary reservoirs in the Gulf Coast (Schumacker and Kennicutt, 1989), remain controversial.

3.1.1.1.3 Oil-oil correlation. Oil-oil correlation requires parameters that: (1) distinguish oils from different sources, and (2) are resistant to secondary processes such as biodegradation and thermal maturation. In many cases, oil-oil correlations can be accomplished using only a few simple "bulk parameters" such as gas chromatographic "fingerprints," carbon or sulfur stable isotope ratios, or V/Ni content. For example, Clayton et al. (1987) distinguished three types of Paleozoic oils in the Northern Denver Basin based mainly on pristane/phytane and stable carbon isotope ratios. Additional data from gas chromatograms and biomarker analyses supported the classification. In cases such as this, where exploration questions are answered using rapid and inexpensive analyses, the interpreter may choose to omit the additional analyses (including biomarkers), thereby saving time and money. However, a complete biomarker analysis invariably adds additional information to any geochemical study of oils and source rocks.

The strength of biomarkers is in their potential to provide detailed information on source, depositional environment, and thermal maturity beyond that available using only bulk parameters. For example, bulk parameters might be used to delineate several families of oils in a basin. Supplementary biomarker analyses on selected oils might then be used to determine organic facies variations in the source rock for each family, resulting in the recognition of regional oil subgroups. Another example is the supplementary use of biomarkers to correlate and describe the level of thermal maturity of biodegraded oils.

Light oils and condensates represent a special correlation problem. The high maturity of many condensates results in a dominance of gasoline-range components. Higher molecular-weight components, including the biomarkers, are low or absent in some condensates (Sec. 2.5.5.2). Thus, low concentrations of biomarkers in condensates can obscure relationships between condensates and less mature, "normal" oils derived from the same source rock. In addition, condensates may solubilize (pick up) biomarkers from less mature rocks during migration (Sec. 4.1). Because biomarker concentrations are already low in most condensates, contaminating biomarkers may adversely affect various interpretations, including correlation, source organic matter input, and thermal maturity.

A critical problem in geochemistry is to distinguish the effects of source (including organic matter input and depositional environment) from those of thermal maturity on petroleum composition. One approach is to apply principal component (eigenvector) analysis to a large geochemical database (e.g., Hughes et al., 1985). As might be expected, geochemical parameters show a range of sensitivities to source and maturity effects. Variations in some parameters are clearly dominated by maturity [e.g., 20S/(20S + 20R) steranes], others are dominated by source input (e.g., C_{27}, C_{28}, or C_{29} steranes versus total C_{27} to C_{29} steranes), and many are affected by both source and maturity [e.g., Ts/(Tm + Ts) or diasteranes/steranes ratios].

3.1.1.1.4 Solid bitumens. Solid bitumens complicate correlation studies because of their highly variable, sometimes refractory character. They consist of a wide variety of organic materials with common names like grahamite, anthraxolite, gilsonite, and albertite. These names are based on an antiquated classification scheme requiring solubility, fusibility, and hydrogen/carbon atomic ratio data (e.g., as described in Hunt, 1979). This generic classification is of little use in correlation or describing the origin of these materials.

Curiale (1986) recommends that this generic scheme be discarded in favor of a classification that divides solid bitumens into pre-oil and post-oil products using biomarkers and supporting data. Most pre-oil bitumens are intimately associated with their source rocks, while post-oil bitumens have undergone extensive migration prior to alteration. By carefully accounting for differences caused by maturation and biodegradation, many solid bitumens can be related to oils and bitumens using biomarkers and supporting methods.

3.1.1.2 Organic Input and Depositional Environment. Under certain conditions, large populations of a single or restricted group of organisms can produce abundant supplies of one or a few diagnostic biomarkers. For example, unusually high concentrations of dinosterol in sediments appear to result from dinoflagellate blooms in the overlying nutrient-rich, upwelling marine waters. Similarly, blooms of purple halophilic bacteria may account for the abundance of carotene in the associated lake-bottom sediments. The presence of unusual concentrations of the saturated analogs of these biomarkers in bitumens or oils can be used to infer environmental conditions that existed at the time of deposition of the source organic matter.

Compared to other methods, biomarker studies are particularly useful when only oils are available, but information on the source rock is desired. For various reasons, proposed source rocks are generally more difficult to obtain than oils for biomarker studies. Nonetheless, biomarkers in the oils can be used to infer the depositional environment, organic input, thermal maturity, and in some cases can even be used to place age constraints on the source rocks (Sec. 4.5). These data can be used to suggest likely source rocks that might be sampled later for detailed oil-source rock correlations.

3.1.2 Nonbiomarker parameters for correlation, source, and depositional environment. Nonbiomarker and biomarker parameters are used together to provide the most reliable interpretation of source organic matter input, depositional environment, and the relationship between samples. For this reason, the following brief discussion of nonbiomarker parameters and their interrelationships with biomarker parameters is included. References describing procedures for many of the nonbiomarker methods are given in each section.

3.1.2.1 Gas Chromatography. Gas chromatography (GC) using high resolution capillary columns is a widely used method (Sec. 2.5.1) for screening and correlating oils and bitumens because

1. it is less expensive and more versatile than many other analytical methods,
2. it requires little sample,
3. little sample preparation is necessary,
4. high-resolution capillary GC columns can be used to generate a reproducible "fingerprint" of petroleum consisting of hundreds of peaks in the range C_2 to C_{40},
5. a large database of peak ratios for each oil or bitumen can be digitally recorded and statistically compared using computers.

Gas chromatograms are sensitive to organic matter input (Fig. 3.1) and secondary processes such as biodegradation (Fig. 3.2) and thermal maturation (Fig. 3.3).

Kaufman et al. (1990) describe such a GC approach for correlation. The method has been effectively used to determine reservoir continuity and mixing relationships between different reservoirs in the same field (Sec. 3.1.2.1.3). Thompson (1983) describes various light hydrocarbon ratios from GC analyses useful for assessment of correlation, maturity, biodegradation, and water washing.

Biodegraded petroleums are particularly difficult to correlate using gas chromatography alone because resolution of a substantial proportion of the hydrocarbons in these samples is not possible. This **unresolved complex mixture (UCM)** of compounds or "hump" rises significantly above the baseline, and is especially pronounced in biodegraded petroleums (Fig. 3.2; Milner et al., 1977; Rubinstein et al., 1977; Killops and Al-Juboori, 1990).

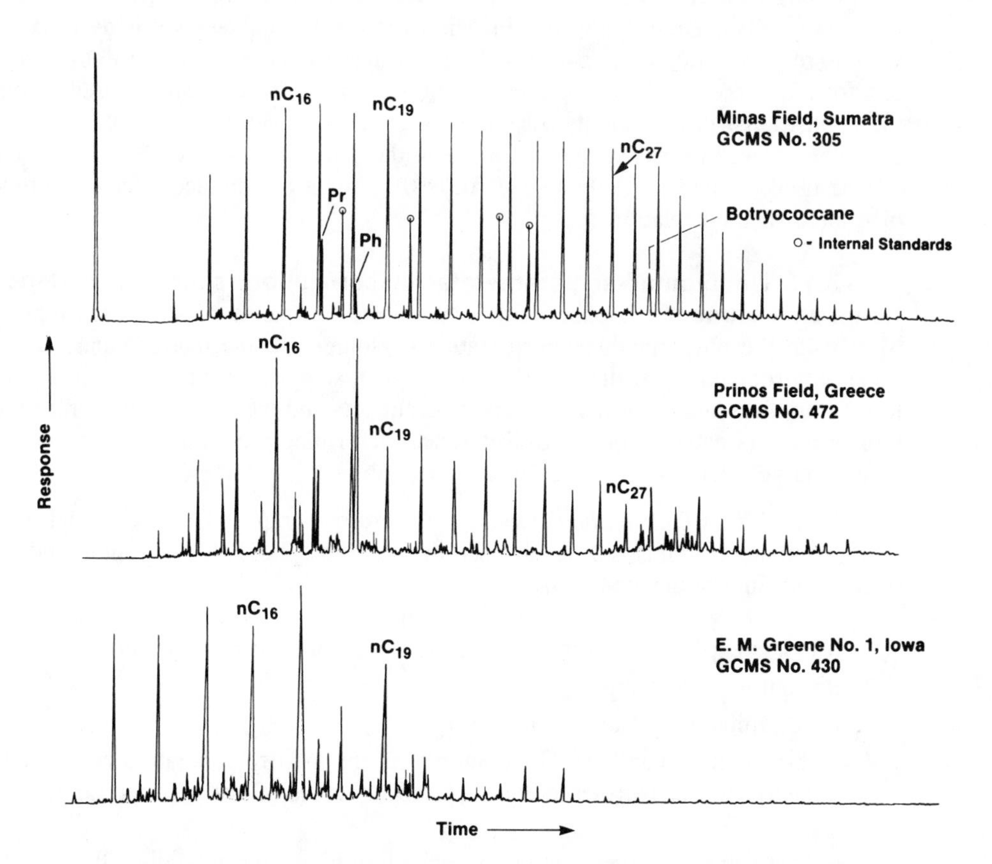

Figure 3.1 (Top) Bimodal *n*-paraffin distribution in a gas chromatogram of Minas oil from Sumatra. The presence of botryococcane indicates that the source rock contains remains of the alga *Botryococcus braunii*. Unlike the typical case where the high molecular-weight *n*-paraffins (near nC_{27} on figure) are derived from terrestrial higher plants, these compounds appear to have originated from lipids within the algae (Gelpi et al., 1970; Moldowan et al., 1985). Pyrolyzates of the nonmarine algae *Tetraedron* and of laminated Messel Shale (Germany) containing densely packed *Tetraedron* algae remains show similar distributions of *n*-alkanes and *n*-alkenes (Goth et al., 1988).

(Middle) Even numbered *n*-paraffin predominance in the gas chromatogram of an oil from Prinos, Greece, generated from a carbonate source rock (Moldowan et at., 1985).

(Bottom) Odd numbered *n*-paraffin predominance in the gas chromatogram of a bitumen from a Middle Ordovician rock from the Greene No. 1 well, Iowa, dominated by input from the microorganism *Gloeocapsamorpha prisca* (Jacobson et al., 1988).

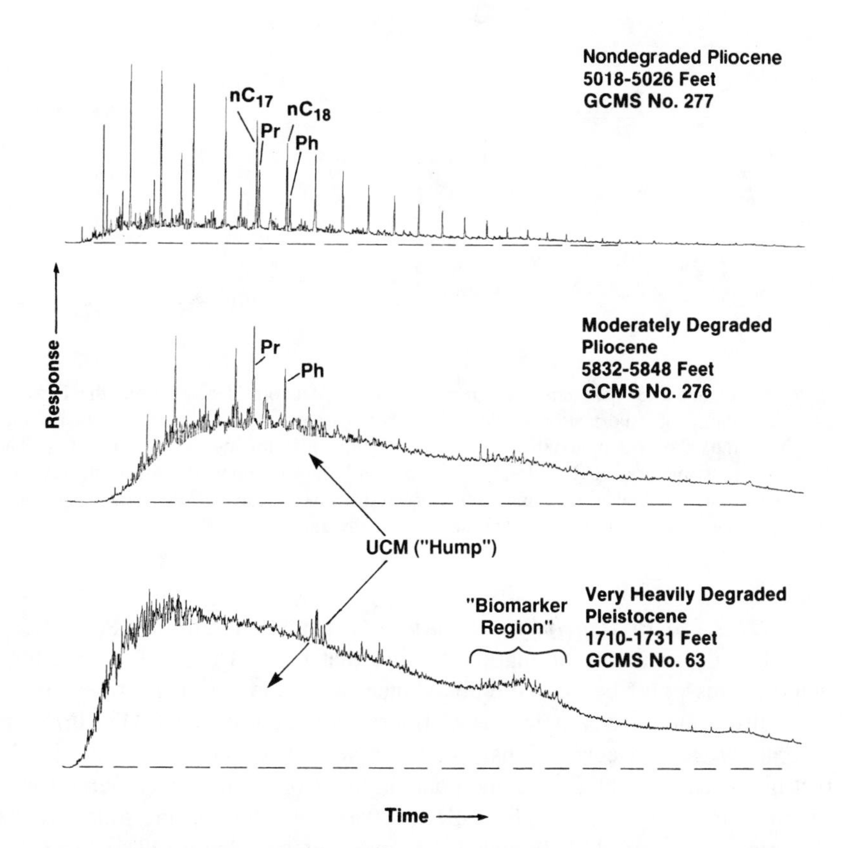

Figure 3.2 Gas chromatograms for three related Gulf Coast oils which differ in the extent of biodegradation. Compared to the nonbiodegraded oil (top), the moderately biodegraded oil (middle) shows loss of *n*-paraffins while the more biodegraded sample (bottom) shows loss of both *n*-paraffins and acyclic isoprenoids. Pr = pristane, Ph = phytane, UCM = unresolved complex mixture, dashed line = baseline.

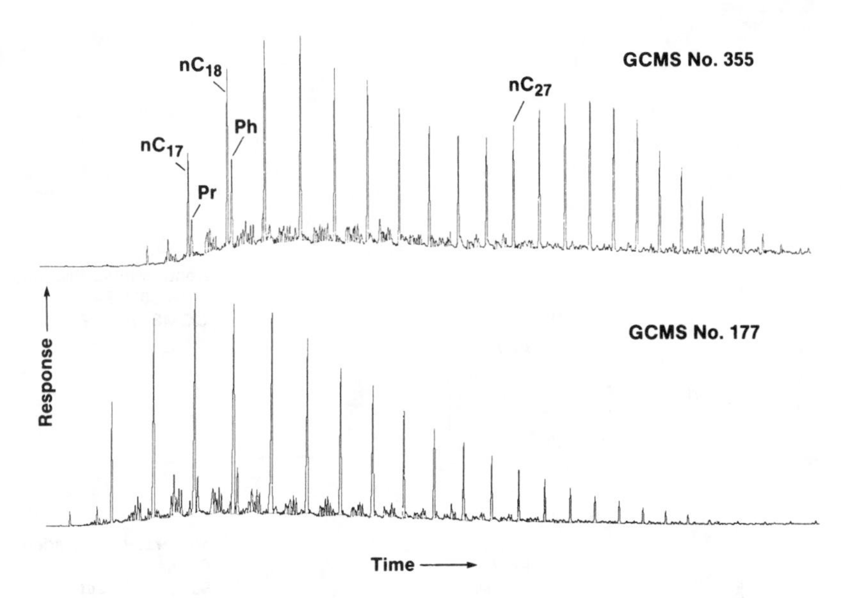

Figure 3.3 Gas chromatograms of two related oils derived from the Permian Phosphoria Formation in Wyoming. The least mature oil (Dillinger Ranch Field; top) shows a bimodal *n*-paraffin distribution maximizing at nC_{20} and nC_{30}. The most mature oil (Dry Piney Field; bottom) shows a unimodal *n*-paraffin distribution. The higher molecular-weight *n*-paraffins have been cracked to lighter products during maturation. Pr = pristane, Ph = phytane.

3.1.2.1.1 Pristane/phytane ratio. Pristane/phytane (Pr/Ph) ratios are discussed in detail in the biomarker section that follows (Sec. 3.1.3.1), but are also noted at this point because the abundance of Pr and Ph in most petroleums allows their direct measurement from GC traces without use of GCMS. Pr/Ph ratios are commonly used in correlations. For example, Powell and McKirdy (1973) showed that high-wax Australian oils and condensates from nonmarine source rocks had Pr/Ph ratios in the range 5 to 11, while low-wax oils from marine source rocks had Pr/Ph ratios from 1 to 3. Although Pr/Ph ratios of petroleum reflect the nature of the contributing organic matter, this ratio should be used with caution. Pr/Ph ratios typically increase with thermal maturation (e.g., Alexander et al., 1981) and some pristane and phytane may be derived from sources other than phytol during diagenesis (ten Haven et al., 1987).

3.1.2.1.2 Isoprenoid/n-paraffin ratios. Pristane/nC_{17} and phytane/nC_{18} are sometimes used in petroleum correlation studies. For example, Lijmbach (1975) noted that oils from rocks deposited under open-water conditions showed Pr/nC_{17} ratios less than 0.5, while those from inland peat-swamps had ratios greater than 1. These ratios should be used with caution for several reasons. For example, both Pr/nC_{17} and Ph/nC_{18} decrease with thermal maturity of petroleum. Alexander et al. (1981) suggested use of the ratio (Pr + nC_{17})/(Ph + nC_{18}) because it is less affected by variations in thermal maturity than Pr/nC_{17} or Ph/nC_{18}. These ratios are also readily affected by secondary processes such as biodegradation. The *n*-paraffins are generally attacked by aerobic bacteria prior to the isoprenoids (Sec. 3.3.1).

3.1.2.1.3 Gas chromatographic "fingerprints." Some GC fingerprints are indicative of certain types of source organic matter input. Bimodal *n*-paraffin distributions, and those skewed toward the range nC_{23} to nC_{30}, are usually associated with terrestrial higher plant waxes. However, this interpretation is complicated by certain algae (e.g., *Botryococcus braunii*) that also contain higher molecular-weight *n*-paraffins (Fig. 3.1, top). Bitumens and oils related to carbonate source rocks commonly show an even carbon number *n*-paraffin predominance (Fig. 3.1, middle), while those related to argillaceous (shaly) rocks show a predominance of odd-numbered *n*-paraffins below nC_{20} (Fig. 3.1, bottom), which is typical of rocks and oils of Middle Ordovician age containing the alga *Gloeocapsamorpha prisca* (Reed et al., 1986; Rullkötter et al., 1986; Jacobson et al., 1988). An odd predominance in *n*-paraffins is common in many lacustrine and marine oils derived from shaly source rocks.

Grouping of oils and bitumens and interpretation of source input using GC "fingerprints" are affected by several limitations. Because of their high concentrations in most oils compared to other compounds, *n*-paraffins and acyclic isoprenoids dominate the general appearance of the corresponding gas chromatograms. These compounds are readily altered by secondary processes, including biodegradation, maturation, and migration. For example, biomodal *n*-paraffin distributions and even- or odd-carbon number predominance are lost with increasing thermal maturity. Thus, oils distinguished by GC might differ because they are from different source rocks, or because they have experienced different histories during migration and/or in the reservoir.

Figure 3.2 shows gas chromatograms for three oils from the Gulf Coast that are known to be related based on detailed biomarker and isotopic compositions. Despite their common origin, the gas chromatograms are dissimilar because of differing degrees of biodegradation (Sec. 3.3). Compared to the nonbiodegraded oil in the figure, the moderately biodegraded oil shows loss of *n*-paraffins, while the more heavily biodegraded oil shows loss of both *n*-paraffins and isoprenoids.

Figure 3.3 shows gas chromatograms for two related oils derived from the Permian age Phosphoria Formation in Wyoming, based on biomarker and isotopic compositions. As for the biodegraded oils discussed, the gas chromatograms of the related Wyoming oils are dissimilar, but in this case differential maturation accounts for the variations in the GC fingerprints.

Figure 3.4 shows gas chromatograms for three oils from Trinidad that are known to be related based on biomarker and isotopic compositions. As in the previously mentioned example, the gas chromatograms are dissimilar. In this case, some of the differences are caused by thermal maturity and others appear due to weathering of the highly mature No. 3 oil after production.

Low resolution, packed column chromatography of petroleum results in coelution of many compounds. These types of columns are rarely used in modern geo-

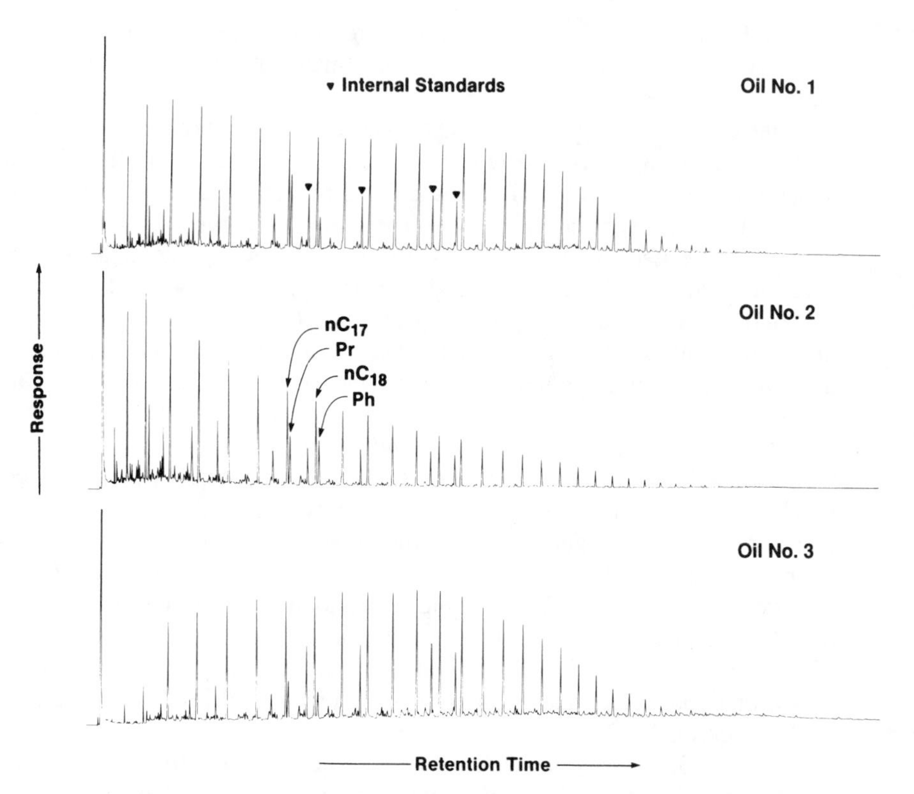

Figure 3.4 *n*-Paraffin distributions can be affected by factors other than source organic matter input. The gas chromatograms for three oils from Trinidad are dissimilar despite biomarker and isotopic evidence indicating they are related (Peters et al., unpublished). Oil No. 1 (top; 27° API) is the least mature of the samples and has reached the early to middle oil-generative window. This oil shows a bimodal *n*-paraffin distribution that is absent in the more mature Oil No. 2 (middle; 38° API). The heavier *n*-paraffins in Oil No. 1 appear derived from terrestrial higher plants (however, see Fig. 3.1). Although slight differences in organic matter input may also be important, the lack of a bimodal *n*-paraffin distribution in the two more mature oils appears to have resulted from thermal maturation. Oil No. 3 (bottom) is the most mature and has reached the late stages of the oil-generative window based on the thermal maturation-dependent biomarker parameters. Despite its high thermal maturity, the oil is depleted in compounds below about nC_{20}, possibly because of evaporation loss or weathering after sampling. This may account for the low gravity of Oil No. 3 (20° API) compared to the other two oils.

chemical studies. Because the relative contributions of different compounds to each GC peak are unknown, comparison of peak-height or peak-area ratios for correlation of different samples becomes problematic. However, grouping of oils by high resolution GC is a useful step during preliminary screening of samples prior to detailed biomarker analysis.

For some applications, especially related to production problems, gas chromatography is a more appropriate analytical approach than detailed biomarker analysis. Because of their high-boiling range, and resistance to water washing, biodegradation, evaporative fractionation, and other processes, biomarkers are less susceptible to reservoir processes than light hydrocarbons. Nonetheless, statistically significant heterogeneities in biomarkers within the Gullfaks Field, Norwegian North Sea are attributed to slight differences in biodegradation, maturity, and source (Horstad et al., 1990).

Oils in different reservoirs within the same field commonly derive from the same source rock, but have experienced slightly different histories. Sealing faults may separate a single accumulation into two reservoirs which are subsequently affected by slightly different secondary processes or filling histories. England (1990) provides evidence that both lateral and vertical differences in petroleum composition exist *within* individual reservoirs because of slow intra-reservoir mixing. Lateral variations are attributed to the way in which the field filled from its source rock(s) and vertical differences are explained as caused by gravitational segregation.

Some examples of the use of GC in solving production problems include:

1. Reservoir continuity (e.g., do the compositions of oils in the same field, but on opposite sides of a mapped fault, support the existence of a seal resulting in two distinct reservoirs?),
2. Production allocation (e.g., what is the relative production from each interval in a well where the total production is co-mingled? Are certain intervals nonproductive?),
3. Detection of drilling fluid contamination or leaks (e.g., is a sample composed of petroleum from the formation or a drilling additive?).

Kaufman et al. (1990) describe a useful GC approach for correlation. The procedure used in this approach will be explained. For a typical production problem, samples might be obtained from the wellhead, drill stem tests (DST), repeat formation tests (RFT), swab runs, or reversed circulation. Large-volume samples from the flowing wellhead are less susceptible to contamination than others. A carefully constructed baseline that is repeatable from one sample to the next is used to derive accurate peak heights and/or areas from the chromatogram. Because performance varies between different chromatographic columns and for the same column over long periods, analyses for each study are completed promptly using the same column and operating conditions. Identified peaks with similar GC retention times are used to calculate the ratios.

Specific ratios used in this approach may differ from one study to the next. The ratios are chosen to maximize differences among the samples that are consistent with geologic and engineering information. For reservoir applications, this GC approach is always used with other information such as structural geology, wire line logs, and pressure tests. Typically no more than a dozen ratios are used. The ratios can be plotted on polar coordinate paper (star diagrams) which facilitate visual comparisons. Cluster analysis or other multivariate techniques can be used to generate dendrogram plots for showing relationships among large numbers of samples.

3.1.2.1.4 TLC-FID. Iatroscan thin-layer chromatography/flame ionization detection (TLC-FID) is used to separate petroleum into fractions of various polarities. The method shows promise as a rapid screening tool to define lateral or vertical changes in the gross composition of petroleum in reservoirs and to assist in selection of samples for more detailed geochemical work. Karlsen and Larter (1989) used TLC-FID to separate up to 70 samples per day into saturate, monoaromatic, diaromatic, polyaromatic, and polar fractions. Differences in composition were used to distinguish petroleum populations and predict barriers to reservoir continuity such as carbonate-cemented horizons or asphalt-rich zones.

3.1.2.2 Stable Isotope Ratios. Stable isotope compositions for carbon, sulfur, and hydrogen are used with biomarkers to group oils and bitumens. Isotopes are atoms whose nuclei contain the same number of protons, but different numbers of neutrons. For example, Fig. 3.5 shows the subatomic composition of the two stable isotopes of carbon. Carbon-12 (^{12}C) and carbon-13 (^{13}C) are called the "light" and "heavy" stable isotopes and account for about 98.89 and 1.11 weight percent of all carbon, respectively.

> ***Note:*** *The abundance of stable isotopes in biomarkers affects their mass spectra and can be used to help in structural elucidation. During mass spectrometry, biomarkers are fragmented into ions (Sec. 2.5.2.1). The probability that one of the atoms in any ion is a ^{13}C isotope increases with the number of atoms. For example, the molecular ion for cholestane (m/z 372) contains 27 carbon atoms and shows a much greater probability of containing a ^{13}C atom than a molecule containing one carbon atom (e.g., $27 \times 1.1 = 29.7\%$). The factor of 1.1 percent per carbon atom varies slightly depending on the organic matter source ($\sim$2% relative; Sec. 3.1.2.2.2). The mass spectrum of cholestane (Fig. 2.18) shows a significant m/z 373 after the m/z 372 molecular ion resulting from the contribution of ^{13}C. If this mass spectrum were for an unknown compound, the maximum number of carbon atoms in the compound could be estimated by comparing the two isotope peaks for the molecular ion (McLafferty, 1980).*
>
> *Isotope peaks can sometimes complicate the interpretation of mass chromatograms. For example, the Beatrice oil contains abundant benzopyrenes characterized by a mass spectral base peak at m/z 252 (Fig. 2.11). Because of their high concentrations in this oil and their isotope peak at m/z 253, the benzopyrenes interfere with monoaromatic steroid mass chromatograms (m/z 253) for this oil (Sec. 2.5.5.8).*

> ***Note:*** *Trace amounts of the unstable (radioactive) carbon-14 are present in organic matter less than about 50,000 years old. Unless introduced as a contaminant, carbon-14 is absent in petroleum.*

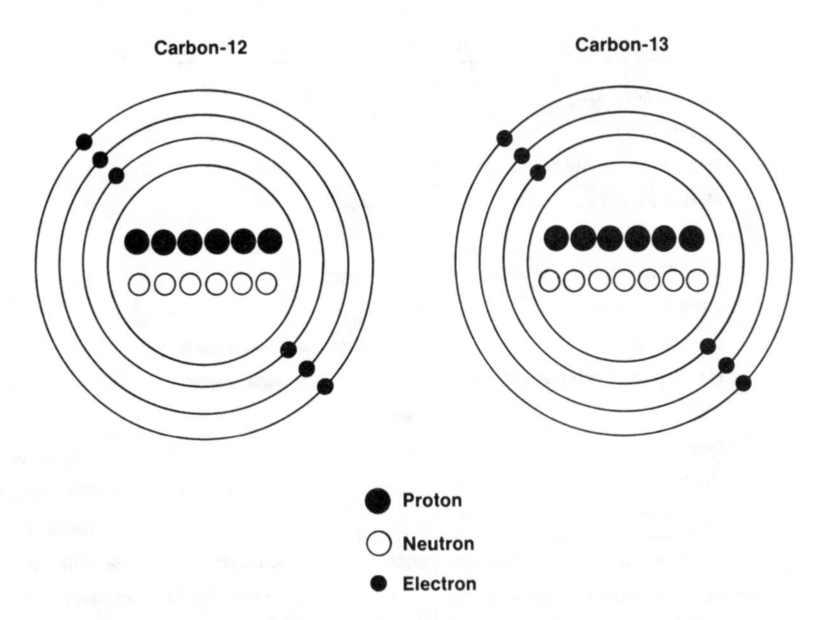

Figure 3.5 Comparison of proton, neutron, and electron configurations in the stable isotopes of carbon. The ^{12}C atom contains six protons and six neutrons in the nucleus (indicated by the inner circle) and accounts for about 98.89 percent of all carbon. The ^{13}C atom contains an additional neutron compared to ^{12}C and accounts of about 1.11 percent of all carbon. The unstable ^{14}C atom (not shown) contains six protons and eight neutrons in the nucleus. ^{14}C is radioactive and accounts for only a trace of naturally occurring carbon. ^{14}C is absent in petroleum that has not been contaminated by carbon from recently living photosynthetic organisms.

The difference in mass between isotopes of the same element results in measurable isotopic fractionation during physical and chemical processes. These fractionations are more pronounced for light elements because their isotopes (e.g., hydrogen vs. deuterium) show proportionally larger differences in mass than heavier elements (e.g., ^{12}C vs. ^{13}C). The main mechanisms for fractionations include:

1. exchange reactions,
2. kinetic effects associated with chemical reactions.

For example, exchange reactions during evaporation and condensation result in isotopic fractionations of the hydrogen and oxygen in water. These fractionations are caused by differences in the vapor pressures of water molecules containing light versus heavy isotopes of hydrogen and oxygen. Another example of chemical exchange occurs for carbon between atmospheric carbon dioxide and bicarbonate dissolved in water. The associated fractionation leads to enrichment of ^{13}C in the bicarbonate compared to carbon dioxide (see Fig. 3.6). Organic matter derived from atmospheric carbon dioxide or oceanic bicarbonate can usually be distinguished by its ^{13}C content (however, see the following).

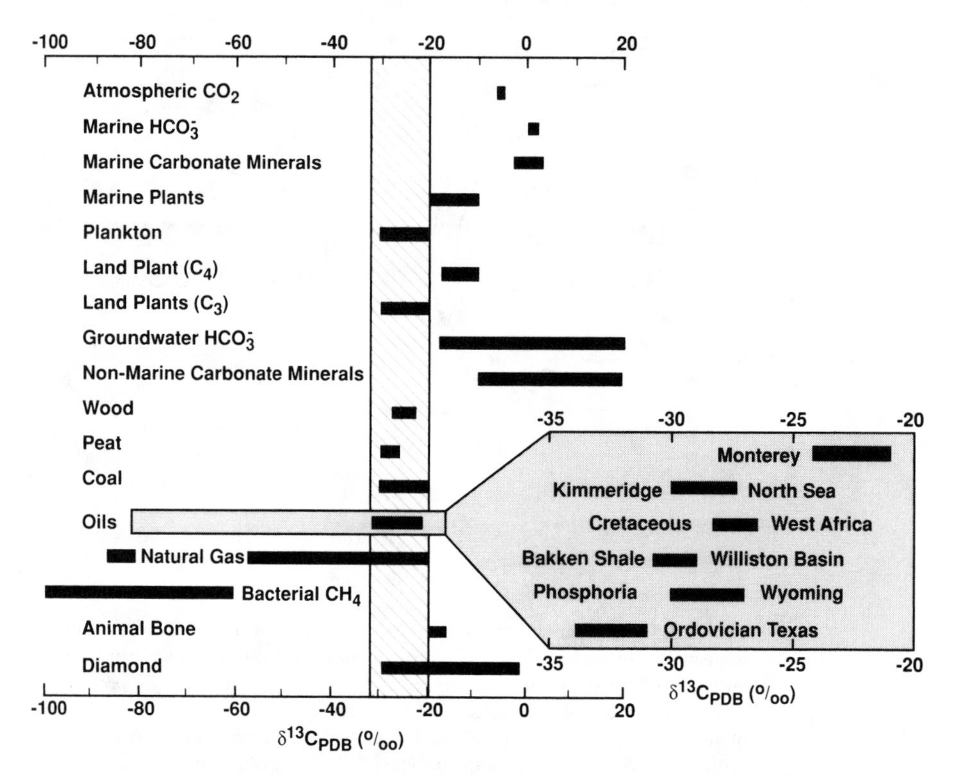

Figure 3.6 Variations in ranges of stable carbon isotope ratios (versus PDB standard, Sec. 3.1.2.2.1) for different organic and inorganic compounds. C_3 and C_4 plants are discussed in Sec. 3.1.2.2. The expanded scale shows ranges of isotopic values for various oils. Stable isotope ratios can be used in conjunction with biomarker data to show relationships between petroleums and their source organic matter. *Concept for figure courtesy of M. Schoell (pers. comm.) and W. G. Mook (1984).*

An example of an important kinetic isotope effect is that associated with photosynthesis, which leads to selective incorporation of ^{12}C into organic matter. The magnitude of this fractionation is believed to depend on various factors, including the specific photosynthetic pathway used for carbon fixation, the rate of diffusion of carbon dioxide into the plant, and the amount of metabolic carbon available (e.g., Hoefs, 1980 and references therein). [However, Galimov (1973) suggests a different explanation for isotopic compositions of organic matter based on intramolecular equilibrium isotope effects.]

Most plants and plankton fix carbon using the C_3- or Calvin pathway. Two additional photosynthetic pathways include the C_4- or Hatch-Slack pathway and the CAM (Crassulacean Acid Metabolism) pathway. C_4 plants are mainly tropical grasses, desert plants, and salt marsh (marine) plants. CAM plants are mainly succulents. Plants using these two photosynthetic pathways are enriched in ^{13}C compared to C_3 plants (see Fig. 3.6).

Detailed discussions of the use of stable isotopes in geochemistry are in Kaplan (1975), Fuex (1977), Hoefs (1980), and Schoell (1984).

3.1.2.2.1 Standards and notation. Stable isotope data are presented as "delta"-values (δ) representing the deviation in parts per thousand (per mil) from an accepted standard

$$\delta\,(\text{per mil}) = \frac{(\text{Rsample} - \text{Rstandard})}{\text{Rstandard}} \times 1000 \tag{3.1}$$

where R represents the isotope abundance ratio, such as $^{13}C/^{12}C$, $^{18}O/^{16}O$, $^{34}S/^{32}S$, $^{15}N/^{14}N$, or D/H ($^{2}H/^{1}H$). The δ-value for carbon, for example, is a convenient means to portray small variations in the relative abundance of the ^{13}C in organic matter (Sec. 3.1.2.2.2). A negative δ-value implies the sample is "light" or depleted in the heavy isotope relative to the standard. A positive value means the sample is isotopically "heavy" or enriched in the heavy isotope relative to the standard. Table 3.1.2.2.1 shows the most commonly used isotopic standards.

TABLE 3.1.2.2.1 Commonly Used Isotopic Standards.

Element	Standard	Abbreviation
Hydrogen	Standard Mean Ocean Water	SMOW
Carbon, primary	*Belemnitella americana* from Cretaceous Peedee Formation	PDB
Carbon, secondary	National Bureau of Standards Oil Sample	NBS-22
Nitrogen	Atmospheric Nitrogen	-
Sulfur	Canyon Diablo Troilite	CD

Converting δ-values reported using different standards. Unfortunately, not all δ-values are given relative to a single standard. For example, some stable carbon isotope measurements are reported relative to NBS-22, while other analyses are relative to PDB (Table 3.1.2.2.1). The PDB and NBS-22 standards are both defined as "zero" for their respective δ-value scales. On the PDB scale, however, NBS-22 oil measures about −29.81 per mil (Schoell, 1984). Because of the peculiar form of Eq. 3.1, conversion of δ-values between different standards is not straightforward. A δ-value for a sample relative to PDB cannot be converted to the NBS-22 scale by simply adding 29.81 per mil to the PDB value. The following equations must be used to convert δ-values for carbon from one standard to another.

PDB to NBS-22

$$\delta^{13}C_{NBS\text{-}22} = 1.03073(\delta^{13}C_{PDB}) + 30.73 \tag{3.2}$$

NBS-22 to PDB

$$\delta^{13}C_{PDB} = 1.9702(\delta^{13}C_{NBS\text{-}22}) - 29.81 \quad (3.3)$$

3.1.2.2.2. Stable carbon isotopes. Stable carbon isotope ratios are used to describe small variations in ^{13}C abundance in organic matter. Figure 3.6 shows an overview of the variations in the carbon isotopic composition for various organic and inorganic components. The expanded portion of the figure shows the range of isotopic values for oils from different localities. Differences in isotopic compositions between natural products are useful for various purposes, including oil and source rock correlation. These variations are controlled by isotopic fractionation, which occurs when carbon transfers from one chemical species to another, or from one phase to another (Hoefs, 1980; Deines, 1980). For example, ^{12}C is preferentially assimilated by plants during photosynthesis compared to ^{13}C. Bacterial decomposition of plants produces methane (marsh gas) containing considerably less ^{13}C than the decaying plants. Similarly, thermal maturation results in loss of ^{12}C-enriched methane and concentration of ^{13}C in the residual kerogen, as modeled in the laboratory (Peters et al., 1981).

Correlation. Several rules of thumb related to stable carbon isotopes can be applied to oil and bitumen correlations.

1. A positive correlation is supported, but not proven, when oils of similar maturity differ by no more than 1 per mil. Based on our experience, maturity differences among related oils can account for isotopic variations of up to about 2 to 3 per mil.
2. Oils which differ by more than about 2 to 3 per mil are usually from different sources, although there are exceptions to this rule. Chung et al. (1981) indicate that the 3.6 per mil range from the least to most mature oils in the Big Horn Basin, Wyoming, is due solely to differences in thermal maturity. Large variations can occur for oils derived from widespread source rocks showing major changes in organic facies (Hwang et al., 1989, discuss examples from Angola). For oils from source rocks deposited in certain restricted depositional environments, large isotopic variations occur, apparently without major differences in the type of contributing organic matter. An extreme example is the 8.1 per mil variation for Middle Ordovician organic matter (Hatch et al., 1987) which is reflected in the isotopic composition of related oils. This variation appears as a result of limited water circulation and variable organic productivity and their effects on carbon cycling by living organisms, rather than changes in organic matter type.
3. Bitumens are generally about 0.5 to 1.5 per mil depleted in ^{13}C compared to their source kerogens. Similarly, oils are depleted in ^{13}C by about 0 to 1.5 per mil compared to their corresponding bitumens. (These general relationships assume approximately equivalent levels of maturity for the compared oil, bitumen, and kerogen.)

Note: *Petroleums from Eocene and Miocene rocks from offshore California can be reliably distinguished by their respective ^{13}C-poor versus ^{13}C-rich carbon isotope compositions (Jones, 1987; Peters et al., 1992b). These differences are reflected in the isotope compositions of kerogens above and below the Neogene boundary.*

The bonds formed by the heavy isotope of an element require more energy for cleavage than bonds of the light isotope. This is the basis for the kinetic isotope effect. During an irreversible thermal reaction, early products (such as methane and other light gases) are thus enriched in the light isotope relative to the reactants (such as kerogen). Oils can become isotopically heavier (more enriched in ^{13}C) with maturation (e.g., Sofer, 1984) or isotopically lighter (e.g., Hughes et al., 1985) depending on which fraction of the evolving oil is sampled. A series of oils might consist of: (1) the volatile products (^{12}C-enriched) derived from progressive maturation and migration from an original oil, or (2) the residual oil (^{13}C-enriched) remaining after removal of more volatile products.

Secondary processes can cause variations in the compound class composition of related oils (e.g., percent saturates versus aromatics) and can, therefore, result in different isotopic compositions for related oils. Care must be used when comparing the stable carbon isotope compositions of whole oils showing large differences in gravity, or when comparing oils with bitumens. Gasoline-range hydrocarbons are enriched in ^{13}C compared to the whole oil (Silverman, 1971) and are prominent in high API gravity oils and condensates. If high API gravity petroleum is not "topped" by distillation to remove these gasoline-range components, comparison to a bitumen may be misleading. Most gasoline-range hydrocarbons originally present in bitumens are lost during sample preparation. (Bitumens are extracted from ground rock using organic solvents. Removal of the solvent by rotoevaporation results in loss of gasoline-range and other components below about nC_{15}.)

Kerogens may contain large amounts of inert or gas-prone macerals that do not contribute to oil generation, but overwhelm the isotopic signature of the oil-prone macerals. Likewise, the isotopic composition of indigenous bitumen may be obscured by migrated oil or contaminants (Sec. 3.1.1.1.2). When either of these problems occur, oil-source rock studies based on stable carbon isotope ratios of bitumens and kerogens may be of limited value. Bailey et al. (1990) avoid this problem by measuring isotopic compositions of the liquid pyrolyzates obtained from pre-extracted rocks. By using rapid anhydrous pyrolysis, they were able to generate pyrolyzates from a large number of closely-spaced rock samples (pyrolyzate isotopic profiling) for isotopic and biomarker comparison with oils in the North Sea. An added benefit of this approach is that geochemically heterogenous intervals within thick source rock units are less likely to be overlooked because of inadequate sampling. The same approach was used by Burwood et al. (1990) to distinguish organic facies within Pre-Salt source rocks from Angola.

Quantitative estimate of oil cosources. Because stable carbon isotope ratios represent a bulk property of the entire oil or bitumen, they can be used in quantitative estimates of source inputs to mixed oils once the co-sources have been established. We are unaware of any attempts to estimate the contributions of more than two sources to an oil.

Peters et al. (1989) show how widely differing carbon isotopic compositions of a whole oil (Beatrice) and bitumens from two co-sources for the oil (M. Jurassic and Devonian) were used to calculate the approximate contributions from each to the oil (however, see Bailey et al., 1990). Because the mixture involved only two oils, they used the following simple equation.

$$\delta^{13}C_{Oil}(100) = \delta^{13}C_{Source\ 1}(x) + \delta^{13}C_{Source\ 2}(100 - x)$$

The contribution (wt %) of source 1 (x) can only be estimated if the $\delta^{13}C$ values of the mixture and the contributing sources are known. Such quantitations are only estimates and require that the samples of mixed oil and contributing oils or bitumens be of approximately the same level of thermal maturity.

Stable carbon isotope type-curves. **The shapes and trends of stable carbon isotope type-curves** (Galimov, 1973; Stahl, 1978) **are used in the same manner as gas chromatograms to fingerprint relationships between oils, bitumens, and kerogens.** Figure 3.7 shows examples of type-curves for oils from the Middle Ob region, West Siberian Basin, Russia. The application of type-curves is based on the heterogeneous distribution of ^{13}C in crude oil fractions. Oils and bitumens show a general enrichment in ^{13}C for fractions of increasing polarity and boiling point (e.g., Galimov, 1973; Chung et al., 1981).

> **Note:** *The reason for this variation is unclear, but could be explained as follows: If the fractions are generated during disproportionation reactions by thermal cracking of kerogen, then the kerogen will become isotopically enriched in ^{13}C (heavy), while the fractions will become relatively depleted (light) in the order of decreasing polarity: asphaltenes > NSO-compounds > aromatics > saturates. Isotope type-curves in the literature show various components on the y-axis, but they are always in the order of decreasing polarity with increasing distance from the origin of the figure. Because of mass balance considerations, the whole oil or bitumen generally shows an isotopic composition between that of the saturates and aromatics.*

Extrapolation of stable carbon isotope type-curves for oils or bitumens and their fractions can be used to predict the approximate isotopic composition of bitumen and kerogen in the source rock. A proposed source rock that contains bitumen or kerogen with an isotopic composition inconsistent with this extrapolation can be eliminated. Further, migration of oil into a rock is indicated when the "bitumen" shows an isotopic composition inconsistent with the associated kerogen (Sec. 4.3.1).

This approach should be applied with caution because some kerogens contain abundant inert or gas-prone macerals with isotopic compositions that differ from the associated oil-prone macerals (e.g., Bailey et al., 1990). Thus, an isotopic relationship between the oil-prone macerals in such a kerogen and an oil might be obscured.

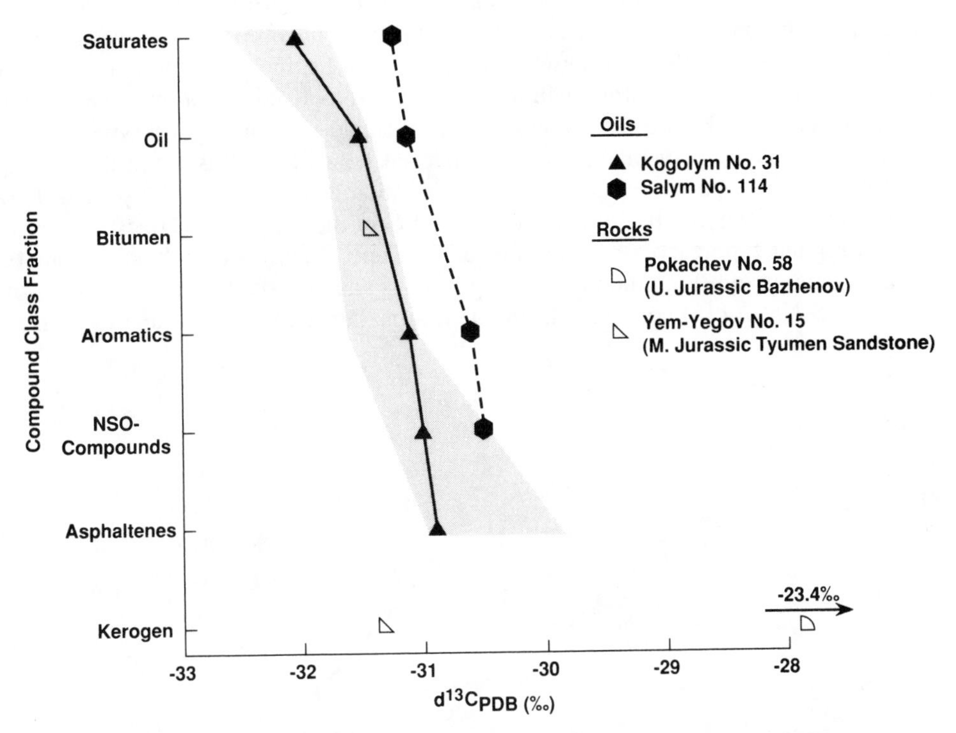

Figure 3.7 Stable carbon isotope type-curves can be used to show relationships between oils or between oils and source rock organic matter. The Kogolym No. 31 oil from the Middle Ob region of West Siberia shows a type-curve similar to other oils from the same area (stipled trend), supporting biomarker results indicating they are related (Peters et al., 1992a). Biomarkers for the Salym No. 114 oil indicate a relationship with Kogolym No. 31 and the other West Siberian oils. Because of its higher thermal maturity (late oil window) compared to the other oils (early to peak oil generation) the Salym No. 114 oil shows isotopically "heavy" (^{13}C enriched) compound class fractions. The trend indicated by all of the type-curves suggests that the kerogen in the source rock for the oils shows a stable carbon isotope value in the range of about −29 to −31 per mil. Kerogen isolated from a prospective U. Jurassic Bazhenov Formation source rock in the Pokachev No. 58 well shows an isotope value (−31.4 per mil) that is consistent with an oil-source rock relationship. Similarly, the bitumen from this rock (−31.5 per mil) falls within the stipled trend established by the related oils, again supporting a relationship between the oils and rock. The bitumen also correlates with the oils based on biomarkers. The carbon isotope composition of kerogen isolated from siltstone clasts in a Middle Jurassic Tyumen Formation sandstone (−23.4 per mil) confirms the biomarker results, indicating that it is unrelated to the oils.

Organic petrography can be used to help identify kerogens with maceral compositions that may present this type of problem (Sec. 3.1.2.5).

The shapes of stable carbon isotope type-curves are a useful characteristic in correlation studies. The trends of many stable carbon isotope type-curves are irregular (Chung et al., 1981) because secondary processes, including thermal maturation, migration, and deasphalting, can influence the isotopic composition of each fraction (as discussed). Figure 3.8 shows irregular stable carbon isotope type-curves for three asphaltene-poor oils from the Timan-Pechora Basin, Russia. In this example, note that the type-curves do not include bitumen or kerogen on the y-axis. Two of the Timan-Pechora oils show depletion of ^{13}C in their asphaltene fractions resulting in irregular type-curves that are similar in shape. Anomalously "light" asphaltenes are common in oils containing low asphaltenes and abundant saturates. The third oil (GCMS No. A115) contained insufficient asphaltenes for isotopic analysis. In cases

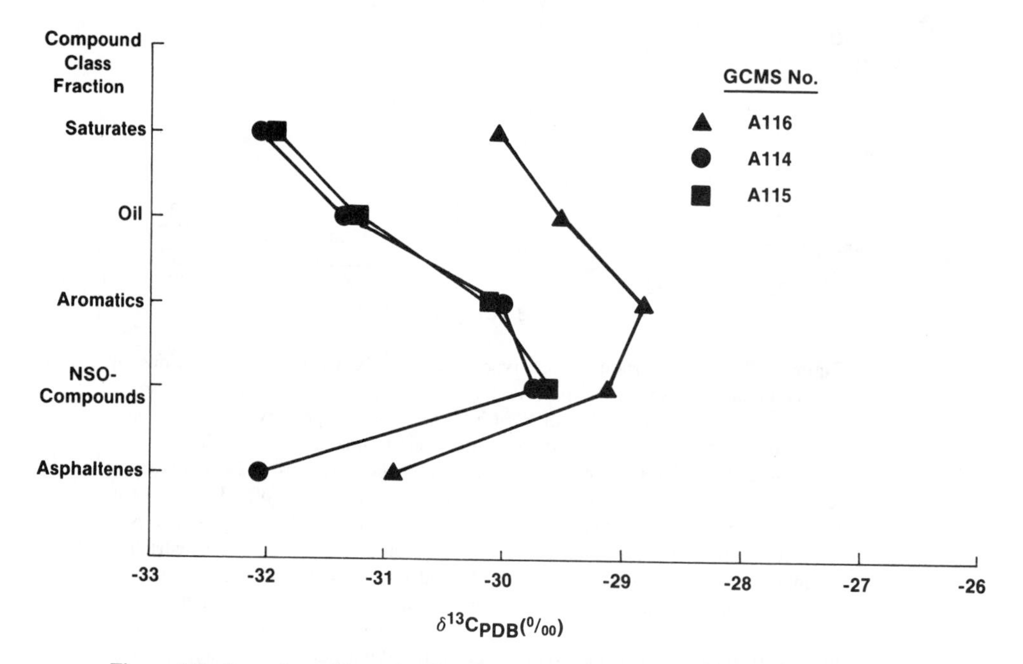

Figure 3.8 Irregular stable carbon isotope type-curves for three related oils from the Timan-Pechora Basin, Russia. In addition to trends, the shapes of type-curves are considered in evaluating oil-oil and oil-source rock correlations. The anomalously "light" (^{13}C depleted) asphaltenes in two of the oils could be caused by coprecipitation of saturates with the small quantities of asphaltenes in these samples during sample preparation. The third oil (GCMS No. A115) contained insufficient asphaltenes for analysis. The three oils each contain more than 54 wt % saturates and less than 0.4 wt % asphaltenes. Biomarker results indicate that GCMS No. A116 oil was derived from a different organic facies of the same source rock that generated the other oils. Based on extrapolation of the type-curves without including the anomalous asphaltenes, the predicted stable carbon isotopic composition of kerogen in the source rock for these oils is about −28 per mil.

of anomalous asphaltene isotopic values, we recommend that they be disregarded when extrapolating type-curves to predict a carbon isotopic composition of the source rock kerogen.

Note: *Based on our analyses of waxy oils from Sudan, we suggest that coprecipitation of isotopically light saturates with asphaltenes during preparation may account for this phenomenon. Both the Sudan and Timan-Pechora oils examined are rich in saturates (more than 45 and 54%, respectively), and poor in asphaltenes (less than 3 and 0.4%, respectively).*

Marine versus terrestrial input. Several publications describe classifications of "marine" and "nonmarine" oils using stable carbon isotope ratios of the whole oil (Silverman and Epstein, 1958) or the saturate and aromatic fractions (Sofer, 1984; 1988). However, statistical studies show that these methods do not completely differentiate oils by source. For example, Peters et al. (1986) used two-group discriminant analysis to show that only 66 percent of the oils in a suite of marine and nonmarine oils were correctly separated using stable carbon isotope analyses of the saturates and aromatics. However, marine and nonmarine oils were successfully distinguished using combined isotope and biomarker data.

Age determination and depositional environment. Although general variations in the bulk stable carbon isotopic compositions of organic matter and carbonate have been described, the variability among petroleums or kerogens of the same age indicates that isotopic composition cannot be used to supply reliable information for age dating. For example, organic matter of Precambrian (Hayes et al., 1983) and Ordovician (Hatch et al., 1987) age is commonly depleted in ^{13}C compared to younger organic matter, but these differences are *not* diagnostic. Schoell and Wellmer (1981) show that some Precambrian organic matter shows unusually low (negative) carbon isotope values, but the average Precambrian organic matter shows values identical to that from the Phanerozoic. Arthur et al. (1988) note a common enrichment in ^{13}C of marine organic matter of about 3 to 4 per mil corresponding to the Cretaceous/Tertiary boundary at several worldwide locations. Similarly, "heavy" stable carbon isotope values are often associated with organic matter produced in hypersaline environments (Schildowski et al., 1984). However, these differences are again not diagnostic.

For these mentioned reasons, isotopic data cannot be used alone to indicate the age or environment of deposition of organic matter. Similarly, stable carbon isotope ratios must be used with caution as indicators of source input because of fractionations occurring both during and after formation of the organic matter (Fuex, 1977; Deines, 1980). However, combined use of stable carbon isotopes and biomarkers allows more confident assessment of organic matter input as exemplified in the following.

Note: *Work on an oil from Oman (GCMS No. A24) is an example of the combined use of biomarkers and isotopes to describe the depositional environment in the source rock for the oil. The oil shows many characteristics suggesting that it is derived from Pre-*

Ordovician rocks. McKirdy et al. (1983) indicate that Pre-Ordovician carbonate-sourced oils commonly show low carbon isotope values, pristane/phytane ratios, and Ts/Tm (<1; Sec. 3.2.4.1), and a predominance of C_{28}- or C_{29}-steranes. The oil shows a "very negative" carbon isotope value of −33.14 per mil, Pr/Ph of 0.85 and Ts/Tm of 0.89, but is not particularly enriched in C_{28}- or C_{29}-steranes. Compared to other oils in the area, this oil shows a higher gammacerane index (Sec. 3.1.3.1) and a lower C_{30}-sterane index (Sec. 3.1.3.2). The low C_{30}-sterane index suggests that the depositional environment of the source rock for the oil may have been more isolated from marine conditions (possibly a starved euxinic basin or lagoonal carbonate-evaporite) than the others. Gammacerane has been tentatively suggested as a marker for hypersaline episodes of source rock deposition, which may occur in alkaline lakes or lagoonal carbonate-evaporite deposits (see the following). The combined biomarker, isotopic, and geologic information on the Omani oil allowed us to infer that it was generated from Infra-Cambrian source rocks found nearby in the basin.

3.1.2.2.3 Sulfur and hydrogen isotopes. Crossplots of the stable isotope ratios for carbon versus sulfur (e.g., Orr. 1974), or carbon versus hydrogen (e.g., Schoell, 1984; Peters et al., 1986) can be used to distinguish petroleum groups.

Burwood et al. (1990) distinguished two families of Angolan oils using carbon-hydrogen isotopic crossplots. Tertiary reservoired "Accretionary Wedge oils" are depleted in deuterium and were previously shown to contain abundant 18α(H)-oleanane (e.g., Riva et al., 1988). These oils are believed to be derived from the Late Cretaceous or Paleogene age, marine Iabe-Landana source rock intervals. Pinda Formation and some Pre-Salt reservoired "Carbonate Platform oils" are enriched in deuterium, suggesting deposition under aquatic conditions with high evaporative loss. These oils show many features in common with lacustrine oils from Brazil, including prominent tricyclic terpanes, 28,30-bisnorhopane, and gammacerane. Many of these Angolan oils show a demethylated hopane tentatively identified as 25,30-bisnorhopane. The Carbonate Platform oils and their Brazilian analogs are believed to be derived from the Pre-Salt, Lower Cretaceous age, lacustrine Bucomazi Formation, which was deposited prior to rifting of South America from Africa. Three oil-prone organic facies were recognized within the Bucomazi Formation based primarily on stable carbon isotopes and the type of organic matter. Interestingly, the 25,30-bisnorhopane marker observed in many of the Carbonate Platform oils was found exclusively in one of these organic facies that was enriched in ^{13}C and contained Type II organic matter.

Because sulfur can be incorporated into oils during secondary processes (Orr, 1974), sulfur isotope compositions may not be directly related to the original organic matter input. Nonetheless, sulfur isotope ratios show potential for oil-oil correlations (Gaffney et al., 1980). Premuzic et al. (1986) used sulfur isotopes to: (1) support source groupings for oils from Prudhoe Bay, Alaska, which were previously established by biomarkers (Seifert et al., 1980, 1983) and (2) distinguish the Prudhoe Bay oils from those in similar formations in the Point Barrow area described by Magoon and Claypool (1981, 1983, 1984).

3.1.2.3 Total Organic Carbon and Pyrolysis. Total organic carbon (TOC) can be measured using various methods, each with limitations.

The "direct" combustion method is the most common and involves acidification of the ground rock sample with 6 N HCl in a filtering crucible to remove carbonate, removal of the filtrate by washing/aspiration, drying at about 55°C, and combustion with accelerator at about 1200°C. The CO_2 generated during combustion is trapped and analyzed. Although the direct method is rapid, it is not accurate for either organic-poor, carbonate-rich rocks or for many immature sediment and rock samples. For example, immature organic matter is susceptible to acid hydrolysis and loss during filtering (Peters and Simoneit, 1982). Diagenetically immature Deep Sea Drilling Project sediments were analyzed using a "modified" direct TOC method (Peters and Simoneit, 1982), which employs nonfiltering crucibles so that hydrolyzate is not lost. The modified direct method showed an average of more than 10 percent higher TOC for the sediments compared to the direct method.

The "indirect" TOC method is usually applied to organic-poor, carbonate-rich rocks. Total carbon (including carbonate carbon) is determined on one aliquot of the sample, while carbonate carbon is determined on another aliquot by coulometric measurement of the CO_2 generated by acid treatment. Organic carbon is determined by the difference between total carbon and carbonate carbon. This method is more time-consuming than the direct method and requires two separate analyses of the sample.

The "Rock-Eval II plus TOC" (Delsi, Inc.) determines TOC by summing the carbon in the pyrolyzate with that obtained by oxidizing the residual organic matter at 600°C. For small samples (100 mg), this method provides more reliable TOC data than the methods previously discussed, which require about 1 to 2 g of ground rock. However, mature samples, where vitrinite reflectance (Sec. 3.2.2.1.2) is more than about 1 percent, yield poor TOC data when determined by this method because the temperature is insufficient for complete combustion.

Tables 3.1.2.3(a) and (b) show how Rock-Eval pyrolysis and other geochemical measurements are used to assess the quantity and quality of source rock organic matter, respectively (more details on this method are in Sec. 3.2.2.1.1). The organic matter in a source rock can be characterized on a hydrogen index versus oxygen index (HI versus OI) diagram (Fig. 3.9) which provides a crude assessment of hydrocarbon generative potential (Espitalié et al., 1977; Peters et al., 1983, Peters, 1986).

TABLE 3.1.2.3(a) Geochemical Parameters Describing Source Rock Generative Potential (Quantity).[1]

Quantity	Wt % TOC	Rock-Eval S2 (mg HC/g rock)	Bitumen, wt %	Hydrocarbons, ppm[2]
Poor	<0.5	<2.5	<0.05	<300
Fair	0.5–1	2.5–5	0.05–0.1	300–600
Good	1–2	5–10	0.1–0.2	600–1200
Very Good	>2	>10	>0.2	>1200

[1] Assumes a level of thermal maturation equivalent to R_o = 0.6%.

[2] 10,000 ppm = 1 wt %.

TABLE 3.1.2.3(b) Geochemical Parameters Describing Type of Hydrocarbon Generated (Quality).[1]

Type	Hydrogen index (mg HC/g TOC)	Rock-Eval S2/S3	Atomic H/C
Gas	50–200	1–5	0.7–1.0
Gas and Oil	200–300	5–10	1.0–1.2
Oil	>300	>10	>1.2

[1] Assumes a thermal maturation equivalent to R_o = 0.6%.

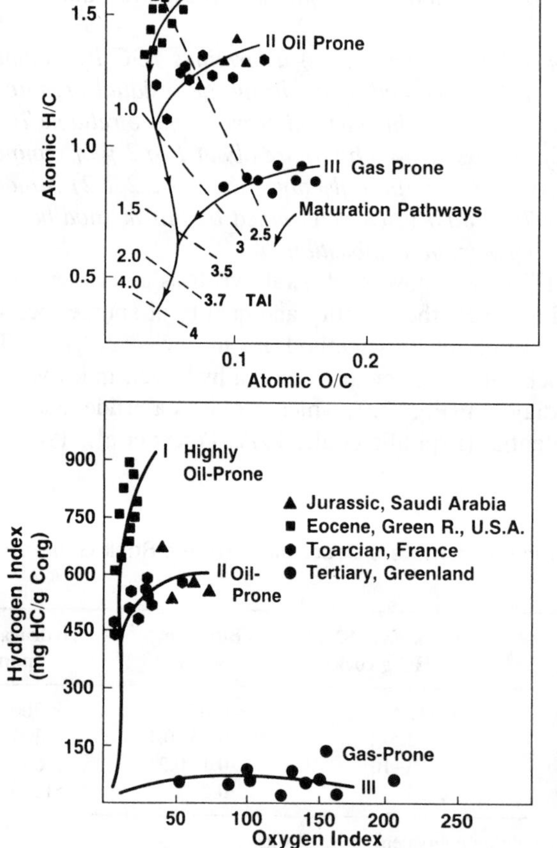

Figure 3.9 Hydrogen index versus oxygen index (Rock-Eval pyrolysis of whole rock) and atomic H/C versus O/C (elemental analysis of kerogen) diagrams can be used to describe the type of organic matter in source rocks (*from Peters,* 1986). R_0 = vitrinite reflectance, TAI = Chevron thermal alteration index (Jones and Edison, 1978). The Type IV (inertinite) pathway is not shown.

3.1.2.4 Elemental Analysis. Oils and kerogens are routinely analyzed for elemental composition. For example, CHN- and sulfur analyzers can be used to provide carbon, hydrogen, nitrogen, and sulfur content of kerogen (Durand and Monin, 1980). Analysis for organic oxygen in kerogen is much more complex, and is commonly not attempted. Sulfur content is the principal elemental measurement determined on oils.

3.1.2.4.1 van Krevelen diagrams, types of organic matter. The most familiar method of classifying organic matter type is the atomic H/C versus O/C or "van Krevelen" diagram. This diagram was originally developed to characterize coals (van Krevelen, 1961; Stach et al., 1982) during their thermal maturation or "coalification." Tissot et al. (1974) extended the use of the van Krevelen diagram from coals to include the kerogen dispersed in sedimentary rocks.

> ***Note:*** *In this work, "coal" is defined as any rock containing more than 50 wt % organic matter. Both coals and sedimentary rocks can contain any combination of macerals and the term "coal" does not imply anything about maceral composition. Unfortunately, no universally accepted classification for the various maceral (Sec. 3.1.2.5) and kerogen types exists in the literature at this time. In this work we will use the Type I, II, III (Tissot et al., 1974) and IV (Demaison et al., 1983) nomenclature to describe kerogens.*

The three principal maceral groups in coals and sedimentary rocks (liptinites, vitrinites, and inertinites; Sec. 3.1.2.5) mature along three different evolutionary paths on van Krevelen diagrams (Fig. 3.9). As a given sedimentary rock becomes more mature during burial, the kerogen becomes more depleted in hydrogen and oxygen relative to carbon. At very high levels of catagenesis, all kerogens approach graphite in composition (pure carbon) near the lower left portion of the diagram.

An understanding of the type of organic matter in a rock sample based on elemental analysis is useful for corroborating biomarker and other geochemical analyses. The four principal types of kerogen in both coals and sedimentary rocks include Type I (very oil prone), Type II (oil prone), Type III (gas prone), and Type IV (inert) (Fig. 3.9) as will be described.

Type I: Immature Type I kerogen shows high atomic H/C (near 1.5) and low atomic O/C (<0.1) ratios. Petrographically Type I kerogens are dominated by liptinite macerals, although vitrinites and inertinites can be present in lesser amounts. The kerogen is dominated by aliphatic structures suggesting major contributions from lipids during diagenesis. Sulfur is low in Type I kerogens. Laboratory pyrolysis (see following) or burial maturation of this kerogen result in higher yields of hydrocarbons than the other kerogen types. Paraffins dominate the pyrolysis products. Type I kerogens appear to be derived by extensive bacterial reworking of lipid-rich algal debris, particularly in lacustrine settings. *Botryococcus* (Sec. 3.1.3.1) and similar lacustrine algae and their marine equivalents such as *Tasmanites*, appear to represent major contributors to Type I kerogens.

Although Type I kerogens are less common than the others, they account for many important petroleum source rocks and oil shales (Hutton et al., 1980), including the organic-rich Green River shale (actually a marl) from Utah, Colorado, and Wyoming, and Chinese oil shales, boghead coals, torbanites from Scotland, and coorongite from South Australia.

Type II: Immature Type II kerogen shows high atomic H/C (1.2 to 1.5) and low O/C ratios compared to Types III and IV. The kerogen is dominated by liptinite macerals, but like Type I kerogens, vitrinites and inertinites can be present in lesser amounts. Sulfur is typically higher in Type II compared to other kerogen types. Unusually high sulfur in certain Type II kerogens such as that in the Permian age Phosphoria Formation (Lewan, 1985) and the Phosphatic Member of the Monterey Formation (Peters et al., 1990; Baskin and Peters, 1992) may explain the tendency of these kerogens to generate petroleum at lower levels of burial maturation than others. Orr (1986) describes "Type II-S" kerogens and procedures for their isolation from the Monterey Formation in the Santa Maria Basin and Santa Barbara coastal area, California. Type II-S kerogens contain unusually high organic sulfur (8–14 wt%, atomic $S/C \geq 0.04$) and appear to begin to generate oil at lower thermal exposure than typical Type II kerogens with less than 6 wt% sulfur. Pyrolysis or burial maturation of Type II kerogen results in higher yields of hydrocarbons than the other kerogens, except Type I.

Type II kerogen originates from mixed phytoplankton, zooplankton, and bacterial debris, usually in marine sediments. Type II kerogens account for most petroleum source rocks, including those of Jurassic age in Saudi Arabia, the North Sea, and West Siberia, the Cretaceous in Venezuela, and the Miocene in California.

Type III: Immature Type III kerogen shows low H/C (less than 1.0) and high O/C (up to ~0.3) ratios. Type III organic matter yields less hydrocarbons than Types I and II during pyrolysis or burial maturation. This type of organic matter is common in Tertiary rocks, is usually derived from terrestrial plants, and is dominated by vitrinite and lesser amounts of inertinite macerals.

Type IV: Type IV kerogen is "dead carbon" showing very low atomic H/C (~0.5–0.6) and low to high O/C ratios (up to ~0.3). This type of kerogen is dominated by inertinite macerals and does not generate significant hydrocarbons. Type IV kerogen can be derived from other kerogen types that have been reworked and oxidized.

Estimation of atomic O/C. Because of difficulties in the accurate measurement of organic oxygen in kerogen, we plot kerogens on the H/C versus O/C diagram using the method of Jones and Edison (1978). The measured atomic H/C defines a horizontal line on the plot (e.g., $H/C = 1.5$). Using independent data on maturation *or* organic matter type allows definition of a point on the diagram which characterizes the particular kerogen and infers an atomic O/C ratio. Elemental compositions of kerogens in Fig. 3.9 have been calibrated to thermal maturity measurements based on microscopy (vitrinite reflectance and thermal alteration index).

3.1.2.4.2 Sulfur content. Sulfur content is a bulk parameter commonly used to support relationships between petroleums. For example, simple plots of API gravity or stable carbon isotope ratio versus wt % sulfur can be used to show tentative relationships between oils as part of sample screening.

Understanding the origin of sulfur in petroleum and kerogen is necessary to make reliable interpretations regarding source input and depositional environment. Some sulfur may be derived from amino acids in the original contributing organic matter in sediments. However, most primary sulfur in oils and bitumens originates from early diagenetic reactions between the deposited organic matter and aqueous sulfide species (S^{2-}) such as hydrogen sulfide (H_2S) or polysulfides (e.g., Francois, 1987). Details of the pertinent pore water and microbial chemistry in shallow sediments are described in Claypool and Kaplan (1974).

Sulfides are produced by sulfate-reducing bacteria, such as *Desulfovibrio*, primarily in highly reducing to anoxic (low Eh) marine sediments (Fig. 3.10). If H_2S migrates upward from an anoxic into an oxic environment in sediment or the water column, it is rapidly oxidized back to sulfate by aerobic bacteria, such as *Chlorobium* or *Thiobacillus* (Fig. 3.10) (Orr and Gaines, 1974). Even in anoxic environments, sulfide ions produced by sulfate reducing bacteria can be oxidized back to sulfate during periodic mixing of oxic and anoxic water layers. However, because of the poor water circulation typical of anoxic basins (Demaison and Moore, 1980), H_2S may increase. H_2S is toxic to aerobic organisms.

Two sinks compete for excess sulfide in anoxic sediments: metals and organic matter. Although some H_2S may be oxidized under anoxic conditions by photosynthetic purple bacteria, the major sink is the reaction of sulfide with iron to form hydrotroilite, troilite, and eventually pyrite. Raiswell and Berner (1985) show how plots of organic carbon versus pyrite sulfur for rocks and sediments can be used to distinguish deposition under normal marine (oxygenated) versus euxinic (anoxic, H_2S-containing, usually deep water) conditions.

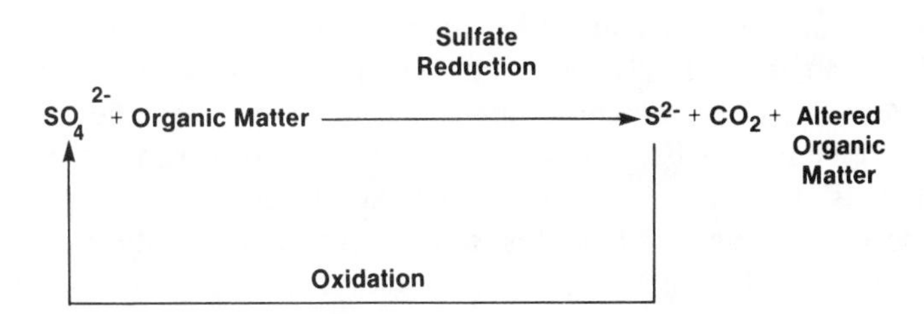

Figure 3.10 Generalized reactions of sulfur species during diagenesis of sediments. Under highly reducing to anoxic conditions, excess sulfides are generated by sulfate-reducing bacteria. Metals, such as iron, compete with organic matter as sinks for the sulfides. Clay poor, carbonate sediments contain few metals and under these conditions, excess sulfides will become incorporated into the immature kerogen.

High- and low-sulfur crude oils are derived from high- and low-sulfur kerogens, respectively (Gransch and Posthuma, 1974). Clay-poor lime muds contain insufficient iron and other metals to scavenge all available sulfide (Tissot and Welte, 1984). Under these conditions, much of the sulfide becomes incorporated into the kerogen. Thus, many high-sulfur kerogens and oils originate from clay-poor, marine rocks (e.g., carbonates or anhydrites) deposited under highly reducing to anoxic conditions.

Metals associated with clays in marine siliclastic rocks (e.g., most shales) may outcompete organic matter for reduced sulfur, leading to low-sulfur kerogens and oils. Most lacustrine kerogens and oils are also low in sulfur, but for a different reason. Lacustrine sediments usually do not contain sufficient sulfate for strong enrichment of sulfur to occur in the organic matter. However, certain salt lake sediments of China contain elevated sulfur (Fu Jiamo et al., 1986) as do the lacustrine-sourced oils from Rozel Point, Utah (up to 14 wt %) (Meissner et al., 1984; Sinninghe Damsté et al., 1987).

High-sulfur kerogens appear to generate petroleum at significantly lower thermal exposure than other kerogens (Lewan, 1985; Orr, 1986; Baskin and Peters, 1991). Because high-sulfur kerogens form in this type of environment, a correlation exists between high-sulfur oils derived from these kerogens and biomarker ratios indicating anoxic depositional conditions such as high C_{35}-homohopane indices or low Pr/Ph ratios (Sec. 3.1.3.1). Many sulfur-rich oils and bitumens also show low diasteranes/steranes ratios, typical of clay-poor source rocks (Section 3.1.3.2). The same factors that control the distribution of sulfur in dispersed kerogens from marine and nonmarine source rocks also apply to the macerals in coals (Casagrande, 1987).

At temperatures greater than 150°C, sulfur can be incorporated into organic matter in contact with sedimentary anhydrite or gypsum (Orr, 1974). Biodegradation can result in increased sulfur content in oils by preferential removal of saturated hydrocarbons.

Schmid et al. (1987) and Sinninghe Damsté et al. (1987) describe long-chain dialkylthiacyclopentanes in oils and suggest a mechanism for diagenetic or thermal incorporation of sulfur based on the structures of these compounds. Payzant et al. (1986) describe terpenoid sulfides in petroleum. Sinninghe Damsté et al. (1987) and Valisolalao et al. (1984) have tentatively identified steroid and hopanoid thiophanes, respectively, in petroleum. Abundant benzothiophenes and alkyldibenzothiophenes in petroleum have been proposed as indicators of carbonate-evaporite source environments (Hughes, 1984). Sinninghe Damsté et al. (1990) describe various highly branched isoprenoid thiophenes in sediments and immature oils. These compounds appear to result from selective incorporation of sulfur into isoprenoid alkenes during diagenesis.

Recent progress in detailed sulfur-compound geochemistry as applied to source rock characterization, paleoenvironmental assessment, maturation, and correlation is described by Sinninghe Damsté and de Leeuw (1990), Kohnen et al. (1991), and Guadalupe et al. (1991). These authors show that biomarker distributions differ between free and sulfur-bound (resin) fractions of petroleum, apparently due to selective preservation of certain compounds by sulfur linkage during diagenesis. These sulfur-bound biomarkers can be released by treatment with Raney nickel. For example, aromatic carotenoids derived from photosynthetic bacteria and/or marine sponges are absent in the free hydrocarbon fraction, but are major compounds in the desulfurized resin fraction of some samples. Thus, analysis of desulfurization products from resins may allow more accurate assessment of the original biomarkers in the depositional environment. Further, distributions of biomarkers released by Raney nickel treatment can be used as "fingerprints" in correlation studies (Sinninghe Damsté and de Leeuw, 1990). Distributions of sulfur-containing compounds derived by flash pyrolysis of kerogens can be used to distinguish organic matter from different Ordovician rocks, including the Guttenberg oil rock from the Decorah Formation, which is composed predominantly of remains of the alga *Gloeocapsamorpha prisca* (Douglas et al., 1991).

Kohnen et al. (1991) used MeLi/MeI for selective chemical degradation of di- and polysulfide linkages in polar and asphaltene fractions of extract from an immature Italian bituminous shale. They showed that a major portion of the total bound biomarkers, including acyclic, branched, isoprenoid, steroid, hopanoid, and carotenoid carbon skeletons, are linked by di- or polysulfide bonds.

X-ray absorption near-edge spectroscopy (XANES) has recently been applied to identify and quantify classes of sulfur-containing compounds in oils and bitumens (Waldo et al., 1991). This information appears useful for describing petroleum source rock depositional environment and thermal maturity. For example, high-sulfur oils show distinct sulfide-rich or thiophene-rich XANES profiles, which appear to be related to clastic vs. carbonate source rocks, respectively.

3.1.2.5 Maceral Analysis. Because multiple lines of evidence increase the reliability of all interpretations, we recommend the use of microscopy to substantiate biomarker and other geochemical analyses. Optical microscopy of kerogen preparations can be used to describe the quantity, quality, and thermal maturity of organic matter. Analyses include organic yield, the relative percentages of oil-prone and other macerals, TAI, vitrinite reflectance, and qualitative fluorescence.

Three different techniques are applied to the microscopy of coals and petroleum source rocks.

1. Strew-mount slides of kerogen in transmitted light,
2. Slides of polished kerogen in reflected light,
3. Polished surfaces of whole rock or coal in reflected light.

Kerogen is defined as the organic matter in rocks and coals that is insoluble in organic solvents and survives acid digestion of the mineral matrix (Durand, 1980). **Macerals** are individual components in the kerogen that show petrographically distinct properties. Macerals consist of individual organic particles or "phytoclasts." For Techniques 1 and 2, kerogen is concentrated by means of acid maceration and/or density flotation (Bostick and Alpern, 1977) and prepared as slides (Baskin, 1979). Technique 3 is typically used only for very organic-rich source rocks or coals.

Unfortunately, no universally accepted maceral classification for the finely-dispersed kerogen in source rocks currently exists. Tissot and Welte (1984, p. 498) attempt to show the approximate relationships between some of the many classifications used to describe kerogens. The principal macerals in coals and sedimentary rocks can be categorized into three groups (Stach et al., 1982) defined as follows:

Liptinites: "Oil-prone" macerals showing low reflectance, high transmittance, and intense fluorescence at low levels of maturity. Many liptinite phytoclasts can be identified by their characteristic shape, e.g., algae (such as *Tasmanites*), resin (impregnating voids), or spores.

Vitrinites: "Gas-prone" macerals showing angular shapes, sometimes with cellular structure. The reflectance of vitrinite phytoclasts is used as an indicator of the thermal maturity for rock samples. Vitrinite macerals show intermediate reflectance ("low-gray"; Bostick, 1979) and transmittance, usually with no fluorescence, unless impregnated by liptinites.

Inertinites: "Inert" macerals showing angular shapes, typically with cellular structure. Inertinite phytoclasts show high reflectance, no fluorescence, and are opaque in transmitted light.

Much of the organic matter observed under the microscope appears amorphous. Not all amorphous organic debris is oil prone. However, immature to mature oil-prone amorphous organic matter typically fluoresces under ultraviolet light while other types of organic matter do not.

3.1.2.6 Nickel and Vanadium Content. Absolute concentrations and ratios of nickel/vanadium can be used to classify and correlate petroleums. These metals exist in petroleum largely as porphyrin complexes (Sec. 3.1.3.4). Oils from marine carbonates or siliciclastics show low wax content, moderate to high sulfur, high concentrations of nickel and vanadium, and low nickel/vanadium (≤ 1) ratios (Barwise, 1990). The dominance of vanadium over nickel in these oils appears to be due to low Eh conditions associated with sulfate reduction in the depositional environment of their marine source rocks and greater relative stability of vanadyl versus nickel porphyrin complexes (Lewan, 1984). Oils from lacustrine source rocks show high wax, low sulfur, moderate quantities of metals, and high nickel/vanadium (>2). Nonmarine oils derived from higher plant organic matter show high wax, low sulfur, and very low metals.

Note: *Comparisons of oils using bulk parameters such as total nickel and vanadium content, sulfur, or API gravity require that the thermal maturity of the samples be taken into account. Metal content and sulfur decrease with increasing maturity (e.g., API) for a given oil type.*

3.1.3 Biomarker parameters for correlation, source, and depositional environment. Tables 3.1.3(a) and -(b) summarize some of the more useful acyclic and cyclic biomarkers, respectively, which serve as indicators of organic matter input or depositional environment of the source rock. These tables must be used with caution. Although a compound or group of compounds may support a particular biological origin or paleoenvironment, exceptions are common. Volkman (1988) summarizes the use of biomarkers in petroleums for reconstructing source rock depositional environment and organic matter input.

TABLE 3.1.3(a) Acyclic Biomarkers as Indicators of Biological Input or Depositional Environment (Assumes High Concentrations of Component).

Compound	Biological origin	Environment	Example
nC_{15}, nC_{17}, nC_{19}	Algae	Lacustrine, Marine	Gelpi et al. (1970), Tissot and Welte (1984)
nC_{15}, nC_{17}, nC_{19}	*Gloeocapsamorpha prisca(?)*	~Ordovician	Reed et al. 1986, Rullkötter et al. (1986), Jacobson et al. (1988), Longman and Palmer (1987)
nC_{27}, nC_{29}, nC_{31}	Higher plants	Terrestrial	Tissot and Welte (1984)
nC_{23}-nC_{31} (Odd)	Nonmarine algae	Lacustrine	Gelpi et al. (1970), Moldowan et al. (1985)
2-methyldocosane	Bacteria?	Hypersaline	Connan et al. (1986)
Pristane/Phytane (low)	Phototrophs, Archaebacteria	Reducing to anoxic, high salinity	Fu Jiamo et al. (1986, 1990)
2,6,10,15, 19-penta-methyleicosane	Archaebacteria	Hypersaline	Brassell et al. (1981), Risatti et al. (1984)
2,6,10,trimethyl-7-(3-methyl-butyl)-dodecane	Green algae	Hypersaline	Yon et al. (1982), Kenig et al. (1990)
Botryococcane	Green algae (*Botryococcus branuii*)	Lacustrine/Brackish	Moldowan and Seifert (1980), McKirdy et al. (1986)
16-demethyl-botryococcane	Green algae (*Botryococcus braunii*)	Lacustrine/Brackish	Seifert and Moldowan (1981), Brassell et al. (1986)
Mid-chain monomethyl alkanes	Cyanobacteria	Hot springs, marine	Shiea et al. (1990)

TABLE 3.1.3(b) Cyclic Biomarkers as Indicators of Biological Input or Depositional Environment (Assumes High Concentrations of Component).

Compound	Biological origin	Environment	Example
Saturates			
C_{15}-C_{23} (Odd) cyclohexyl alkanes	*Gloeocapsamorpha prisca(?)*	~Ordovician, Marine	Reed et al. (1986) Rullkötter et al. (1986)
β-carotane	Bacteria	Arid, Hypersaline	Jiang and Fowler (1986)
Phyllocladanes	Conifers	Terrestrial	Noble et al. (1985a, 1985b, 1986)
C_{27}-C_{29} Steranes	Algae (C_{27}), Algae and higher plants (C_{29})	Various	Moldowan et al. (1985), Volkman (1986)

TABLE 3.1.3(b) *(cont.)*

Compound	Biological origin	Environment	Example
C_{30} 24-*n*-propylcholestanes (4-desmethyl)	Chrysophyte algae	Marine	Moldowan et al. (1985), Peters et al. (1986), Moldowan et al. (1990)
4-methylsteranes	Dinoflagellates/ some bacteria	Lacustrine or Marine	Brassell et al. (1986), Wolff et al. (1986)
Diasterenes	Algae or higher plants	Clay-rich rocks	Rubinstein et al (1975)
Dinosteranes	Dinoflagellates	Marine, Triassic or Younger	Summons et al. (1987), Goodwin et al. (1988)
25,28,30-trisnorhopane	Bacteria?	Anoxic Marine, Upwelling?	Grantham et al. (1980) Volkman et al. (1983c)
28,30-bisnorhopane	Bacteria?	Anoxic Marine, Upwelling?	Seifert et al. (1978), Grantham et al. (1980)
C_{35} $17\alpha,21\beta$(H)-hopane	Bacteria	Reducing to Anoxic	Peters and Moldowan (1991), Moldowan et al. (1992)
2-methylhopanes	Bacteria	Carbonate rocks	Summons and Walter (1990)
23,28-bisnorlupanes	Higher Plants?	Terrestrial	Rullkötter et al. (1982)
4β(H)-eudesmane	Higher Plants	Terrestrial	Alexander et al. (1983a)
Gammacerane	Protozoa? Bacteria	Hypersaline	Kleemann et al. (1990), Moldowan et al. (1985), Fu Jiamo et al. (1986), ten Haven et al. (1988)
18α(H)-oleanane	Higher plants (angiosperms)	Cretaceous or Younger	Ekweozor et al. (1979), Ekweozor and Udo (1988), Riva et al. (1988)
Hexahydrobenzohopanes	Bacteria	Anoxic Carbonate-anhydrite	Connan and Dessort (1987)
Pregnane, homopregnane	Unknown	Hypersaline	ten Haven et al. (1986)
C_{24} tetracyclic terpane	Unknown	Hypersaline	Connan et al. (1986)
Squalane	Archaebacteria	Hypersaline	ten Haven et al. (1986)
Norhopane (C_{29} hopane)	Various	Carbonate/ evaporite	Clark and Philp (1989)
C_{31}-C_{40} head-to-head isoprenoids	Methanogens	Unspecified	Risatti et al. (1984)
C_{19}-C_{30} tricyclic terpanes	*Tasmanites*?	Unspecified	Aquino Neto et al. (1989), Volkman et al. (1989)
Aromatics			
Benzothiophenes, alkyldibenzothiophenes	Unknown	Carbonate/ evaporite	Hughes (1984)
Aryl isoprenoids (1-alkyl, 2,3,6,trimethylbenzenes)	Green sulfur bacteria	Hypersaline	Summons and Powell (1987), Clark and Philp (1989)
Trimethylated 2-methyl-2-trimethyldecylchromans	Unknown	Saline	Schwark and Püttman (1990)

Conclusions on correlations, source, and depositional environment should always be based on a thorough evaluation of all available geochemical information, including other biomarker, isotopic, and supporting data. A few of many examples of integrated geochemical correlations include: Shi Ji-Yang et al. (1982), Palmer (1984a), Grantham et al. (1987), Peters et al. (1989), Moldowan et al. (1992). Further discussion is in the following text.

Caution should be applied when comparing biomarker parameters derived from different laboratories. The mass chromatograms and peaks used to determine each biomarker parameter described in the following sections may differ between laboratories. For this reason, publications that provide details on measurement procedures for each parameter (e.g., Fu Jiamo et al., 1990) are very useful.

3.1.3.1 Terpanes

m/z 191 Fingerprint

Often applied, specific for correlation/depositional environment when examined in detail but affected by interfering peaks, measured using MID GCMS.

Many of the terpanes in petroleum originate from bacterial (prokaryotic) membrane lipids (Ourisson et al., 1982). These bacterial terpanes include several homologous series, including acyclic, bicyclic (drimanes), tricyclic, tetracyclic, and pentacyclic (e.g., hopanes) compounds. The following is a brief overview of more detailed discussions of these compounds found later in the text.

Tricyclics, tetracyclics, hopanes, and other compounds contribute to the terpane fingerprint (m/z 191) commonly used to relate oils and source rocks (Seifert et al., 1980). Examples of terpane fingerprints for oils from different types of source rocks are shown in Fig. 3.11 (also see other figures showing m/z 191 mass chromatograms). Terpane fingerprints reflect source rock depositional environment and organic matter input. Because bacteria are ubiquitous in sediments, terpanes are found in nearly all oils, and oils from different source rocks deposited under similar conditions may show similar terpane fingerprints. For example, although the C_{29} (norhopane) and C_{30} (hopane) 17α(H)-hopanes represent the dominant triterpanes in many oils, their relative abundance is not very useful for separating oils into genetically related families. Most petroleums show C_{30}-hopane/C_{29}-hopane m/z 191 peak ratios greater than 1. The ratio measured using m/z 191 is more sensitive to C_{30}-hopanes than C_{29}-hopanes because the C_{30}-hopanes undergo two fragmentations yielding m/z 191 fragments, while C_{29}-hopanes only undergo one. Petroleums from organic-rich evaporite-carbonate rocks generally show enhanced relative concentrations of the C_{29}- compared to the C_{30}-hopane (Zumberge, 1984; Connan et al., 1986; Clark and Philp, 1989). However, others have observed that high C_{29} hopanes characterize many petroleums from source rocks rich in terrestrial organic matter (e.g., Brooks, 1986).

The tricyclic terpanes (Connan et al., 1980; Aquino Neto et al., 1983) extend from C_{19} to at least C_{45} because of their isoprenoid side chains (Moldowan et al.,

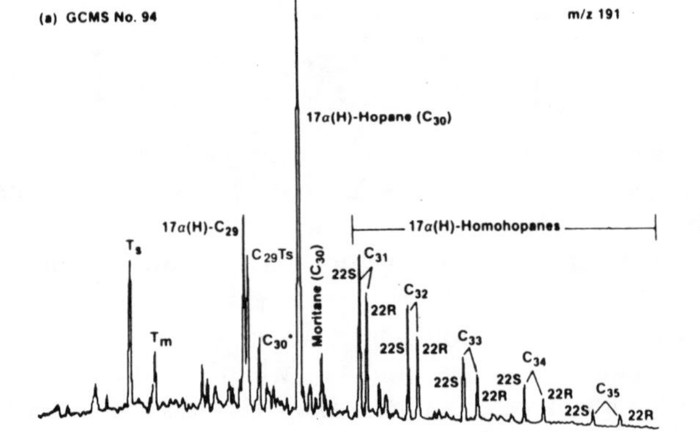

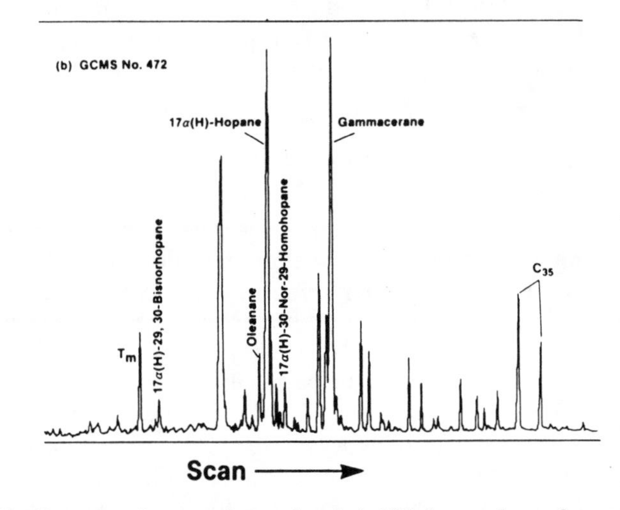

Figure 3.11 Examples of terpane fingerprints (m/z 191) for petroleums from various types of source rock depositional environments:

(a) An oil (GCMS No. 94) from the northwestern shelf of Australia shows a mass chromatogram typical of source rock deposited under marine shelf conditions. The regular stair-step progression of C_{31} to C_{35} homohopanes is consistent with suboxic bottom waters during deposition. The C_{30}^{*} peak represents a rearranged hopane [17α(H)-diahopane] probably related to hopanes which resulted from clay-mediated acid catalysis reactions on hopenes. Absence of oleanane in an oil from a shelf or deltaic source rock suggests, but does not prove, that the source is older than Cretaceous.

(b) Prinos oil (GCMS No. 472) from Greece (Seifert et al., 1984) is probably derived from a carbonate source rock deposited in an anoxic basin as indicated by strong preservation of the C_{35}-homohopanes. The 30-norhopane series, typified by 17α(H)-29,30-bisnorhopane and 17α(H)-30-nor-29-homohopane, is present as in many carbonate-derived oils. The strong gammacerane peak may signify hypersaline water during deposition of the source rock. The presence of oleanane indicates a Tertiary to Cretaceous age.

(c) Ravni Kotari-3 oil (GCMS No. 558) from Yugoslavia (Moldowan et al., 1992) is another oil from an anoxic to slightly suboxic hypersaline source rock. Its terpane fingerprint features gammacerane and a predominance of the C_{34}- homohopanes.

(d) Miocene oil from the Kenai reservoir (GCMS No. 618), Swanson River Unit at Cook Inlet, Alaska (Peters et al., 1986) shows relatively little response in the C_{27} to C_{35} terpane range. The only identified terpanes are labeled. This unusual liquid is thought to be derived from a resinitic coal source.

(e) An oil from offshore Angola (GCMS No. 887) is derived from a Lower Cretaceous lacustrine source rock. High gammacerane is typical of saline to hypersaline lake deposition. The C_{28} to C_{30} tricyclic terpanes (cheilanthanes) are also present in this m/z 191 fingerprint.

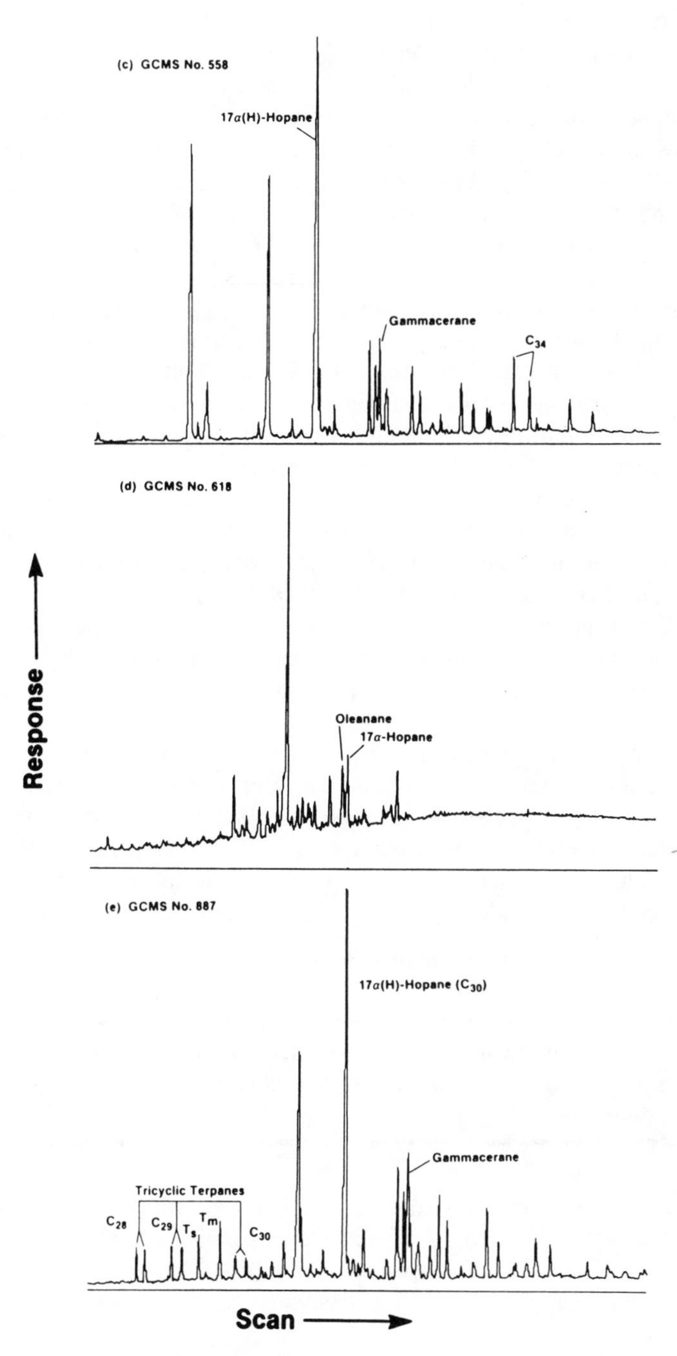

Figure 3.11 *(cont.)*

1983). The C_{28} and C_{29} tricyclics have been used extensively in correlations of oils and bitumens (Seifert et al., 1980; Seifert and Moldowan, 1981). The tricyclic terpanes ($<C_{30}$) appear to be derived from a regular C_{30} isoprenoid such as tricyclohexaprenol (Aquino Neto et al., 1983), and could be a constituent in prokaryote membranes (Ourisson et al., 1982). However, recent work shows that high concentrations of tricyclic terpanes and their aromatic analogs correlate with *Tasmanite*-rich rocks, suggesting they may be related to these primitive algae (Aquino Neto et al., 1989; Volkman et al., 1989; Azevedo et al., 1992). These compounds are unrelated to the hopanes (discussion follows) because of their differing point of attachment of the side chain. Some tricyclic terpanes (Fig. 3.24) are terrigenous indicators (Noble, 1986). Tricyclic diterpanes (C_{19} to C_{20}) are believed to be derived from diterpenoids such as abietic acid, which are produced by vascular plants (Barnes and Barnes, 1983). Walters and Cassa (1985) show that the ratio of tricyclic diterpanes to the sum of sesterterpane and triterpane tricyclics appears to be a sensitive indicator of source input for offshore Gulf Coast oils. Figure 1.10 shows an example of the tricyclic terpanes (C_{29} extended *ent*-isocopalane; Aquino Neto et al., 1983).

Bicyclic terpanes of the drimane series are ubiquitous in sediments and crude oils, and for this reason are thought to be of microbial origin (Alexander et al., 1983a). Noble and Alexander (1989) proposed a mechanism for their formation by oxidation of $\Delta 11(12)$-bacteriohopanetetrol during diagenesis.

Based on structure studies (Trendel et al., 1982), the C_{24}-C_{27} tetracyclic terpanes appear to be degraded hopanes (17,21-secohopanes, Fig. 1.10). These tetracyclic terpanes appear to be more resistant to biodegradation and maturation than the hopanes.

Hopanes are pentacyclic triterpanes commonly containing 27 to 35 carbon atoms in a naphthenic structure composed of four six-membered rings and one five-membered ring (Van Dorsselear et al., 1977). Hopanes are derived from precursors in bacterial membranes (Ourisson et al., 1979). Hopanoids in these membranes, such as bacteriohopanetetrol (Figs. 1.10 and 1.20), are believed to be synthesized by cyclization of squalene precursors (Fig. 1.26) (Rohmer, 1987).

Homohopane index or distribution

Often applied as an indicator of oxicity (redox potential) of marine sediments during diagenesis, but also affected by thermal maturity, measured on m/z 191 chromatograms.

The homohopanes (C_{31}-C_{35}) are believed to be derived from bacteriohopanetetrol and other polyfunctional C_{35} hopanoids common in prokaryotic microorganisms (Fig. 1.20) (Ourisson et al., 1979, 1984; Rohmer, 1987). Abundant C_{35} homohopane may be related to extensive bacterial activity in the depositional environment.

The relative distribution of C_{31} to C_{35} $17\alpha(H),21\beta(H)$,22S and 22R homohopanes in marine petroleum is used as an indicator of the redox potential (Eh) during and immediately after deposition of the source sediments. Although high C_{35} homohopanes are commonly associated with marine carbonates or evaporites (Boon et al., 1983; Connan et al., 1986; Fu Jiamo et al., 1986; ten Haven et al., 1988; Mello et al., 1988 a and b; Clark and Philp, 1989), we interpret this phenomenon as a general indicator of highly reducing (low Eh) marine conditions during deposition (Peters and Moldowan, 1991). The **C_{35}-homohopane index** (or simply homohopane index) is the ratio $C_{35}/(C_{31}$ to $C_{35})$ homohopanes, usually expressed as a percentage. For example, the m/z 191 chromatogram for oil standard GCMS No. 81-3 from Hamilton Dome, Wyoming, in Fig. 2.12 (top-center, peak numbers 34 and 35) shows elevated C_{35} homohopanes (22S + 22R doublet) compared to the C_{34} homologs. Figure 3.12 (left) shows several examples of homohopane distributions for petroleums from the central Adriatic Basin, Italy, and Yugoslavia.

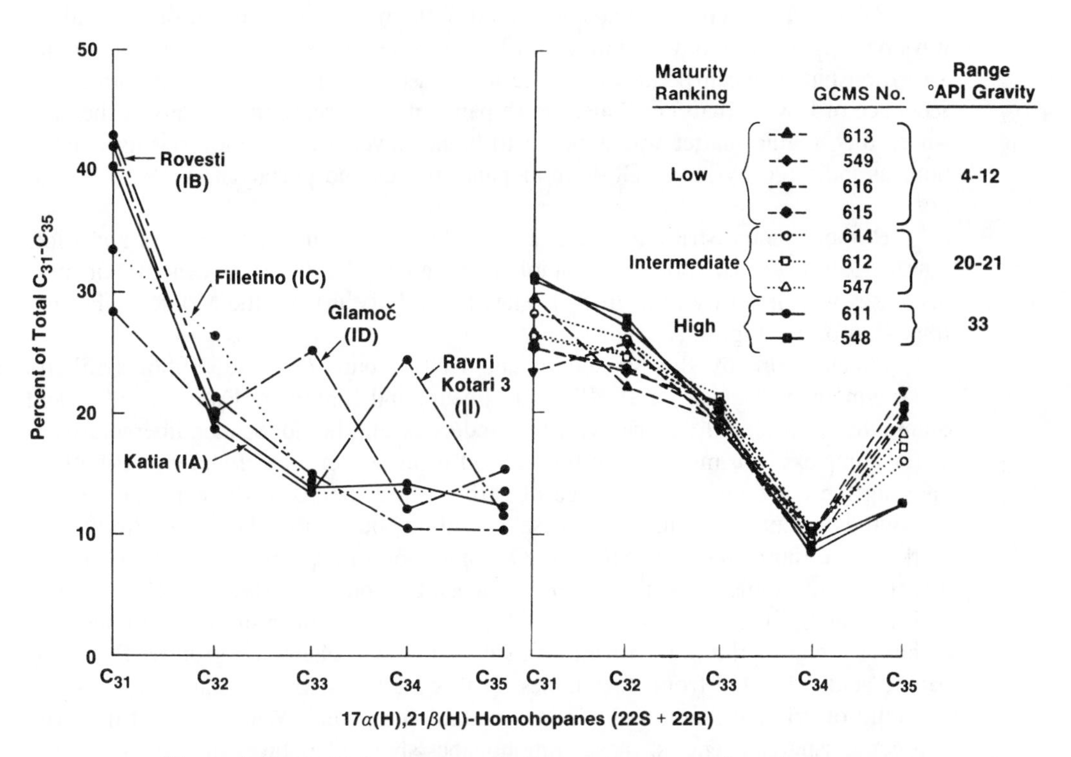

Figure 3.12 (Left) Homohopane distributions for several oils and seeps from the central Adriatic Basin (Italy and Yugoslavia) show variations that were used to divide the samples into genetically different groups (Moldowan et al., 1992). (Right) A series of related Monterey oils from offshore California show similar enrichment in the C_{35} homohopane, typical of organic matter derived from anoxic depositional settings (Peters and Moldowan, 1991). The example shows that a wide range of thermal maturity can affect homohopane distributions. The C_{35} homohopane index $[\%C_{35}/(C_{31}$ to $C_{35})]$ decreases, while $\%C_{31}$ homohopane increases with increasing maturity. Section 3.3.3 discusses another example of a unique, heavily biodegraded oil where the homohopanes show a "reversed" distribution in the range C_{31} to C_{35}.

Oils and bitumens of similar maturity showing high concentrations of C_{33}, C_{34}, or C_{35}-homohopanes compared to lower homologs are believed to indicate a highly reducing (low Eh) marine environment of deposition with no available free oxygen. When free oxygen is available, the precursor bacteriohopanetetrol is oxidized to a C_{32} acid, followed either by loss of the carboxyl group to C_{31}, or, if all the oxygen is used, preservation of the C_{32} homolog. The latter environment is called suboxic or dysoxic, while the former may be oxic or suboxic depending on the amount of oxygen and its accessibility to the organic matter (Demaison et al., 1983). This situation may be complicated by the bacteriohopane precursors having different (as yet unreported) side chain hydroxyl substitution or side chain lengths. Thus, C_{33} and C_{34}-homohopane predominances might reflect different types of bacterial input. Alternatively, preservation of intermediate homologs could indicate mildly suboxic exposure at the time of deposition followed by partial oxidation of the bacteriohopanetetrol side chain.

Paleoenvironmental conclusions based on homohopane distributions should always be supported by other parameters. For example, Moldowan et al. (1986) found a relationship between pristane/phytane and diasteranes/regular steranes ratios in a sequence of Lower Toarcian shales. Both parameters increase in sections of the core where the organic matter was exposed to higher levels of oxidation during deposition, as indicated by lower C_{35}-homohopane indices and porphyrin V/(V + Ni) ratios.

Homohopane distributions are affected by thermal maturity (Peters and Moldowan, 1991). For example, the homohopane index [$C_{35}/(C_{31}$ to $C_{35})$ homohopanes] decreases with maturity in a suite of related oils derived from the Monterey Formation, California (Fig. 3.12, right).

Recent work by the Strasbourg and Delft groups (Schmid, 1986; Trifilieff, 1987; Sinninghe Damsté et al, 1988; de Leeuw and Sinninghe Damsté, 1990; and Sinninghe Damsté and de Leeuw, 1990) indicates that homohopanes liberated from sulfur complexes in immature sulfur-rich oil using Raney nickel show distributions favoring the C_{35} homolog. Cleavage of carbon-sulfur bonds in the early oil-generative window can result in the release of sulfur-bound homohopanes. Additional work on the rhuthenium tetroxide (RuO_4) oxidation of asphaltenes from sulfur-rich oils shows that some homohopanes are attached to aromatic systems by C—C bonds (Trifilieff et al., 1992). Oxidation by RuO_4 disrupts aromatic systems by producing a carboxyl group at the point of attachment of the side chains or groups. Homohopanoic acids liberated from asphaltenes by this method carry the carboxyl group at the point of original functionality of the natural compound. When cracked from the kerogen during catagenesis, these homohopanes show a "mature" distribution favoring the C_{31} homolog. Thus, catagenesis of a source rock containing bitumen with a C_{35} homohopane predominance (typical of sulfur-bound homohopane species liberated during early maturation) may lead to cracking of kerogen-bound homohopanes and a relative increase in the concentration of C_{31} homologs later in the oil window.

Diastereomeric pairs (22S and 22R) of side-chain-extended $\alpha\beta$-homohopanes with carbon numbers ranging up to C_{40} have been observed (Rullkötter and Philp,

1981). The C_{36} to C_{40} homologs are in low concentrations compared to their C_{31} to C_{35} counterparts. They are believed to originate during thermal cracking of bacteriohopane precursors that are bound to the kerogen by covalent carbon-carbon bonds.

Pristane/phytane ratio

Often applied, low specificity for redox conditions in the source rock as a result of interference by thermal maturity and source input, measured using gas chromatogram or reconstructed ion chromatogram.

Pristane/phytane (Pr/Ph) ratios of oils or bitumens have been used to indicate the redox potential of the source sediments (Didyk et al., 1978) (Fig. 3.13). According to these authors, Pr/Ph ratios less than unity indicate anoxic deposition, particularly when accompanied by high porphyrin and sulfur contents. Oxic conditions are indicated by Pr/Ph > 1. Pr/Ph ratios are commonly applied, partly because Pr and Ph are easily measured using simple gas chromatography.

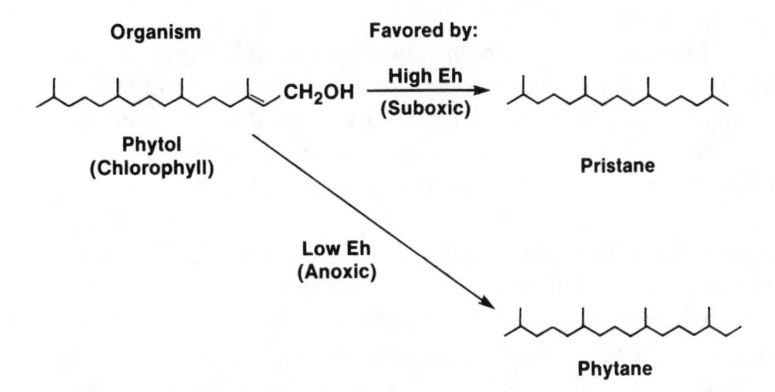

Figure 3.13 Diagenetic origin of pristane and phytane from phytol (derived from side chain of chlorophyll-a). See also Fig. 1.27. Other sources of acyclic isoprenoids having 20 carbon atoms or less include chlorophyll-b, bacteriochlorophyll-a, α-and β-tocopherols, carotenoid pigments, and archaebacterial membrane components. The pristane/phytane ratio of a petroleum provides information on the redox potential of the depositional environment for the source rock, but must be used with caution.

Pristane (C_{19}), phytane (C_{20}), and smaller isoprenoids are primarily derived from the phytyl side chain of chlorophyll in phototrophic organisms (Sec. 1.3.3), although other sources, such as archaebacteria, are described by Chappe et al. (1982), Illich (1983), Goosens et al. (1984) and Rowland (1990). Under anoxic conditions in sediments, the phytyl side chain is cleaved to yield phytol, which is reduced to dihydrophytol and then phytane. Alternately, dihydrophytol has been shown to be a component in archaebacterial cell membranes and a building block for kerogen

(Chappe et al., 1982). Under oxic conditions, phytol is oxidized to phytenic acid, decarboxylated to pristene and then reduced to pristane (Fig. 3.13).

In samples of low thermal maturity, Pr/Ph ratios are not recommended to describe paleoenvironment (Volkman and Maxwell, 1986). **For samples within the oil-generative window, high Pr/Ph ratios (>3.0) indicate terrestrial organic matter input under oxic conditions and low values (<0.6) typify anoxic, commonly hypersaline environments. For samples showing Pr/Ph in the range 0.8 to 2.5, we do not recommend that Pr/Ph be used as an indicator of paleoenvironment without corroborating data.** For example, in a study of oils generated from lacustrine source rocks in Angola, a relationship was found between Pr/Ph and the gammacerane index (Fig. 3.14). Considered in isolation, neither of these parameters could be used with confidence to assess the salinity of the depositional environment. However, the supporting relationship between Pr/Ph and gammacerane index reinforces the inferred salinity relationship. Similarly, Schwark and Püttmann (1990) note a general decrease in Pr/Ph during deposition of the Permian age Kuperschiefer sequence, Germany, which they interpret to indicate increasing paleosalinity. This interpretation is supported by a parallel increase in trimethylated 2-methyl-2-(trimethyltridecyl)chromans, aromatic compounds believed to be markers of salinity.

Inferences from Pr/Ph ratios on the redox potential of the source sediments should always be supported by other geochemical data and by geologic information. Typically, conditions of source rock deposition inferred from Pr/Ph ratios of oils agree with other indicators, such as sulfur content or the C_{35}-homohopane index. For example, one oil from Oman (GCMS No. A28) shows a low Pr/Ph (<1), high sulfur (1.6 wt %), and high C_{35} homohopane index (8.1), typical of anoxic depositional conditions in the source rock. Many high-sulfur Gulf Coast oils show low Pr/Ph ratios and vice versa (Walters and Cassa, 1985).

Unfortunately, the Pr/Ph ratio is a crude parameter and its use to describe redox of the environment of deposition has been criticized. Ten Haven et al. (1987) indicate that it is impossible to draw valid conclusions on the oxicity of the environment of deposition from Pr/Ph alone. They show that low Pr/Ph (<1) ratios are typical of hypersaline environments where the ratio appears controlled by the effects of variable salinity on halophilic bacteria. Some problems related to Pr/Ph ratios are listed.

1. In addition to chlorophyll, archaebacterial lipids (from methanogens or halophiles) appear to be a source for Pr and Ph and tocopherols are a source for Pr (Goosens et al., 1984). Thus, differences in populations of organisms contributing to the sediment can affect Pr/Ph ratios.
2. A branched, irregular isoprenoid in petroleum [2,6,10-trimethyl-7(3-methylbutyl)-dodecane] (Sec. 1.2.7, Table 3.1.3(1)] coelutes with Pr on most capillary columns (Volkman and Maxwell, 1986). Mass chromatograms of m/z 168 can be used to show this compound, even in the presence of large amounts of pristane.

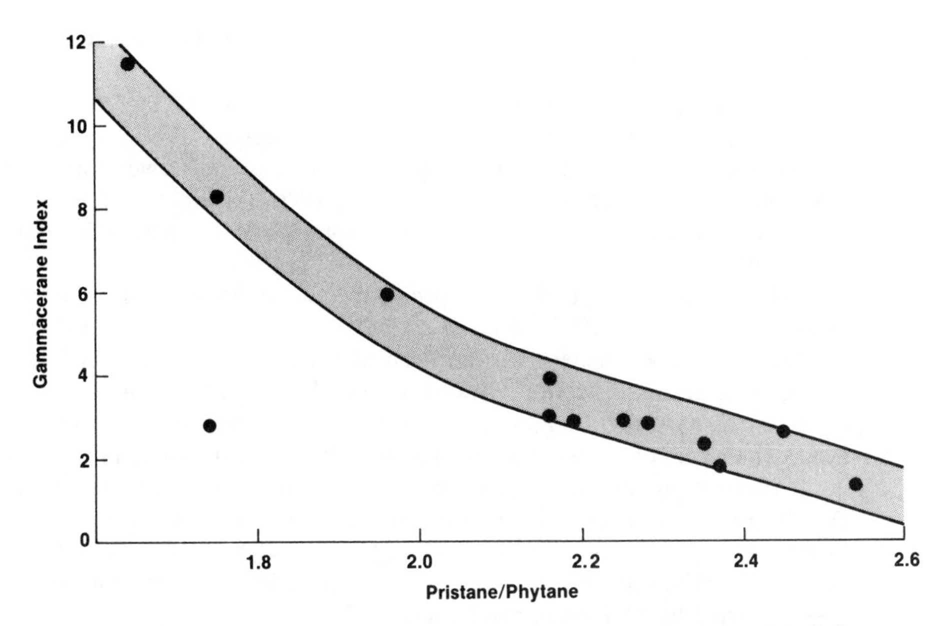

Figure 3.14 Variations in Pr/Ph (redox) and gammacerane index (salinity?) for oils derived from lacustrine sources rocks in Angola. Increased water salinity in the source rock depositional environment results in higher gammacerane indices. Higher salinity is typically accompanied by density stratification and reduced oxygen content in bottom waters (i.e., lower Eh), which results in lower Pr/Ph ratios. *Figure courtesy of B. J. Huizinga.*

3. Thermal maturity changes the Pr/Ph ratio. Typically the Pr/Ph ratio of petroleum increases with maturity (Connan, 1974). Ten Haven et al. (1987) indicate an increase in Pr/Ph and a decrease in Ph/nC_{18} with thermal maturation. For a relatively uniform sequence of argillites from the Douala Basin, Cameroon, Albrecht et al. (1976) found that the Pr/Ph ratio first increases with depth to 4.9 in the principal zone of oil formation, but then decreases to 1.5 at higher levels of maturity. Similarly, Radke et al. (1980), Brooks et al. (1969), and Connan (1984) found an increase in Pr/Ph to a maximum (corresponding to vitrinite reflectance values of 0.88, 1.0, and 0.7 percent, respectively) for different coals, beyond which the ratio decreased. Unlike the above, Burnham et al. (1982) show decreases in Pr/Ph for oils of increasing maturity generated by pyrolysis of Green River oil shale. This may be caused by preferential release of pristane- compared to phytane-precursors from the kerogen during very early catagenesis.

Acyclic isoprenoids (>C_{20})

Rarely used, highly specific for certain archaebacteria and as a correlation tool, measured using m/z 183, m/z 253, or a variety of other fragments (Fig. 2.11).

The higher molecular-weight acyclic isoprenoids include the regular isoprenoids (head-to-tail linked) ranging up to C_{45} (Albaiges, 1980), the head-to-head isoprenoids with a parent C_{40} compound (bisphytane) and lower pseudohomologs (Moldowan and Seifert, 1979; Chappe et al., 1980; Albaiges, 1980), and various single compounds like squalane (C_{30}, tail-to-tail) and lycopane (C_{40}, tail-to-tail) (Fig. 1.10).

Head-to-head isoprenoids are specific markers of archaebacterial input to sediments. Petrov et al. (1990) found a series of C_{28} to C_{39} head-to-head linked isoprenoids, which are believed to be derived from archaebacterial cell walls, in a lightly biodegraded oil from northwest Siberia. Most of these compounds consist of a pristane unit coupled to a second isoprenoid unit in the range C_{10} to C_{20}. The C_{12} and C_{17} members of the second unit are absent because this would require the breaking of two carbon-carbon bonds at a branch position during maturation. Petrov et al. (1990) also found a range of components consisting of a phytane unit coupled to lower isoprenoids (i.e., C_{20}-C_{19}, C_{20}-C_{18}, C_{20}-C_{16}, C_{20}-C_{13}, C_{20}-C_{11}). These results extend the carbon number range and structural variety of these compounds previously reported by Moldowan and Seifert (1979).

The acyclic isoprenoids are often ignored in correlation studies because their analysis by GCMS requires removal of the *n*-paraffins which have the same fragment ions and often dominate in intensity. Further, special gas chromatographic conditions may be required to separate the parent compounds (C_{40} homologs) of each series. This may be one reason why lycopane (tail-to-tail, C_{40}; Figs. 1.10 and 2.11) has only rarely been identified in oils (Albaiges et al., 1985). Another reason could be that lycopane is derived from an unsaturated to polyunsaturated C_{40}-isoprenoid. These alkenes appear to be preserved as sulfur-bound species under anoxic marine conditions (de Leeuw and Sinninghe Damsté, 1990). For example, large concentrations of lycopane were found in hydrocarbons liberated from the sulfur-rich "red band" fraction isolated from immature, high-sulfur oils (Adam et al., 1992).

Several C_{40} isoprenoids, which appear to serve as membrane constituents in archaebacteria, contain five-membered rings within the chain (Fig. 1.10) and have been identified in Spanish Amposta and Tarraco crude oils (Albaiges et al., 1985) and in Indonesian Mahakam oil (Chappe et al., 1982). Thermoacidophilic archaebacteria produce related compounds containing three or four rings within the chain (DeRosa et al., 1977a and b) which have not yet been found in petroleum, but the cyclized lipids may be paleoclimatic markers in sediments (P. Albrecht, pers. comm.).

Because lacustrine-sourced oils are typically very low in steranes, which are common source correlation parameters, the acyclic isoprenoids are very important for correlation of these types of oils. For example, in a study of Sumatran oils, distributions and relative concentrations of C_{33}-C_{40} head-to-head isoprenoids compared with those of C_{19}-C_{31} regular isoprenoids and botryococcanes were useful in grouping lacustrine-derived oils (Fig. 3.15) (Seifert and Moldowan, 1981).

Certain acyclic isoprenoids may be used as biological markers for methanogenic bacteria (e.g., C_{31}-C_{40} head-to-head isoprenoids; Risatti et al., 1984; Row-

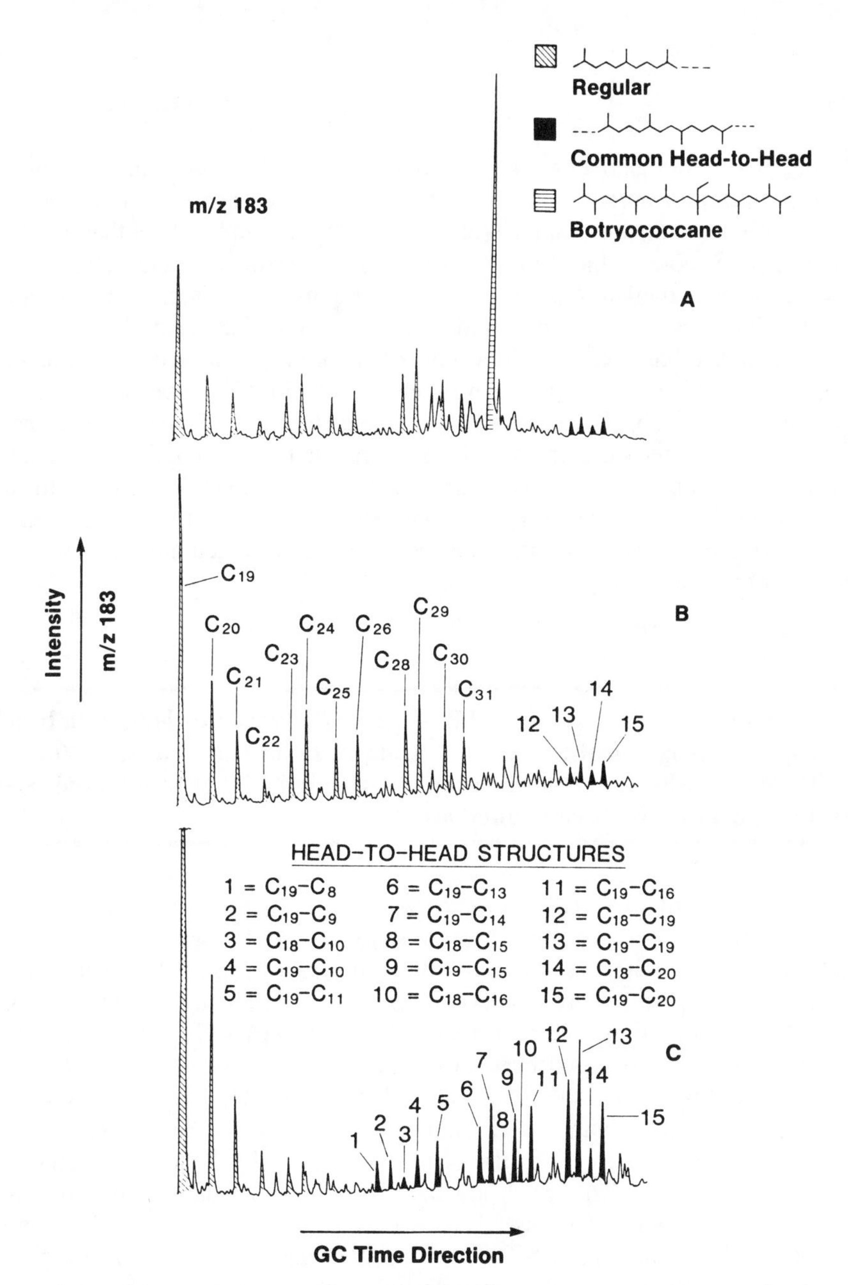

Figure 3.15 Variations in acyclic isoprenoids revealed by (MID) GCMS m/z 183 of saturate fractions proved useful in grouping Sumatran oils (Seifert and Moldowan, 1981). Like many other oils generated from lacustrine source rocks, these oils contain few steranes, which are commonly used for correlation. The isoprenoids were concentrated by urea adduction of the saturate fractions of the oils, thus removing the otherwise abundant *n*-paraffins which interfere with the analysis (Sec. 2.5.5).

(A) Minas oil (GCMS No. 305) shows botryococcane and low amounts of head-to-head isoprenoids compared to the other samples.

(B) Petapahan oil (GCMS No. 306) shows a regular and head-to-head isoprenoid distribution similar to Minas (A) but lacks botryococcane.

(C) Damar oil (GCMS No. 307) shows higher concentrations and more diverse head-to-head isoprenoids than either Minas (A) or Petapahan (B).

land, 1990). Methanogens are only found under highly reducing conditions. They use the anaerobic process of fermentation to derive energy from various oxidized forms of organic matter. Brassell et al. (1981) for example, show that 2,6,10,15,19-pentamethyleicosane and squalane (Fig. 1.10) in sediments and petroleum indicate past biological methanogenesis. Squalane (Figure 1.10) is common in petroleum, but has not been used for correlations (Gardner and Whitehead, 1972).

Squalane has been used as a biomarker for archaebacteria and hypersaline depositional environments (ten Haven et al., 1988). The precursor for squalane, squalene (Fig. 1.26), is ubiquitous in sediments because it is the starting material for all polycyclic triterpenes and sterols. Squalene and its saturated analog squalane are acyclic and symmetrical, containing two farnesyl residues linked tail-to-tail (Fig. 1.10). Squalene has been isolated from asphaltenes in petroleum (Samman et al., 1981). High concentrations of squalenes have been detected in living halophilic bacteria (Tornabene, 1978).

Botryococcane

Always applied when found, highly specific (*Botryococcus braunii* in brackish or lacustrine environment), measured using m/z 183 chromatogram (or m/z 238, 239, 294, 295; Fig. 2.11) of the saturate urea nonadduct (isoprenoids separated from *n*-paraffins by urea adduction).

This irregular C_{34}-isoprenoid appears to be derived from botryococcene known in only one living organism, *Botryococcus braunii*, a fresh/brackish water alga (Moldowan and Seifert, 1980; McKirdy et al., 1986) (Fig. 3.16). Thus, studies utilizing this compound represent some of the most specific applications of biomarkers to paleoreconstruction. The occurrence of botryococcane (Fig. 3.15) and closely related 16-demethylbotryococcane in petroleum is limited to a few geographic regions, including Sumatra (Seifert and Moldowan, 1981), Australia (McKirdy et al., 1986), and the Maoming Shale, China (Brassell et al., 1986). Some extracts of *B. braunii* consist of a mixture of C_{30} to C_{37} branched alkenes and botryococcenes (Metzger et al., 1985a), which are the probable precursors for the more complex mixtures of botryococcane pseudohomologs in a Sumatran oil (Seifert and Moldowan, 1981). *B. braunii* is believed to be the major organism contributing to boghead coals, such as torbanite, some Tertiary age oil shales, and rubbery deposits of coorongite from Australia (Maxwell et al. 1968).

Two modern races of *B. braunii* (PRBA and PRBB) showing different hydrocarbon compositions have been identified. Only race B produces botryococcene (Metzger et al., 1958b). Hence, the occurrence of botryococcane proves the presence of *B. braunii* in the depositional environment, but lack of botryococcane does not prove that *B. braunii* was absent (Derenne et al., 1988). During its scenescent stage, *B. braunii* contains from 70 to 90 wt % of botryococcene and related hydrocarbons (Maxwell et al., 1968; Knights et al., 1970).

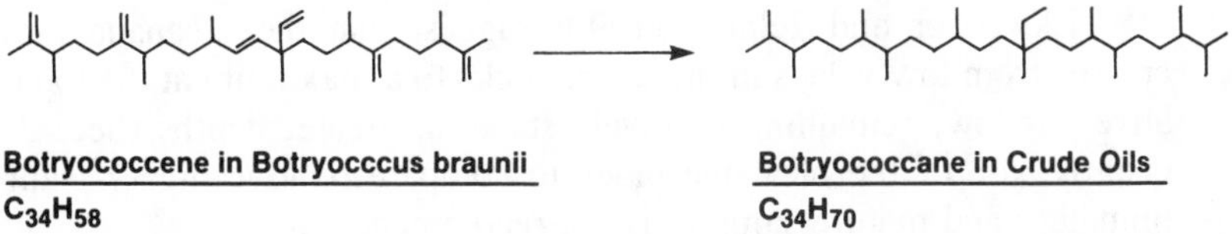

Figure 3.16 Botryococcane is a marker for a specific lacustrine alga, *Botryococcus braunii*. Botryococcane is identified on the m/z 183 mass chromatograms of oils from Sumatra in Fig. 3.15.

Oleanane/C_{30} hopane (oleanane index)

Often applied, highly specific for higher plant input of Cretaceous or younger age, measured using m/z 191 chromatograms (Fig. 2.11).

18α(H)-oleanane is thought to be a Cretaceous or younger higher plant marker. Oleananes are probably derived from betulins (Grantham et al., 1983) and other pentacyclic triterpenes in angiosperms (Whitehead, 1973, 1974; ten Haven and Rullkötter, 1988). Angiosperms first became prominent in the Late Cretaceous (<100 m.y.). No examples of oleananes have been found in oils known to be older than Cretaceous. (Note however, that the absence of oleanane does not prove an oil to have been generated from Cretaceous or older rocks.)

Crude oils from the Tertiary Niger delta contain large amounts of oleananes (Ekweozor et al., 1979) and there is a correlation between the abundance of higher plant macerals (e.g., vitrinite and resinite) and the oleanane index (Udo and Ekweozor, 1990). Marine shales of the Tertiary Akata Formation, which received substantial input from terrigenous higher plants, are considered to be the main source of the Niger delta oils. The widespread occurrence of oleananes is now recognized (Grantham et al., 1983; Hoffman et al., 1984; Riva et al., 1988; Zumberge, 1987a; Czochanska et al., 1988; Ekweozor and Udo, 1988; Zeng et al., 1988; Fu Jiamo and Sheng Guoying, 1989).

Use of the oleanane index (oleanane/C_{30} hopane) when comparing samples showing major differences in maturity can be complicated by the presence of 18β(H)-oleanane in immature bitumens. Although both 18α(H)- and 18β(H)-oleanane are found in immature bitumens and oils, the latter is thermally less stable (Fig. 3.51) (Riva et al., 1988). Equilibrium between the 18β(H)-isomer probably occurs before peak oil generation. Thus, the sum of 18α(H)- and 18β(H)-isomers should be used in the oleanane/C_{30} hopane ratio for purposes of correlation. From the relative retention time and mass spectrum, we believe compound "J" identified in earlier literature (Hills and Whitehead, 1966; Grantham et al., 1983) is 18β(H)-oleanane. Figure 2.19 shows the provisional identification of 18α(H)- and 18β(H)-oleanane in a bitumen by coinjection of authentic standards.

Ekweozor and Telnaes (1990) suggest that the oleanane ratio generally increases from low values in immature rocks to a maximum at the top of the oil-generative window, remaining relatively stable at greater depths (Sec. 3.2.4.1). Consequently, use of the oleanane index to compare relative higher plant input between immature and mature samples is not recommended.

We recommend that interpretations of terrestrial versus marine input using oleanane in mature samples be complemented using other parameters. For example, plots of oleanane versus C_{35}-homohopane index or C_{30}-sterane index (marine input) are often useful (Fig. 3.32).

Oleanane is difficult to identify using conventional GCMS because of possible interference by other triterpanes with similar retention times and mass spectra. For example, the mass spectra of 17α(H)-hopane and oleanane are similar. In most published work, oleanane is identified by GMCS in MID mode using m/z 191 and m/z 412 mass chromatograms and by co-elution experiments with 18α(H)- and 18β(H)-oleanane standards. Unfortunately, both 17α(H)-hopane and oleanane undergo cleavage of the C-ring during ionization, yielding both A/B-ring and D/E-ring fragments that carry the main ion current of the spectrum at m/z 191. The molecular ion, m/z 412, and the M − CH_3 (molecular ion minus methyl group) fragment, m/z 397, are only slightly larger for the oleananes than for 17α(H)-hopane. These characteristics are also typical of other known sterioisomeric C_{30}-hopanes. The most important difference between the hopanes and oleananes is a small m/z 369 hopane fragment, which is absent for oleanane.

We identify oleanane using a combination of traits determined by MRM-GCMS.

- Oleanane occurs as a recognizable doublet at the proper retention times, where the earlier-eluting 18α(H)-peak is larger than the 18β(H)-peak in mature oils,
- Metastable transitions m/z 412 → 397, 412 → 412, and 414 → 191 decrease slightly in that order relative to those of 17α(H)-hopane,
- Metastable transition m/z 412 → 369 shows no significant peak [a commonly occurring, but unidentified interfering compound shows a singlet peak at m/z 412 → 369 (Fig. 3.17)].

28,30-Bisnorhopanes (BNH) and 25,28,30-trisnorhopanes (TNH), BNH/(BNH + TNH)

Rarely applied, highly specific as a correlation tool, probable bacterial markers associated with some anoxic depositional environments, measured on m/z 191, 177, and 163 fragmentograms.

BNH and TNH (Fig. 1.10) are desmethylhopanes which occur as $17\alpha,18\alpha,21\beta$(H)-, $17\beta,18\alpha,21\alpha$(H)-, and $17\beta,18\alpha,21\beta$(H)-epimers. High concentrations of BNH and TNH are typical of petroleum from highly reducing to

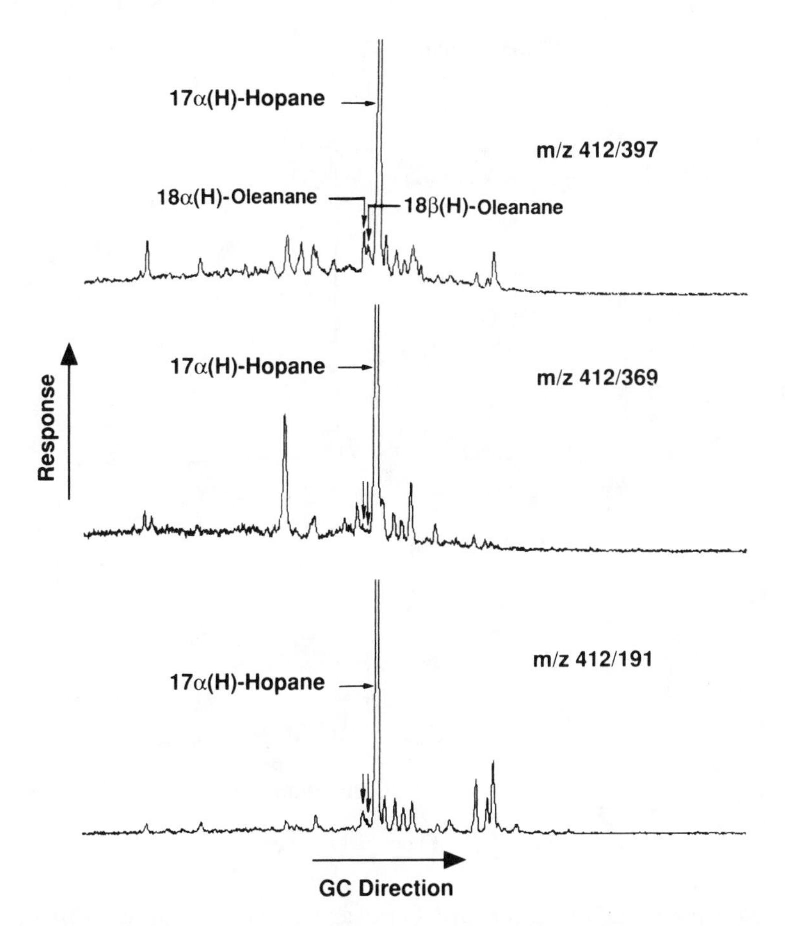

Figure 3.17 Oleananes are (A) present in the Upper Cretaceous (Campanian) source rock from South America and (B) not detected in an Upper Cretaceous (Santonian) source rock from Wyoming. The 18α(H)- and 18β(H)-oleanane doublet (indicated by arrows) in A was confirmed by coinjection of authentic standards. Note that the oleanane doublet is present in m/z 412 $\rightarrow$ 191 and 412 $\rightarrow$ 397, but absent in m/z 412 $\rightarrow$ 369 in A. The chromatograms in B show a single peak with retention time between 18α(H)- and 8β(H)-oleanane (indicated by two arrows). It is present in all three chromatograms. Chromatographic displays are cut at 50 percent of peak height of highest peak, 17α(H)-hopane, in order to visualize small amounts of oleananes.

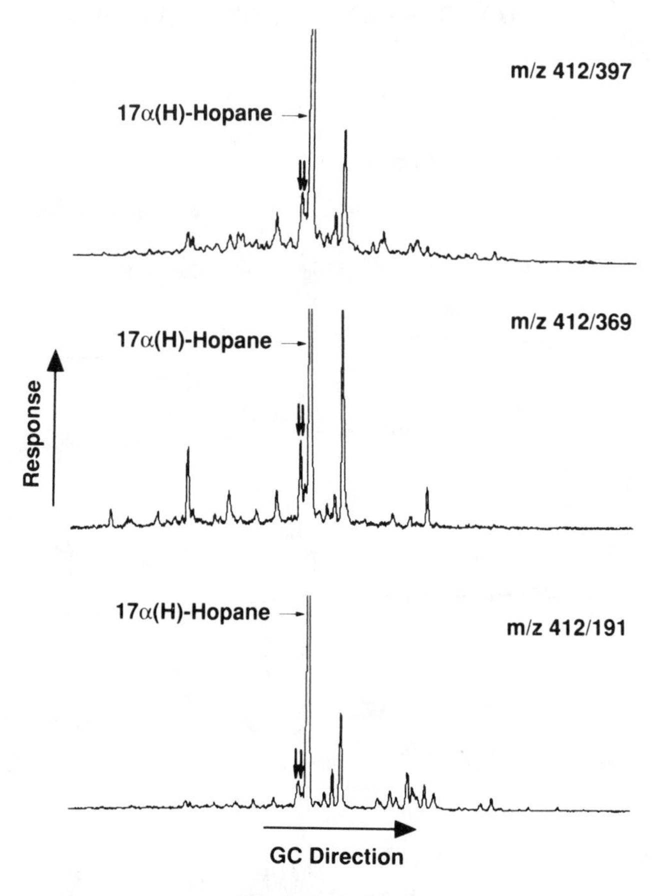

Figure 3.17 (*cont.*)

anoxic depositional environments. For example, Mello et al. (1990) note high BNH and TNH (up to 130 ppm), a predominance of phytane over pristane, and high concentrations of porphyrins (up to 5700 ppm) associated with bitumens from Late Cretaceous age rocks deposited under anoxic conditions on the Brazilian continental margin.

The BHN/TNH parameter is based on all epimers and was reported by Moldowan et al. (1984) for six different California oil-source rock groups. The BNH/TNH ratio was useful in separating some groups and was unaffected by thermal maturity. BNH and TNH do not appear to be generated from kerogen, but are passed on from the original free bitumen in the source rock to the oil (Moldowan et al., 1984; Noble et al., 1985c; Tannenbaum et al., 1986). Therefore, the concentrations of BNH and TNH drop as source rocks generate oil during maturation (e.g., note the relative decrease in BNH in Fig. 3.45 between Kimmeridge bitumen and related, but more mature Piper oil). Because BNH and TNH are destroyed at about the same rate, the BNH/TNH ratio remains approximately constant during maturation, until one or both are depleted.

Heavy biodegradation may invalidate the BNH/TNH ratio because, like the degradation of 17α(H)-hopanes to 17α(H)-25-norhopanes, BNH may be converted to TNH by microbes.

When abundant in petroleum, BNH and TNH indicate deposition of the source rock under anoxic conditions (Katz and Elrod, 1983; Mello et al., 1988a and b; Curiale and Odermatt, 1989). However, absence of these compounds does not exclude anoxia.

Abundant BNH in the Miocene Monterey Formation is hypothesized to be derived from bacterial mats (Williams, 1984). Abundant fossilized bacterial mats in the Monterey Formation led Katz and Elrod (1983) to suggest the sulfur-oxidizing bacterium *Thioploca* as a possible source for the diagenetic precursor of BNH in many Monterey oils and bitumen extracts (Philp, 1985; Curiale and Odermatt, 1989). However, the lack of hopanoids in lipids extracted from *Thioploca* (McCaffrey et al., 1989) indicates that this organism is probably not the source for bisnorhopane.

Gammacerane index

Often applied, highly specific for hypersaline conditions (?), measured on m/z 412 or 191 fragmentograms (Fig. 2.11).

Gammacerane is a C_{30}-triterpane (Fig. 1.10) that appears to represent a marker for highly saline marine and nonmarine depositional environments. Our results for oils from lacustrine source rocks in Angola (Fig. 3.14) indicate that increased water salinity during deposition of the source rock results in higher gammacerane indices (see the following text), and lower pristane/phytane ratios. Similar results were obtained for petroleum from Tertiary rocks, offshore China (Mann et al., 1987). In addition to β-carotane and related carotenoids (see the following text), gammacerane is a major biomarker in many lacustrine oils and bitumens, including the Green River marl in Utah and oils from northwest China (Hills et al., 1966; Moldowan et al., 1985; Jiang and Fowler, 1986; Fu Jiamo et al., 1986, 1988; Brassell et al., 1988). The Green River marl was deposited in a widespread arid or semiarid lacustrine setting where the organic matter was derived primarily from algae and bacteria. Gammacerane is also abundant in certain marine petroleums from carbonate or evaporite source rocks (Rohrback, 1983; Moldowan et al., 1985; Mello et al., 1988a and b; Moldowan et al., 1992).

Gammacerane may be derived by reduction of tetrahymanol (Venkatesan, 1989; ten Haven et al., 1989), a lipid thought to replace steroids in the membranes of certain protozoa (Caspi et al., 1968; Nes and McKean, 1977; Ourisson et al.,

1987), phototrophic bacteria (Kleemann et al., 1990), and possibly other organisms. Although present in at least trace amounts in most crude oils, large amounts of gammacerane indicate highly reducing, hypersaline conditions during deposition of the contributing organic matter (Moldowan et al., 1985; Fu Jiamo et al., 1986). These conditions may favor the organisms which produce tetrahymanol. **Although oils and bitumens with high gammacerane ratios (e.g., gammacerane/$\alpha\beta$-hopane) can often be traced to hypersaline depositional environments, these environments do not always result in high gammacerane ratios** (Moldowan et al., 1985). Late Proterozoic (~850 m.y.) rocks of the Chuar Group in the Grand Canyon, Arizona, represent the oldest known occurrence of high concentrations of gammacerane (Summons et al., 1988a). This compound is more resistant to biodegradation than the hopanes (Zhang Dajiang et al., 1988).

Gammacerane is also useful in distinguishing families of petroleum. For example, Poole and Claypool (1984) used gammacerane to distinguish oils and bitumens from different source rocks in the Great Basin. Bitumens from the oil shale unit of the lacustrine Elko Formation were related to Elko-derived oils and distinguished from other samples based on high gammacerane and supporting geochemical data. Palmer (1984a) distinguished organic facies *within* the Elko Formation using gammacerane and other data such as the distribution of C_{27} to C_{29} steranes (e.g., Fig. 3.31). Gammacerane was present in the oil shale, but absent in the lignitic siltstone facies.

Care must be taken to accurately quantify gammacerane using the m/z 191 mass chromatogram. Because of its high degree of symmetry (Fig. 1.15), two identical m/z 191 fragments are generated in the mass spectrometer from gammacerane (Fig. 2.11). Thus, a sizable peak on the m/z 191 mass chromatogram represents a low concentration of gammacerane compared to other terpanes, requiring a sensitivity correction (Seifert and Moldowan, 1979). Further, under conditions of poor column performance (Sec. 2.5.5.1), gammacerane can nearly coelute with the 22R epimer of C_{31}-homohopane (M+ = m/z 426). The most reliable quantitation of gammacerane is made using the m/z 412 (molecular ion) mass chromatogram because it reduces interference from other terpanes with the gammacerane peak on the m/z 191 chromatogram (Seifert and Moldowan, 1986).

17α(H)-diahopane/18α(H)-30-norneohopane ($C_{30}^{*}/C_{29}Ts$)

Rarely applied, probable relationship between C_{30}^{*} and oxic-suboxic/clay-rich depositional environments, measured on m/z 191 chromatograms and by parents of m/z 191 using GCMSMS.

We have isolated two new rearranged hopanes from a Prudhoe Bay oil: 17α(H)-diahopane (C_{30}^{*}) was characterized by X-ray crystallography and 18α(H)-30-norneohopane ($C_{29}Ts$) by advanced NMR methods (Moldowan et al., 1991c). The peak for C_{30}^{*} is commonly detected in the m/z 191 chromatogram of the saturate fraction of petroleums (Fig. 3.18). For example, Volkman et al. (1983a) found what

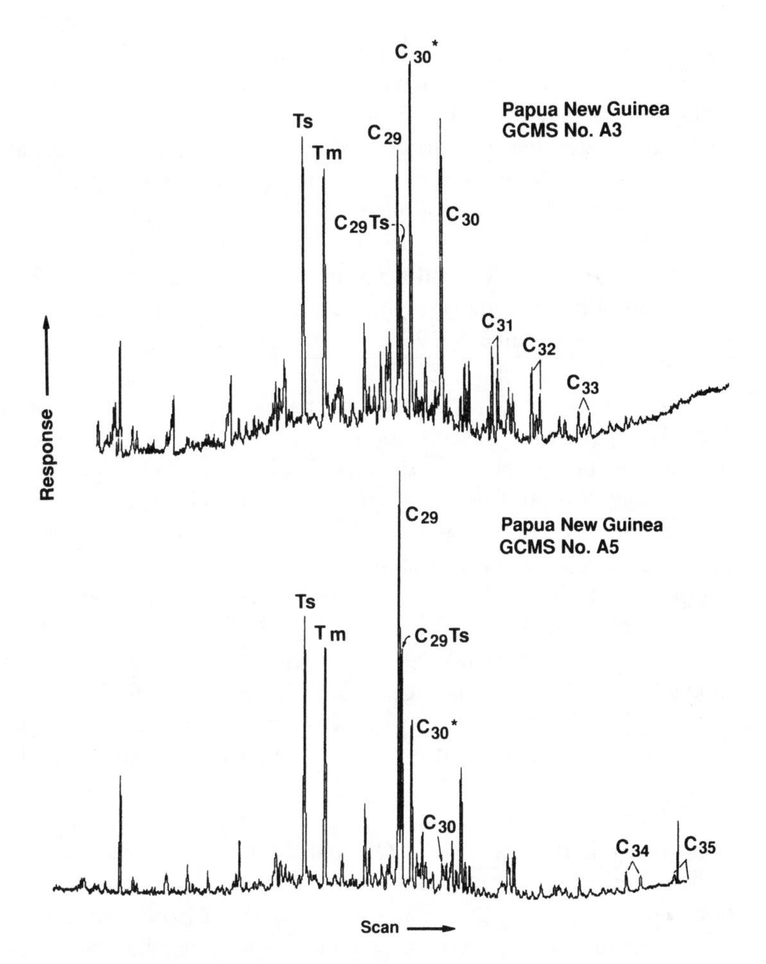

Figure 3.18 Two groups of oils and seep oils from Papua New Guinea represented by GCMS No. A3 and GCMS No. A5, respectively, can be distinguished by the ratio of the two rearranged hopanes, 18α(H)-30-norneohopane and 17α(H)-diahopane (C_{29} Ts and C_{30}^*, respectively). Sample No. A5 is a seep oil which showed C_{30}-C_{33} 17α(H)-hopane removal by biodegradation, but C_{29} Ts and C_{30}^* were resistant, increasing their utility for correlation of biodegraded oils.

appears to be C_{30}^* in oils and rock extracts from the Barrow Subbasin of Western Australia, and Philp and Gilbert (1986) used an unknown terpane they called X-C_{30}, which appears to be the same compound, for correlations of oils from many basins. Both papers regard compound C_{30}^* as a possible terrestrial marker because of its presence in coals and terrestrially sourced oils.

However, the structure of 17α(H)-diahopane suggests that it is formed by rearrangement of a hopanoid carrying functionality on the D-ring (Corbett and Smith, 1969). We also find that this rearranged hopane has nearly the same stable carbon isotope ratio as the C_{27}-C_{30} 17α(H)-hopanes, Ts and C_{29}Ts in the Prudhoe Bay oil, suggesting they are all derived from precursors in the same or similar organisms. Thus, the occurrence of C_{30}^* in oils may be related to bacterial hopanoid precursors that have undergone oxidation in the D-ring and rearrangement by clay-mediated acidic catalysis. Many source rocks rich in terrestrial organic input were deposited under oxic to suboxic conditions and are clay-rich. Thus, our results suggest that C_{30}^* is derived from bacterial input into a sediment containing clays and that deposition under oxic or suboxic conditions favors its formation. This interpretation is also consistent with the type of highly terrestrial oils where it is found in greatest abundance (Philp and Gilbert, 1986; Volkman et al., 1983a).

On the other hand, C_{29}Ts shows virtually the same geochemical behavior as Ts, and is probably as widespread in crude oils, except that poor resolution from C_{29} 17α(H)-hopane may obscure its analysis (e.g., Fig. 3.18). For example, it can be used as a maturity marker in the ratio C_{29}Ts/[C_{29} 17α(H)-hopane + C_{29}Ts], which is similar to Ts/(Ts + Tm). [See Sec. 3.2.4.1, C_{29}Ts/(C_{29} 17α(H)-hopane + C_{29}Ts)].

Summons et al. (1988a, 1988b) discovered a pseudohomologous series of peaks in the C_{27}-C_{34} range that appears to include C_{30}^* in Proterozoic petroleums. A comparable GCMSMS analysis of an oil from Jordan shows what is probably the same series of peaks (starred series, Fig. 3.19).

Because these peaks fall in chromatographic succession with Ts, they were thought to be pseudohomologs of Ts. However, the finding that C_{29}Ts is the peak eluting immediately after C_{29} 17α(H)-hopane, plus the X-ray structure of C_{30}^*, leave little doubt that the starred peaks in the figure are pseudohomologous 17α(H)-diahopanes.

Although application of the C_{30}^*/C_{29}Ts parameter is not well established, it appears that relative amounts of C_{30}^* and C_{29}Ts depend most strongly on environment of deposition, and that oils derived from shales deposited under oxic-suboxic conditions will show higher ratios than those derived from source rocks deposited under anoxic conditions. Thus, in a correlation study of oils and rocks from Papua New Guinea, the C_{30}^*/C_{29}Ts ratio was effective in distinguishing oil source groups (Fig. 3.18). Both oils in the figure show relatively high C_{30}^*/C_{29}Ts ratios, suggesting that their source rocks were deposited under oxic-suboxic conditions.

Application of molecular mechanics calculations to these structures indicates that compounds of the 17α(H)-diahopane series should be more stable than compounds of the 18α(H)-neohopane series, which, in turn, should be more stable than those of the 17α(H)-hopane series. For example, the calculated heat for formation for C_{30}^* (76.1 kcal/mole, D. S. Watt, personal communication) is 3.7 kcal less than the hypothetical compound 18α(H)-neohopane (C_{30}Ts) and 6.1 kcal less than 17α(H)-hopane (Kolaczkowska et al., 1990). Thus, the C_{30}^*/C_{29}Ts ratio should increase with increasing thermal maturity, particularly in the late oil window.

While depositional environment appears to exert the principal control on relative C_{30}^* concentrations, molecular mechanics calculations indicate that 17α(H)-diahopanes are more stable than 18α(H)-neohopanes, which are more stable than 17α(H)-hopanes (Moldowan et al., 1991b). Thus, increasing maturity should result

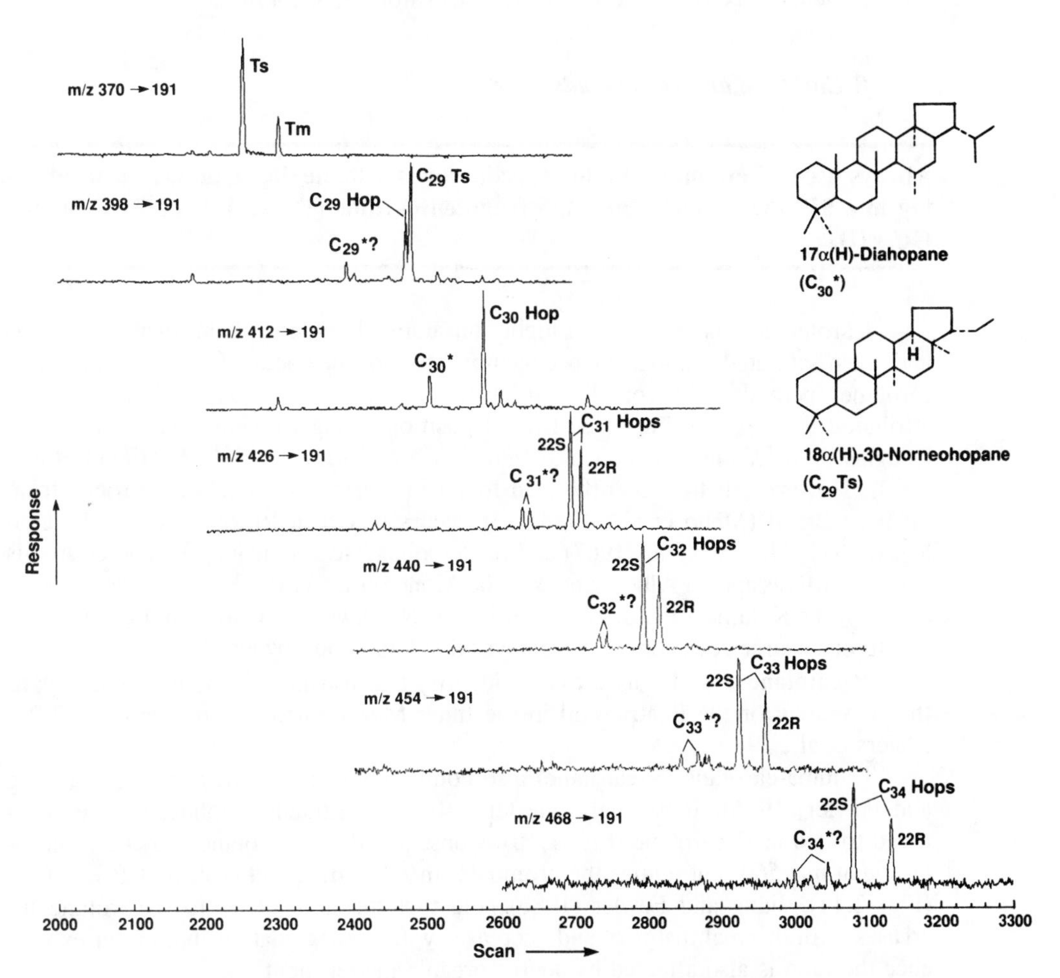

Figure 3.19 Parent ion analysis of m/z 191 daughter ions by GCMSMS in the saturate fraction of a Jordanian oil (GCMS No. 896) shows a probable pseudohomologous series of C_{29}-C_{34} 17α(H)-diahopanes (C_{29}-C^{*}_{34}). The C_{27} and C_{29} 18α(H)-neohopanes (Ts and C_{29} Ts) and C_{27} (Tm) and C_{29}-C_{34} 17α(H)-hopanes are also evident. Each chromatogram is normalized internally to the highest peak. GCMSMS data were recorded on a Finnigan MAT TSQ-70 triple quadrupole system interfaced to a Varian 3400 gas chromatograph, using a 60 m DB-1 (J & W Scientific) fused silica capillary column, 0.25 mm I.D., 0.25 mm film thickness, H_2 carrier gas, 150°C programmed at 2°/min. to 300°C. Scans are cycled at 1.5 seconds.

in increased ratios of 17α(H)-diahopane to either 18α(H)-30-norneohopane or 17α(H)-hopane. Horstad et al. (1990) used the ratio $C_{30}^{*}/[C_{30}^{*}$ + 17α(H)-hopane] along with sterane isomerization [C_{29} $\alpha\alpha\alpha$ 20S/(20S + 20R)] in oils to map maturity gradients in North Sea oil fields (cf., Cornford et al., 1986).

β-Carotane and carotenoids

Always used when found, highly specific for lacustrine deposition, measured using m/z 558 (M^+) and/or m/z 125 fragmentograms (Fig. 2.11), or by capillary GC/FID.

Carotenoids include certain highly unsaturated compounds in many organisms and their saturated analogs in petroleum. Most occurrences of beta-carotane (β-carotane; perhydro-β-carotene; Fig. 1.10), a fully saturated C_{40}-dicyclane, are attributed to anoxic, saline, lacustrine deposition of algal organic matter (Hall and Douglas, 1983; Jiang and Fowler, 1986; Irwin and Meyer, 1990; Fu Jiamo et al., 1990). However, it has recently been found in marine oils and source rocks from offshore Brazil (Mello et al., 1988a). It occurs in Green River shale (marl) (Gallegos, 1971; Murphy et al., 1967) and related oils. β-carotane has been noted in oils from the Mississippian Albert Shale in the Moncton Basin (R. M. K. Carlson, pers. comm.), the Kelamayi Field, Zhungeer Basin, Northwestern China, and from a Carboniferous source rock for the Kelamayi oils (Jiang and Fowler, 1986).

β-carotane was also used to help identify a lacustrine Devonian source rock as the co-source for the Beatrice oil in the Inner Moray Firth, North Sea (Fig. 3.20) (Peters et al., 1989).

Gamma-carotane (γ-carotane) also commonly occurs with β-carotane (Jiang and Fowler, 1986). It has only one tetra-alkyl substituted cyclohexane ring compared to two in β-carotane (Fig. 1.10). Consequently, γ-carotane shows a molecular ion at m/z 560, but retains the prominent m/z 125 fragment typical of β-carotane (Fig. 2.11). Jiang and Fowler (1986) note that the ratio of γ- to β-carotane increases with thermal maturity and decreases with biodegradation, but in our experience the ratio is also affected by source organic matter input.

Recent work by the group at Strasbourg (Adam et al., 1992) sheds some light on the fate of β-carotene in anoxic marine and some hypersaline nonmarine sediments. They liberated β-carotane from sulfur complexes in high-sulfur oils that show no β-carotane in the saturate fraction. Apparently the double bonds in β-carotene react with sulfur species in these environments becoming part of a sulfur cross-linked system, while in low-sulfur anoxic lacustrine systems β-carotene is reduced to β-carotane. Lycopane (Fig. 1.10), which is rare in oils, was also liberated from the sulfur complexes. Like β-carotane, lycopane is also derived from a highly unsaturated precursor, in this case lycopene.

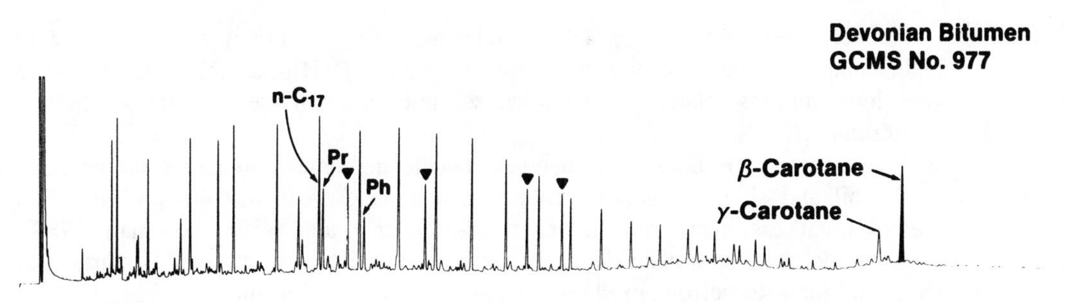

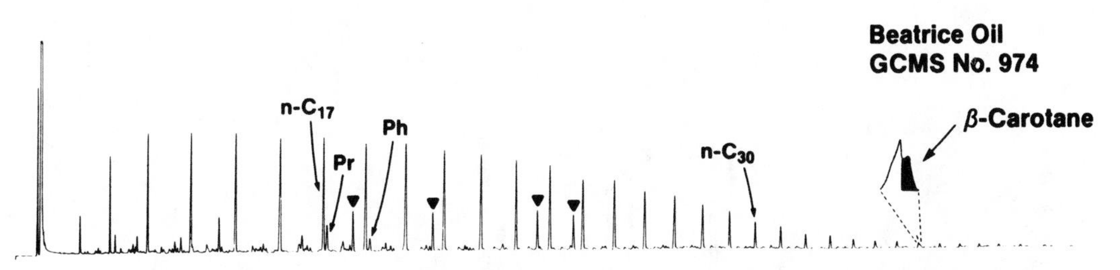

Figure 3.20 Gas chromatograms showing β-carotane in bitumen from a Devonian age flagstone (top) and Beatrice oil (bottom) from the Moray Firth, U.K. The compound was identified by coinjection of the authentic standard and by the similarity of the mass spectrum of the peak in the oil with that of the standard. γ-carotane is also a prominent peak in the gas chromatogram of the Devonian sample. The small concentrations of β-carotane and other evidence helped to show that the Devonian flagstones are a co-source for the Beatrice oil (Peters et al., 1989).

Bicyclic sesquiterpanes (eudesmane and drimane)

Rarely used, specificity unknown, sometimes reported as terrestrial markers, measured using GCMS m/z 123 (Fig. 2.11).

Several reports show the occurrence of bicyclic sesquiterpanes in crude oils (Bendoraitis, 1974; Seifert and Moldowan, 1979; Philp et al., 1981). The significance of most of these compounds remains unknown because of a lack of precise structure assignments. However, two compounds, 4β(H)-eudesmane (Fig. 1.10) and 8β(H)-drimane were identified using synthetic standards in an Australian oil of terrestrial origin (Alexander et al., 1983a). Recently, van Aarssen and de Leeuw (1989) reported a group of stereoisomeric cadinanes, probably related to resins, in southeast Asian oils (see "Cadinanes" section).

The carbon skeleton of 4β(H)-eudesmane (Fig. 2.11) is clearly related to higher plant terpenes. Despite its terrestrial origin, 4β(H)-eudesmane is present in very low amounts relative to the other sesquiterpanes in the terrestrially derived Australian oil.

In contrast to eudesmane, drimane occurs in higher abundance in the Australian oil and shows structural features and a widespread distribution similar to many biomarkers of prokaryotic origin (Alexander et al., 1983a; Volkman, 1988). Thus, we do not recommend the general use of bicyclic sesquiterpanes as markers of terrestrial input to petroleum without rigorous structure identification (Fig. 3.21).

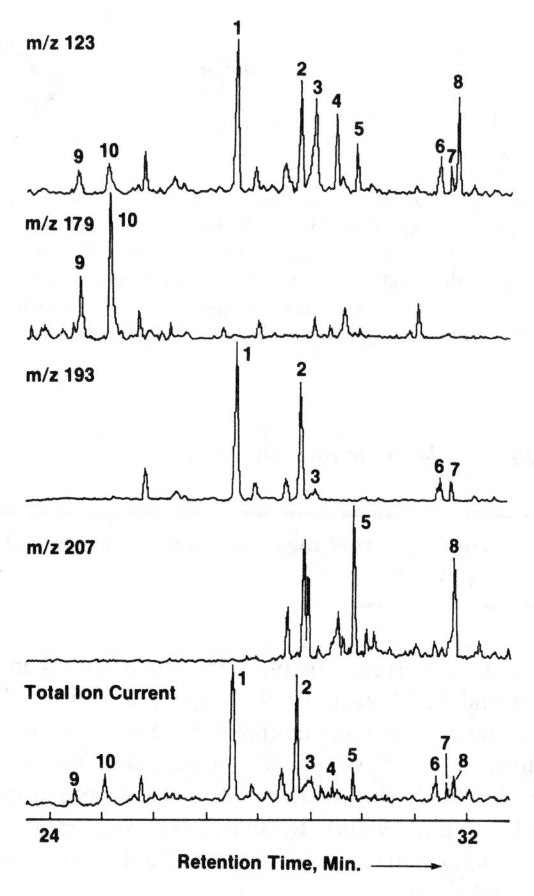

Figure 3.21 The figure shows bicyclic sesquiterpanes identified by Noble (1986) using synthesized standards and mass spectral studies. The chromatograms show the elution order and relative intensity of these compounds in Miandoum oil (Chad) using a 50 m 0.2 mm I.D. methylsilicone capillary column, temperature programmed at 4°C/minute. The figure also shows structures of the important mass spectral fragments used in identification of these compounds.

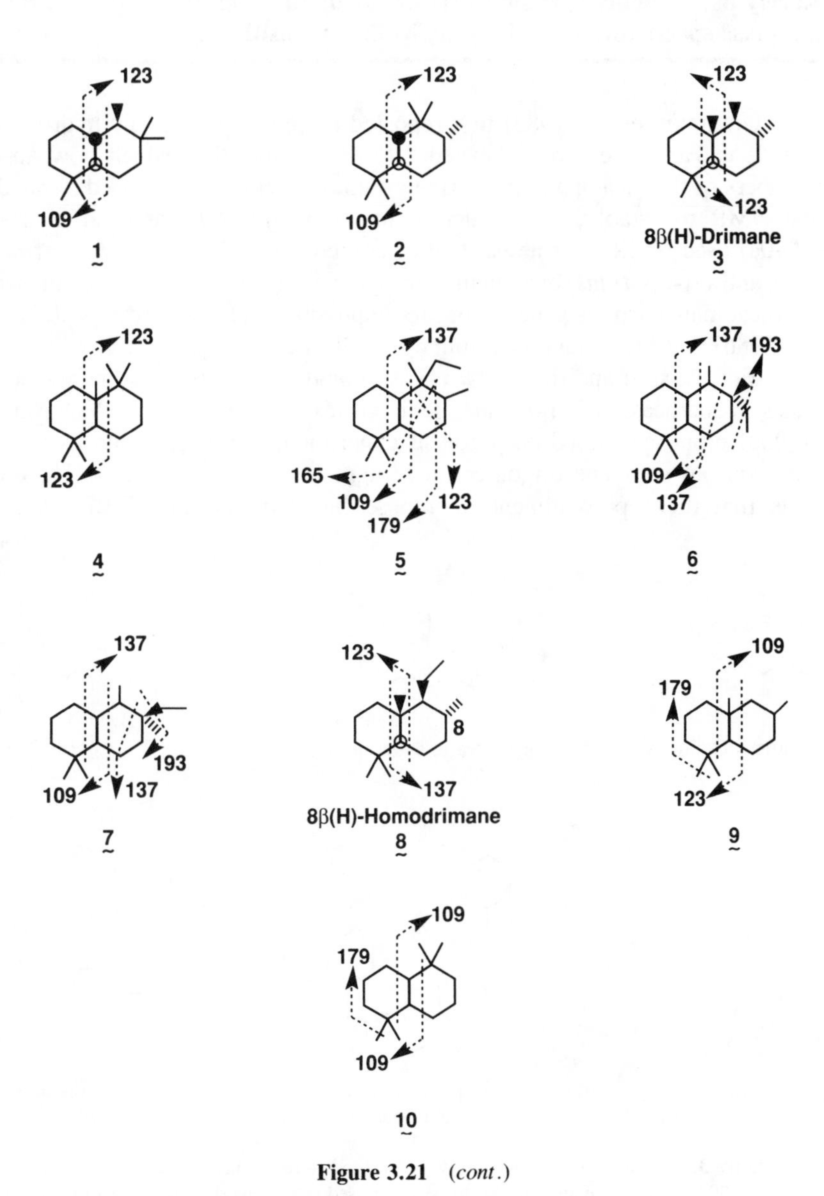

Figure 3.21 (*cont.*)

Cadinanes (mono-, bi-, and tricadinanes)

Rarely used, highly specific for resinous input from higher plants, detected using mass spectrum, or various GCMSMS transitions.

Grantham et al. (1983) first reported three C_{30}-pentacyclic hydrocarbon compounds in Far Eastern oils, labeled "W," "T," and "R," which now appear to be members of the bicadinane triterpane group. Cox et al. (1986) used X-ray diffraction and NMR to establish the structure for compound "T" and van Aarssen et al. (1990a) used NMR to establish that of compound "W" as *trans-trans-trans-*bicadinane and *cis-cis-trans-*bicadinane, respectively (Fig. 3.22). Direct biosynthesis of the bicadinanes from squalene appears impossible and it was thus postulated (Cox et al., 1986) that bicadinane is a dimer of cadinene.

Van Aarssen and de Leeuw (1989) found cadinanes, bicadinanes, and tricadinanes in southeast Asian oils and sediment extracts (Fig. 3.22). Van Aarssen et al. (1990b) proposed, based on pyrolysis experiments with fossil and fresh resins, that these various cadinene oligomers are fragments of polycadinenes. Their work suggests that these polycadinenes compose biopolymers (Fig. 3.22) that form an-

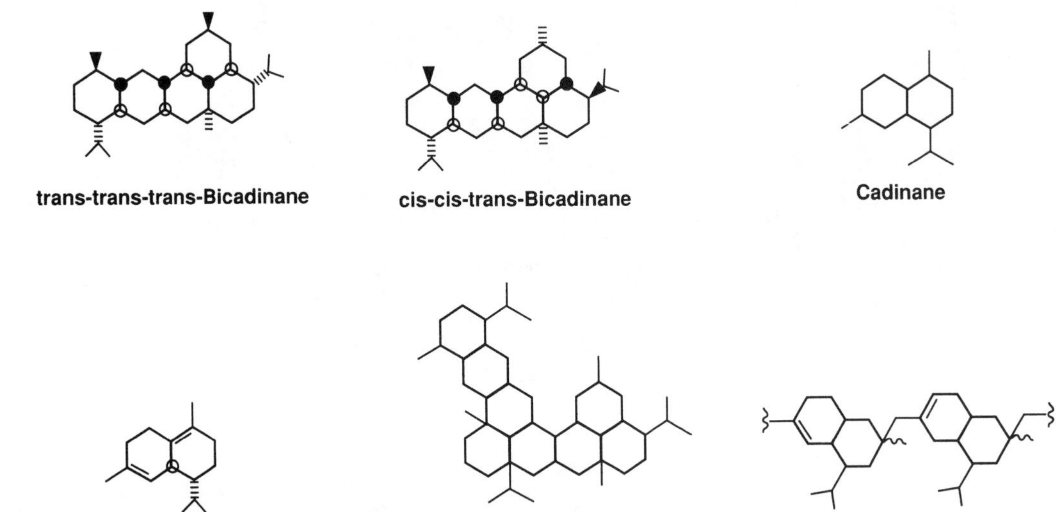

Figure 3.22 The figure shows the two identified stereoisomeric bicadinanes (van Aarssen et al., 1990b). The cadinanes appear to be comprised of various stereoisomers of cadinane. Cadinane shows five asymmetric carbons resulting in 32 (or 2^5) possible stereoisomers. A possible structure for tricadinane is suggested (van Aarssen and de Leeuw, 1989). It is proposed that the various oligomers of cadinane in crude oils are derived from the catagenesis of polycadinene biopolymers that compose angiosperm dammar resins (van Aarssen et al., 1990b).

giosperm dammar resins of some southeast Asian higher plants. Numerous isomers of each oliogomer are typically present when they occur in oils and rock extracts. Bicadinanes are common in land-plant derived Tertiary oils from the Surma Basin in northeast Bangladesh (Alam and Pearson, 1990). Oleanane is abundant, but does not co-vary with bicadinanes in these oils, suggesting separate land plant sources for these two triterpanes.

The mass spectra of bicadinanes contain prominent m/z 191 and 217 fragments and, therefore, peaks can appear in the corresponding chromatograms used in the analysis of hopanes and steranes (e.g., Fig. 3.23). Because the bicadinane mass spectrum has an intense m/z 369 for loss of *iso*-propyl and a strong molecular ion (m/z 412), it can be conveniently monitored with little interference using a chromatogram of the m/z 412 to 369-GCMSMS transition. Only the hopanes (C_{30}), which show the same transition, typically give additional but weaker peaks on this chromatogram. However, the hopanes are easily distinguished because they show a much higher retention time than the bicadinanes under normal analytical GC conditions.

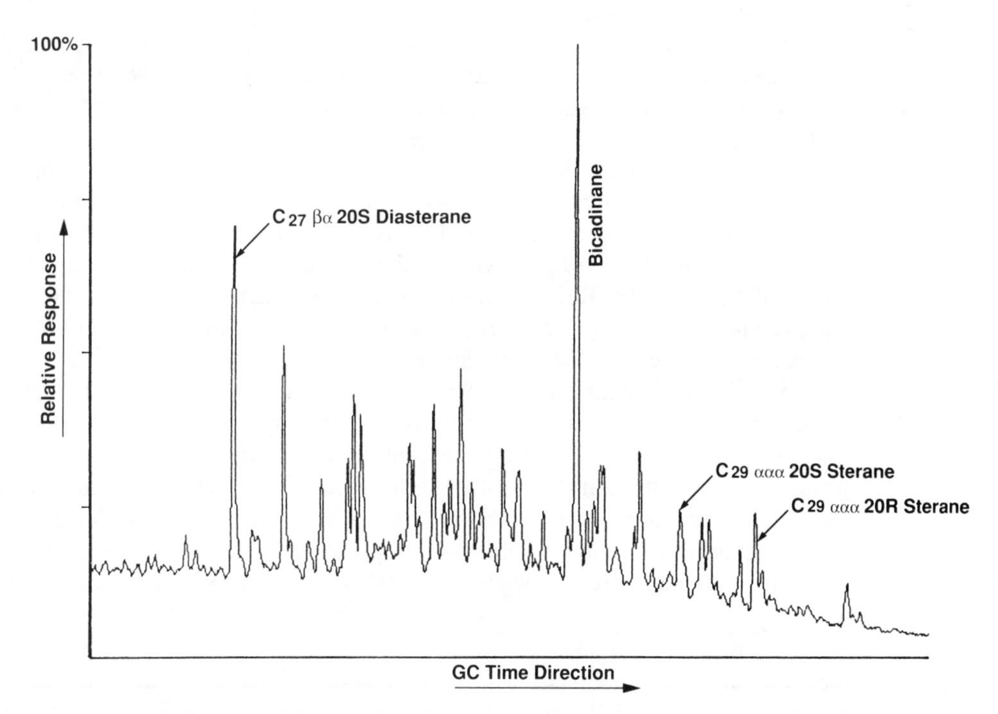

Figure 3.23 The m/z 217 chromatogram for steranes from the saturate fraction of a southeast Asia crude oil shows a dominant peak that was identified by its mass spectrum, and comparison with that in Cox et al. (1986), as bicadinane. Other less prominent bicadinane isomers were detected using a CAD GCMSMS m/z 412 to 369 transition. Although not identified here, these bicadinane isomers contribute to the complexity of this m/z 217 fingerprint.

Bicyclic and tricyclic diterpanes

Rarely used, high specificity for various microbial, gymnosperm, and angiosperm inputs, measured using GCMS m/z 123 (Fig. 2.11).

The tetracyclic (see the following) and tricyclic diterpanes can be used to evaluate terrestrial input to some petroleums. Philp et al. (1981, 1983) analyzed diterpane distributions in various Australian oils from the Gippsland Basin and found a correlation between these distributions and higher plant input. Richardson and Miiller (1982, 1983) recognized diterpanes of terrestrial origin in a crude from southeast Asia and Snowdon (1980) tentatively identified pimarane-type compounds in Canadian crude oils.

Uncertainties in the use of these data arise from lack of structural identification of the compounds. For example, some degraded tricyclohexaprenanes (C_{19}, C_{20}), which may be of algal (*Tasmanites*) origin (Volkman et al., 1989; Aquino Neto et al., 1989), show similar GCMS fragment ions and retention times to the higher plant-derived diterpanes.

Previous studies suggest a resinite source for bi- and tricyclic sesqui- and diterpanes (Snowdon, 1980; Philp et al., 1983). Noble (1986) and Noble et al. (1986) identify several bicyclic sesquiterpanes (Fig. 3.21) and tricyclic diterpanes (Fig. 3.24) and discuss their significance as terrestrial markers. Compounds identified in Australian oils include 8β(H)-labdane (a bicyclic), 4β(H)-19-norisopimarane, rimuane, and isopimarane (Fig. 3.24). Isopimarane [iso(sandaraco)pimarane] was isolated and structurally characterized by Blunt et al. (1988) in seep oils from New Zealand. Weston et al. (1989) used distributions of diterpanes to distinguish among major oil fields in the Taranaki Basin, New Zealand. Fichtelite (see the following text), retene, iosene (phyllocladane), and pimarane are common in lignite. Tricyclic diterpanes generally show a major fragment at m/z 123 which is useful for identification along with the tetracyclic diterpanes (see the following). These diterpanes appear to be markers for gymnosperms (mainly conifers) which occur in Permian and younger sediments. An exception is pimarane, which may be of angiosperm (flowering plant) origin in some cases. Based on a comparison of mass spectra, isopimarane appears to be present in Maui oil from New Zealand (Restle, 1983; Weston et al., 1989). Labdane is reported as a series of C_{15}-C_{24} bicyclic alkanes of probable microbial origin (Dimmler et al., 1984) in Athabasca tar sand.

Fichtelite

Rarely used, suggested specificity for land plant input, measured using GCMS m/z 219 (Fig. 3.24).

Mackenzie (1984) indicates that fichtelite (Fig. 1.10), a tricyclic diterpane, is a potential land plant source indicator. Fichtelite, retene, and iosene have been known for a long time as crystalline deposits in lignites (Noble, 1986; Simoneit, 1986; Wang and Simoneit, 1990).

Tetracyclic diterpanes (beyerane, phyllocladanes, kauranes)

Rarely used, many are highly specific for terrestrial organic matter input, measured using M^+ (m/z 274) in combination with various fragment ions (Fig. 3.24).

Many tetracyclic diterpanes are believed to be age-related markers for conifers. For example, Noble et al. (1985a, 1985b, 1986) report a family of bridged tetracyclic diterpanes in Australian coals, sediments, and oils. They identified com-

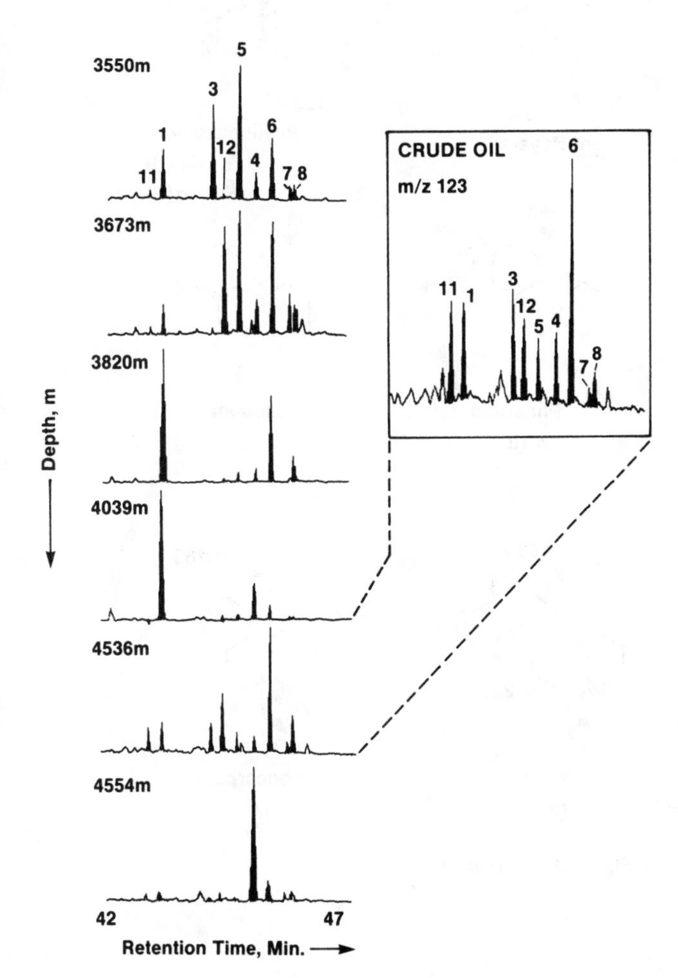

Figure 3.24 Noble (1986) used distributions of tetracyclic and other diterpanes (m/z 123) to correlate a crude oil with one source rock horizon within a 1 km depth interval in a well from the Gippsland Basin, Australia. Structures for the tricyclic diterpanes identified in the mass chromatogram show their principal fragment ions. Peaks for compounds labeled 2, 9, 10, 13, 14, and 15 are not shown.

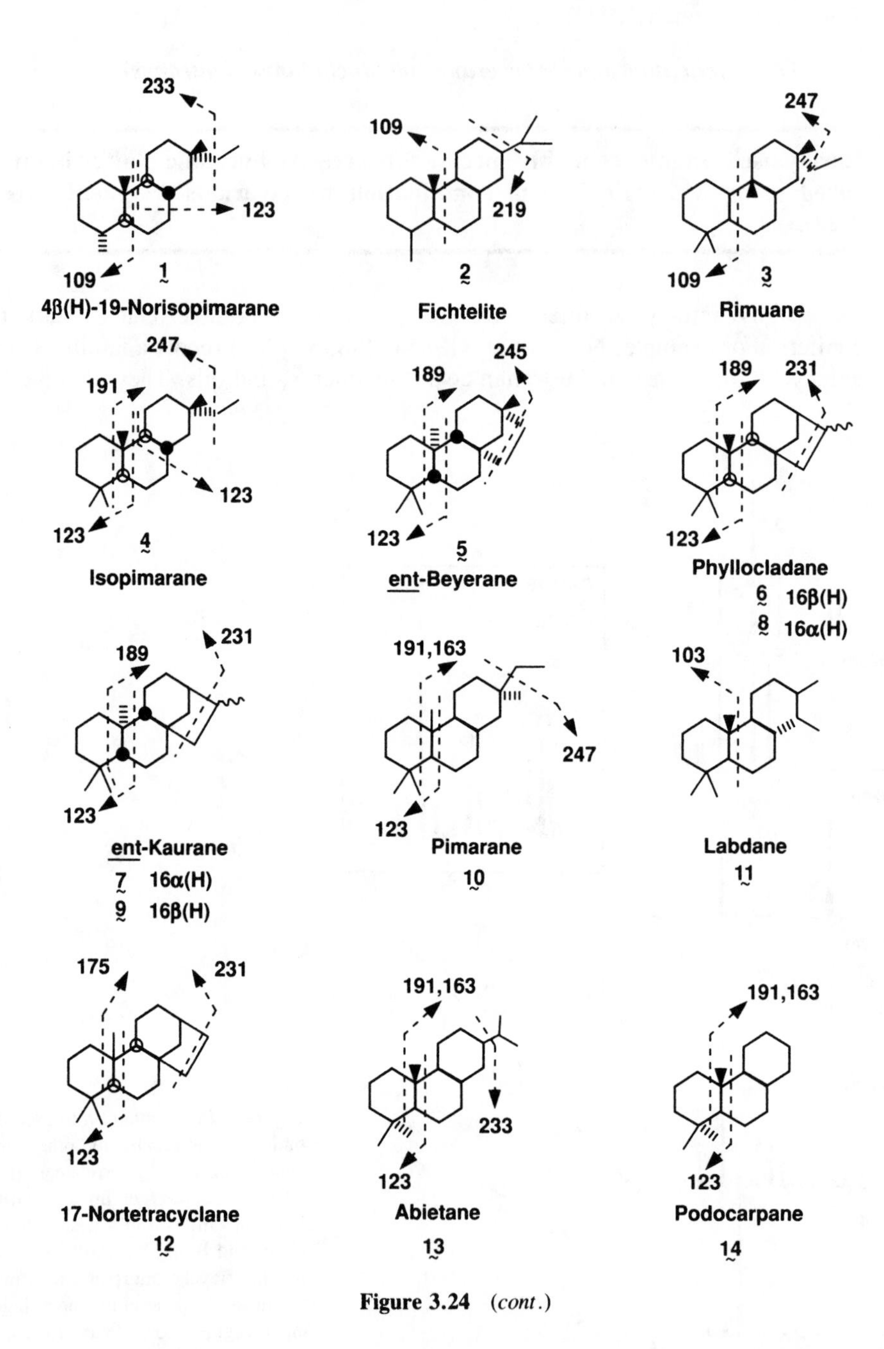

Figure 3.24 (*cont.*)

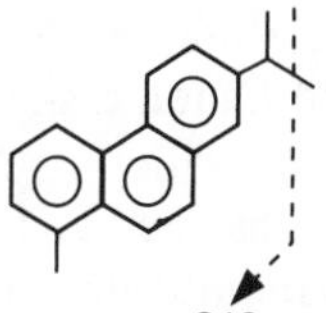

219

Retene
15

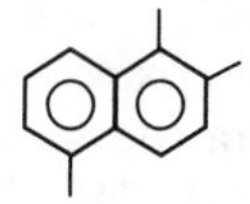

1,2,5-Trimethylnaphthalene
16

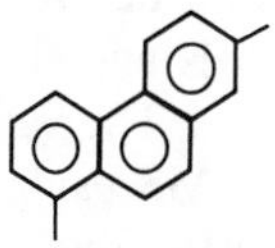

1,7-Dimethylphenanthrene
17

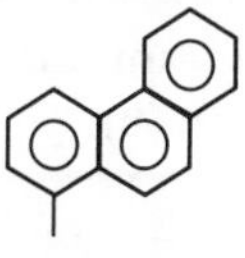

1-Methylphenanthrene
18

Figure 3.24 *(cont.)*

pounds including: *ent*-beyerane, 16α(H)- and 16β(H)-phyllocladane, *ent*-16α(H)- and *ent*-16β(H)-kaurane (Fig. 3.24), and a 17-nortetracyclic diterpane thought to be related to diterpanes abundant in leaf resins of conifers belonging to the *Podocarpaceae, Araucariaceae,* and *Cupressaceae* families. Phyllocladane was identified in resins from *Podocarpus* and *Dacrydium* (species of *Podocarpaceae*), while kaurane was detected in *Agathis* (a species of *Araucariaceae*). Both of these families are currently restricted to the southern hemisphere. Weston et al. (1989) also reported many of these markers in oils from the Taranaki Basin, New Zealand.

Although difficult to distinguish petrographically, limnic coals from the Saar District are enriched in *ent*-beyerane, while paralic marine coals of the Ruhr Area in Germany contain more kauranes (Schulze and Michaelis, 1990; ten Haven et al., 1991). Ten Haven et al. (1992) also found the phyllocladanes, *ent*-kauranes and *ent*-beyerane in a Pennsylvanian age coal from Oklahoma, USA. Because these coals are of Late Carboniferous age and *Cupressaceae, Araucariaceae*, and *Podocarpaceae* did not evolve until Late Triassic time, the authors conclude that the phyllocladane-type compounds in their coals were derived from early conifers, the *Voltziales*, which may have already been able to biosynthesize these compounds. The *Voltziales* appear to have evolved during late Carboniferous time.

Thermodynamic equilibrium is reached before the beginning of oil generation where the ratio of isomers $16\alpha(H)/16\beta(H)$ for *ent*-kaurane and *ent*-phyllocladane reach final values of < 0.1 and 0.3, respectively. Simoneit (1986) notes that likely precursors for some diterpanes, including those which form kaurane, abietane, pimarane, and podocarpane (Fig. 3.24), are found, but only as minor components in marine organisms. Noble (1986) used distributions of tetracyclic and other diterpanes to correlate a crude oil to one horizon within a series of Gippsland Basin rocks extending over a 1 km depth range (Fig. 3.24). Villar et al. (1988) identified resin-derived diterpenoids with the phyllocladane and kaurane skeleton in Tertiary coals and shales from Argentina, indicating input from conifers.

The various tetracyclic diterpane isomers can be analyzed and distinguished by GCMS using their mass spectra and by selective ion monitoring of the major diagnostic fragments at m/z 123, 231, 245, 259, and 274 (Fig. 2.11).

Tricyclic terpanes (cheilanthanes) and tricyclics/17α(H)-hopanes

---

Often used, low specificity as a correlation tool, widespread in oils and bitumens, measured using m/z 191 fragmentogram (Fig. 3.49).

The ratio of tricyclics/17α(H)-hopanes is primarily a source parameter that compares a group of bacterial or algal lipids (tricyclics) with markers that arise from different prokaryotic species (hopanes). We measure the sum of four tricyclic terpane peaks, 22R and 22S doublets, representing the C_{28} and C_{29} pseudohomologs of tricyclohexaprenane (29,30-bisnortricyclohexaprenane and 30-nortricyclohexaprenane, respectively), for the numerator of the ratio. The sum of the C_{29}-C_{33} 17α(H)-hopanes comprises the denominator of the ratio. The origin of these two groups of bacterial lipids is discussed in the following text.

A series of tricyclic terpanes [also called 13β(H),14α(H)-cheilanthanes; Fig. 2.11], ranging from C_{19} to C_{30} is found in most oils and bitumens (Seifert et al., 1978; Aquino Neto et al., 1983). Palacas et al. (1984) found that tricyclic terpanes were the most useful series of biomarkers for differentiating potential from effective source rocks in the South Florida Basin. The structures of several homologs of the tricyclic terpanes have been proven by synthesis (Aquino Neto et al., 1982; Ekweozor and Strausz, 1982; Heissler et al., 1984; Sierra et al., 1984). All four isomers at C-13 and C-14 ($\beta\alpha$, $\alpha\alpha$, $\alpha\beta$, and $\beta\beta$) are present in immature rocks with $\beta\alpha$ and $\alpha\alpha$ predominating, but with increasing maturity, the $\beta\alpha$ isomer becomes dominant (Chicarelli et al., 1988). Moldowan et al. (1983) show that the tricyclic terpane series extends to C_{45} in a California oil.

Note: *For tricyclic terpanes containing 25 or more carbon atoms, C-22 is an asymmetric center resulting in two stereoisomers or a doublet on m/z 191 mass chromatograms. For tricyclic terpanes containing 30 or more carbon atoms, C-27 is also an asymmetric center and four stereoisomers are expected for each compound. The tricyclic terpane series lacks or shows low concentrations of C_{22}, C_{27}, C_{32}, C_{37}, and C_{42} homologs because of the isoprenoid structure of the side chain (methyl substitution every fourth carbon atom). For example, to generate the C_{22} tricyclic terpane requires the cleavage of two carbon-carbon bonds rather than one bond at the C-22 position. As described earlier (Sec. 1.2.7), the C_{17} acyclic isoprenoid is low or absent in petroleum for a similar reason.*

The widespread occurrence of tricyclic terpanes and molecular properties of tricyclohexaprenol suggest that they are derived from prokaryotic membranes (Ourisson et al., 1982). These authors propose tricyclohexaprenol as the parent compound for tricyclic terpanes containing 30 or less carbon atoms. However, high concentrations of tricyclic terpanes in *Tasmanite* rock extracts indicates a possible origin from these algae (Volkman et al., 1989; Aquino Neto et al., 1989). Evidence suggests a higher plant source for some C_{19} and C_{20} tricyclic terpanes (Fig. 3.24). Zumberge (1983) indicates that C_{19}-C_{20} tricyclics may also be produced by the thermal cleavage of the alkyl side chain in sester- and triterpanes. The C_{23} homolog is typically the most prominent tricyclic terpane (Connan et al., 1980; Aquino Neto et al., 1983; Sierra et al., 1984).

Oils and bitumens from carbonate rocks appear to show low concentrations of tricyclic terpanes above C_{26} compared to those from other depositional environments where the C_{26}-C_{30} and C_{19}-C_{25} homologs show similar concentrations (Aquino Neto et al., 1983).

Because of their extreme resistance to biodegradation, tricyclic terpanes permit correlation of intensely biodegraded oils (Seifert and Moldowan, 1979; Palacas et al., 1986). They also appear resistant to thermal maturation compared to homohopanes (Sec. 3.2.4.1), although the lower carbon number homologs are favored at high thermal maturity (Peters et al., 1990). Tricyclic terpanes are generated from the kerogen at relatively higher thermal maturity than homohopanes (Aquino Neto et al., 1983; Peters et al., 1990). Sofer (1988) used tricyclic terpanes, stable carbon isotopes of saturates and aromatics, and other data to classify groups of Gulf Coast oils showing wide variations in thermal maturity.

Pentacyclic triterpenoids (Fig. 1.10), including precursors of the hopanes, are found in prokaryotes and higher plants, but appear absent in algae. Bacteria appear to be the major source for sedimentary hopanoids. The extended hopanes (C_{31} or more) are related to specific bacteriohopanepolyols found in bacteria such as bacteriohopanetetrol (Fig. 1.20), while the lower pseudohomologs (C_{30} or less) may also be related to C_{30} precursors such as diploptene or diplopterol (Fig. 3.25) found in nearly all hopanoid-producing bacteria (Rohmer, 1987).

Tetracyclic terpanes

Rarely used, specificity unknown, measured using m/z 191; abundant C_{24} tetracyclic may indicate carbonate or evaporite depositional environment.

Tetracyclic terpanes of the 17,21-secohopane series (Fig. 2.11) (Trendel et al., 1982) occur in most oils and bitumens. Aquino Neto et al. (1983) show that tetracyclic terpanes range from C_{24} to C_{27} with tentative evidence for homologs up to C_{35}. Some of these compounds are identified as peaks on the m/z 191 trace in Fig. 2.12 (peaks 4, 5, and 10). They are thought to be derived by thermal or microbial

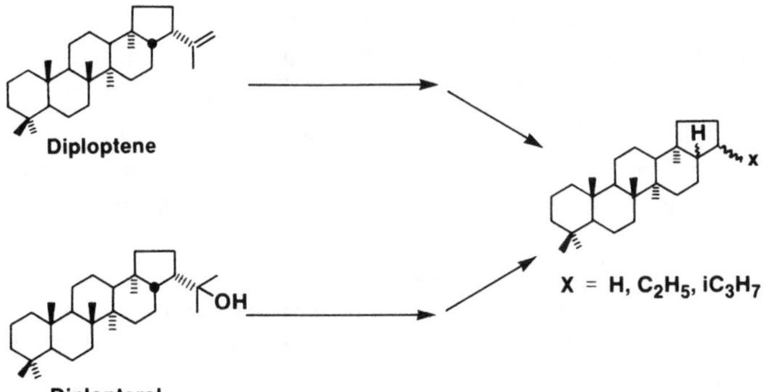

Diploptene

Diplopterol

X = H, C_2H_5, iC_3H_7

Figure 3.25 Diploptene and diplopterol are common in nearly all hopanoid-producing bacteria and represent likely sources for hopanes containing 30 carbon atoms or less. The extended hopanes (C_{31} or more) appear most likely derived from bacteriohopanepolyols (Fig. 1.20).

rupture of the five-membered E-ring (Fig. 1.12) in hopanes or precursor hopanoids, although an independent biosynthetic route to the tetracyclic terpanes may exist in bacteria. Ratios of tetracyclic terpanes/hopanes increase in more mature source rocks and oils, indicating greater stability of the tetracyclic terpanes. Tetracyclic terpanes also appear more resistant to biodegradation than the hopanes (Aquino Neto et al., 1983). For these reasons, they are occasionally used in correlations of altered petroleums (Seifert and Moldowan, 1979).

The C_{24} tetracyclic terpane (Fig. 2.11; X = H) shows the most widespread occurrence, followed by C_{25} to C_{27} homologs. Abundant C_{24} tetracyclic terpane in petroleum (e.g., Fig. 1.10 and 3.49) appears to be a marker for carbonate and evaporite depositional environments (Palacas et al., 1984; Connan et al., 1986; Connan and Dessort, 1987; Mann et al., 1987; Clark and Philp, 1989). However, this compound is also present in Australian oils believed to be generated from terrigenous organic matter (Philp and Gilbert, 1986). The C_{25} to C_{27} tetracyclic terpanes have also been reported in carbonate and evaporite samples (Connan et al., 1986).

Hexahydrobenzohopanes

Rarely used, specificity unknown, but appear diagnostic of carbonate-anhydrite depositional environments, measured using m/z 191 or molecular ions (Fig. 2.11).

Hexahydrobenzohopanes (Fig. 2.11) range from C_{32} to C_{35} and are probably formed in a reducing (low Eh) depositional environment by cyclization of the side chains on extended hopanoids. They appear diagnostic of oils and bitumens from sulfur-rich carbonate-anhydrite source rocks (Connan and Dessort, 1987; Rinaldi et al., 1988). Although hexahydrobenzohopanes (also called hexacyclic hopanoids) can be analyzed using a m/z 191 fragmentogram of the saturate fraction, they are usually in low concentration relative to the extended hopanes. These compounds can be monitored using the molecular ions at m/z 438, 452, 466, and 480 (Fig. 2.11). Alternately, the high selectivity offered by the specific parent to daughter transitions ($M^+ \rightarrow$ m/z 191) allows use of linked-scan GCMSMS approaches (Fig. 3.26).

Note: *Unlike most other biomarkers, a cyclohexyl ring (Ring D) in hexahydrobenzohopanes is in the "boat" rather than the usually preferred "chair" conformation. (See Note in Sec. 1.2.6).*

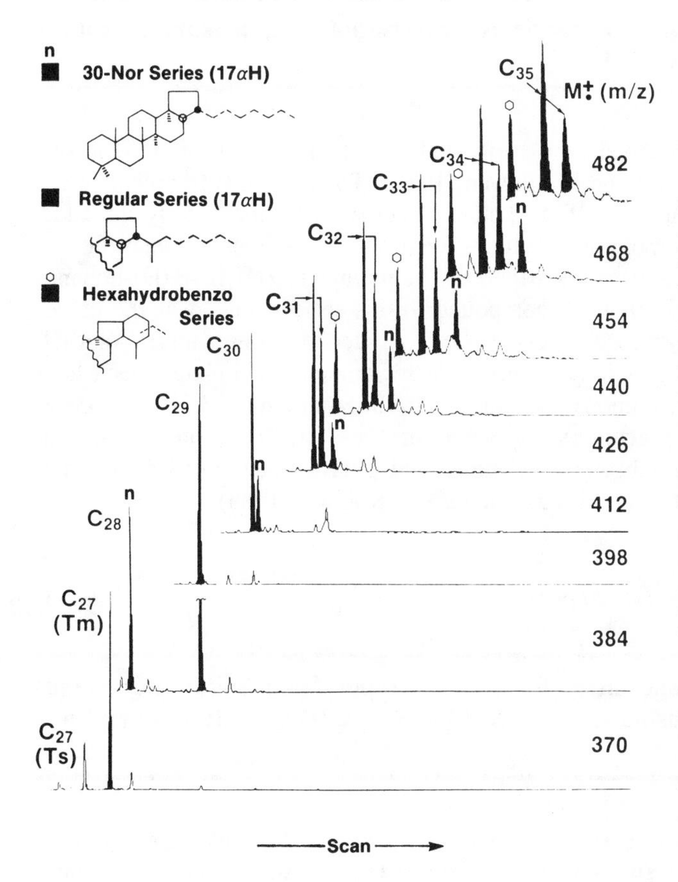

Figure 3.26 MRM GCMS analysis of pentacyclic triterpanes in Santa Maria-3 oil, Italy (Moldowan et al., 1992). Pentacyclic hopanes with 27 to 35 carbon atoms are analyzed on separate $M^+ \rightarrow$ m/z 191 chromatograms, with M^+ ranging every 14 mass units from m/z 370 to m/z 482, respectively. Both the regular 17α(H)-hopane series and the 30-nor-17α(H)-hopane series (Sec. 3.3) are identified in this sample. Peaks corresponding to the series of C_{32} to C_{35} hexahydrobenzohopanes are also found. The m/z 191 fragments from their molecular ions, m/z 438, 452, 466, and 488, respectively, are recorded here because of the low resolution of the MRM method (Sec. 2.5.3.3) compared to triple quadrupole GCMSMS. The same transitions recorded at higher resolution (i.e., with a triple quadrupole system) do not show the hexahydrobenzohopane series. The precise stereochemical structures of these compounds have not been determined.

Lupanes

Rarely used, specificity unclear but appear to indicate terrestrial input, measured using m/z 177 (bisnorlupanes) and m/z 191.

Lupane is believed to indicate terrigenous organic matter and is common in coals (e.g., Wang and Simoneit, 1990).

The 23,28-bisnorlupanes appear to be derived from lupanes or other higher plant precursors during diagenesis (Rullkötter et al., 1982). A relationship between the concentrations of bisnorlupanes and oleanane (a higher plant marker) in petroleums from the Mackenzie Delta, Canada, supports their terrigenous origin (Brooks, 1986).

Methylhopanes

Rarely used, potentially highly specific as a correlation tool, prokaryotic source input, measured using m/z 205.

Ring A and B methylhopanes were first detected in a series of Jurassic oils from the Middle East (Seifert and Moldowan, 1978). The ring A/B fragment of the hopane series, normally m/z 191 (Fig. 2.11), is increased to m/z 205 by the additional methyl group. The most prominent series of these compounds identified in some Precambrian oils has been identified as 2α-methyl-17α(H),21β(H)-hopanes (Summons and Jahnke, 1992). Each compound in this series elutes near the corresponding 17α(H)-hopane with one less carbon. A second series, the 3β-methyl-17α(H),21β(H)-hopanes, generally occurs in lesser amounts and elutes much later than the 2α-compounds (Summons and Walter, 1990; Summons and Jahnke, 1992) (Fig. 3.27). The likely precursors for some of these hydrocarbons, 2β-methyldiplopterol and 3β-methylbacteriohopane polyols, have been found in methylotrophic bacteria (Bisseret et al., 1985; Zundel and Rohmer, 1985).

3.1.3.2 Steranes

Regular steranes/17α(H)-hopanes

Often used, moderate specificity for relative input from eukaryotes versus prokaryotes, measured using GCMSMS ($M^+ \rightarrow$ m/z 217) for steranes and m/z 191 for hopanes.

The regular steranes/17α(H)-hopanes ratio reflects input of eukaryotic (mainly algae and higher plants) versus prokaryotic (bacteria) organisms to the source rock.

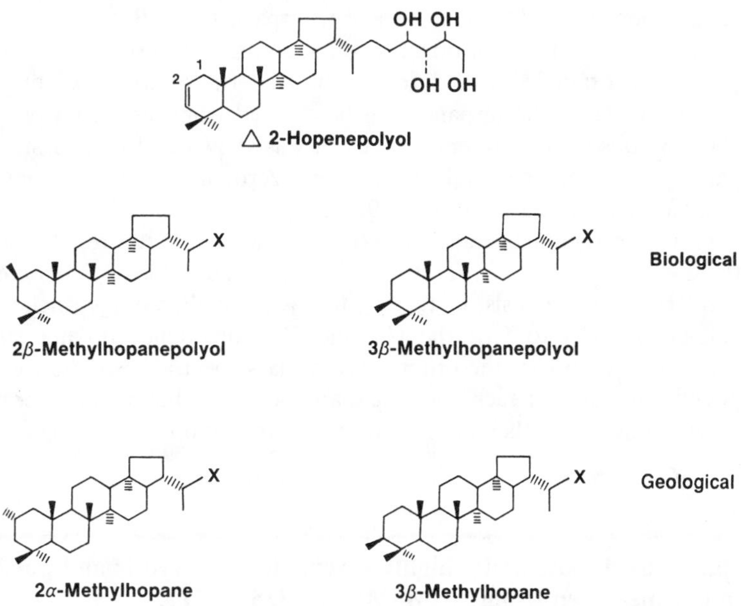

Δ 2-Hopenepolyol

2β-Methylhopanepolyol

3β-Methylhopanepolyol

Biological

2α-Methylhopane

3β-Methylhopane

Geological

Figure 3.27 Origin of 2- and 3-methylhopanes from a proposed Δ2-hopenepolyol precursor. The X-group in the "biological" methylhopanepolyols is probably similar to the polyol side chain in Δ2-hopenepolyol. The X-group in the "geological" methylhopanes is a saturated *n*-alkyl side chain.

Thus, related oils of differing thermal maturity typically fall along a line on plots of sterane versus hopane concentration. Unrelated oils may or may not fall on this line. Because organisms vary widely in their sterol and hopanoid contents, only very large differences in this ratio allow confident assessment of eukaryote versus prokaryote input.

Some work indicates a possible maturity effect on the ratio (Seifert and Moldowan, 1978), but in a study of oils from Oman the regular steranes/17α(H)-hopanes ratio remained relatively constant for a group of related oils of widely different thermal maturities.

In general, we have observed that high concentrations of steranes combined with high sterane/hopane ratios (≥ 1) seem to typify marine organic matter with major contributions from planktonic and/or benthic algae (e.g., Moldowan et al., 1985). Conversely, low steranes and low sterane/hopane ratios are more indicative of terrigenous and/or microbially reworked organic matter (e.g., Tissot and Welte, 1984).

Some workers use the ratio steranes/triterpanes as an indicator of organic matter input, assuming that steranes are derived from algae and higher plants, while triterpanes come mainly from bacteria. For example, Connan et al. (1986) used low sterane/triterpane ratios (<0.05) and other molecular data for bitumen from anoxic, carbonate-anhydrite facies as evidence of high microbial input. We prefer to use the sterane/hopanes rather than sterane/triterpane ratio. Both ratios are of limited use because of the variety of organisms that contribute to steranes, and especially triterpanes. For example, in addition to the bacterial and thermal origins as discussed, C_{19} and C_{20} tricyclic terpanes can be derived from terrestrial plants. In a study of about 40 oils from different sources, it was shown that regular steranes/17α(H)-hopanes ratios were generally lower (near zero) in nonmarine-compared to marine-sourced oils (Moldowan et al., 1985).

In the regular steranes/17α(H)-hopanes ratio, the regular steranes consist of the C_{27}, C_{28}, and C_{29} $\alpha\alpha\alpha$(20S + 20R) and $\alpha\beta\beta$(20S + 20R) compounds and the 17α(H)-hopanes consist of the C_{29} to C_{33} pseudohomologs (including 22S and 22R epimers for C_{31} to C_{33}; the C_{29} and C_{30} compounds do not show 22S or 22R epimers). Because of the differences in mass spectral response for the various compounds, amounts of each compound are measured first in parts per million of the saturate fraction, and then combined to give the ratio (Sec. 2.5.4).

C_{26} steranes

Rarely used, potentially highly specific for eukaryotic and possibly prokaryotic input, measured using GCMSMS m/z 358 → 217.

Three series of C_{26}-steranes are known, 21-, 24- and 27-norcholestanes (Moldowan et al., 1991a) (Fig. 3.28). The only practical analysis for these C_{26}-steranes is by GCMSMS, as a result of strong interference by more abundant higher molecular-weight steranes in MID GCMS. The 5α,14α,17α(H) 20S + 20R and

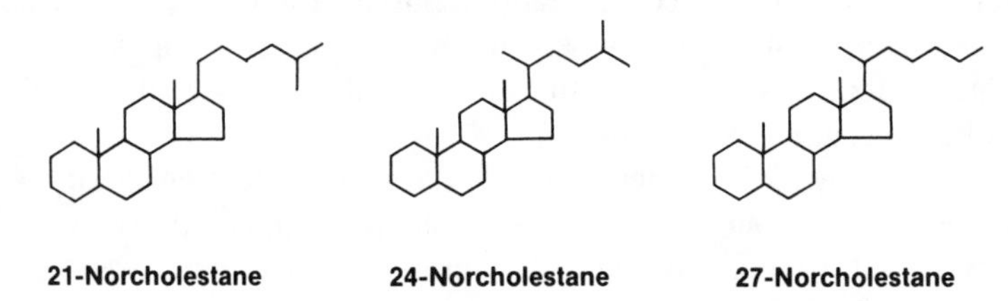

Figure 3.28 Three C_{26}-sterane structures have been identified (Moldowan et al., 1991a).

5α,14β,17β(H) 20S + 20R compounds have been identified in the 24- and 27-norcholestanes, but the 20S and 20R isomers do not occur in the 21-norcholestanes because of lack of the methyl group at C-20. Under normal GC conditions 5α,14α,17α(H)- and 5α,14β,17β(H)-21-norcholestanes coelute (Fig. 3.29).

The 21- and 27-norcholestanes appear to have no direct sterol precursors, but may be derived through bacterial oxidation or thermally induced cleavage and loss of a methyl group from larger steroids ($>C_{26}$). On the other hand, traces of 24-norcholesterols occur in living marine algae and invertebrates, suggesting an origin in

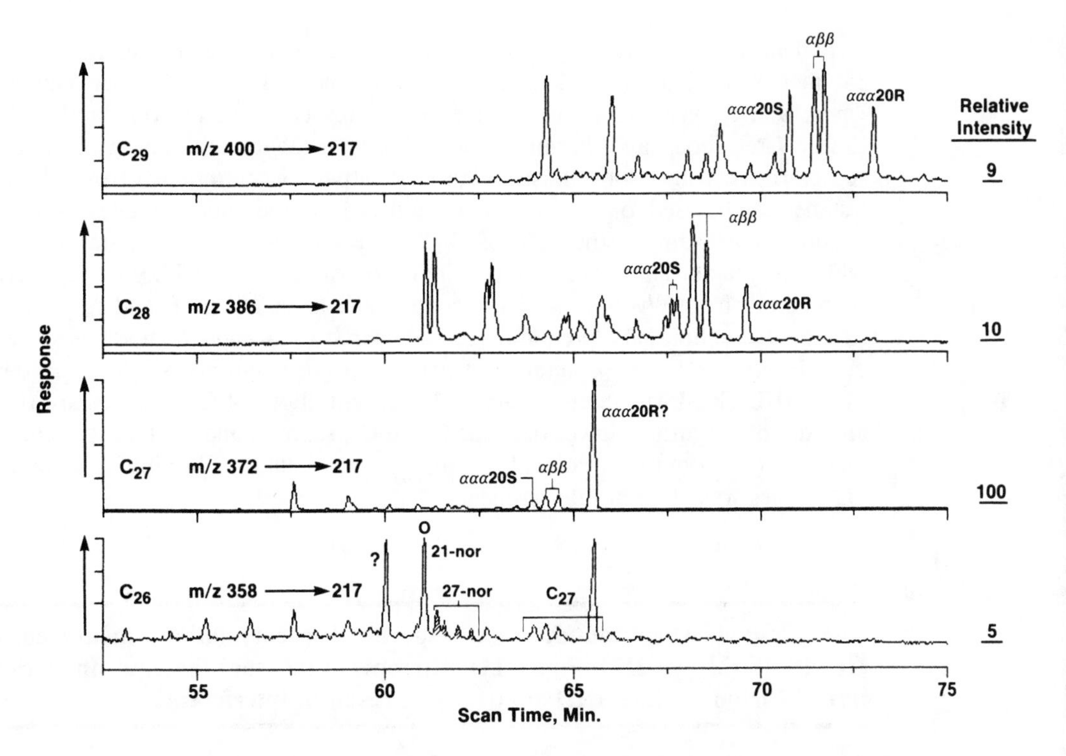

Figure 3.29 MRM GCMS analysis of steranes in a condensate from Angola (GCMS No. A159) shows a strong predominance of 21-norcholestanes among the C_{26}-steranes. The MRM method gives lower resolution than triple quadruple GCMSMS, allowing C_{27}-steranes to interfere on m/z 358 → 217 traces. Triple quadrupole data (Fig. 3.30) are not affected by this interference. The mature sterane epimer patterns for the C_{28} and C_{29} steranes are in accordance with this sample being a condensate. However, the pattern of the C_{27} steranes shows a strong $\alpha\alpha\alpha$20R peak, which could be interpreted as indicating immaturity. Because this pattern of C_{27} steranes is characteristic of many mature petroleums, it is suggested that a compound with similar properties may coelute with $5\alpha,14\alpha,17\alpha$(H), 20R-cholestane.

eukaryotes with or without prokaryotic symbionts (Goad and Withers, 1982). All three series of C_{26}-steranes have been found in both marine and nonmarine petroleums.

> ***Note:*** *Symbionts are organisms that live together, sometimes to their mutual benefit. For example, lichens consist of algae and fungi which grow together as symbionts. Root nodules of leguminous plants, such as soybeans, peas, or alfalfa, commonly contain symbiotic bacteria of the genus* Rhizobium. *These nodules are sites of nitrogen fixation, where atmospheric N_2 is converted to combined organic nitrogen, as found in the chlorophyll and amino acids that are critical to sustain plant life. Without the ability to fix nitrogen, legumes would be less likely to survive in unfertilized soils. Neither legumes nor* Rhizobium *alone are able to fix nitrogen.*

The ratio of $C_{24}/(C_{24} + C_{27})$-norcholestanes is an effective source correlation parameter. It has been used to distinguish marine and nonmarine petroleums from Upper and Lower Cretaceous source rocks, respectively, in Angola (Fig. 3.30), Lower Cretaceous and Permian sourced oils in Wyoming, and two nonmarine source rocks in the Bohai Basin, China. Relative concentrations of the 21-norcholestanes can be used to gauge thermal maturity from the middle to late oil-generative window. For example, the ratio of $C_{21}/(C_{21} + C_{24} + C_{27})$-norcholestanes increases with API gravity and other maturity indicators in a series of Phosphoria (Permian) sourced oils from Wyoming. Mature condensates from Wyoming and Angola show a strong 21-norcholestane predominance compared to 24- and 27-norcholestanes.

Information on C_{26} steranes in petroleum is seldom accessible using conventional MID GCMS methods because: (1) concentrations of C_{26} steranes are typically an order of magnitude lower than the C_{27} to C_{29} steranes and (2) their gas chromatographic retention times coincide with the early eluting C_{27} to C_{29} steranes and diasteranes, resulting in interference.

C_{27}-C_{28}-C_{29} steranes

Always used when found, highly specific for correlation, measured using GCMSMS ($M^+ \rightarrow 217$) (Fig. 2.11); attempts to measure these parameters using m/z 217 from routine MID analyses can result in interference.

Based on a study of recent marine and terrestrial sediments, Huang and Meinshein (1979) show that the ratio of cholest-5-en-3β-ol to 24-ethylcholest-5-en-3β-ol is a source parameter. They extended this approach by suggesting that the distributions of C_{27}-, C_{28}-, and C_{29}-homologous sterols on a ternary diagram can be used to differentiate ecosystems. Similarly, the relative abundances of C_{27}-, C_{28}-, and C_{29}-sterane homologs in oils reflect the carbon number distribution of the sterols in the organic matter in the source rocks for these oils (Mackenzie et al., 1983b; Moldowan et al., 1985).

Figure 3.31 shows a C_{27}-C_{28}-C_{29} sterane ternary diagram that represents a composite of data for oils from various depositional environments (Moldowan et al., 1985). There is so much overlap on this figure that the analysis is seldom used to differentiate depositional environments of the source rocks for petroleums, with the possible exception of certain samples containing predominantly higher plant organic matter (e.g., nonmarine shales in area B, Fig. 3.31). Monoaromatic steroid ternary diagrams (see following text) are more useful for this purpose (Moldowan et al., 1985).

Nevertheless, sterane ternary diagrams are used extensively to show relationships between oils and/or source rock bitumens (e.g., Peters et al., 1989). Based on our experience, plot locations on these diagrams do not change significantly throughout the oil-generative window. However, some workers contend that a slight relative

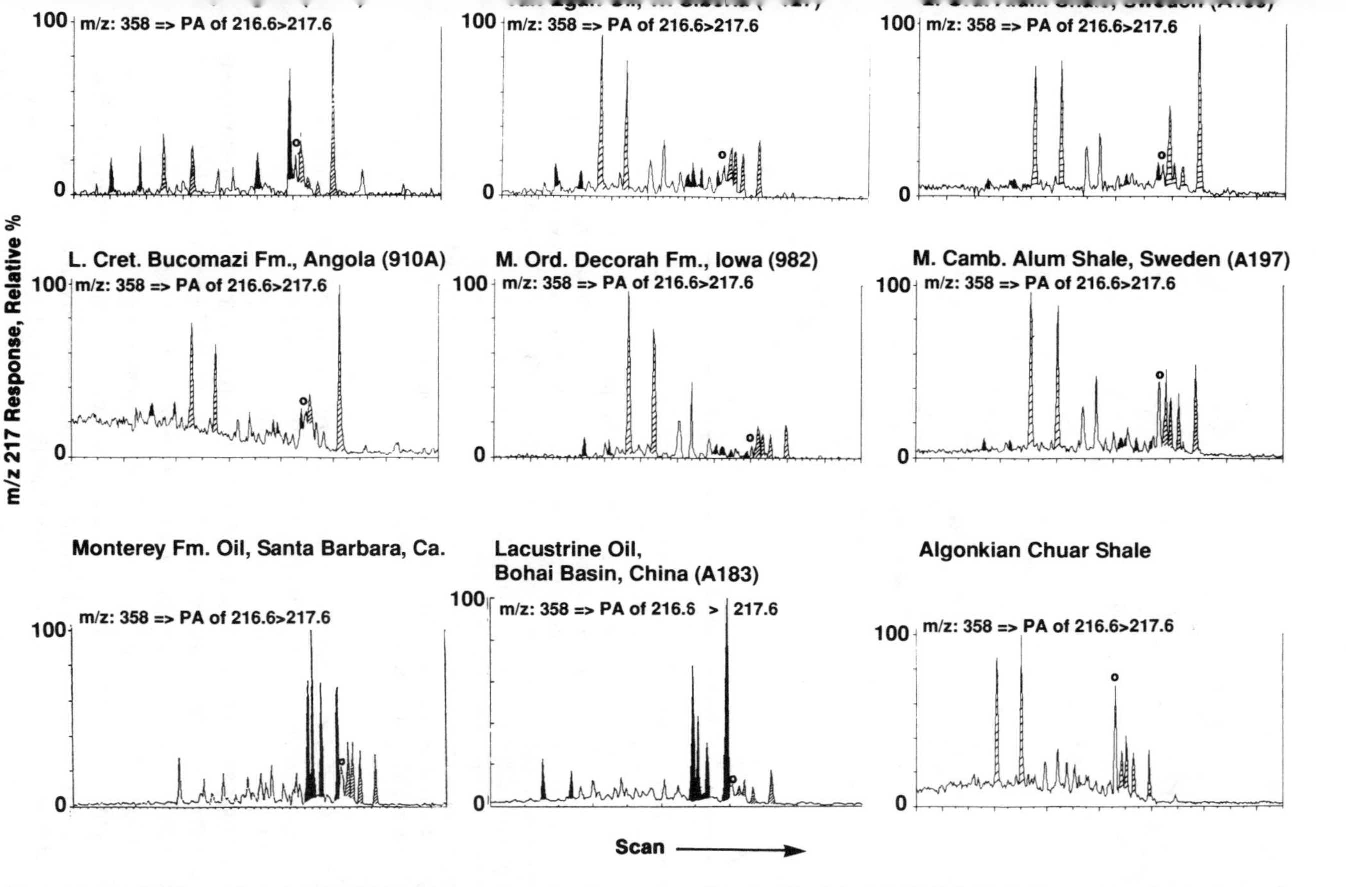

Figure 3.30 Triple quadrupole GCMSMS of nine oils and rock extracts using the Finnigan MAT TSQ-70 shows various distributions of C_{26} steranes (m/z 358 → 217). The Paleozoic and older rocks show little or no 24-norcholestanes. A lacustrine rock (Bucomazi Formation, Angola) shows little 24-norcholestane, whereas another (Bohai Basin, China) shows a strong predominance of 24-norcholestanes. The Monterey Formation sample also shows predominant 24-norcholestanes, but this is unusual. Most oils show a predominance of 27-norcholestanes.

Black peaks = 24-norcholestane (due to low concentrations, identifications are uncertain in some samples); Hachured peaks = 27-norcholestanes; Circle over peak = 21- norcholestanes; Refer to Fig. 3.28 for structures.

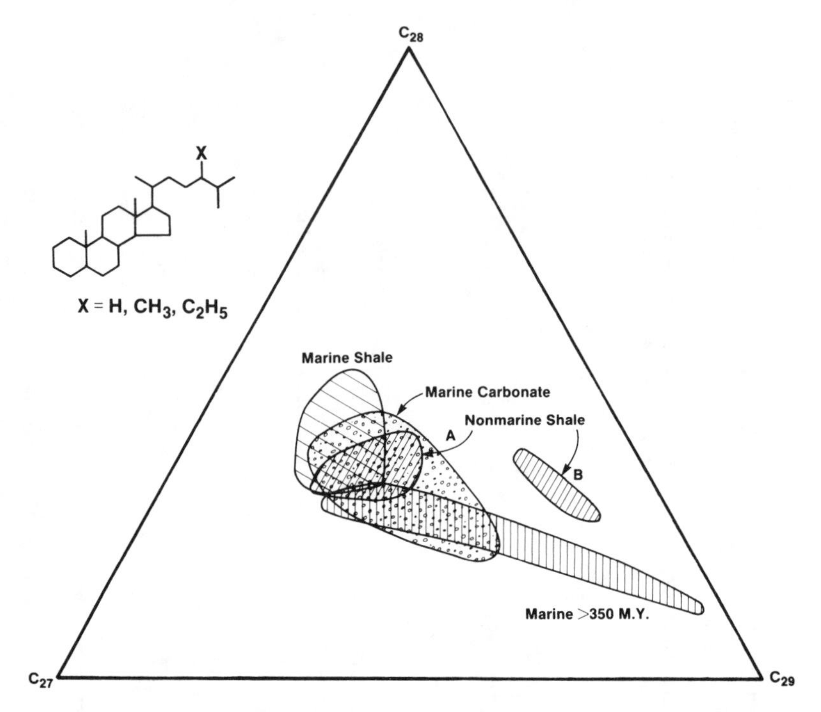

Figure 3.31 Ternary diagram showing the relative abundances of C_{27}-, C_{28}-, and C_{29}-regular steranes [5α(H),14α(H),17α(H),20S + 20R and 5α(H),14β(H),17β(H),20S + 20R] in the saturate fractions of petroleums determined by GCMSMS ($M^+ \rightarrow 217$). The labeled areas represent a composite of data for oils where the source rocks are known (Moldowan et al., 1985). The extensive overlap between the different oil source types on the diagram limits the use of sterane hydrocarbon distributions in describing the depositional environment of organic matter contributing to the petroleum. However, the diagram is still useful for showing relationships among oils and bitumens. Nonmarine shale areas A and B refer to petroleums sourced mainly from nonmarine algal organic matter or terrestrial (higher plant) organic matter, respectively.

depletion of ααα 20R-ethylcholestane may occur relative to the other epimers with increasing maturity (Mackenzie, 1984; Curiale, 1986; Sakata et al., 1988; Curiale, 1992).

The principal use of C_{27}-C_{28}-C_{29} sterane ternary diagrams is to distinguish groups of petroleums from different source rocks or different organic facies of the same source rock. For example, Grantham et al. (1987) used a sterane ternary diagram, stable carbon isotope ratios, and other supporting data to classify five groups of Oman oils and relate most of them to source rock extracts. Palmer (1984a) used a sterane ternary diagram to differentiate two major lacustrine organic facies in the Eocene-Oligocene Elko Formation from northeastern Nevada: a lignitic siltstone and an oil shale. Bitumens from the siltstones showed a predominance of C_{29} steranes, high pristane/phytane (>1), abundant C_{19} and C_{20} tricyclic diterpanes, and other characteristics indicating higher plant input. The oil shale bitumens showed more C_{27} and C_{28} steranes than those from the siltstones, and other characteristics of oil-prone organic matter, including low pristane/phytane (<0.5), high C_{28}, C_{29}, and C_{30} 4-methyl steranes, and the presence of gammacerane.

Because the relative concentration of specific isomers may depend on more than the initial input of steroids, some ternary diagrams of steroid composition use the sum of several isomers. We use the sum of the $5\alpha,14\alpha,17\alpha$ 20S and 20R and $5\alpha,14\beta,17\beta$ 20S and 20R peaks obtained from GCMSMS analysis of the four major epimers of the C_{27}, C_{28}, and C_{29} steranes in petroleum.

The accuracy of the $\%C_{27}/(C_{27}$ to $C_{29})$, $\%C_{28}/(C_{27}$ to $C_{29})$, and $\%C_{29}/(C_{27}$ to $C_{29})$ sterane ratios used in the ternary diagrams depends on separating the individual carbon numbers from interfering peaks. It is possible to obtain information on C_{27}, C_{28}, and C_{29} steranes in petroleum using conventional MID mode GCMS, but the specificity of sterane analysis improves using GCMSMS. We do not recommend that the reader construct ternary diagrams using data from different instruments (e.g., benchtop GCMS versus GCMSMS) or from the same instrument analyzed at different times unless the data are corrected using standards.

High concentrations of C_{29} steranes (24-ethylcholestanes) compared to the C_{27}- and C_{28}-steranes may indicate a land-plant source (e.g., Czochanska et al., 1988). This interpretation is based on the work of Huang and Meinshein (1979), who observed high C_{29}-sterol predominances in higher plants and sediments. However, as recommended by Volkman (1986, 1988), caution should be applied in interpreting C_{29}-sterol predominances in recent sediments. For example, Volkman et al. (1981) show that 24-ethylcholest-5-en-3β-ol (a C_{29} sterol) is a significant component in a mixed diatom culture. The validity of the C_{29} sterol as a terrestrial marker is thus, questionable. Furthermore, many Paleozoic and older oils and some oils from carbonate source rocks contain high C_{29} steranes, but little or no higher plant input (e.g., Moldowan et al., 1985; Grantham, 1986b; Rullkötter et al., 1986; Vlierbloom et al., 1986; Fowler and Douglas, 1987; Buchardt et al., 1989). Land plants were absent and cannot account for C_{29} steranes in rocks or oils older than Devonian in age. In crude oils of probable Precambrian origin from southern Oman, Grantham (1986a) attributed strong C_{29}-sterane predominances to algae.

Data indicate a general increase in the relative content of C_{28} steranes and a decrease in C_{29} steranes in marine petroleum through geologic time (Moldowan et al., 1985; Grantham and Wakefield, 1988). The increase in the C_{28} steranes may be related to increased diversification of phytoplankton assemblages including diatoms, coccolithophores and dinoflagellates in the Jurassic and Cretaceous. Although this approach is not sufficiently accurate to determine the age of the source rock for an oil, it is possible to distinguish Upper Cretaceous age and Tertiary oils from Paleozoic or older oils (Grantham and Wakefield, 1988). These authors observed that the C_{28}/C_{29} sterane ratio is less than 0.5 for Lower Paleozoic and older oils, 0.4 to 0.7 for Upper Paleozoic to Lower Jurassic oils, and greater than about 0.7 for Upper Jurassic to Miocene oils. We do not recommend such an age differentiation without additional supporting data. The age relationship between the relative contents of C_{28} and C_{29} steranes described by Grantham and Wakefield (1988) applies only to samples from marine source rocks. Furthermore, we have observed many exceptions to this relationship, even for petroleums that are clearly of marine origin.

$C_{30}/(C_{27}$ to $C_{30})$ steranes (C_{30}-sterane index)

Often used, highly specific for marine organic matter input, measured using GCMSMS of M+ (414) → 217.

The presence of 4-desmethyl C_{30} steranes as measured by GCMSMS is the most powerful parameter for identifying input from marine organic matter to the source rock (Moldowan et al., 1985; Peters et al., 1986). These C_{30} steranes are identified as 24-*n*-propylcholestanes (Fig. 1.10), which are related to 24-*n*-propylcholesterols (Moldowan et al., 1990). The latter are biosynthesized by marine Chrysophyte algae of the order *Sarcinochrysidales* and are common in marine invertebrates, presumably via a dietary origin (Raederstorff and Rohmer, 1984). Analysis of the 24-*n*-propylcholestanes requires GCMSMS (m/z 414 → 217) because they are usually low compared to related compounds showing similar GC retention times, such as the C_{27}- to C_{29}-steranes and some C_{30} 4-methyl steranes.

Figure 3.32 shows GCMSMS mass chromatograms of the C_{29} and C_{30} steranes for a series of oils and bitumens from the North Sea (Peters et al., 1989). The C_{30} steranes in the oils or bitumens with input from marine source rock (Beatrice oil, Piper oil, Kimmeridge bitumen, Middle Jurassic bitumen) all elute at predicted retention times after their C_{29} homologs. The two Devonian bitumens lack C_{30} steranes, consistent with geologic, paleontologic, and geochemical evidence that they are lacustrine in origin. Similar application of C_{30} sterane data supported conclusions on the origins of various Brazilian petroleums (Mello et al., 1988a and b).

Where C_{30} steranes show low concentrations in petroleum, questions may arise regarding their correct identification. For example, Bailey et al. (1990) analyzed two Beatrice oil samples and concluded that C_{30} steranes were too minor to identify with confidence and were effectively absent. The striking difference in signal-to-noise ratio between mass chromatograms for C_{30} steranes obtained by Peters et al. (1989) and Bailey et al. (1990) for Beatrice oil is difficult to explain. Both groups of workers used MRM GCMS, although the specific instruments may differ in sensitivity. Peters et al. (1989) analyzed a saturate fraction, which concentrates components relative to the whole oil. However, Bailey et al. (1990) indicate that in addition to the whole oil used for their published mass chromatogram, analyses of the saturate fraction of Beatrice oil failed to detect C_{30} steranes.

Ratios of $C_{30}/(C_{27}$ to $C_{30})$ steranes plotted against oleanane/hopane give a better assessment of marine versus terrestrial input to petroleum than either parameter alone (Fig. 3.33). Based on a plot of $C_{30}/(C_{27}$ to $C_{30})$ steranes versus C_{34} or $C_{35}/(C_{31}$ to $C_{35})$ 17α(H)-homohopanes, Moldowan et al. (1992) propose that oils derived

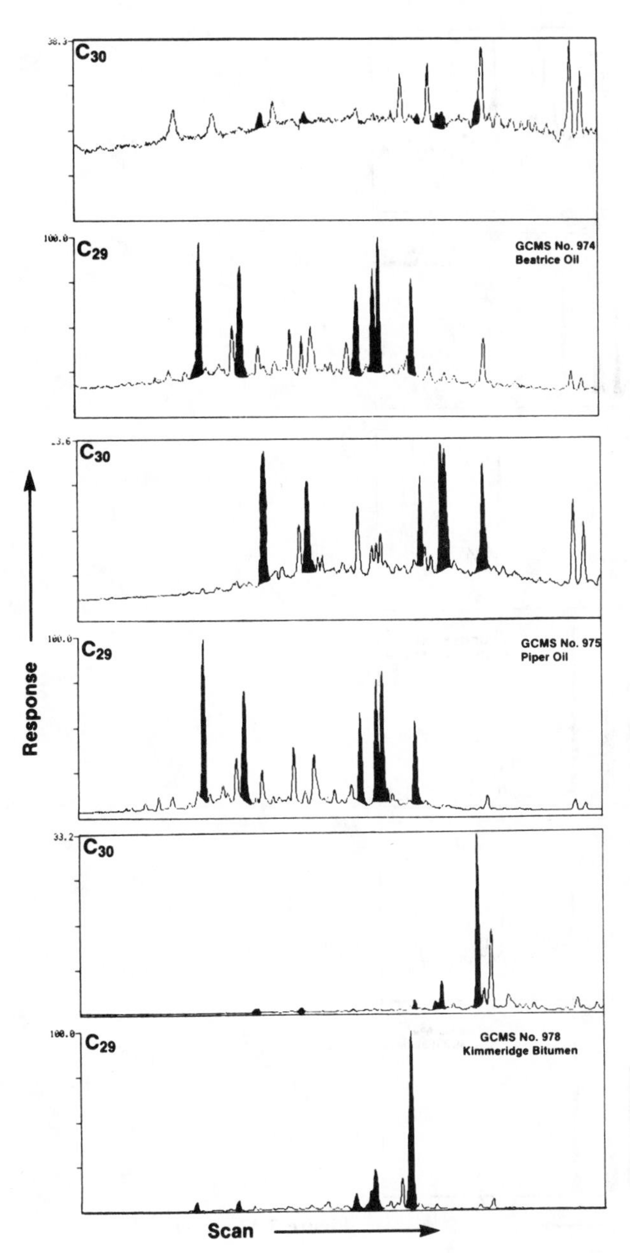

Figure 3.32 C_{29}- and C_{30}-sterane distributions for saturate fractions of oils and bitumens obtained by MRM GCMS. The distributions were obtained by monitoring parent to daughter ion transitions, $M+ \rightarrow m/z\ 217$, where M+ corresponds to molecular ions at m/z 400 and m/z 414 and C_{29}- and C_{30}-steranes, respectively. Note the interference of terpanes on the $m/z\ 414 \rightarrow 217$ (C_{30}) chromatograms. The higher resolution of a triple quadrupole mass spectrometer (monitoring CAD ions) compared to the magnetic sector instrument (monitoring metastable transitions) used here prevents this interference.

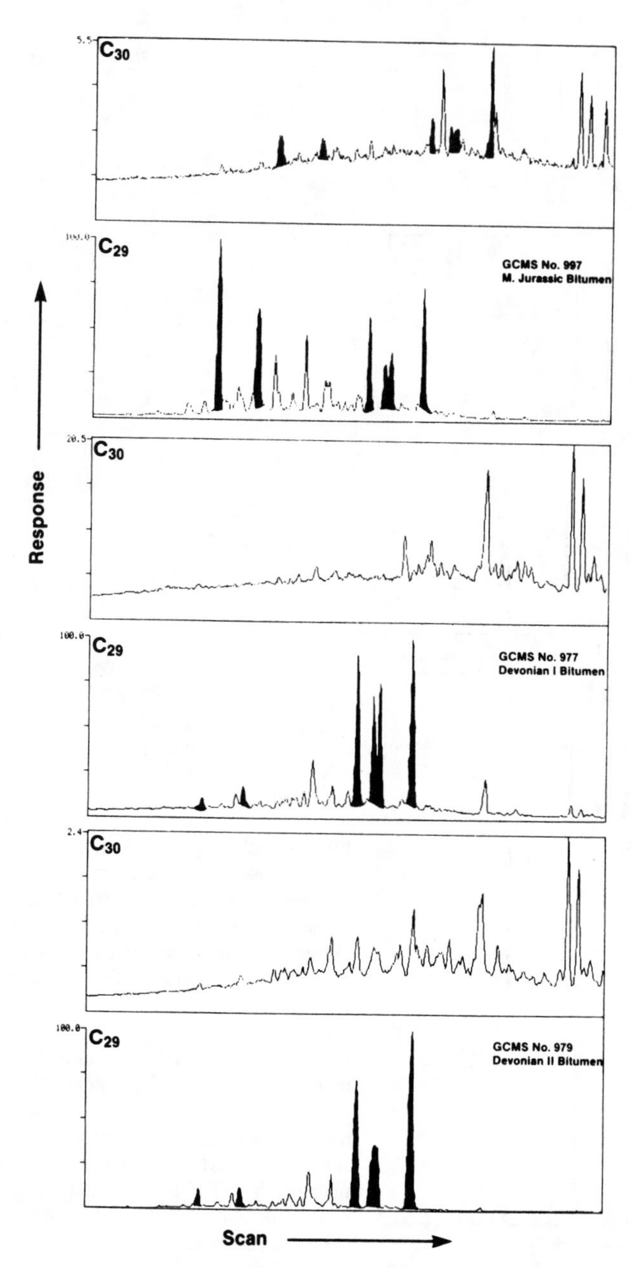

Figure 3.32 (*cont.*)

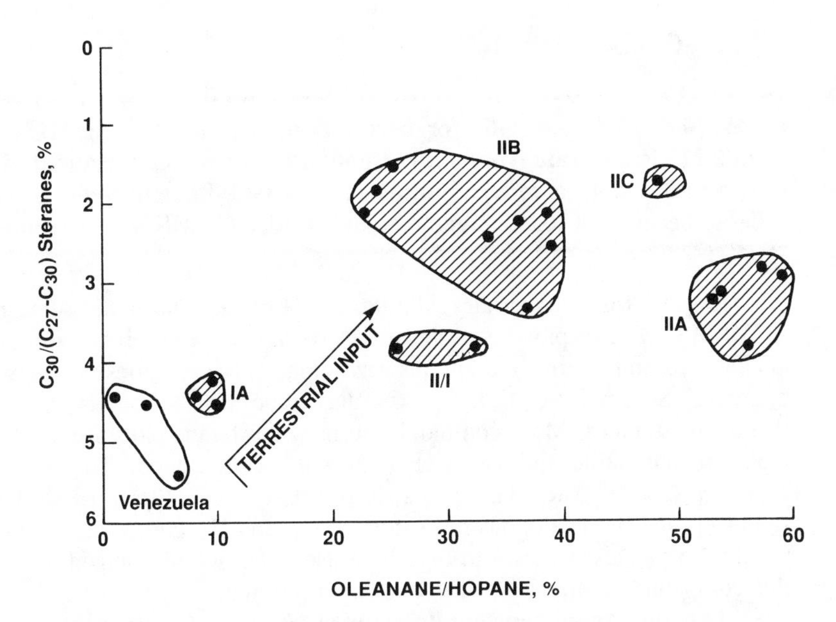

Figure 3.33 Oils and seep oils from Colombia are grouped and compared with oils from the Maracaibo Basin, Venezuela, using relative abundances of marine (24-n-propylcholestanes) and terrestrial (oleanane) markers. Oil group IIA, IIB, and IIC have relatively high oleanane and lower C_{30}-steranes suggesting a deltaic source with strong terrestrial input. Oil group IA has higher C_{30}-sterane and lower oleanane contents, like the Venezuela oils (probably sourced from the La Luna Formation), suggesting they have a marine source which is more remote from terrestrial influences.

from source rocks deposited under restricted saline to hypersaline lagoonal conditions show lower $C_{30}/(C_{27}$ to $C_{30})$ sterane ratios than those from open marine systems. They show that the C_{30} sterane ratio varies inversely with C_{34} 17α(H)-homohopanes in Cretaceous and Liassic-Triassic-sourced oils from the Adriatic Basin. Low C_{30}-sterane ratios and high C_{34} 17α(H)-homohopanes appear to indicate a restricted, evaporitic source rock depositional environment.

Values of zero for the C_{30}-sterane ratio generally correspond to nonmarine oils. Most of the small number of Cambrian and Precambrian oils thus far examined (Moldowan et al., 1985; Peters et al., 1986) show no detectable C_{30} steranes. The absence of C_{30} steranes in petroleum older than about 500 m.y. was interpreted as an evolutionary lag in the appearance of C_{30} sterols in marine organisms or domination of the marine biota by a few species that did not contain C_{30} sterols (Moldowan et al., 1985). However, we have recently detected C_{30} steranes in bitumens extracted from Precambrian Chuar rock in Arizona.

C_{27}-C_{28}-C_{29} diasteranes

Often used, highly specific for correlation, measured using GCMSMS of $M^+ \rightarrow m/z$ 217 in saturate fraction. (Although the principal fragment of diasteranes shown in Fig. 2.11 is m/z 259 under routine GCMS conditions, m/z 217 is more reliable because of its stronger response under GCMSMS conditions.)

Results for the ternary diasterane plots are normally determined using GCMSMS. These plots can be used to support interpretations based on the analogous sterane ternary diagrams regarding source relationships between oils and bitumens. Specificity of C_{27}, C_{28}, C_{29}-diasterane distributions is about the same as that of the steranes. More commonly, ternary diasterane plots are used when sterane plots are unreliable and vice versa, as will be discussed. The data include [C_{27} $13\beta,17\alpha$ (20S + 20R) diasteranes]/[C_{27} + C_{28} + C_{29} $13\beta,17\alpha$ (20S + 20R) diasteranes], and the analogous ratios for the C_{28} and C_{29} compounds. The structure of C_{27} diacholestanes is shown in Fig. 1.10. Note the two rearranged methyl groups for this compound compared to the steranes in the figure.

The most important applications of C_{27}, C_{28}, C_{29}-diasterane plots are for: (1) heavily biodegraded oils where steranes are altered, but diasteranes remain intact, and (2) some highly mature oils and condensates that show low steranes, but more abundant diasteranes. On the other hand, some oils from clay-poor source rocks show high steranes, but the diasteranes are not useful for correlation because of low concentrations. Figure 3.34 shows triple quadrupole GCMSMS results for diasteranes in a biodegraded seep oil from Papua New Guinea.

Diasteranes/regular steranes

Often used, moderately specific for source rock mineralogy and oxicity, interference caused by thermal maturation, GCMSMS analysis of $M^+ \rightarrow m/z$ 217.

The conversion of sterols to diasterenes during diagenesis (Fig. 3.35) is believed to be catalyzed by acidic sites on clays (Rubinstein et al., 1975; Sieskind et al., 1979). Diasterenes are ultimately reduced to diasteranes (rearranged steranes) showing $13\beta,17\alpha$(H) 20S and 20R (major isomers) and $13\alpha,17\beta$(H) 20S and 20R (minor isomers) stereochemistries. The diasteranes/steranes ratio used at Chevron is based on [$13\beta,17\alpha$(H) 20S + 20R]/{[$5\alpha,14\alpha,17\alpha$(H) 20S + 20R] + [$5\alpha,14\beta,17\beta$(H) 20S + 20R]} for the C_{27}, C_{28}, and C_{29} steranes obtained from GCMSMS. Occasionally only one carbon number is used, for example C_{29}, as specified.

Diasteranes/steranes ratios are commonly used to distinguish petroleum from carbonate versus clastic source rocks (e.g., Mello et al., 1988b). **Low diasteranes/steranes ratios (m/z 217) in oils indicate anoxic, clay-poor, carbonate source rock.** During diagenesis of these carbonate sediments, bicarbonate and

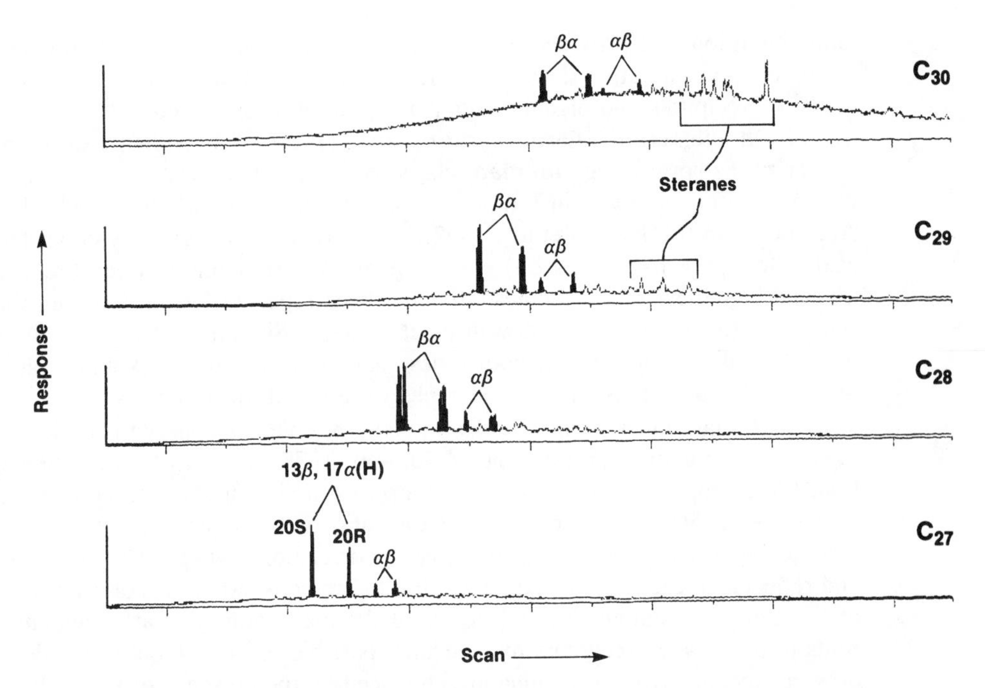

Figure 3.34 Seep oil (GCMS No. A6) from Papua New Guinea shows loss of most regular steranes caused by heavy biodegradation. However, some C_{29} steranes remain, showing their relative resistance to biodegradation compared to C_{28} and C_{27} steranes. The full suite of C_{30} steranes (24-*n*-propylcholestanes) remains intact, demonstrating their still greater resistance to biodegradation compared to the C_{27}-C_{29} steranes. The biodegradation ranking of this oil (level 8 in Fig. 3.62) is consistent with partial removal of 17α(H)-hopanes and all *n*-paraffins and isoprenoids. Despite biodegradation of the steranes, the unaltered C_{27}-C_{29} diasterane distribution in this oil was used in a ternary diagram to support its correlation with other samples.

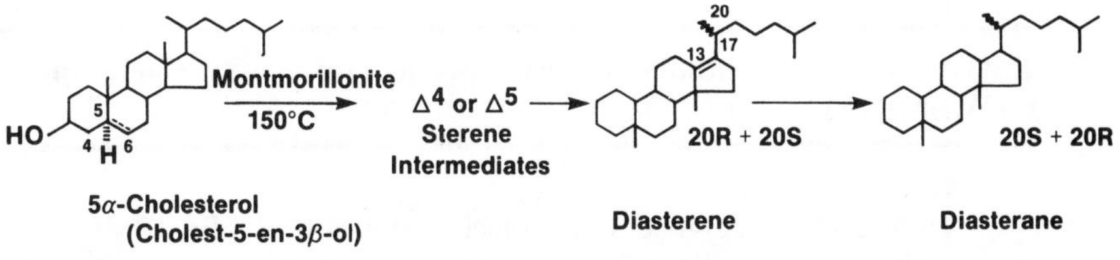

HO
5α-Cholesterol (Cholest-5-en-3β-ol)
Montmorillonite 150°C
Δ^4 or Δ^5 Sterene Intermediates
20R + 20S
Diasterene
20S + 20R
Diasterane

Figure 3.35 Diasterenes (rearranged sterenes) are believed to result from clay catalyzed rearrangement of sterols or sterenes during diagenesis and early catagenesis (Rubinstein et al., 1975). Saturation of the double bond (Δ13-17) in diasterenes during catagenesis results in diasteranes.

ammonium ions are provided by bacterial activity (Berner et al., 1970) resulting in increased water alkalinity. Under these conditions of low Eh and high pH, calcite tends to precipitate and organic matter preservation is improved.

High diasteranes/steranes ratios are typical of petroleum derived from source rocks containing abundant clays. However, high diasteranes/steranes ratios have also been observed in bitumens from organic-lean carbonate rocks from the Adriatic basin (Moldowan et al., 1992). These rocks were probably deposited in an acidic (low pH), oxic (high Eh) environment. A correlation between low pH, high Eh, and high diasteranes/steranes ratios has been reported for the Toarcian shales of southwestern Germany (Moldowan et al., 1986). Similarly, Palacas et al. (1984) noted high diasteranes in clay-poor limestones from Florida. Clark and Philp (1989) list several publications where diasteranes were found in carbonates.

High diasteranes/steranes ratios in some petroleums appear to result from: (1) high thermal maturity (Seifert and Moldowan, 1978), and/or (2) heavy biodegradation. For example, burial maturation of a series of similar shaly carbonate rocks increases both vitrinite reflectance and the ratio of C_{27} diasteranes/(C_{27} diasteranes plus steranes) (Goodarzi et al., 1989). Such correlations between the diasterane ratio and reflectance can only be applied to limited regions, where lithology and organic matter types are similar. At high levels of thermal maturity, rearrangement of steroids to diasterane precursors may become possible, even without clays, due to hydrogen exchange reactions which are enhanced by the presence of water (Rullkötter et al., 1984a). Alternatively, diasteranes may simply be more stable and survive thermal degradation better than steranes. The diasteranes/steranes ratio is useful for distinguishing source rock depositional conditions only when the samples show comparable levels of thermal maturity (Sec. 3.2.4.2).

Heavy biodegradation can result in selective destruction of steranes relative to diasteranes (Sec. 3.3). This interpretation can be supported by other evidence of heavy biodegradation, including depletion of *n*-paraffins and isoprenoids, or the presence of 25-norhopanes. However, it is possible that a nonbiodegraded oil might mix with a heavily biodegraded oil showing a much higher diasteranes/steranes ratio. In such cases, only careful quantitative assessment of each biodegradation-sensitive parameter can lead to a correct interpretation.

3-Alkyl steranes

Rarely used, potentially highly specific for correlation, measured using GCMSMS ($M^+ \rightarrow m/z$ 231).

Two recently assigned series of cholestanes (e.g., Fig. 3.36), ergostanes, and stigmastanes include the 2α-methyl steranes and 3β-alkyl steranes that show homologation up to at least 3β-*n*-pentyl (Dahl et al., 1992). Summons and Capon (1988, 1991) characterized the 2α-methyl and 3β-methyl and 3β-ethyl steranes using synthetic standards. In oils derived from source rocks of differing age and deposited under different environmental conditions 3β-methyl steranes predominate

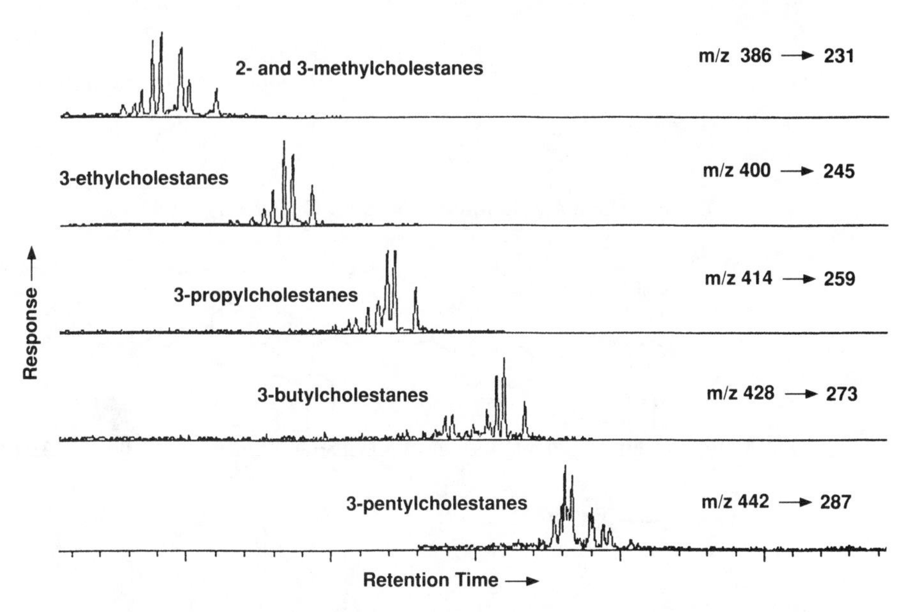

Figure 3.36 Homologous series of cholestane isomers with alkylation at the 3-position are detected in mass chromatograms constructed from GCMSMS parent-daughter transitions of the saturate fraction of the 81-3 Hamilton Dome oil from Wyoming. 3-Ethylcholestanes were confirmed by coinjection of authentic standards. Parent to m/z 262 transitions (see Fig. 2.11) confirmed alkylation in the A-ring for all homologs. Similar homologous series are observed for 3-alkyl ergosteranes and 3-alkyl stigmastanes. Intensities are normalized to the largest peak within each chromatogram.

over 2α-methyl steranes and their ratio remains constant. The distribution of 3β-methylcholestanes, 3β-methylergostanes and 3β-methylstigmastanes is the same as that of cholestanes, ergostanes and stigmastanes in the same oil sample. The same relationship holds for the 2α-methyl steranes. This evidence suggested to Summons and Capon (1988, 1991) that 3β-methyl, 3β-ethyl and 2α-methyl steranes originate from stenols via Δ^2-sterenes through a bacterial process, possibly methylation, and that 2α-methyl and 3β-methyl steranes have a common Δ^2-sterene precursor. However, Dahl et al. (1992) found that *n*-pentylcholestane is preferentially preserved in an anoxic marine environment (e.g., Monterey oil). This preservation is most pronounced in the alkanes released from the N, S, O fraction by Raney nickel. Thus, these compounds may be formed by bacteria, in close analogy to the biosynthesis of bacteriohopanetetrol (Flesch and Rohmer, 1988; see Sec. 1.2.10.2), by combining of a Δ^2-sterene and a C_5 sugar. No natural product analogs for 2- or 3-alkyl steranes are known in modern organisms.

The 2- and 3-methyl steranes, when found in the absence or near absence of 4-methyl steranes, are potentially age-related biomarkers for pre-Mesozoic petroleum. The 2- and 3-methyl steranes are the only nuclear methylated steranes in a suite of Paleozoic and Precambrian oils examined by Summons and Capon (1988). Surveys for 2- and 3-methyl steranes in Mesozoic and Tertiary oils are incomplete, but 4α-methyl steranes were usually predominant in the samples studied.

Analysis of 2- and 3-methyl steranes can be difficult because of the large numbers of diastereomers and isomers (i.e., rearranged isomers and 4-methyl steranes). For C_{30} methyl steranes, four series are known (Fig. 3.37).

Cholestane Alkylation Patterns for C30-Methyl Steranes

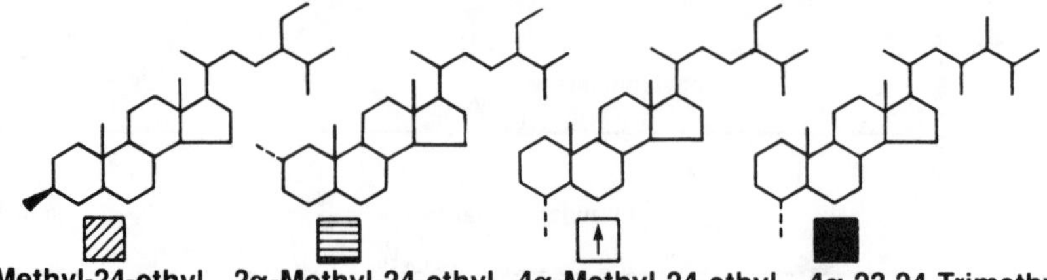
3β-Methyl-24-ethyl
2α-Methyl-24-ethyl
4α-Methyl-24-ethyl
4α,23,24-Trimethyl

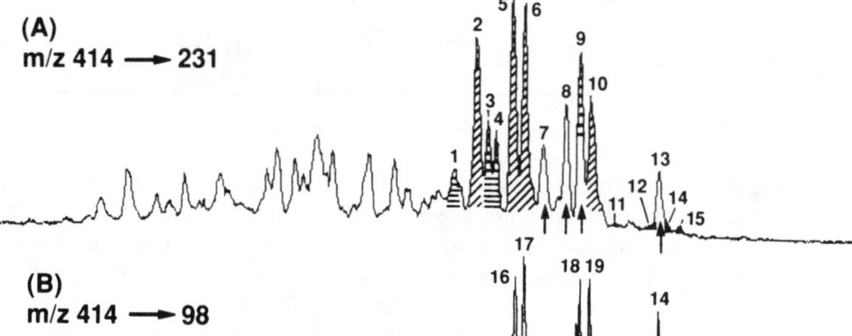
(A)
m/z 414 → 231
Response

(B)
m/z 414 → 98

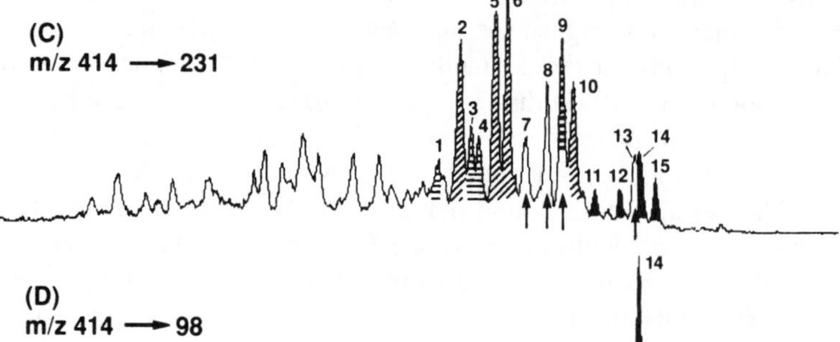
(C)
m/z 414 → 231

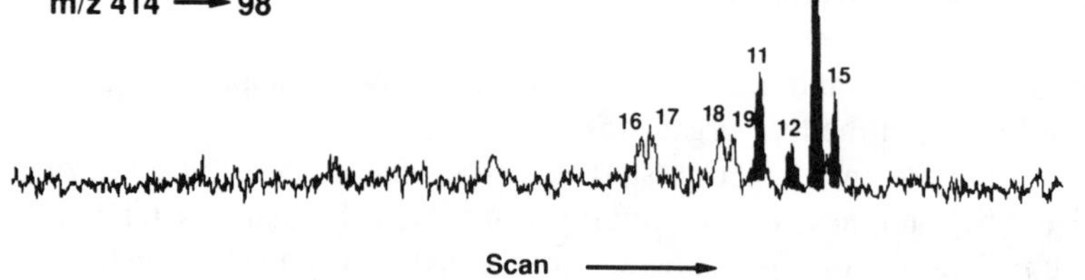
(D)
m/z 414 → 98
Scan

Figure 3.37

Figure 3.37 Four series of methylsteranes are detected in an Asian oil by GCMSMS of the saturate fraction using a 60 m DB-1 (J and W Scientific) fused silica capillary column. (A): Relative retention times of 2α-methyl-24-ethylcholestanes (horizontally striped peaks), 3β-methyl-24-ethylcholestanes (hachured peaks), 4α-methyl-24-ethylcholestanes (arrows), and 4α,23,24-trimethylcholestanes (dinosteranes, black peaks) are indicated on the m/z 414 $\rightarrow$ 231 chromatogram. (B): Relatively small concentrations of dinosteranes are shown by m/z 414 $\rightarrow$ 98 peaks that are insignificant for the other methylsteranes. (C) and (D): A coelution experiment using four dinosteranes confirms their presence in the oil. The four 2α-methyl-24-ethylcholestanes and four 3β-methyl-24-ethylcholestanes were identified by coelution of 2α-methyl-24-ethylcholestane and 3β-methyl-24-ethylcholestane (*Courtesy of R. Summons*) that had been isomerized over Pd/C catalyst at 260° (Seifert et al., 1983). The four dinosterane stereoisomers were prepared by reduction of dinosterol isolated from a gorgonian (*Courtesy of R. M. K. Carlson and D. S. Watt*), with stereochemical assignments by NMR (*Courtesy of R. M. K. Carlson*). Assignments of 4α-methyl-24-ethylcholestanes are by comparison of relative retention times with the literature (Summons et al., 1987). *Analyses were completed by P. A. Lipton; peak identifications courtesy of B. J. Huizinga*. Identity of peaks:

1. 2α-methyl-24-ethylcholestane 20S
2. 3β-methyl-24-ethylcholestane 20S
3. 2α-methyl-24-ethylcholestane 14β,17β(H),20R*
4. 2α-methyl-24-ethylcholestane 14β,17β(H),20S*
5. 3β-methyl-24-ethylcholestane 14β,17β(H),20R*
6. 3β-methyl-24-ethylcholestane 14β,17β(H),20S*
7. 4α-methyl-24-ethylcholestane 20S
8. 4α-methyl-24-ethylcholestane 14β,17β(H),20R*
9. 2α-methyl-24-ethylcholestane 20R + 4α-methyl-24-ethylcholestane 14β,17β(H),20S*
10. 3β-methyl-24-ethylcholestane 20R
11. 4α,23S,24S-trimethylcholestane 20R
12. 4α,23S,24R-trimethylcholestane 20R
13. 4α-methyl-24-ethylcholestane 20R
14. 4α,23R,24R-trimethylcholestane 20R
15. 4α,23R,24S-trimethylcholestane 20R
16. 4α,23,24-trimethylcholestane diastereomer**
17. 4α,23,24-trimethylcholestane diastereomer**
18. 4α,23,24-trimethylcholestane diastereomer**
19. 4α,23,24-trimethylcholestane diastereomer**

* 20S and 20R designations for 14β,17β(H) compounds may be reversed.

** Based on m/z 414 $\rightarrow$ 98 response and relative retention time.

4-Methyl steranes

Rarely used, but applications are rapidly increasing (e.g., 4-methyl steranes are abundant in certain lacustrine rocks, such as in China), potentially highly specific for marine or nonmarine dinoflagellates or bacteria, monitored using m/z 231 and 232 fragmentograms or preferably by GCMSMS.

The 4-methyl steranes can be divided into two major classes: (1) C_{28}-C_{30} analogs of the steranes substituted at positions 4 and 24 (e.g., the C_{30} compound is 4α-methyl-24-ethylcholestane), and (2) C_{30} dinosteranes (e.g., 4α,23,24-trimethylcholestanes). The origin of the 4,24-substituted methyl steranes is unclear. The 4-methyl steranes in petroleum may be mainly derived from 4α-methylsterols from living dinoflagellates (Wolff et al., 1986). However, 4α-methylsterols have also been found in prymnesiophyte microalgae of the genus *Pavlova* (Volkman et al., 1990). Bird et al. (1971) propose an additional source from certain bacteria, notably *Methylococcus capsulatus,* but these 4-methylsterols are not alkylated at C-24. Dinosterane is derived from dinosterol or dinostanol and appears to be particularly specific for input from dinoflagellates (Withers, 1983). Both marine and nonmarine petroleums contain 4-methyl steranes (Moldowan et al., 1985; Summons et al., 1987; Goodwin et al., 1988). Several nonmarine dinoflagellate species have been identified (Curiale, 1987 and references therein). In marine dinoflagellates, 4α-methylsterols are more abundant than the desmethylsterols.

Fu Jiamo et al. (1990) show that bitumens from nonmarine rocks deposited under hypersaline conditions contain less C_{28} vs. C_{29} and C_{30} 4-methyl steranes than those from freshwater and brackish water lacustrine environments in China. Further, samples from freshwater settings contain more total 4-methyl steranes than the others. They conclude that species differences among dinoflagellates in freshwater, brackish, and hypersaline lacustrine settings account for the different 4-methyl compositions, and that dinoflagellate blooms occur mainly in the freshwater settings.

Abundant C_{30} 4-methyl steranes are found in both freshwater lacustrine and marine sediments. Both 4α-methyl-24-ethylcholestanes and 4α,23,24-trimethylcholestanes (dinosteranes) are found in marine dinoflagellate-rich sediments, but the only C_{30} 4α-methyl sterane found to date in lacustrine sediments is 4α-methyl-24-ethylcholestane (Goodwin et al., 1988; Summons et al., 1987). The 4α-methyl-24-ethylcholestanes are less specific for dinoflagellates than dinosteranes because they may have an origin from 4α-methyl-24-ethylcholesterols in prymnesiophyte algae (Volkman et al., 1990).

Dinosterane has only been reported in petroleums younger than Triassic age (Summons et al., 1987). This age relationship corresponds with the earliest widespread fossil evidence for dinoflagellates during the Triassic, although sporadic occurrences of possible dinoflagellate fossils are indicated into the Paleozoic with the oldest suspected species from rocks of Silurian age (Tappan, 1980). Black shales in the Norian/Rhaetian (Upper Triassic) sequence from Watchet in North Somerset,

U.K., contain abundant cysts of the earliest positively identified dinoflagellate, *Rhaetogonyaulax rhoetica,* and abundant 4-methylsteranes, including 4,23,24-trimethylcholestanes (Thomas et al., 1989).

The 4α-methyl steranes are an analytical challenge (Fig. 3.37) because of their complexity and because their major fragments (except m/z 231 and 232) are the same as the steranes (Fig. 2.11). The major homologs are 4α-methyl analogs of cholestane (C_{28}), ergostane (C_{29}), and stigmastane (C_{30}) plus dinosterane (4,23,24-trimethylcholestane). The 4β- methyl group occurring in immature rocks doubles the potential number of stereoisomers compared to the steranes, however the 4α(CH_3)-isomers strongly dominate in thermally mature bitumens (Wolff et al., 1986). Rubinstein and Albrecht (1975) have identified 4β-methyl steranes in the more immature parts of the Toarcian Formation in the Paris basin. Epimerization at C-23 and C-24 in dinosterane results in four stereoisomers. The large number of 4-methyl sterane isomers suggests that these compounds will eventually be as useful as the steranes for oil-oil and oil-source rock correlation. The same as the steranes, the complexity of 4-methyl sterane distributions typically necessitates the use of GCMSMS for the most reliable results. The 4-methyl steranes can be analyzed by MID GCMS using a benchtop quadrupole system such as the GC-MSD (Sec. 2.5.6), but only when concentrations of these compounds are high, as in certain immature lacustrine oils (Hwang et. al., 1989). However, analysis of 4-methyl steranes in many samples where these compounds are low may be difficult even using GCMSMS, particularly for the C_{30} compounds. For example, in some oils m/z 414 $\rightarrow$ 231 shows 2-, 3-, and 4-methyl-24-ethylcholestanes and their many stereoisomers plus all the epimeric dinosteranes (Fig. 3.37). Dinosterane has an additional m/z 98 fragment (Fig. 2.11) from the side chain which can be used for its identification (i.e., m/z 414 $\rightarrow$ 98; Summons et al., 1987).

3.1.3.3 Aromatic Steroids and Hopanoids. Aromatic biomarkers can provide valuable information on organic matter input. For example, aromatic hopanoids originate from bacterial precursors, while tetra- and pentacyclic aromatics with oleanane, lupane, or ursane skeletons indicate higher plant input (Garrigues et al., 1986; Loureiro and Cardoso, 1990). However, more detailed research appears necessary before many of these aromatic compounds can be used in routine exploration applications. At this time, aromatic biomarkers are primarily used for oil-to-oil and oil-to-source rock correlations (see the following) and as supporting evidence for assessments of thermal maturity (Sec. 3.2.4.3).

C_{27}-C_{28}-C_{29} C-ring monoaromatic steroid hydrocarbons

---

Often used, high specificity as a correlation tool, eukaryotic species input, measured using m/z 253 fragmentograms (Fig. 2.11).

Plot locations of C-ring monoaromatic (MA)-steroids on C_{27}-C_{28}-C_{29} ternary diagrams (Fig. 3.38) have been related to various types of source input in a manner

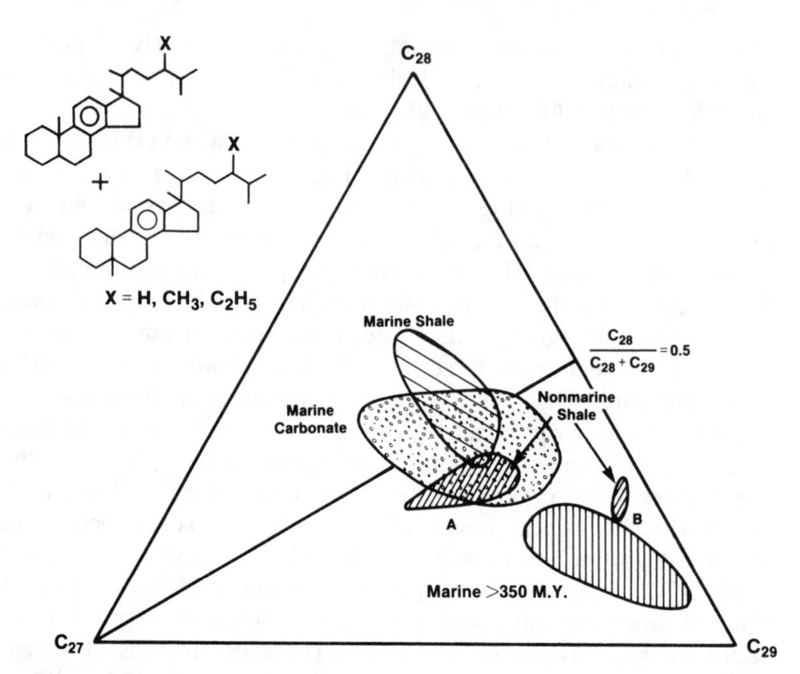

Figure 3.38 Monoaromatic steroids in oils provide information on source rock characteristics. The ternary diagram shows the relative abundance of C_{27}-, C_{28}-, and C_{29}- monoaromatic steroids in aromatic fractions of petroleums determined by GCMS. The labeled areas represent a composite of data for oils where the source rocks are known (Moldowan et al., 1985). MA-steroid distributions are more variable than those for steranes (Fig. 3.31), which enhances their usefulness in describing the depositional environment of the source rock for petroleums. The most apparent use of the MA-steroid triangular diagram is in distinguishing oils derived from nonmarine versus marine shales. Samples from marine shale sources generally contain less C_{29} MA-steroids than nonmarine oils. Typically, more terrigenous organic matter is deposited in nonmarine than marine source rocks, and the nonmarine rocks thus contain more C_{29} sterols. Alternately, nonmarine algae may contain relatively higher amounts of C_{29}-sterols (e.g., Moldowan et al., 1985; Volkman, 1986). A = oils derived from nonmarine algal input; B = oils derived from higher plant input.

similar to the early sterol work of Huang and Meinshein (1979). C-ring MA-steroids may be derived exclusively from sterols with a side chain double bond during early diagenesis (Riolo et al., 1986; Moldowan and Fago, 1986). In this respect, C-ring MA-steroids are more precursor-specific than steranes. Thus, **ternary diagrams of both MA-steroids and steranes provide more powerful evidence for correlations than either one alone because they represent compounds of differing origins and thus, provide independent evidence for correlation purposes. Furthermore, plot locations on these diagrams do not change significantly throughout the oil-generative window** (e.g., Peters et al., 1989).

MA-steroids in petroleum plot on the C_{27}-C_{28}-C_{29} diagram in fields associated with terrestrial, marine, or lacustrine input, although overlaps in the distributions occur (Moldowan et al., 1985). The field for carbonate-derived marine oils extends to higher C_{29} MA-steroids than shale-derived marine oils, although there is considerable overlap. Terrestrial input is poor in C_{27} and C_{28} MA-steroids. Nonmarine shales show MA-steroid $C_{28}/(C_{28} + C_{29})$ ratios less than 0.5 (however, see Peters et al., 1989).

Limited data for Lower Paleozoic oils show an increase in the $C_{29}/(C_{27}$ to $C_{29})$ MA-steroid ratio with age. The same tendency has been observed for steranes (Moldowan et al., 1985; Grantham and Wakefield, 1988). However, several Lower Paleozoic and Precambrian oils have recently been reported which show higher C_{27} and/or C_{28} steranes than expected based on this model (R. Summons and D. McKirdy, 1988, pers. comm.).

Accurate measurement of C-ring MA-steroids requires high resolution capillary GC columns (narrow I.D., > 50 m length) and authentic standards to identify peaks. In mature petroleum, monoaromatic steroids are dominated by those with an aromatic C-ring showing a few different combinations of rearranged methyl groups (Fig. 3.39). A typical distribution of MA-steroids in a marine petroleum is shown in Fig. 3.40(a); however, MA-steroid distributions dominated by the $5\beta(CH_3)$, 10β(H)-isomers are also common [Fig. 3.40(b)].

For ternary (triangular) diagrams as described, the ratios of $C_{27}/(C_{27}$ to $C_{29})$, $C_{28}/(C_{27}$ to $C_{29})$, and $C_{29}/(C_{27}$ to $C_{29})$ MA-steroids are measured. For each carbon number, six isomeric compounds are used in these ratios including 5α (20S + 20R),

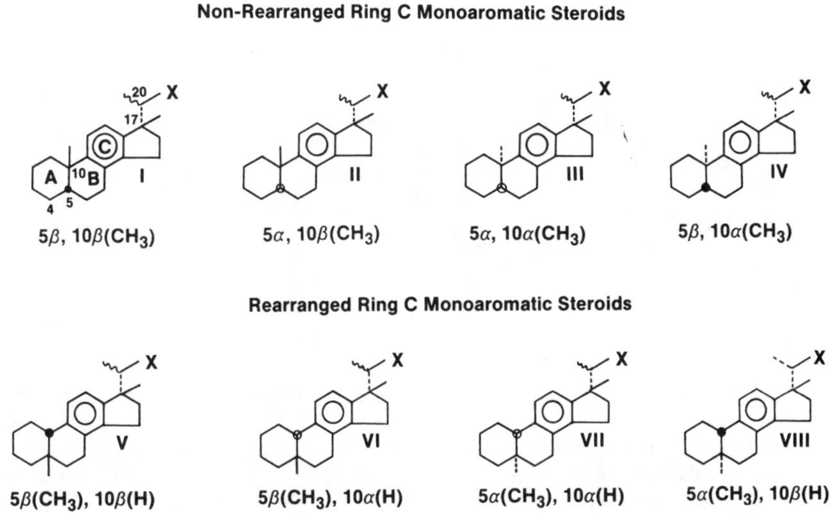

Non-Rearranged Ring C Monoaromatic Steroids

Rearranged Ring C Monoaromatic Steroids

Figure 3.39 These monoaromatic steroid structures are known to occur in petroleum. Structures I, II, and V are the most common and are routinely quantified for MA-steroid ternary diagrams (Fig. 3.38). Structures V and VII have been noted as particularly important in anhydrites from sabkha environments while Structure VII is much less significant in pure carbonates (Riolo et al., 1986; Connan et al., 1986). The other isomers generally occur in lower relative amounts than Structures I, II, V, and VII, and more polar GC columns (e.g., OV-73 and OV-1701) have been used for their analysis.

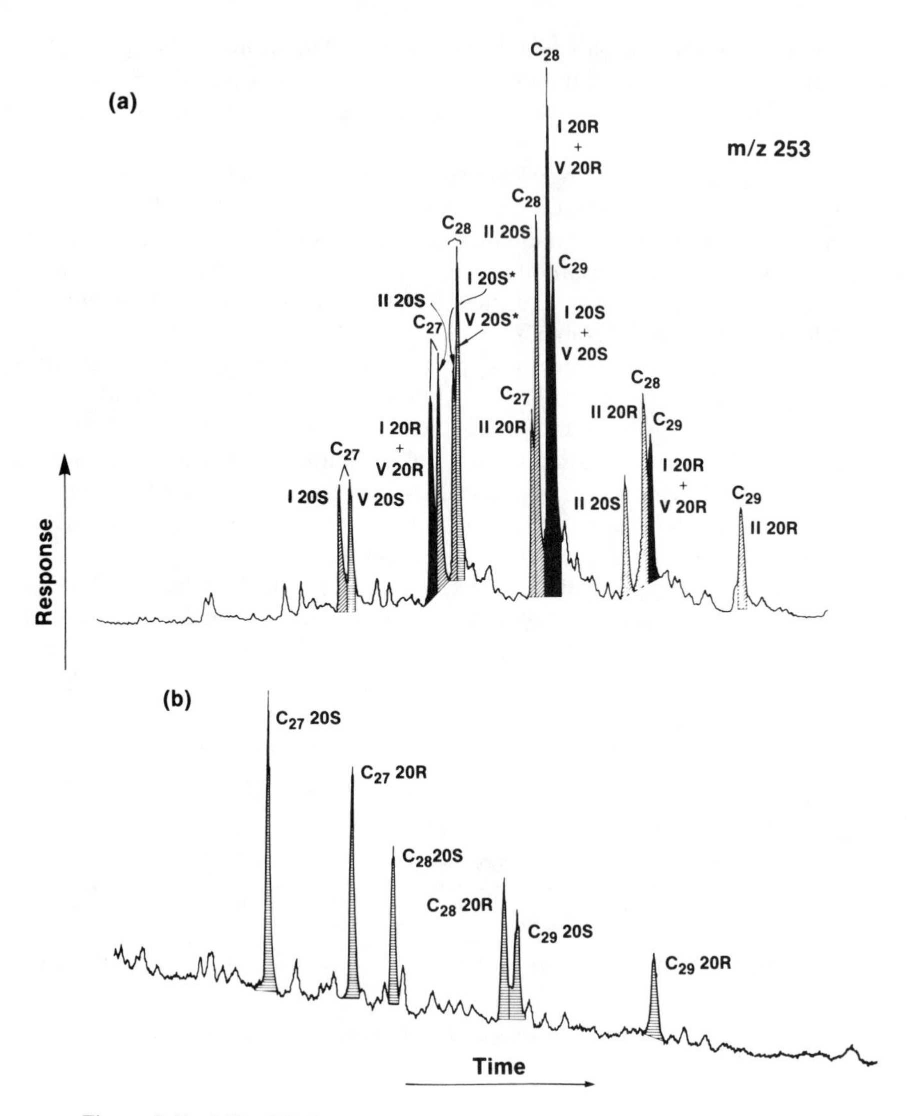

Figure 3.40 MID GCMS m/z 253 trace of (a) Carneros, California, oil (GCMS No. 257), and (b) Neiber Dome, Wyoming, oil (GCMS No. 358), exemplify two types of monoaromatic steroid distributions commonly seen in oils. The pattern in (a) shows about equal proportions of regular (Fig. 3.39, Structures I and II) and rearranged (Structure V) monoaromatic steroids (see also Fig. 2.13). The pattern in (b) is dominated by rearranged (Structure V) monoaromatic steroids with no detectable regular compounds. GC conditions: (a) OV-101 coated 50 m, fused silica, Hewlett-Packard capillary column programmed from 150 to 320°C at 2°C/minute; (b) 60 m DB-1 J and W Scientific, thick phase column, programmed as in (a) (Moldowan and Fago, 1986). Diagonal hachure = Structure I or II (Fig. 3.39); Horizontal hachure = Structure V; Solid = mixture of I and V; * = assignment may be reversed with adjacent peak.

5β(20S + 20R) and $10\beta \rightarrow 5\beta$ methyl-rearranged 20R and 20S isomers. For C_{28} and C_{29}, the number of compounds is actually doubled (12) because of C-24 R and S isomers. However, these R and S isomers are unresolved, and for practical purposes are ignored.

Dia/(dia + regular) C-ring monoaromatic steroid hydrocarbons

Rarely used, low specificity, related to paleoenvironment of source rock deposition, measured using m/z 253 fragmentogram.

The structures of rearranged monoaromatic (dia-MA-) steroids (Fig. 1.10 and 3.39) have been established as 10-desmethyl 5α- and 5β-methyl (20S and 20R) diastereomers (Riolo et al., 1985; Riolo and Albrecht, 1985; Moldowan and Fago, 1986). Although the mechanism for their formation is not clear, evidence from a study of Guatemalan samples suggests an influence of clay catalysis on diasterane formation in the source rock (Riolo et al., 1986). Anhydrites formed in evaporitic sabkhas show a large predominance of dia-MA-steroids, although some clastic source rocks also show such a predominance. A study of organic facies in the Toarcian shales (Moldowan et al., 1986) shows a correlation between C_{27} dia/(dia + regular) MA-steroids and C_{27} dia/(dia + regular) steranes. The dia/(dia + regular) MA-steroid ratio can also be increased by thermal maturity (Moldowan and Fago, 1986).

At Chevron we use the C_{27} $5\beta(CH_3)/[5\beta(CH_3) + 5\beta(H)]$ ratio for the 20S isomers to represent the dia/(dia + regular) MA-steroid ratio. The 5α- and $5\beta(CH_3)$,20S (dia) and 5β(H),20S (regular) C_{27} MA-steroids show the best resolution by chromatography on the m/z 253 fragmentograms (Figure 2.13). The same isomers for the C_{28} and C_{29} MA-steroids are not generally as well resolved as those for C_{27} (Moldowan and Fago, 1986).

C_{26}-C_{27}-C_{28} triaromatic (TA) steroid hydrocarbons

Rarely used, specificity unknown, measured using m/z 231 or GCMSMS of $M^+ \rightarrow$ m/z 231 (Fig. 2.11).

Triaromatic (TA-) steroids can be generated by aromatization and loss of a methyl group (—CH_3) from monoaromatic steroids (e.g., the C_{29} monoaromatic can be converted to the C_{28} triaromatic, Fig. 3.57). The ratios of C_{26}/(C_{26} to C_{28}), C_{27}/(C_{26} to C_{28}), and C_{28}/(C_{26} to C_{28}) TA-steroids are potentially effective source parameters similar to those described for the C_{27}-, C_{28}-, and C_{29}- MA-steroids. The structures of the TA-steroids used in these parameters are given in Fig. 3.41 (e.g., Ludwig et al., 1981). The TA-steroid ratios should be more sensitive to thermal maturation than those for MA-steroids or steranes because the TA-steroids appear to be maturation products from aromatization of MA-steroids (Mackenzie et al.,

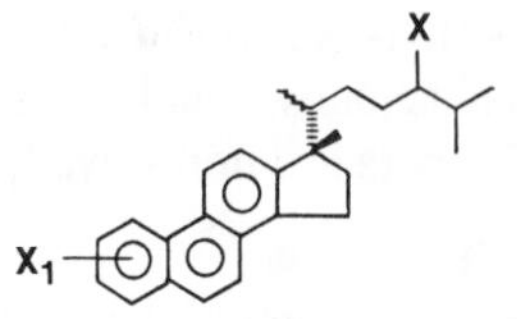

$X = H, CH_3, C_2H_5, nC_3H_7$

$X_1 = H, CH_3$

Figure 3.41 Generalized structure for triaromatic steroids common in petroleum. Other side chains and substitution patterns are possible, but have not yet been characterized. If X = H, the major mass spectral fragment is m/z 231 (Fig. 2.11); while if X = CH_3, the major mass spectral fragment is m/z 245. Note that unlike sterols (Fig. 1.22) and steranes (Fig. 1.12), triaromatic steroids lack methyl groups at C-10 and C-13, but show a methyl group at C-17.

1982b). As aromatization proceeds in the early part of the oil window, there may be changes in the TA-steroid ratios reflecting the relative ease of aromatization of various precursors and possible additional precursors other than MA-steroids. For example, the ratio of C_{27}/C_{29} MA-steroids does not correlate with the ratio of C_{26}/C_{28} (20S) TA-steroids in a study of immature to mature oils and seeps from Greece (Seifert et al., 1984).

Because the C_{26}20R isomer coelutes with the C_{27}20S isomer under all reported GC conditions using GCMS m/z 231 chromatograms (Fig. 2.14), TA $C_{27}/(C_{26}$ to $C_{28})$ and TA $C_{26}/(C_{26}$ to $C_{28})$ cannot readily be measured. Alternately, a GCMSMS approach using $M^+ \rightarrow 231$ (Fig. 3.42) can be applied. For example, we used GCMSMS for TA-steroids to show relationships between oils from the Eel River basin successfully (unpublished data). The TA-steroid results for these samples are consistent with other correlation parameters, including C_{27}, C_{28}, and C_{29} sterane distributions. This approach suffers from reduced sensitivity for TA-steroids, because TA-steroid molecular ions are only in low abundance in TA-steroid electron impact spectra, leading to weak metastable or collision spectra. For this reason, low-voltage electron impact, field ionization, or chemical ionization methods (Sec. 2.5.3) may be useful in TA-steroid analysis. The TA-steroids are related to the common desmethyl sterols, probably through MA-steroid intermediates. There are additional families of TA-steroids having the mass spectral base peak m/z 245 most likely related to the various ring A methyl sterols. However, little work on their structures or applications has been published (Riolo et al., 1986) and they are difficult to identify without synthetic standards for coelution experiments.

Benzohopanes

Rarely used, moderate specificity for evaporite or carbonate depositional environments, measured using m/z 191 (Fig. 2.11) or GC-FID of the aromatic fraction.

Benzohopanes are probably formed by cyclization of extended hopanoid side chains followed by aromatization (Hussler et al., 1984a and 1984b). Benzohopanes

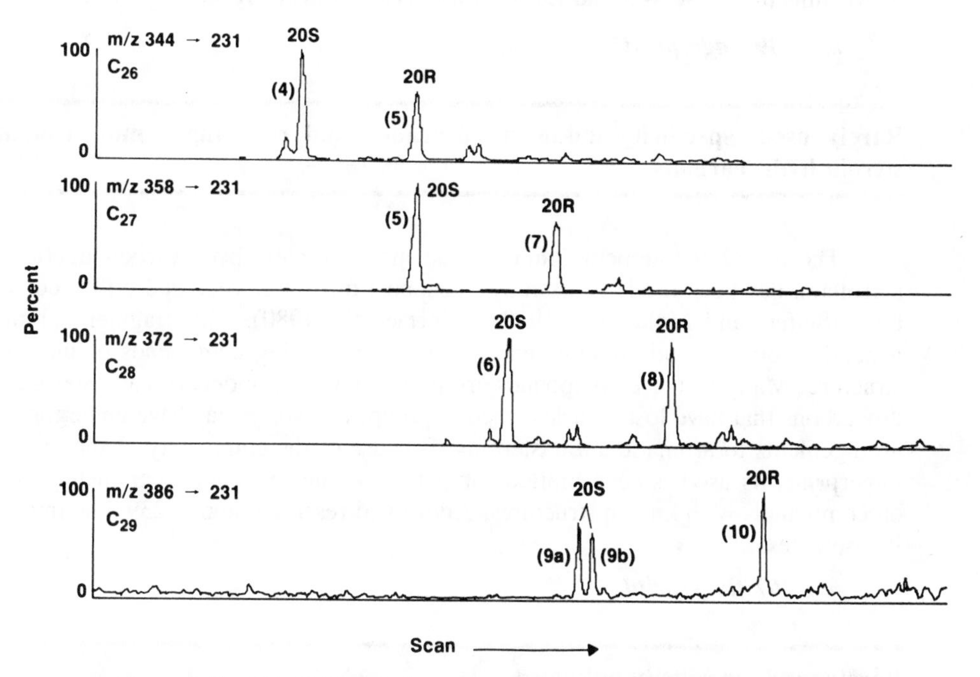

Figure 3.42 GCMSMS analysis of triaromatic steroid hydrocarbons (Fig. 3.41; X = H) in the aromatic fraction of oil from the Carneros Formation, California (GCMS No. 257) using the triple quadrupole (Finnigan MAT TSQ-70) set to record parents of the m/z 231 fragment. Compared to MID m/z 231 mass fragmentography (Fig. 2.14), this method facilitates quantitation of all TA-steroid epimers. For example, the C_{26} 20R and C_{27} 20S compounds are analyzed separately on the m/z 344 → 231 and m/z 358 → 231 chromatograms, respectively, thus eliminating most interference problems. Application of ternary diagram relationships and 20S/(20S + 20R) ratios among the C_{26}-C_{28} TA-steroid homologs is possible using this method.

Small amounts of the C_{29} homologs (peaks 9 and 10) are also more easily recognized and quantified using the GCMSMS technique. The C_{29} 20S compound is split into a doublet (peaks 9a and 9b) consisting of 24S and 24R epimers (Fig. 1.12 shows the steroid numbering system) of the *n*-propyl group. Other 24S and 24R epimers coelute in this analysis. Peak numbers 4-10 refer to identifications in Fig. 2.14. Figure modified from Gallegos and Moldowan (1992).

range in carbon number from C_{32} to C_{35}, consistent with their proposed origin by cyclization of the homohopanoid side chain during early diagenesis. Oils and bitumens from evaporitic and carbonate source rocks show the highest concentrations of benzohopanes, although they are found in trace amounts in most source rocks and petroleums (e.g., He Wei and Lu Songnian, 1990 and references therein).

m/z 239 fingerprint

Rarely used, specificity unknown, degraded and rearranged monoaromatic steroid hydrocarbons.

The m/z 239 fragmentogram for the monoaromatic hydrocarbon fraction of petroleum has been used successfully to support oil-source rock and oil-oil correlations (Seifert and Moldowan, 1978; Seifert et al., 1980). This fragmentogram is generally complex and contains many peaks representing compounds of unknown structure. Many of these compounds are probably C-ring monoaromatic steroid hydrocarbons that have lost a nuclear methyl group, and some may have undergone rearrangements to aromatic anthrasteroids. Because of the complexity of the m/z 239 fingerprint, its use has been limited compared to other biomarker parameters based on compounds with known structures. Additional research on m/z 239 may improve its usefulness.

m/z 267 fingerprint

Rarely used, specificity unknown.

The m/z 267 fragmentogram appears dominated by a series of C-ring monoaromatic steroid hydrocarbons (Fig. 2.11), probably derived in part from 2-, 3-, and 4-methylsterols. This fingerprint has found little application, but has similar potential to the m/z 239 fingerprint (see previous text).

Perylene

Rarely used, suggested specificity for land plant input, measured using m/z 252 (Fig. 2.11).

The precursor for perylene must have a widespread distribution and its formation requires deposition in highly reducing sediment (Gschwend et al., 1983). Perylene (Fig. 2.11) may be a land plant source indicator (Aizenshtat, 1973), although its occurrence in Walvis Bay sediment, a site thought to be largely free of terrestrial organic matter input, led Wakeham et al. (1979) to question whether the precursor must be terrestrial.

Degraded aromatic diterpanes

Rarely used, suggested specificity for land plant input, measured using GC-MS-MID or GC-FID of aromatic fractions.

Alexander et al. (1992) found that saturated diterpane biomarkers are in extremely low concentrations in oils sourced from coal measures in the Cooper and Eromanga Basins, Australia. In contrast, the degraded aromatic compounds, 1,2,5-trimethylnaphthalene, 1,7-dimethylphenanthrene, 1-methylphenanthrene, and retene (Fig. 3.24), derived from natural product resin precursors are particularly abundant. These are related mostly to Araucariaceae conifer remains in Jurassic to Lower Cretaceous source rocks which generated the oils in the Eromanga Basin. Different distributions of the same compounds occur in Cooper Basin oils derived mostly from remains of pteridosperms (seed ferns) in Permian source rocks there. While the more degraded aromatics, 1,2,5-trimethylnaphthalene, 1,7-dimethylphenanthrene, and 1-methylphenanthrene (*16, 17,* and *18,* Fig. 3.24) could conceivably be derived from pentacyclic triterpanes, their collective presence and their abundance relative to other aromatics such as 1,3,6-trimethylnaphthalene and 9-methylphenanthrene (lacking the proper alkyl substitution positions to be related to common diterpenes), was given as strong evidence of their diterpenic origin (Alexander et al., 1992).

3.1.3.4 Porphyrins. The following provides a limited discussion of porphyrins to supplement that in Secs. 1.3.3, 3.2.4.5, and 3.3.3. More detailed discussion of porphyrins can be found in Sundararaman et al. (1988a and b); Baker and Louda (1983, 1986); Louda and Baker (1986), and Chicarelli et al., 1987.

Porphyrins are tetrapyrrolic organometallic compounds that account for much of the vanadium and nickel in petroleum (Boduszynski, 1987). Deoxophylloerythroetioporphyrin (DPEP) and etioporphyrin structures are the most common tetrapyrroles (Fig. 1.27). These compounds are particularly resistant to biodegradation (Sec. 3.3.3). More detailed discussions on porphyrins and their use as maturity parameters are in Sec. 3.2.4.5 and Sundararaman et al. (1988a and b).

Routine application of petroporphyrins to geochemical problems has been limited for several reasons. Porphyrins are complex and show very large numbers of isomers (Barwise and Whitehead, 1980). An easily reproducible, practical method of fingerprinting these compounds has been developed by Sundararaman (1985) who showed the feasibility of direct high pressure liquid chromatographic (HPLC) separation of vanadyl porphyrins without demetallation. Demetallation and remetallation can result in selective decomposition of porphyrins (Boreham, 1992). The high molecular weights and low volatilities of these compounds preclude their separation using the GCMS approach. Consequently, structural analysis normally requires that porphyrins be inserted directly into the mass spectrometer. This requires large amounts of sample and provides no isomer-specific information.

New developments in analytical techniques may allow more routine analyses of porphyrins. For example, Gallegos et al. (1991) used commercially available fused silica columns to separate porphyrins chromatographically and obtain complete mass spectra without the need for demetallation. For this new approach, the authors employed high-temperature gas chromatography/electron impact mass spectrometry (HTGC/EIMS) and high-temperature gas chromatography/field ionization mass spectrometry (HTGC/FIMS) to analyze C_{28} to C_{33} etio- and deoxophylloerythroetioporphyrins (DPEP) isolated from Boscan crude and a Monterey source rock.

Porphyrins can be analyzed using liquid chromatography-mass spectrometry (LCMS). Advances in LCMS have been impeded by the problem of delivering the liquid effluent from the column to the vacuum system of the mass spectrometer. A technology called thermospray is the most popular LCMS interface, but it permits only mild chemical ionization of analytes (compound to be analyzed). Such chemical ionization spectra do not show the rich fragmentation patterns of electron impact spectra that make it possible to identify unknown compounds. Newly developed particle-beam LCMS interfaces are more compatible with different ionization sources than other interfaces. Electron impact, chemical ionization, and other ionization techniques can be used.

Supercritical fluid chromatography (SFC) mass spectrometry (MS) is another technique with potential for porphyrin analysis. This technique has been applied using capillary gas chromatographic columns (Campbell et al., 1988) and may allow many of the benefits obtained using GCMS or GCMSMS methods for hydrocarbons.

Many alkyl substitution patterns around the porphyrin nucleus have been elucidated and the structures have been related to either chlorophyll-d in green photosynthetic (*Chlorobiaceae*) bacteria, chlorophyll-c in certain species of eukaryotic algae, or chlorophyll-a, which is widespread in eukaryotic algae and higher plants (Ocampo et al., 1984, 1985a, 1985b). Some of the isolated porphyrins from Messel shale have also been related to specific groups of organisms by stable carbon isotope ratios (Hayes et al., 1987). However, progress in the application of petroporphyrins to source rocks and oils is slow because of the laborious methods necessary for their isolation. Identification of peaks on HPLC chromatograms using authentic standards will be necessary for further progress. New technologies such as liquid chromatography- and capillary supercritical fluid chromatography-mass spectrometry may ultimately allow routine application of porphyrins to geochemical studies.

V/(V + Ni) porphyrins

Often used, related to redox conditions in source rock depositional environment, separation by column chromatography (porphyrin-polar cut), measurement by ultraviolet spectrophotometry.

The proportions of vanadyl and nickel porphyrins are used as a source parameter in oil-to-oil and oil-to-source rock correlations (Lewan, 1984). Vanadium and nickel are the major metals in petroleum (Boduszynski, 1987) but are not part of the original tetrapyrrole pigments in living organisms. These metals enter into the porphyrin structure by chelation during early diagenesis (Fig. 1.27) and their relative proportions are governed by the depositional environment (Lewan, 1984).

Lewan (1984) proposed that in marine sediments, Ni^{2+} and VO^{2+} in solution in the pore waters compete for chelation with free-base porphyrins. Under normal oxic conditions, nickel is favored by a higher equilibrium constant for reaction with the free-base porphyrins than vanadium. However, under low Eh conditions, sulfate-reducing bacteria generate hydrogen sulfide. High sulfide in the pore waters of anoxic sediments causes nickel ion (Ni^{2+}) to precipitate as nickel sulfide, leaving vanadyl ion (VO^{2+}) to complex with available free porphyrins. Low V/(V + Ni) porphyrin ratios in Toarcian rocks of marine origin reflect oxic to suboxic marine conditions, while high ratios reflect anoxic sedimentation (Moldowan et al., 1986).

Nickel porphyrins generally predominate in lacustrine rocks and related oils, while vanadium is low. However, one exceptional lacustrine source rock from the Cretaceous Bucomazi Formation in West Africa shows a predominance of vanadyl porphyrins in the most organic-rich sections.

Sundararaman and Boreham (1991) explain the high V/(V + Ni) porphyrin ratios (up to 0.9) in bitumens from parts of the Bucomazi Formation as due to the combined influence of low Eh and pH. They observed that two porphyrin ratios decrease uphole in a Bucomazi core: V/(V + Ni) and C_{30} 3-norDPEP/(C_{30} 3-nor DPEP + C_{32}DPEP). Low Eh and pH favor vanadyl porphyrins. The decreasing V/(V + Ni) porphyrin ratios are attributed mainly to increasing oxicity during evolution of the lake. However, because the V/(V + Ni) and C_{30} 3-norDPEP/(C_{30} 3-nor DPEP + C_{32}DPEP) ratios show different sensitivities to second-order cycles of deposition, they appear to be controlled by two different factors.

The ratio C_{30} 3-norDPEP/(C_{30} 3-norDPEP + C_{32}DPEP) appears sensitive to pH. Increased pH apparently suppresses the devinylation reaction that leads from C_{32}DPEP to C_{30} 3-norDPEP. The results imply that the pH of the lake increased gradually with time, which is supported by more marls and fewer organic-rich shales and mudstones in the upper parts of the Bucomazi core.

Porphyrin distributions

Rarely used, measured using HPLC methods.

Recent work by the groups at Strasbourg, France, and Bristol, U.K., has identified several porphyrin structures in immature petroleums. However, the complexity of porphyrin fingerprint patterns (Sundararaman, 1985) has largely defied specific application to correlation problems. Some of the structures show specific links to eukaryotes or prokaryotes (Hayes et al., 1987). For example, porphyrins ranging from C_{34} to C_{36} with extended side chains can be identified in HPLC fingerprints and originate from photosynthetic bacteria (Ocampo et al., 1985a and b), while other porphyrins with a rearranged exocyclic 5-membered ring can be related to algae such as dinoflagellates (Ocampo et al.,1984).

Michael et al. (1990) correlated oils, source rocks, and heavily biodegraded tar-sand bitumens from the Ardmore and Anadarko basins using various parameters, including: tricyclic terpane, C_{24}-tetracyclic terpane, hopane, mono- and triaromatic steroid hydrocarbon and porphyrin distributions.

Chicarelli et al. (1987) and Callot et al. (1990) describe sedimentary porphyrins with structures providing clear evidence of specific precursor chlorophylls or bacteriochlorophylls, while others are not obviously related to known pigments. One C_{32} porphyrin isolated from Gilsonite contains a methyl-substituted, five membered exocyclic ring. Although of unknown origin, this compound may be a marker for lacustrine settings.

3.1.4 Geochemical Characteristics of Petroleum from Carbonate versus Shale Source Rocks. In sparsely explored or frontier basins, the petroleum source rocks are usually unknown and only a few crude oil or seep samples may be available for study. Various geochemical methods have been applied to such oils in attempts to provide information on their source rocks. This information is useful to the geologist who plans further exploration within a basin. For example, structural, mineralogical, and organic geochemical studies can help delineate probable migration pathways or the types of petroleum that will be found. The reader is referred to two statistical studies of geochemical data obtained from oils as a means for determining source rock organic matter type and/or mineralogy (Peters et al., 1986; Zumberge, 1987a).

Combined use of the parameters in Sec. 3 can be used to describe the organic matter type, depositional environment, and mineralogy of the source rock from oil composition. Table 3.1.4 is an example showing characteristics of oils derived from carbonate versus shale source rocks. In the text, the term **carbonate rocks** refers to fine-grained sedimentary rocks containing 50 percent or more of carbonate minerals, typically associated with evaporitic, siliceous, and argillaceous components. The data in the table apply only to oils of comparable maturity prior to peak oil generation.

Other environments can be differentiated using biomarkers. For example, Mello et al. (1988b) describe a series of biomarker and nonbiomarker parameters they used to separate offshore Brazilian oils and bitumens from the following depositional environments: lacustrine freshwater, lacustrine saline, marine evaporitic, marine carbonate, marine deltaic, marine calcareous, and marine siliceous lithology. Connan et al. (1986) show that detailed biomarker analysis can be used to distinguish anhydrites from carbonates in a core from a paleo-sabhka in Guatemala.

Oil-prone source rocks, whether carbonates or shales, show many common characteristics, including lamination, high total organic carbon, and hydrogen-rich organic matter. Despite these similarities, oils from carbonate rocks are typically richer in cyclic hydrocarbons and sulfur compared to those from shales. Jones (1984) describes differences in the rock matrix of carbonates and shales that result in different primary migration characteristics. He believes that carbonate source rocks do not have a lower minimum TOC compared to other source rocks to generate and expel petroleum.

TABLE 3.1.4 Some Characteristics of Petroleum from Carbonate Versus Shale Source Rocks.

Characteristics	Shales	Carbonates	Reference
Nonbiomarker Parameters			
API, Gravity	Medium-High	Low-Medium	1, 2, 3
Sulfur, wt %	Variable	High[1]	1, 2, 3, 6, 9
Thiophenic sulfur	Low	High	1
Saturate/aromatic	Medium-High	Low-Medium	1, 2, 3
Naphthenes/paraffins	Medium-Low	Medium-High	1, 3
CPI (C_{22}-C_{32})	≥ 1	≤ 1	1, 2, 6, 9
Biomarker Parameters			
Pristane/phytane	High (e.g; ≥ 1)	Low (e.g; ≤ 1)	1, 2, 6, 9, 10
Phytane/nC_{18}	Low (≥ 0.3)	High (≤ 0.3)	2, 6
Steranes	$C_{27} < C_{29}$	$C_{27} > C_{29}$	1
Steranes/17α(H)-hopanes	High	Low	7, 9
Diasteranes/steranes	High	Low	1
C_{24} Tetra-/C_{26} tricyclic diterpanes	Low-Medium	Medium-High	2, 7
C_{29}/C_{30} Hopanes	Low	High (>1)	10, 11
C_{35}-Homohopane index	Low	High	4, 10
Hexahydrobenzohopanes and benzohopanes	Low	High	5
Dia/(reg + dia) MA-steroids	Low	High	8
Ts/(Ts + Tm)	High	Low	4
C_{29} MA-steroids	Low	High	9

[1] Oils from lacustrine carbonates such as the Green River formation may be low in sulfur.

References: (1) Hughes (1984), (2) Palacas (1984), (3) Tissot and Welte (1984), (4) McKirdy et al. (1983), (5) Connan and Dessort (1987), (6) Connan (1981), (7) Connan et al. (1986), (8) Riolo et al. (1986), (9) Moldowan et al. (1985), (10) ten Haven et al. (1988), (11) Fan Pu et al. (1987).

None of the parameters in Table 3.1.4 should be considered proof of the origin of an oil. Exceptions to the general ranges of values used as a guide in the table are common. For example, very mature oils are characterized by severe alteration of biomarker and other components, making interpretations of source rock type difficult. Rather, the parameters should be used together in a cumulative sense to indicate the origin of the oil.

3.1.5 Marine versus Terrestrial Organic Matter Input. Few systematic studies have addressed the effects of variations in the salinity of the source rock depositional environment on geochemical parameters. Recently, Philp et al. (1989) examined oils thought to be derived from rocks deposited in brackish, saline, and freshwater lacustrine environments in China. They concluded that brackish-derived oils showed high tricyclic terpanes compared to hopanes, and a predominance of 24-methyl- and 24-ethylcholestanes (C_{28} and C_{29}) with few cholestanes (C_{27}). The saline-derived oils showed high gammacerane. The freshwater-derived oils showed low tricyclic terpanes compared to hopanes, an unknown C_{30} pentacyclic terpane (probably C_{30}^{*}, Sec. 3.1.3.1), small amounts of C_{31}-C_{35} homohopanes compared to C_{30} hopane, and large amounts of 24-ethylcholestanes (C_{29}) compared to other steranes.

Fu Jiamo et al. (1990) conclude that the most useful parameters for distinguishing Chinese rocks deposited in freshwater, brackish, and hypersaline lacustrine environments include relative abundances of *n*-alkanes, acyclic isoprenoids, 4-methyl steranes, the hopane/sterane ratio, and gammacerane and homohopane indices. Samples from each of these groups were distinguished using principal component analyses of these data.

Some parameters appear useful for indicating marine versus terrestrial input of organic matter to the source rock. Many of these parameters are described in the previous discussion, although few are diagnostic in every case. For example, oleanane indicates higher plant input, but its absence does not prove lack of that input. Sea grasses such as *Zostera,* are basically vascular plants found in marine rather than terrestrial environments. For these reasons, we recommend against heavy reliance on one or a few of these parameters. The most reliable statements about organic matter input are made based on multiple parameters. For example, Talukdar et al. (1986) used vanadium, sulfur, pristane/phytane, pristane/nC_{17}, sterane distributions, C_{19} and C_{20} diterpanes, oleanane, and hopane/sterane ratios to distinguish marine, terrestrial, and mixed oils in the Maracaibo Basin, Venezuela.

Some of the principal differences between nonbiodegraded petroleum derived from marine, terrestrial, and lacustrine algal input are listed in Table 3.1.5. The table is greatly simplified because of the enormous diversity of environments and types of contributing organisms included within these categories, and because more or less terrestrial organic matter is commonly mixed with either marine or lacustrine organic matter. The reader is urged to consult the detailed discussions of various parameters in the text.

Terrestrial markers are commonly low in marine and lacustrine oils. Sediments deposited in deltaic environments are mixtures that show evidence for both marine and terrestrial input, and in the extreme case generate "terrestrial" oil. Terrestrial oils also originate from lacustrine sediments that are dominated by higher plant input. However, the true "lacustrine" or "nonmarine" oils are those generated from nonmarine algal and bacterial organic matter in lacustrine source rocks. Note that hypersaline lacustrine environments are not included in Table 3.1.5. Absolute concentrations of biomarkers also provide important information. For example, high concentrations of markers for vascular plants might be expected in lacustrine or estuarine sediments, but smaller amounts might be transported by wind or turbidity currents to deep sea sediments.

3.2 Maturation

3.2.1 Concepts. "Thermal maturity" describes the extent of heat-driven reactions which convert sedimentary organic matter into petroleum. For example, kerogen in fine-grained source rocks can be thermally converted to oil and gas,

TABLE 3.1.5 Some Typical Characteristics of Petroleums Derived from Various Types of Organic Matter Input.[1]

Characteristics	Marine	Terrestrial (higher plant)	Lacustrine (Algal)
Nonbiomarker Parameters			
Sulfur	High	Low	Low
C_{21}-C_{35} n-Alkanes (High C_{27}, C_{29}, C_{31})	Low	High	High
Biomarker Parameters			
Pristane/phytane	≤2	≥3	~1-3
Pristane/nC_{17}	Low (<0.5)	High (>0.6)	-
4-Methyl steranes	Moderate	Low	High
C_{27}-C_{29} steranes	High C_{28}	High C_{29}	Low C_{28}
C_{30} steranes (24-n-propylcholestanes)	Present	Absent or Low	Absent
Steranes/hopanes	High	Low	Low
Bicyclic sesquiterpanes (e.g., Eudesmane)	Low	High	Low
Tricyclic diterpanes (e.g., Pimarane, isopimarane, or abietane skeletons)	Low	High	Low
Tetracyclic Diterpanes (e.g., Phyllocladane, beyerane, or kaurane skeletons)	Low	High	Low
Lupanes, bisnorlupanes	Low	High	Low
Oleananes	Low or Absent	High	Low
Botryococcane	Absent	Absent	High (Rare)
V/(V + Ni)	High-Low	Low or Absent	Low or Absent

[1] The terms "marine", "terrestrial", or "lacustrine oil" can be misleading. For example, "marine oil" might refer to: (1) oil produced from reservoir rocks of marine origin, (2) oil generated from a source rock deposited under marine conditions, or (3) oil derived from marine organic matter. The table refers to the origin of the organic matter that generated the oil rather than the environment in which it was deposited.

which migrate to coarser-grained reservoir rocks (Figs. 1.2 and 3.46). Early diagenetic processes convert bacterial and plant debris in sediments to kerogen (insoluble, particulate organic matter) and bitumen (extractable organic matter). Thermal processes generally associated with burial then convert part of this organic matter to petroleum and, ultimately, to gas and graphite. Thus, petroleum is a complex mixture of metastable products which evolve toward greater thermodynamic stability during maturation.

> ***Note:*** *It is generally believed that both kerogen and oil are unstable during catagenesis and progressively decompose to pyrobitumen and gases (e.g., Hunt, 1979; Tissot and Welte, 1984). Mango (1991) shows evidence that hydrocarbons in oil are much*

more thermally stable than their kerogenous precursors. He believes that oil and gas are generated by direct thermal decomposition of kerogen, but that hydrocarbons in oils show no evidence of decomposition to gas in the Earth. This scenario does not exclude some oxidative decomposition of hydrocarbons during thermochemical sulfate reduction (e.g., Orr, 1974; Krouse et al., 1988).

Potential petroleum source rocks are described in terms of the quantity, quality, and level of thermal maturity of the organic matter. A **potential source rock** contains adequate amounts of the proper type of dispersed kerogen to generate significant amounts of petroleum but is not yet thermally mature. A potential source rock becomes an **effective source rock** only at the appropriate levels of thermal maturity (i.e., within the "oil-generative window").

In general terms, **organic matter can be described as immature, mature, or postmature,** depending on its relation to the oil-generative window (Tissot and Welte, 1984). "Immature" organic matter has been affected by diagenesis, including biological, physical, and chemical alteration, but without a pronounced effect of temperature. "Mature" organic matter has been affected by catagenesis, the thermal processes covering the temperature range between diagenesis and metagenesis. As used in this book, catagenesis is equivalent to the oil and wet gas zones. "Postmature" organic matter has been heated to such high temperatures that it has been reduced to a hydrogen-poor residue capable of generating only small amounts of hydrocarbon gases.

Recognizing the need to describe the thermal maturity of sedimentary organic matter accurately, organic geochemists have developed various thermal maturity parameters. Conventional geochemical methods for assessing source rock maturity include Rock-Eval pyrolysis, compound class distributions, vitrinite reflectance (R_0), thermal alteration index (spore coloration or TAI), carbon preference index (CPI), and others. However, few of these parameters can be applied to oils. Within the last two decades, detailed molecular parameters based on ratios and distributions of specific biomarkers are finding increased use in studies of thermal maturity.

Two types of thermal maturity parameters exist: (1) "generation" or "conversion" parameters used as indices of the stage of petroleum generation (independent of the magnitude of thermal stress), and (2) "thermal stress" parameters used to describe relative effects of temperature/time. For example, two rocks containing different types of kerogen might generate equivalent amounts of oil at a given atomic H/C value, but the vitrinite reflectance of the samples may differ. In this hypothetical case, the atomic H/C is linked to hydrocarbon generation while vitrinite reflectance is linked only to thermal stress (however, see Sec. 3.2.2.1.2). For this reason, the reflectance value associated with the threshold of oil generation can vary between different rocks. Vitrinite reflectance values of about 0.6 percent are widely accepted as indicating the start of oil generation in most source rocks (Dow, 1977; Peters, 1986). However, source rock in the Monterey formation appears to generate significant amounts of oil at R_o values as low as 0.3 percent (Isaacs and Petersen, 1988), while comparable levels of petroleum generation in the Green River marl are not reached until R_o values approach 0.7 percent (Tissot et al., 1978).

3.2.2 Nonbiomarker maturity parameters. Reliable assessment of the thermal maturity of organic matter typically requires integrating both biomarker and nonbiomarker maturity data. Fig. 3.46 and Table 2.2.2.1.1 show the general correlation between the biomarker and nonbiomarker maturity parameters described below. Because many of the maturity parameters are related to "thermal stress" rather than "generation" as has been defined, relationships between these parameters and the oil generation window can only be approximate.

3.2.2.1 Rocks

3.2.2.1.1 Pyrolysis. Peters (1986) describes guidelines for evaluating or screening petroleum source rocks using Rock-Eval (programmed temperature) pyrolysis. Figure 3.43 shows a typical Rock-Eval pyrogram and related parameters. A flame ionization detector (FID) senses any organic compounds generated during pyrolysis. The first peak in the figure (S1) represents hydrocarbons that can be thermally distilled from a rock. The second peak (S2) represents hydrocarbons gener-

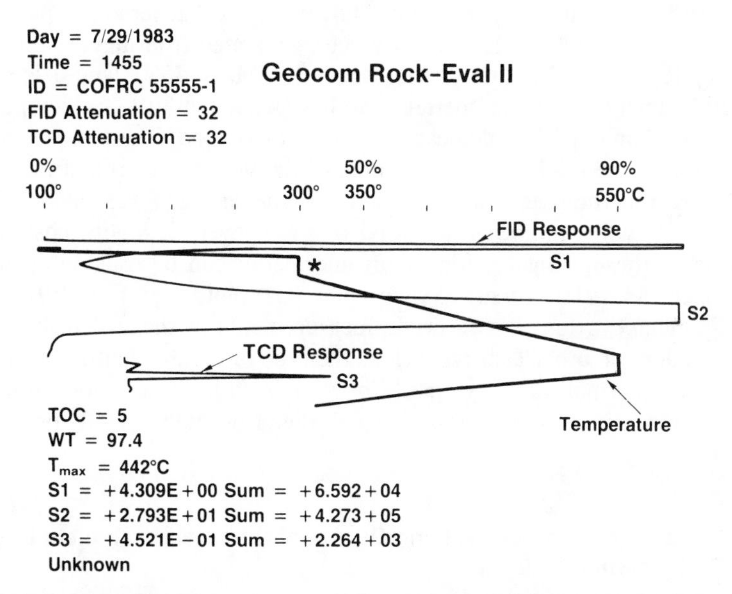

Figure 3.43 Typical Rock-Eval II pyrogram and report showing oven-temperature profile, flame ionization detector (FID; S1 and S2), and thermal conductivity detector (TCD; S3) responses (increasing time from top to bottom). Asterisk near the temperature curve designates isothermal heating period at 300°C. The first number to the right of S1 and S2 represents mg hydrocarbon/g rock. The first number to the right of S3 represents mg CO_2/g rock. The second number to the right of S1, S2, and S3 represents the peak area (area counts) measured by the integrator. For example, the S2 peak consists of 4.273×10^5 area counts, equivalent to 2.793×10^1 mg hydrocarbon/g rock. "Unknown" indicates that the sample is from an active study and is not a reference standard. See Peters (1986) for additional discussion. For reliable Rock-Eval interpretations, we recommend pyrograms every 30 to 60 feet (9–18 m) with depth in each well.

ated by pyrolytic degradation of the kerogen in the rock. (Although the literature expresses S1 and S2 in milligrams of "hydrocarbons" per gram of rock, the FID also detects nonhydrocarbons provided carbon atoms are present.) The third peak (S3) represents milligrams of carbon dioxide generated from a gram of rock during temperature programming up to 390°C, and is analyzed using a thermal conductivity detector (TCD). During pyrolysis, the temperature is monitored by a thermocouple. The temperature at which the maximum amount of S2 hydrocarbons is generated is called T_{max}. The hydrogen index (HI) corresponds to the quantity of pyrolyzable organic compounds from S2 relative to the total organic carbon (TOC) in the sample (mg HC/g TOC). The oxygen index (OI) corresponds to the quantity of carbon dioxide from S3 relative to the TOC (mg CO_2/g TOC). The production index (PI) is defined as the ratio S1/(S1 + S2).

Table 3.2.2.1.1 shows how to use the production index [PI = S1/(S1 + S2)] and T_{max} from Rock-Eval pyrolysis to estimate thermal maturity. Like vitrinite reflectance (R_o) and TAI, T_{max} is a "thermal stress" parameter. Changes in thermal stress parameters depend primarily on time/temperature conditions and can only be approximately related to the stage of petroleum generation for different rock types. The PI and bitumen/TOC ratios, however, are "generation" parameters related to how much petroleum has actually been generated from the organic matter.

Rock-Eval T_{max} and PI values less than about 435°C and 0.1, respectively, indicate immature organic matter that has generated little or no petroleum. A T_{max} greater than 470°C coincides with the wet-gas zone. The PI reaches about 0.4 at the bottom of the oil window (beginning of the wet-gas zone) and increases to 1.0 when the hydrocarbon-generative capacity of the kerogen has been exhausted. Usually some S1 will remain as adsorbed dry gas, even in highly postmature rocks. The level of thermal maturation of organic matter can be roughly estimated from a hydrogen index (HI) versus oxygen index (OI) plot (e.g., Fig. 3.9).

The T_{max} and PI are crude measurements of thermal maturity, but are partly dependent on other factors, such as the type of organic matter. Thus, conclusions regarding thermal maturity should be supported by other geochemical measurements such as vitrinite reflectance, thermal alteration index (TAI), or biomarker parame-

TABLE 3.2.2.1.1 Rock-Eval and Other Geochemical Parameters Describing Source Rock Thermal Maturity.

	Rock-Eval pyrolysis				
Maturation level	PI [S1/(S1 + S2)]	T_{max} (°C)	R_o(%)	TAI	Bit/TOC
Beginning Oil window (Birthline)	~0.1	~435–445[1]	~0.6	1.5–2.6	0.05–0.1
Peak oil window	~0.25	~445–450	~0.9	2.9–3.0	0.15–0.25
End oil window	~0.4	~470	~1.4	>3.2	<0.1

[1] Many maturation parameters (particularly T_{max}) depend on type of organic matter.

ters. For example, Peters et al. (1983) established the thermal maturity of Cretaceous black shales from the Cape Verde Rise in the eastern Atlantic that were intruded by Miocene diabase sills by using pyrolysis T_{max} and PI, kerogen elemental compositions, quantity of bitumen extract, and vitrinite reflectance.

3.2.2.1.2 Vitrinite reflectance. Vitrinite reflectance (R_o) is widely accepted by exploration geologists for measuring the thermal maturity of the kerogen in sedimentary rocks (Bostick, 1979). Although vitrinite reflectance is related more to thermal stress experienced by the vitrinite than to petroleum generation (Sec. 3.2.1), approximate R_o values have been assigned to the beginning and end of oil generation (Table 3.2.2.1.1, Fig. 3.46).

The increase in reflectance of vitrinite continues throughout all of thermal oil generation and appears due to complex, irreversible aromatization reactions. Vitrinite is present in many sedimentary rocks and is largely independent of rock composition. Vitrinite is believed to originate from terrestrial higher plant debris. Thus, vitrinite is not found in rocks older than Devonian because higher plants had not yet evolved. Compared to the thermal alteration index (TAI), vitrinite reflectance is more precise and less subjective.

Maturation lines defined in terms of both TAI and vitrinite reflectance (Fig. 3.9) have been placed on the van Krevelen atomic H/C versus O/C diagram (Jones and Edison, 1978). These workers show that by assuming atomic H/C values for specific macerals at a given level of maturity, a maceral analysis can be used to calculate the atomic H/C of the kerogen for comparison to measured atomic H/C.

Kerogens isolated from rock samples are embedded in epoxy and polished (Bostick and Alpern, 1977; Baskin, 1979) for measurement of the percentage of incident light of a given wavelength (usually 546 nm) reflected from phytoclasts (small particles of organic matter) of the vitrinite maceral group under oil immersion (Stach et al., 1982). Reported R_o (%) values typically represent the average of measurements on about 50 to 100 phytoclasts in each polished kerogen preparation. Each vitrinite reflectance value for a sample is derived from a histogram composed of about 50 to 100 individual measurements (Fig. 3.44).

Probably the major limitation of vitrinite reflectance is that vitrinite group macerals do not contribute significantly to oil generation compared to the liptinite macerals. Some very oil-prone source rocks, such as the Hanifa-Hadriya interval in Saudi Arabia (Ayres et al., 1982), contain little or no vitrinite. Reflectance measurements determined from fewer than about 50 phytoclasts may be unreliable. Furthermore, evidence suggests that large amounts of oil-prone macerals (Hutton and Cook, 1980; Price and Barker, 1985) or bitumen (Hutton et al., 1980) retard the normal progression of vitrinite reflectance with maturity.

Geochemical logs (e.g., Espitalié et al., 1987; Peters and Cassa, 1992) provide some of the most reliable R_o values because of the support gained by trends in R_o using the recommended downhole sample spacing of about 90 m (~300 ft), and because these trends can be corroborated by additional maturity parameters, such as Tmax. One example of a particularly suspect R_o value might be that from an outcrop (possible weathering), where less than 50 to 100 phytoclasts were available for measurement, and supporting maturity measurements were not completed.

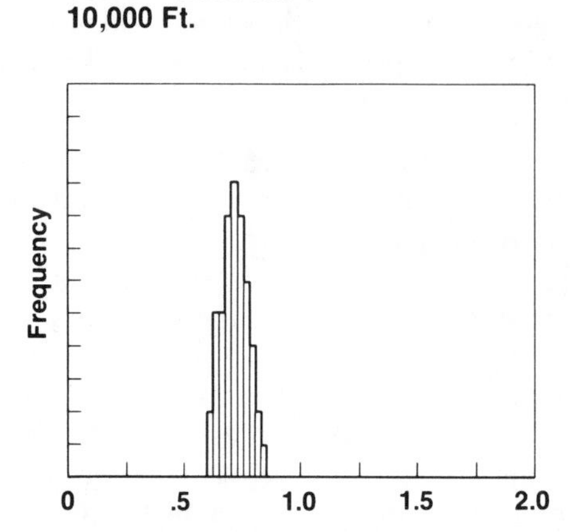

No. of Meas. = 50
Avg. Refl. = .74
Std. Dev. = .05

Reflectance Values

.64	.64	.65	.66	.66
.67	.67	.68	.68	.69
.69	.69	.70	.70	.70
.71	.72	.72	.72	.72
.73	.73	.73	.73	.74
.74	.74	.74	.74	.75
.75	.75	.76	.76	.76
.76	.77	.78	.78	.78
.78	.79	.79	.80	.80
.81	.81	.83	.84	.85

Figure 3.44 Example of a typical vitrinite reflectance histogram for kerogen isolated from rock. Fifty reflectance values (ordered from lowest to highest by the computer; right) were determined on this sample. Each reflectance value represents the percentage of incident light (546 nm) reflected from one vitrinite phytoclast. The quoted reflectance value for a sample typically represents the mean of all the individual measurements (usually 50–100) in a histogram. Reliability of vitrinite reflectance as a maturity indicator increases when numerous reflectance values are available with depth in a well. We recommend sampling intervals of 300 feet (about 90 m) or less for vitrinite reflectance.

Vitrinite reflectance is subject to several other problems (Table 3.2.2.1.2), some of which are discussed. For these reasons, R_o measurements of maturity should always be supported by other geochemical evidence, including biomarker maturity parameters.

Compared to most biomarker thermal maturity parameters, vitrinite reflectance shows lower sensitivity and accuracy up to a maturity equivalent to

TABLE 3.2.2.1.2 Some Examples of Potential Problems Affecting Interpretation of Vitrinite Reflectance.

Problem	Effect on R_o value
Caving of uphole cuttings	Lower
Poorly polished vitrinite	Lower
Mud contamination (e.g., lignite additives)	Lower (usually)
Oxidized or recycled vitrinite	Higher (usually)
Natural reflectivity variations within vitrinite subgroups	Higher or lower
Statistical errors (insufficient number of measurements)	Higher or lower
Incorrect maceral identification (e.g., liptinite, solid bitumen)	Higher or lower

the onset of petroleum generation (~R_o = 0.6%; Table 3.2.2.1.1). In this range of low maturity, the biomarker parameters described provide a more accurate assessment of thermal maturity than vitrinite reflectance (Mackenzie et al., 1988a).

Within and beyond the zone of petroleum generation (R_o > 0.6%) the usefulness of vitrinite reflectance improves compared to biomarkers. At these thermal maturities, biomarker maturity parameters are still valuable for supporting vitrinite reflectance results and vice versa. Although vitrinite reflectance can assist in ranking the thermal maturity of organic matter in rocks, it cannot be applied to oils.

3.2.2.1.3 Thermal alteration index (TAI). TAI is a numerical scale based on thermally induced color changes in spores and pollen (from yellow to brown to black) observed under the microscope in transmitted light (Staplin, 1969). The conodont alteration index (CAI) is a similar scale (Epstein et al., 1977), although it is much less commonly used in geochemical studies. These methods are advantageous because they are rapid, inexpensive, and do not require sophisticated instrumentation. Another advantage is that unlike vitrinite reflectance, where identification of vitrinite can be problematic, identification of spores or pollen is comparatively more reliable. Visual examination of samples for TAI (or CAI) using a microscope is also useful because it can result in the detection of reworked organic matter or particulate contamination.

The Chevron TAI scale differs somewhat from that of Staplin (1969). Chevron TAI ranges from 0 (very pale yellow) to 4 (black) and correlates with vitrinite reflectance (Table 3.2.2.1.3) and hydrocarbon generation zones (Fig. 3.9) (Jones and Edison, 1978). The easiest to measure and most important color changes between 2.4 and 3.1 occur during beginning to peak oil generation. TAI is presumably accurate to 0.1 units, although variations can be caused by other factors including measurements on different taxa, palynomorph thickness variations, weathering, or subjective error in color assessment.

Frequent comparison of the sample with TAI standard microslides is necessary. Measurements are best accomplished using a split-stage comparison microscope where the sample and standard can be viewed simultaneously.

Disadvantages of the TAI method include potential errors resulting from the subjective determination of color by an operator, the need for long-ranging micro-

TABLE 3.2.2.1.3 Approximate Relationship between Chevron TAI and Vitrinite Reflectance.

TAI												
1.6	1.8	2.0	2.2	2.4	2.6	2.8	3.0	3.2	3.4	3.6	3.8	4.0
R_o, %												
0.22	0.26	0.30	0.35	0.43	0.60	0.80	1.0	1.2	1.4	1.7	2.7	4.0

fossils of the same taxon for best results, and a limited range of applicability. The method is not effective for measuring maturity of organic matter below or above TAI values of about 2.4 and 3.1, respectively. The TAI shows low sensitivity to changes in maturity near the onset of oil generation compared to certain biomarker ratios, such as the C_{29} sterane 20S/(20S + 20R) ratio (Sec. 3.2.4.2) (Mackenzie et al., 1983b).

3.2.2.1.4 van Krevelen diagrams. In the van Krevelen diagram (Fig. 3.9) of atomic H/C versus O/C, different types of kerogens are shown as Type I (very oil prone), Type II (oil prone), and Type III (gas prone). Type IV (inert) kerogens contain very little hydrogen and plot near the bottom of the figure. The thermal maturation of each kerogen type is described by pathways; the most mature samples are near the lower left corner (little hydrogen or oxygen relative to carbon in the kerogen). Because of mixtures of organic matter, oxidation, and other variables, van Krevelen diagrams provide only a crude estimate of relative maturity and require support by other methods (Peters, 1986).

3.2.2.1.5 Bitumen (transformation) ratios. The ratio of extractable bitumen to total organic carbon (Bit/TOC) in fine-grained, nonreservoir rocks, sometimes called the transformation ratio, ranges from near zero in shallow sediments to about 0.25 (i.e., up to 250 mg/g TOC) at peak oil generation (Table 3.2.2.1.1). At greater depths the Bit/TOC ratio decreases because of conversion of bitumen to gas. Both the Bit/TOC and hydrocarbon/TOC ratios are used as crude methods for estimating the level of maturity of organic matter with depth in wells, especially the threshold of oil generation (Tissot and Welte, 1984, p. 180).

Lithology and the type of solvent used for extraction of the bitumen play critical roles in measured Bit/TOC ratios. Coarse-grained rocks, such as siltstones and sandstones, commonly contain migrated hydrocarbons and show higher Bit/TOC ratios than fine-grained rocks from similar depths. Contamination from migrated oils and drilling fluids can also affect this ratio. However, careful avoidance of coarse-grained, reservoir-type rocks and analysis of many closely-spaced samples can be used to establish Bit/TOC versus depth curves.

3.2.2.2 Oils. A variety of oil characteristics can be used for qualitative assessment of relative levels of thermal maturity. For related oils of increasing thermal maturity, the *n*-paraffin envelope becomes displaced toward lower molecular-weight homologs (Fig. 3.3), API gravity, nC_{19}/nC_{31} and saturate/aromatic ratios increase, while sulfur, nitrogen, and the isoprenoid/*n*-paraffin ratio decrease. For example, thermal maturity is the principal factor controlling the strong inverse correlation between API gravity and sulfur content for Monterey-sourced oils from the Santa Barbara Channel and offshore Santa Maria Basin (Baskin and Peters, 1992). Some of the more commonly applied qualitative maturity parameters are discussed as follows.

3.2.2.2.1 Isoprenoid/n-paraffin ratios. Pristane/nC_{17} and phytane/nC_{18} ratios decrease with increasing thermal maturity as more *n*-paraffins are generated from kerogen by cracking (Tissot et al., 1971). The ratios can be used to assist in ranking the thermal maturity of related, nonbiodegraded oils and bitumens. However, organic matter input (Alexander et al., 1981) and secondary processes, such as biodegradation, can affect these ratios.

3.2.2.2.2 CPI, OEP. The relative abundance of odd versus even carbon-numbered *n*-paraffins can be used to obtain a crude estimate of thermal maturity of petroleum. These measurements include the carbon preference index (CPI; Bray and Evans, 1961) and the improved odd : even preference (OEP; Scalan and Smith, 1970). In practice, the OEP can be adjusted to include any specified range of carbon numbers. Some examples of CPI and OEP variations are shown.

$$\mathrm{CPI} = \left[\frac{C_{25} + C_{27} + C_{29} + C_{31} + C_{33}}{C_{26} + C_{28} + C_{30} + C_{32} + C_{34}} + \frac{C_{25} + C_{27} + C_{29} + C_{31} + C_{33}}{C_{24} + C_{26} + C_{28} + C_{30} + C_{32}}\right] \Big/ 2$$

$$\mathrm{CPI}\ (1) = 2(C_{23} + C_{25} + C_{27} + C_{29})/[C_{22} + 2(C_{24} + C_{26} + C_{28}) + C_{30}]$$

$$\mathrm{OEP}\ (1) = \frac{C_{21} + 6C_{23} + C_{25}}{4C_{22} + 4C_{24}}$$

$$\mathrm{OEP}\ (2) = \frac{C_{25} + C_{27} + C_{29}}{4C_{26} + 4C_{28}}$$

CPI or OEP values significantly above (odd preference) or below (even preference) 1.0 indicate the oil or extract is thermally immature. Values of 1.0 suggest, but do not prove an oil or extract is thermally mature. CPI or OEP values less than 1.0 are unusual and are typically associated with oils or bitumens from carbonate (Table 3.1.4.1) or hypersaline environments (e.g., Fig. 3.1, middle).

Organic matter input affects CPI and OEP values. For example, many Ordovician oils show abundant *n*-paraffins below nC_{20} accompanied by a strong predominance of the odd-numbered homologs in the range nC_{10} to nC_{20} (Hatch et al., 1987) (Fig. 3.1, bottom). This odd-predominance of *n*-paraffins is common, even in Ordovician oils showing high API gravities. Thus, we see traits normally considered to indicate low (odd predominance) and high (high API) thermal maturity in the same Ordovician oils. This is an example of the effect of organic matter input on these parameters. The estimation of thermal maturity of these samples can be improved using thermally dependent biomarker parameters.

3.2.2.2.3 API gravity. API gravity is a bulk physical property of oils that can be used as a crude indicator of thermal maturity. API gravity is inversely related to specific gravity.

$$^{\circ}\mathrm{API} = \frac{141.5^{\circ}}{\text{Specific Gravity At } 60^{\circ}\mathrm{F}} - 131.5^{\circ}$$

Generalized relationships between API and various parameters, such as GOR (gas-oil ratio), reservoir depth, percent sulfur, and trace metal contents are described in Tissot and Welte (1984), and Hunt (1979). These relationships are only approximate and should be used with extreme caution, because of many exceptions.

During thermal maturation, the heavy components in oil, NSO-compounds, asphaltenes, and heavy saturate and aromatic compounds undergo increased cracking, resulting in increased API gravity. However, API gravity is also affected by other factors including original organic matter input, biodegradation, water washing, migration (phase separation, molecular partitioning), and inspissation (evaporation). For example, Walters and Cassa (1985) noted no correlation between API gravity and reservoir depth for a suite of offshore Gulf of Mexico oils. Some of the shallow oils were biodegraded while other oils migrated into reservoirs from deeper source rocks of varying maturities.

3.2.2.2.4 Methylphenanthrene index. Various isomer ratios among methylated aromatic hydrocarbons and sulfur heterocycles have been developed as thermal maturity indicators (Radke, et al., 1982; Radke and Welte, 1983; Radke et al., 1986; Radke, 1987; Radke et al., 1990). For example, the methylphenanthrene index (MPI 1) appears equivalent or superior to vitrinite reflectance as a maturity parameter in some field studies (Radke et al., 1982; Radke, 1988; Farrington et al., 1988). Calibrations between methylphenanthrene indices and vitrinite reflectance are available (e.g., Radke and Welte, 1983; Boreham et al., 1988). Radke et al. (1990) use various methylphenanthrene indices for bitumens from coals in a well in Indonesia to estimate the present depth of the source rock interval for oils from the same area.

Several difficulties limit the use of methylphenanthrene indices:

1. samples of different maturity can show identical methylphenanthrene ratios,
2. our experience shows that variations in the type of organic matter in the source rock can adversely affect methylphenanthrene ratios in petroleums,
3. they require different separation methods than those used for our isolation of biomarkers.

3.2.3 Criteria for ideal biomarker maturity parameters. The ideal molecular maturity parameter is based on measuring the relative concentrations of reactants (A) and products (B) in the following reaction:

$$A \underset{r2}{\overset{r1}{\rightleftharpoons}} B$$

where r1 and r2 represent the rates of the forward and reverse reactions, respectively. In the ideal case, the following conditions are satisfied:

1. A and B are single compounds and the reaction from A to B is irreversible (r1 much greater than r2),
2. The concentration of B is zero while easily measured amounts of A are present in all samples before heating,

3. A is transformed by heat only to B,
4. B is thermally stable and is only formed from A,
5. Conversion of A to B occurs in the range of maturity of interest for petroleum generation.

A convenient maturity parameter for expressing the extent of this reaction is the ratio of the concentration of B to the sum of A and B. Unlike B/A, **the ratio B/(A + B) can only range from 0 to 1 (or 0 to 100%) with increasing thermal maturity.**

In actuality, most biomarker maturity parameters cannot satisfy the ideal conditions as listed. In many known reactions, A or B can be derived from or degraded to other reactants and products. Some B may be present before heating begins and the conversion of A to B may never reach 100 percent even at very high maturity. Further, few biomarker maturity parameters can be used throughout the oil-generative window and all biomarkers are eventually thermally degraded to simpler hydrocarbons.

Proposed equilibrium reactions are commonly employed as biomarker maturity parameters. In contrast to the ideal irreversible reaction of A to B as discussed, the rates of forward and reverse reactions (r1 and r2) become equal (reach equilibrium) at a given level of maturity. Thus, after reaching equilibrium, no further information on maturity is available because the ratio B/(A + B) remains constant with further heating.

Compounds not directly related as precursor and product are also used in relative maturity assessments. Their ideal relationship is shown by the following reaction:

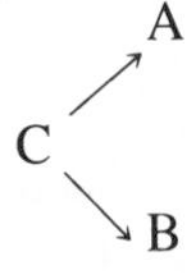

where A and B are biomarkers related to a common precursor C. During maturation, A and B are degraded or changed into other compounds at different rates and B is more stable toward these conversions than A. Although relative maturity can still be expressed using the ratio B/(A + B), there is greater potential for interference with the ratio of products A and B by variables unrelated to maturity than in the simpler case where A and B are the precursor and product, respectively. For example, precursor C might be a family of compounds and the initial formation of A and B might vary according to source organic matter input, conditions in the depositional environment, and early diagenesis.

3.2.4 Biomarker maturity parameters. As recommended earlier (Sec. 3.1.3), caution must be applied when comparing biomarker parameters derived from different laboratories. The mass chromatograms and peaks used to determine biomarker thermal maturity parameters may differ.

Assessment of the level of thermal maturity of bitumens and oils assists in correlation studies. Migrated oil in a sandstone that is more mature than indigenous bitumen from surrounding shales clearly could not have originated in the shales. Such sandstones may show anomalously high bitumen/TOC, production index, and S1 values (Sec. 3.2.2.1). Tests to determine whether bitumen is indigenous are described in Sec. 4.3.1. For example, biomarker maturity parameters for shales in the Shengli oilfield in China show good correlation with depth, while oils do not (Shi Ji-Yang et al., 1982). This suggests that the maturity of the oils is related more to maturation of the source rock (deeper in the section) than the extent of reservoir maturation.

Reduced biomarker concentrations of all types indicate increased maturity (e.g., Mackenzie et al., 1985) and signal the interpreter that special care should be taken in interpreting results (e.g., Fig. 2.21). Especially where biomarkers are low, mass chromatograms should be examined to verify quantified biomarker peak ratios. A general idea of the relative thermal maturity of oils or bitumens can be gained by plotting total steranes versus terpanes or 17α(H)-hopanes (measured in ppm, Sec. 2.5.4), especially when the samples are related.

> ***Note:*** *Very few oils are so mature that biomarkers are absent. One possible example is an oil seep from the billion-year old Nonesuch Shale in the White Pine copper mine, Michigan, which contained no steranes or terpanes (Hoering, 1978; Imbus et al., 1988). However, other oils from the Nonesuch Shale show biomarkers (Pratt et al., 1991). These observations may be explained by oxidative degradation of the biomarkers during ore deposition. Local precipitation of copper sulfides, native copper, and silver in the White Pine mine appears to have occurred by reduction of hot, metalliferous brines and the associated oxidation of organic matter (Ho et al., 1990).*

Some of the more reliable biomarker parameters, such as C_{29}-sterane isomerization ratios, are not useful for samples showing maturities beyond peak oil generation ($R_o \sim 0.9\%$) because the reactions they represent have reached equilibrium. Biomarker ratios that can be used for assessment of high levels of maturity include the "side chain cleavage" ratios for the mono- and triaromatic steroids (Sec. 3.2.4.3). Some parameters that are useful at high levels of thermal maturity (e.g., up to and above the equivalent of vitrinite reflectance 1.0%) are described by Van Grass (1990). These include absolute concentrations of steranes and hopanes, tricyclics/17α(H)-hopanes, diasteranes/steranes, and Ts/(Ts + Tm).

Biomarkers are typically measured using selected ion monitoring-GCMS (SIM), which is also called multiple ion detection-GCMS (MID) (Sec. 2.5.3.1). The major ions used in this approach are indicated in the discussion that follows. Figure 3.45 shows examples of sterane and terpane mass chromatograms for an immature bitumen and a mature, related oil. The MID approach allows a rapid, general assessment of the thermal maturity of samples. The immature Kimmeridge bitumen in the figure shows simple mass chromatograms compared to the mature Piper oil. This is mainly due to thermally induced stereoisomerization (Sec. 1.2.9) of the comparatively simple distribution of "biological" epimers of various biomarkers inherited from organisms to a more complex mixture including "geological" epimers. For example, the C_{27} to C_{29} steranes in the bitumen (Fig. 3.45, page 223) are dominated

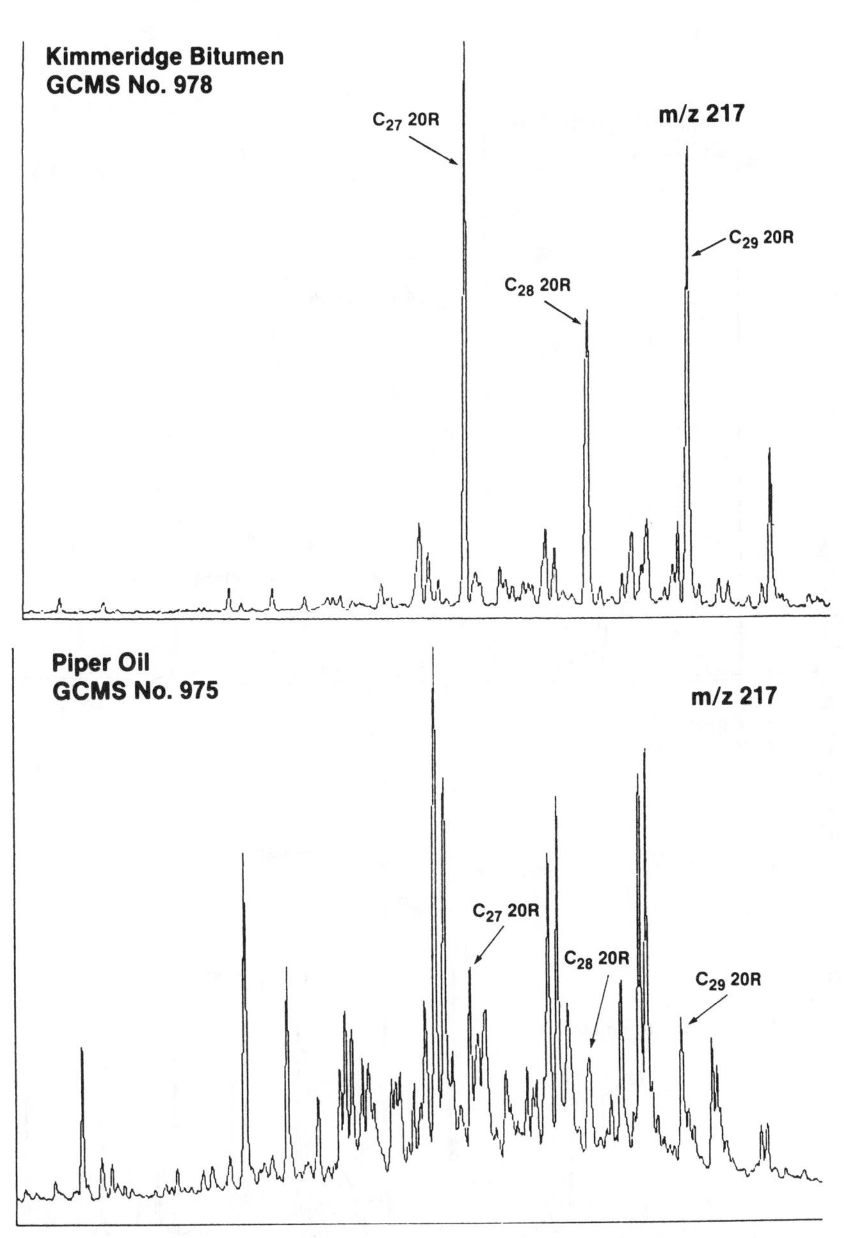

Figure 3.45 Sterane (m/z 217) and terpane (m/z 191) mass chromatograms for an immature bitumen from the Kimmeridge shale and those for a mature, related oil from the Piper Field in the North Sea. Despite their differing thermal maturities, biomarker and isotopic analyses allowed correlation of these samples (Peters et al., 1989).

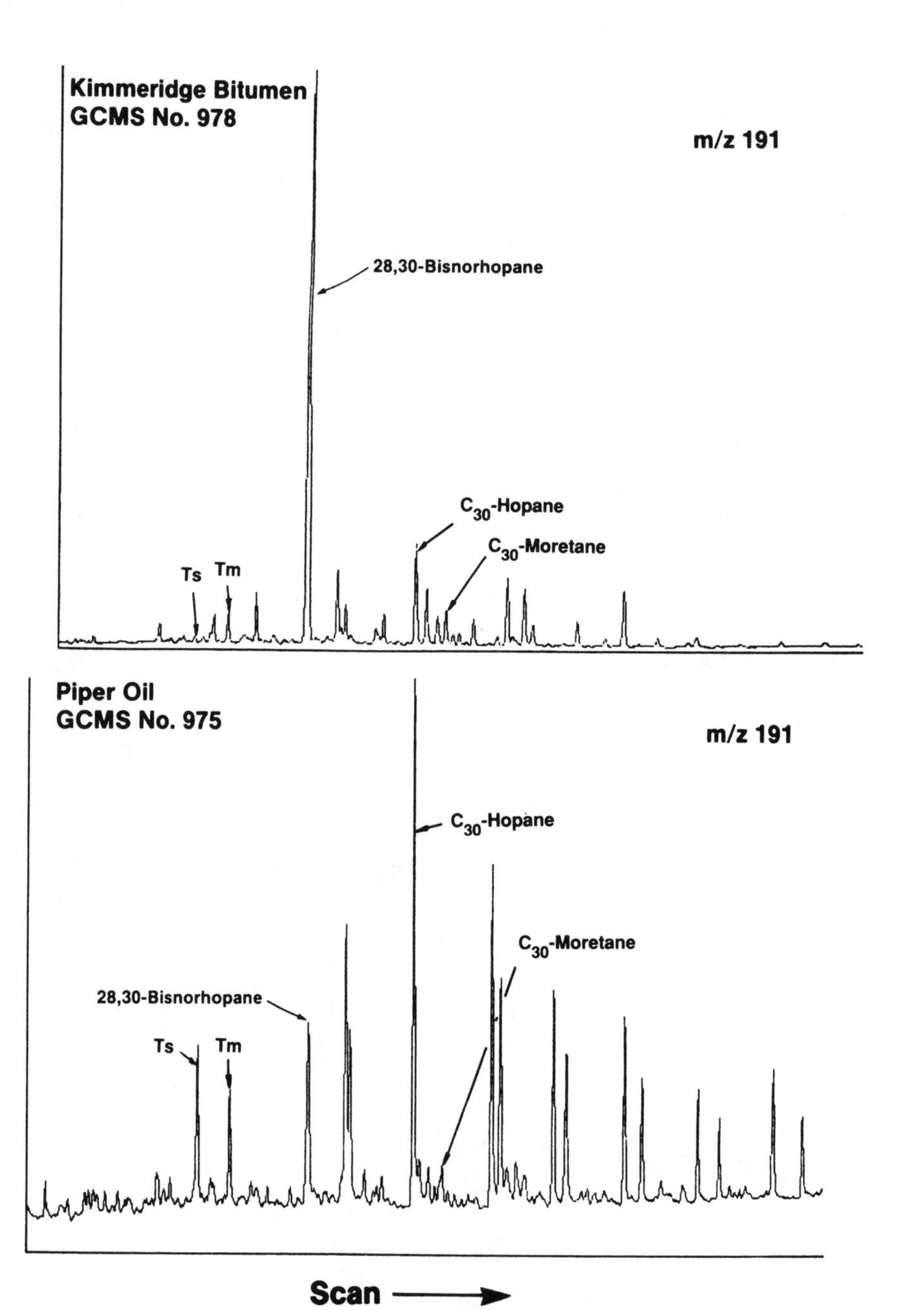

Figure 3.45 (*cont.*)

by 20R stereochemistry. Maturity results in a complex mixture of epimers (page 224).

However, biomarker compound classes can also be analyzed using some form of GCMSMS (Sec. 2.5.3.4), which reduces interferences and allows detection and quantitation at lower concentrations than the SIM approach. Linked-scanning is a form of GCMSMS that can be completed using

1. a double focusing magnetic sector instrument,
2. a multisector instrument, such as the triple sector quadrupole (TSQ) by monitoring collision-activated decompositions (CAD),
3. a hybrid instrument combining magnetic and quadrupole sectors (Sec. 1.5.3.2).

Ratios of certain saturated and aromatic biomarker compounds are some of the most commonly applied thermal maturity indicators. These indicators result from two types of reactions: (1) cracking reactions (including aromatizations), or (2) isomerizations at certain asymmetric carbon atoms. While both types of indicators are used, isomerizations are more commonly applied. For example, one of the more reliable cracking reactions is the conversion of monoaromatic to triaromatic steroids (Sec. 3.2.4.3). This reaction is less commonly applied for maturity assessment than isomerizations because of the more tedious quantitation procedures necessary for measurement of aromatic steroids. The two most commonly used isomerizations are those involving hydrogen atoms at the C-22 position in the hopanes and the C-20 position in the steranes, as discussed below (Sec. 3.2.4.1).

Several other biomarker thermal maturity parameters supplement the configurational isomerizations. These include the moretane/17α(H)-hopane [17β(H),21α(H)/17α(H),21β(H)] ratio, the 18α(H)/17α(H)-trisnorhopane ratio [Ts/Tm; usually presented as Ts/(Ts + Tm)] (Sec. 3.2.4.1), and ratios of aromatic steroids with short side chains to those with long side chains (Sec. 3.2.4.3) (side chain cleavage reactions). In general, these ratios are used to support the more commonly used biomarker ratios or when other maturity indicators are unavailable or unreliable.

Figure 3.46 shows the approximate ranges of various biomarker thermal maturity parameters relative to the oil-generative window. Because many biomarker thermal maturity parameters depend primarily on temperature and time (thermal stress) rather than the amount of petroleum produced (generation), all plots of this type are approximate.

> ***Note:*** *To simplify this discussion, variations in many biomarker maturity ratios are attributed to "cracking" or "isomerization." However, in many cases, the use of these terms is not strictly correct. For example, changes in the ratio TA(I)/TA(I + II) are commonly described as caused by side chain cleavage or cracking of long- [TA(II)] to short- [TA(I)] chained triaromatic steroids. It is now known that the relative abundances of TA(I) and TA(II) are at least partly controlled by differential thermal stability*

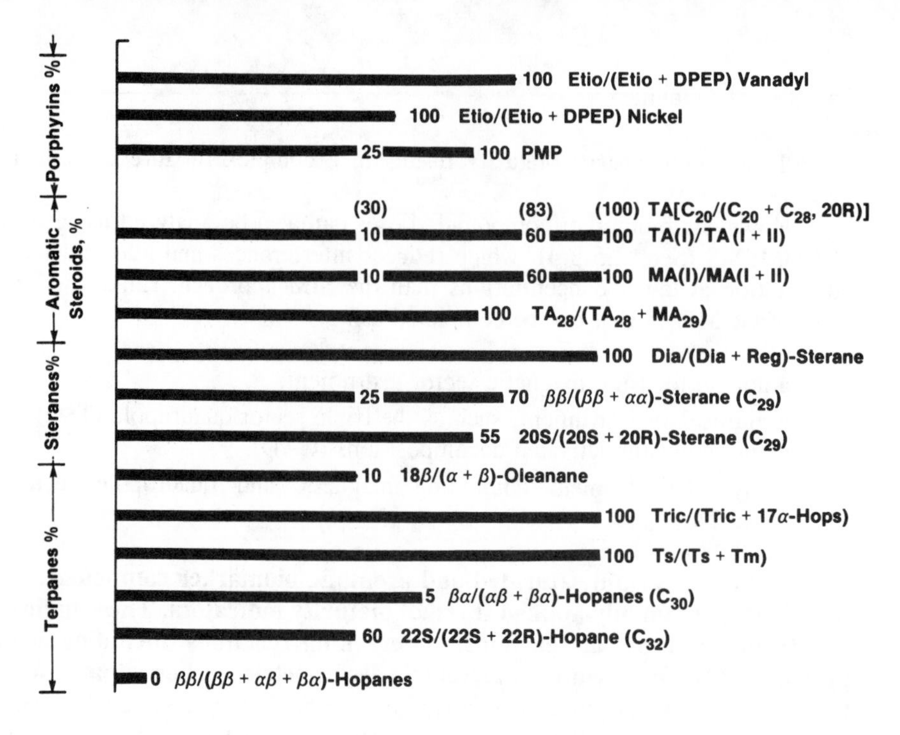

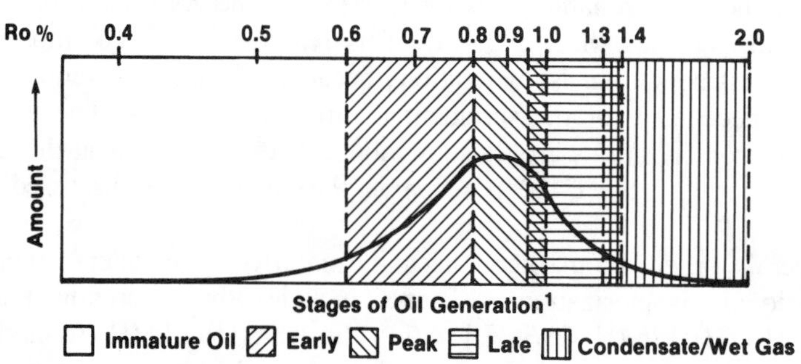

Figure 3.46 Biomarker maturation parameters respond in different ranges of maturity. Approximate ranges of biomarker maturity parameters are shown versus vitrinite reflectance and a generalized oil generation curve (after Mackenzie, 1984). Variability in stages of oil generation is shown by overlaps of shaded areas. Part of this variation results from comparing generation versus maturation parameters, as discussed in the text. Variations of ± 0.1 percent reflectance for biomarker ratios are common, and even greater variations can occur. Values for vanadyl and nickel porphyrins are from Mackenzie (1984) and the porphyrin maturity parameter is from Sundararaman et al., (1988a). The other values are based on experience of the authors and published papers. (See text.)

The solid black bars indicate the range of applicability for each ratio with respect to the stages of oil generation. The ratio reaches a constant value indicated by the number at the end of the bar and remains constant at higher maturities. This number is a maximum value for most of the parameters, except $18\beta/(18\alpha + 18\beta)$-oleanane, $\beta\alpha/(\alpha\beta + \beta\alpha)$-hopanes ($C_{30}$) and $\beta\beta/(\beta\beta + \alpha\beta + \beta\alpha)$-hopanes, where the numbers 10, 5, 0, respectively, are minimum values. For a few ratios intermediate values are shown within the bars to correspond with earlier stages of oil generation. However, the solid bars do not represent calibrated scales and are not meant to suggest a linear change in each parameter up to its maximum value (e.g., the reader is cautioned against inferring a vitrinite reflectance value from a given biomarker ratio). All values and ranges with respect to oil generation and vitrinite reflectance are approximate and may vary with heating rate, lithofacies, and organic facies of the source rock for the oil being evaluated.

(Beach et al., 1989). The factors controlling each maturity ratio are discussed in the following pages.

3.2.4.1 Terpanes

22S/(22S + 22R); Homohopane isomerization

Often applied; high specificity for immature to early oil generation (Fig. 3.46); measured using m/z 191 chromatogram or GCMSMS.

Isomerization at the C-22 position in the C_{31} to C_{35} 17α(H)-hopanes (Fig. 3.47) (Ensminger et al., 1977) occurs earlier than many biomarker reactions used to assess the thermal maturity of oils and bitumens such as isomerization at the C-20 position in the regular steranes. Schoell et al. (1983) showed that equilibrium for the C_{32} hopanes occurs at a vitrinite reflectance of about 0.5 percent in Mahakam Delta rocks. The biologically produced hopane precursors carry a 22R configuration that is gradually converted to a mixture of 22R and 22S diastereomers. The proportions of 22R and 22S can be calculated for any or all of the C_{31} to C_{35} compounds. These 22R and 22S doublets in the range C_{31} to C_{35} on the m/z 191 mass chromatogram are called **homohopanes** (Fig. 2.12; peaks 22, 23, 25, 26, 29, 30, 32, 33, 34, 35). (For comparison, the reader may wish to identify these peaks on the m/z 191 chromatogram of the mature Piper oil in Fig. 3.45).

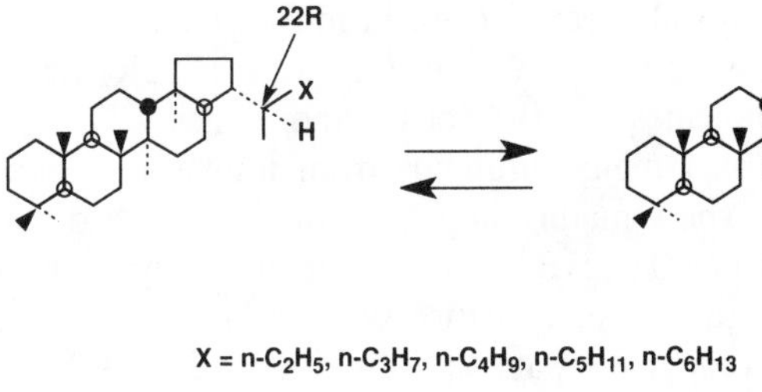

22R 22S

X = n-C_2H_5, n-C_3H_7, n-C_4H_9, n-C_5H_{11}, n-C_6H_{13}

Figure 3.47 Equilibration between 22R (biological epimer) and 22S (geological epimer) for the C_{31} to C_{35} homohopanes.

The 22S/(22S + 22R) ratios for the C_{31} to C_{35} 17α(H)-homohopanes may differ slightly. Typically, the C-22 epimer ratios increase slightly for the higher homologs from C_{31} to C_{35}. For example, Zumberge (1987b) calculated the average equilibrium 22S/(22S + 22R) ratios for 27 low maturity oils at C_{31}, C_{32}, C_{33}, C_{34}, and C_{35} as 0.55, 0.58, 0.60, 0.62, and 0.59, respectively. In some cases interference by coeluting peaks can invalidate certain ratios. For this reason, it is useful to

1. check each reported 22S/(22S + 22R) ratio for a given homohopane versus the other homohopanes,
2. measure the ratio using other important ions, such as m/z 205 for the C_{31} homohopanes,
3. use GCMSMS of the transition for the appropriate molecular ion to m/z 191.

Typically C_{31}- or C_{32}-homohopane results are used for calculations of the 22S/(22S + 22R) ratio. The 22S/(22S + 22R) ratio rises from 0 to about 0.6 (0.57 to 0.62 = equilibrium; Seifert and Moldowan, 1986) **during maturation.** Samples showing 22S/(22S + 22R) ratios in the range 0.50 to 0.54 have barely entered oil generation, while ratios in the range 0.57 to 0.62 indicate that the main phase of oil generation has been reached or surpassed. Some oils exposed to very light thermal stress show 22S/(22S + 22R) ratios as low as about 0.55. Philp (1982) describes a crude oil from the Gippsland Basin, Australia with a 22S/(22S + 22R) ratio less than 0.50 (using the C_{31} homohopane). A probable explanation is leaching or "solubilization" (Sec. 4.1) of homohopanes from immature lignites in contact with the reservoir.

Certain factors such as lithology may affect the rate of 17α(H)-homohopane isomerization. For example, Moldowan et al. (1992) found fully isomerized homohopanes in very immature carbonate rocks from the Adriatic Basin. Laboratory simulations of burial maturation indicate that free homohopanes in the bitumen isomerize more rapidly than those attached to the kerogen (Peters et al., 1990).

After reaching equilibrium at the early oil-generative stage, no further maturity information is available because the 22S/(22S + 22R) ratio remains constant. However, the inflection point in a plot of 22S/(22S + 22R) versus vitrinite reflectance or other maturity/generation parameters can be used to calibrate these parameters to the onset of oil generation for a given source rock in a basin.

Although not a routine maturity parameter, the 22S/(22S + 22R) ratio for the C_{31}-C_{35} moretanes shows variations similar to those for the C_{31}-C_{35} hopanes (Larcher et al., 1987). As with the 17α(H)-hopanes, the C_{31}-C_{35} moretanes with the presumed 22R configuration of the biological precursor are in much greater concentration in immature rocks. With increasing maturity, the 22S/(22S + 22R) ratio approaches about 0.6 for the 17α(H)-hopanes and 0.4 for the moretanes.

Ten Haven et al. (1986) note that many bitumens from immature rocks deposited under hypersaline conditions show mature hopane patterns. These bitumens show hop-17(21)-enes and extended 17α(H),21β(H)-homohopanes fully isomerized at C-22 (50–60% 22S) typical of immature and mature samples, respectively. Apparently unusual diagenetic pathways for the hopanes (and steranes, Sec. 3.2.4.2) in hypersaline environments account for this discrepancy.

βα-Moretanes/αβ-hopanes and ββ-hopanes

Often applied, high specificity for immature to early oil generation range (Fig. 3.46), measured using m/z 191 chromatograms or GCMSMS.

The biological 17β(H),21β(H)-configuration ($\beta\beta$) of hopanoids in organisms is very unstable and is not found in crude oils (unless contaminated by immature sedimentary organic matter). The $\beta\beta$-hopanes readily convert to $\beta\alpha$-(moretane) and $\alpha\beta$-hopane configurations (Fig. 1.20) along a reaction scheme proposed by Seifert and Moldowan (1980). During diagenesis, a temperature is reached where sufficient energy is available to overcome the energy barriers ΔG1 and ΔG2 (Fig. 3.48), allowing conversion of $\beta\beta$-hopanes to either $\beta\alpha$-moretanes or $\alpha\beta$-hopanes. At this low temperature, the conversion of these compounds back to $\beta\beta$-hopanes is not possible because of the high energy barriers, ΔG3 and ΔG4. At higher temperatures, the conversion of moretanes back to $\alpha\beta$-hopanes becomes possible through a $\beta\beta$-hopane intermediate. However, the high ΔG4 energy barrier allows little conversion of $\alpha\beta$-hopanes to $\beta\beta$-hopanes, resulting in an equilibrium mixture favoring $\alpha\beta$-hopanes over $\beta\alpha$-moretanes by about 20 : 1. Recent evidence from Carboniferous coals suggests that moretane/hopane ratios are controlled by differing rates of destruction of the isomers (ten Haven et al., 1992). According to this interpretation, both moretanes and hopanes are formed during diagenesis and the more labile moretanes decrease more rapidly during catagenesis compared to $\alpha\beta$-hopanes.

The ratio of 17β(H),21α(H)-moretanes to their corresponding 17α(H),21β(H)-hopanes **decreases** with increasing thermal maturity (e.g., Fig. 3.45) from about 0.8 in immature bitumens to values of less than 0.15 in mature source rocks and oils to a minimum of 0.05 (Mackenzie et al., 1980; Seifert and Moldowan, 1980). Based on a survey of 234 crude oils, Grantham (1986b) concluded that oils from Tertiary source rocks show higher moretane/hopane ratios (0.1–0.3 with many values between 0.15 and 0.20) than those from older rocks (generally 0.1 or less).

At our laboratory, the C_{30} compounds are most often used for the moretane/hopane ratio; although this ratio is also quantified using C_{29} compounds (e.g., Seifert

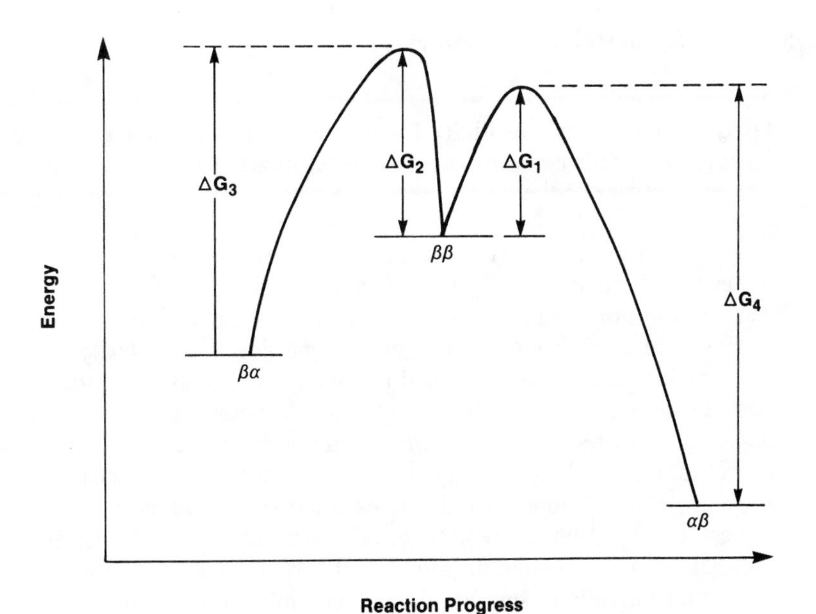

Figure 3.48 Proposed stability relationships among the three major classes of hopanes in petroleum (Seifert and Moldowan, 1980).

and Moldowan, 1980). Others have used both C_{29} and C_{30} compounds together (Mackenzie et al., 1980).

Evidence suggests that the moretane/hopane ratio is partly dependent on source input or depositional environment. For example, Rullkötter and Marzi (1988) noted higher moretane/hopane ratios in bitumens from hypersaline rocks compared to those from adjacent shales.

With the exception of the C_{30} pseudohomologs, the three most commonly encountered hopane stereoisomers can be distinguished by their MS fragmentation patterns in full scan GCMS according to the relative intensities of their A + B and D + E ring fragments (see Fig. 2.11; Ensminger et al., 1974; Van Dorsselaer, 1974; Seifert and Moldowan, 1980). The $\alpha\beta$-hopane isomers all have the m/z 191 fragment for the A/B-ring portion of the molecule that is larger than the m/z 148 + X (side-chain) fragment for the D/E ring portion (Fig. 2.11). For the $\beta\alpha$ and the $\beta\beta$-hopane isomers the situation is reversed, and all have an m/z 148 + X fragment larger than the m/z 191 fragment. The $\beta\alpha$ and $\beta\beta$ stereoisomers are distinguished from each other because the ratio of the fragments m/z (148 + X)/191 = ~2 for the $\beta\beta$ isomers, but is < 1.5 for the $\beta\alpha$ isomers. C_{27} 17β(H)-hopane, which belongs strictly to neither the $\beta\beta$ nor the $\beta\alpha$ series, has a ratio of ~1.5. The D/E-ring and A/B-ring fragments for the C_{30} hopanes are both m/z 191 and cannot be distinguished by this method. While this method is useful for distinguishing the common hopane series, the same reasoning does not extend to all related hopanoids. For example, the $\alpha\beta$ and $\beta\alpha$-28,30-bisnorhopanes have nearly identical mass spectra (Moldowan et al., 1984).

Tricyclics/17α(H)-hopanes

Often applied, low specificity for mature to postmature range caused by interference from source input, measured on m/z 191 chromatograms or GCMSMS.

The tricyclics/17α(H)-hopanes ratio increases systematically for related oils of increasing thermal maturity (Seifert and Moldowan, 1978). The ratio increases because proportionally more tricyclic terpanes (i.e., cheilanthanes, Fig. 2.11) than hopanes are released from the kerogen at higher levels of maturity (Aquino Neto et al., 1983), which is also demonstrated by an observed increase in the tricyclics/17α(H)-hopanes ratio during hydrous pyrolysis of Monterey shale (Peters et al., 1990). Fig. 3.49 compares terpane mass chromatograms for three related Oman-area oils. The tricyclics/17α(H)-hopanes ratio increases with thermal maturity of these oils. The tricyclics/17α(H)-hopanes ratio of expelled oil increases during hydrous pyrolysis (Peters et al., 1990), possibly because tricyclics migrate faster and/or 17α(H)-hopanes show a greater affinity for the rock matrix. Laboratory simulations of petroleum migration using an alumina column show that tricyclic terpanes and $5\alpha,14\beta,17\beta$(H)-steranes elute faster than hopanes and $5\alpha,14\alpha,17\alpha$(H)-steranes (Jiang Zhusheng et al., 1988). Future research will be needed to determine whether this applies to natural expulsion, although Kruge et al. (1990) suggest that more rapid mi-

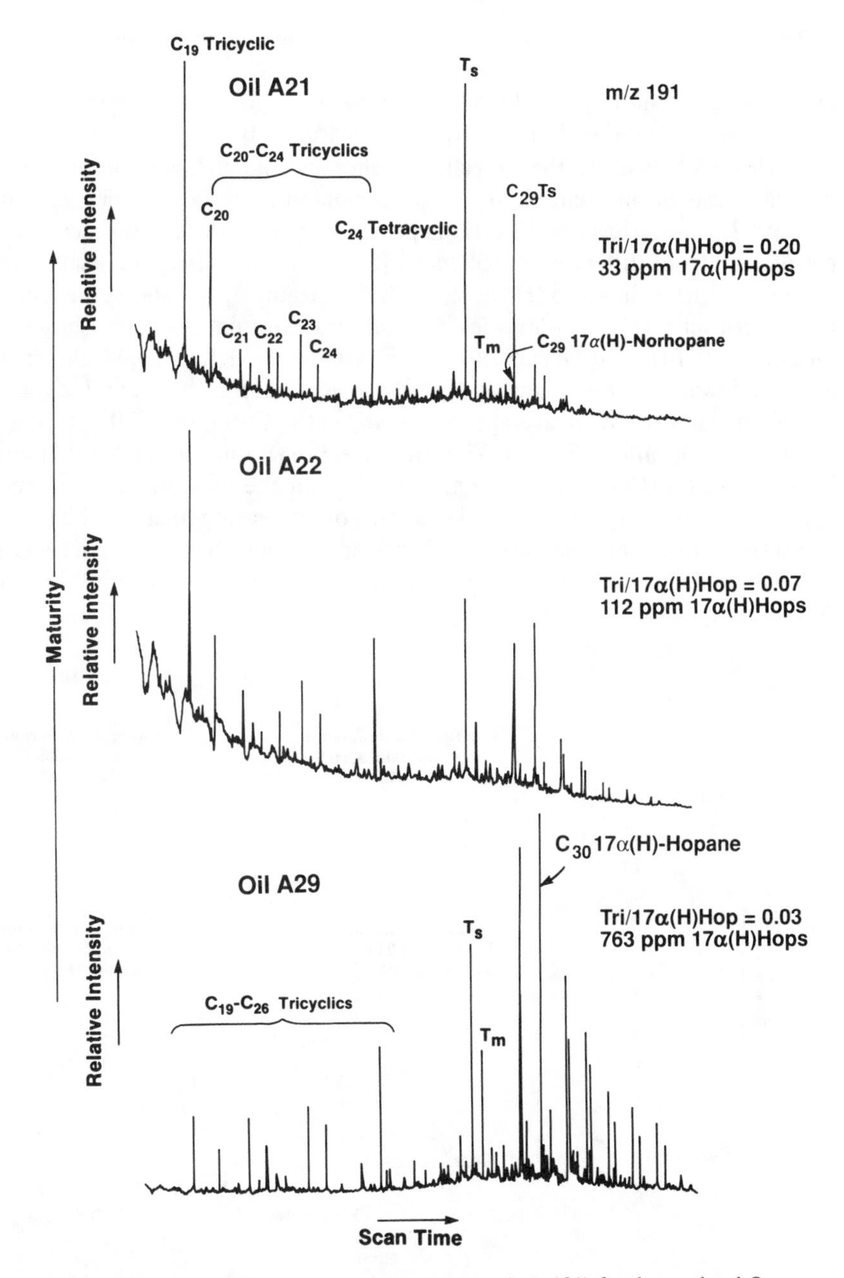

Figure 3.49 Comparison of terpane mass chromatograms (m/z 191) for three related Oman-area oils. Ranking of the oils by thermal maturity was difficult because all are highly mature and contain few biomarkers. Nonetheless, several biomarker parameters were used successfully for this purpose. For example, the tricyclics/17α(H)-hopanes ratio increases and the absolute amounts of 17α(H)-hopanes (C_{29}-C_{33}) decrease with maturity. The figure also demonstrates the use of C_{19}-C_{29} tricyclic terpanes to solve a difficult correlation problem. Based on numerous biomarker and supporting parameters, Oils A22 and A29 are closely related. However, the highly mature Oil A21 shows few if any steranes or hopanes. The similar pattern of the thermally resistant C_{19}-C_{29} tricyclic terpanes for Oils A21 and A22 (the oil showing the most similar maturity to Oil A21) and other supporting data were used to show that Oil A21 is related to the other two oils.

gration of tricyclic terpanes than hopanes partly explains biomarker distributions in black shales of the East Berlin Formation, Hartford Basin, Connecticut.

However, because the tricyclic terpanes and the hopanes appear to result from the diagenesis of different biological precursors in bacterial membranes (Ourisson et al., 1982), the tricyclics/17α(H)-hopanes ratio can differ considerably between petroleums from different source rocks [Sec. 3.1.3.1, "Tricyclic terpanes (cheilanthanes) and tricyclics/17α(H)-hopanes"]. For example, the source-dependent effect on the tricyclics/17α(H)-hopanes ratio allows differentiation of Carneros (0.17), Phacoides (0.14), and Oceanic (0.09) oil groups in McKittrick, California (Seifert and Moldowan, 1978). Maturity-dependent biomarker ratios show that these oils increase in maturity with average reservoir depth: Carneros (6300 ft) < Phacoides (8500 ft) < Oceanic (8870 ft). The tricyclics/17α(H)-hopanes ratio is largest for the least mature oil (Carneros; 0.17) and smallest for the most mature (Oceanic; 0.09); opposite of that expected for related oils of increasing maturity. The tricyclics/17α(H)-hopanes ratio was also used to help distinguish two Jurassic source rocks (Kingak and Shublik) as possible sources for various oils from Prudhoe Bay, Alaska, (Fig. 3.50) (Seifert et al., 1980).

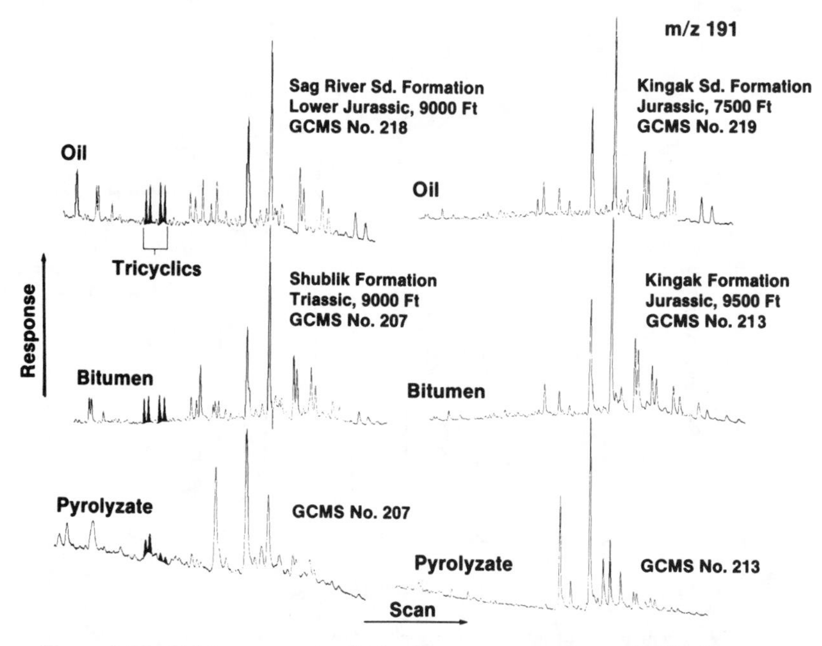

Figure 3.50 Relative amounts of tricyclic terpanes on these m/z 191 fragmentograms (black peaks) support a genetic relationship between the Sag River oil (GCMS No. 218) and the Shublik shale (GCMS No. 207). Absence of tricyclic terpanes in the Kingak oil (GCMS No. 219) and the Kingak shale (GCMS No. 213) supports a relationship between these samples (Seifert et al., 1980). The indigenous nature of the bitumen in the source rocks is confirmed by comparing m/z 191 fragmentograms for pyrolyzates of these rocks, which show the presence and absence of tricyclic terpanes in the Shublik and Kingak shales, respectively.

Ts/(Ts + Tm)

Often applied; some interference from depositional environment; immature to mature or postmature range (Fig. 3.46); measured on m/z 191 or GCMSMS.

During catagenesis, C_{27} 17α(H)-trisnorhopane (Tm or 17α(H)-22,29,30-trisnorhopane) shows lower relative stability than C_{27} 18α(H)-trisnorhopane II (Ts or 18α(H)-22,29,30-trisnorneohopane) (Fig. 3.51) (Seifert and Moldowan, 1978). This observation has been substantiated using molecular mechanics calculations for the formation of various hopanes, including Ts and Tm (Kolaczkowska et al., 1990). It is unknown whether conversion of Tm to Ts may also occur. Figures 3.49 and 3.45 show examples of maturity effects on the relative amounts of Ts and Tm on m/z 191 mass chromatograms.

The Ts/(Ts + Tm) ratio (sometimes reported as Ts/Tm) is both maturity- and source-dependent. Moldowan et al. (1986) show that the ratio can vary depending on organic facies. Hong et al. (1986) studied the thermal maturity of rocks in the Linyi Basin, China. Because of the low maturity of the organic matter in the shallow rocks in this basin, they included a third compound related to Tm and Ts, but showing even lower thermal stability, C_{27} 22,29,30-trisnor-17β(H)-hopane. They found a systematic increase in the relative abundance of Ts and a decrease in C_{27} 17β(H)-trisnorhopane compared to Tm with depth. The concentration of 17β(H)-trisnorhopane approaches zero at maturities equivalent to the earliest oil window. 17β(H)-trisnorhopane is more stable than 17β,21β(H)-hopane, but less stable than moretane as shown by a series of Cretaceous shales from the Wyoming Overthrust Belt (Seifert and Moldowan, 1980).

The Ts/(Ts+Tm) ratio should be used with caution. Tm and Ts commonly coelute with tricyclic or tetracyclic terpanes on the m/z 191 mass chromatogram, resulting in spurious Ts/(Ts+Tm) ratios. For example, Rullkötter and Wendisch (1982) show coelution of a C_{30} tetracyclic triterpane with Tm on their capillary columns. If interference is suspected, quantifying Tm and Ts from the m/z 370 (molecular ion or

Tm (Less Stable) Ts (More Stable)

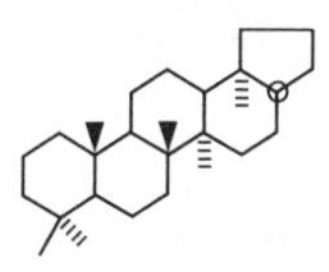

17α(H)-22,29,30-Trisnorhopane 18α(H)-22,29,30-Trisnorneohopane

Figure 3.51 Structures of 17α(H)-22,29,30-trisnorhopane (Tm) and 18α(H)-22,29,30-trisnorhopane II (Ts). Ts is more stable to thermal maturation than Tm.

M^+) mass chromatogram may be helpful (Volkman et al., 1983a). Measurement of the transition from m/z 370 to m/z 191 by GCMSMS gives the most reliable results.

The Ts/(Ts + Tm) ratio is most reliable as a maturity indicator when evaluating oils from a common source of consistent organic facies. The relative importance of lithology or oxicity of the depositional environment in controlling this ratio remains unclear. The Ts/(Ts + Tm) ratio appears to be sensitive to clay-catalyzed reactions. For example, oils from carbonate source rocks appear to have anomalously low Ts/(Ts + Tm) ratios compared to those generated from shales (McKirdy et al., 1983, 1984; Rullkötter et al., 1985; Price et al., 1987; Sec. 3.1.4). Bitumens from many hypersaline source rocks show high Ts/(Ts + Tm) ratios (Rullkötter and Marzi, 1988). The Ts/(Ts + Tm) ratio increases at lower Eh and decreases at higher pH for the depositional environment of a series of Lower Toarcian marine shales from southwestern Germany (Moldowan et al., 1986), but also decreases in an anoxic (low Eh) carbonate section.

C_{29}Ts/(C_{29} 17α(H)-hopane + C_{29}Ts)

Rarely applied, unknown specificity, measured on m/z 191 chromatograms.

Described for many years as the "unknown C_{29} terpane" or "C_{29}X" eluting immediately after C_{29} 17α(H)-hopane on m/z 191 chromatograms (Figures 3.11, 3.18, and 3.19), 18α-(H)-30-norneohopane (C_{29}Ts), has recently been identified by advanced NMR methods (Moldowan et al., 1991b). Most reports suggest the abundance of this compound (e.g., compared to the C_{29} 17α(H)-hopane) on m/z 191 chromatograms is related to thermal maturity (Hughes et al., 1985; Sofer et al., 1986; Sofer, 1988; Cornford et al., 1988; Riediger et al., 1990). Molecular mechanics calculations indicate that C_{29}Ts should be more stable than 17α(H)-30-norhopane by 3.5 kcal/mole, compared with the C_{27} analog Ts, which is more stable than Tm by 4.4 kcal/mole (Kolaczkowska et al., 1990). Thus, the thermal maturity effect on the C_{29}Ts/(C_{29} 17α(H)-hopane + C_{29}Ts) ratio should be comparable, but slightly less than that on Ts/(Ts + Tm). This inference is supported by a study from the Jeanne d'Arc Basin, eastern Canada (Fowler and Brooks, 1990). In this basin, trends in the C_{29}Ts/C_{29}-hopane ratio paralleled those for Ts/Tm, and both parameters increased with thermal maturity as indicated by other biomarker parameters. The highly mature condensate A21 (Fig. 3.49), for example, shows a very high C_{29}Ts/(C_{29} 17α(H)-hopane + C_{29}Ts) ratio.

Ts/C_{30} 17α(H)-hopane

Rarely applied, unknown specificity for mature to postmature range, measured on m/z 191 chromatograms.

Volkman et al. (1983a) propose this ratio as a maturity parameter for very mature oils and condensates (e.g., Fig. 3.49). Cleavage of the side chain of C_{29} and higher carbon-number hopanoids is believed to produce precursors of C_{27} hopane II (Ts) by a rearrangement mechanism (Seifert and Moldowan, 1978).

18α/(18α + 18β)-oleananes and oleanane index

Rarely applied, immature to early mature range (Fig. 3.46), measured on m/z 191 chromatograms (Fig. 2.11).

Apart from being a reliable marker for higher plant (angiosperm) input in rocks of Cretaceous or younger age (Sec. 3.1.3.1), oleanane epimerization can be used as a maturity parameter (Riva et al., 1988; Ekweozor and Udo, 1988). Oleananes appear to be derived from oleanenes and other pentacyclic triterpenes (i.e., taraxerenes) by rearrangements and double-bond isomerizations through a $\Delta 13(18)/\Delta 18$ intermediate (ten Haven and Rullkötter, 1988). Reduction of the double bond gives a mixture of 18α(H)- and 18β(H)-oleananes. Increased maturation favors the 18α(H) configuration, suggesting it is more stable than 18β(H). Riva et al. (1988) show correlations between the oleanane epimer ratio and other maturity parameters, including Ts/Tm, vitrinite reflectance, and T_{max} with depth based on data from several wells.

Figure 3.52 compares the relative peak areas for 18α(H)- and 18β(H)-oleanane for two groups of petroleum samples from the Eel River area, California. These two compounds were provisionally identified by coinjection of authentic standards (Fig. 2.19). The relative maturity of the samples in Fig. 3.52 was established using other biomarker parameters (see Fig. 3.60).

Ekweozor and Telnaes (1990) suggest that the oleanane index (oleanane/C_{30} hopane) can be used to delimit the top of the oil-generative window (boundary between diagenesis and catagenesis). In wells from the Niger delta, they observed maxima in the oleanane index at the top of the oil-generative window, corresponding to vitrinite reflectance values of about 0.55 percent. Because this approach has not been extensively studied, we recommend that maturity estimates based on the oleanane index be cross-checked using other maturity parameters, such as the hopane isomerization ratio [22S/(22S + 22R)].

(28,30-Bisnorhopanes + 25,28,30-trisnorhopanes)/hopanes, (BNH + TNH)/hopanes

---

Often applied when BNH and TNH are present, highly specific for immature to mature range, measured on m/z 191 and 163 fragmentograms (Fig. 2.11).

The (BNH + TNH)/hopanes ratio is defined as (28,30-bisnorhopanes + 25,28,30-trisnorhopanes)/(C_{29} + C_{30} 17α(H)-hopanes) (Moldowan et al., 1984), but can be expressed in other ways, such as BNH/[17α(H)-hopane + BNH]. A nonzero value for this ratio indicates a contribution from a source rock containing

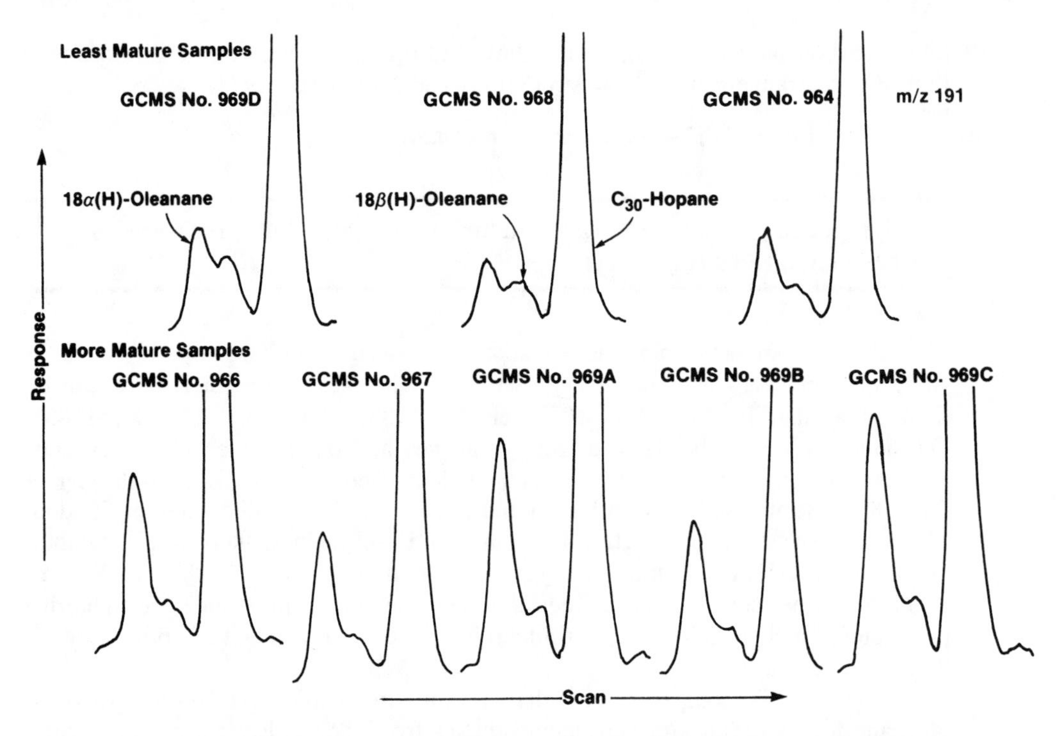

Figure 3.52 Terpane mass chromatograms (m/z 191) for oils and bitumens from the Eel River Basin, California, show that the relative proportions of 18α(H)- and 18β(H)-oleanane change with thermal maturity. The less mature samples (top) show proportionally more 18β(H)-oleanane than more mature samples (bottom). Maturity relationships were established using other biomarker parameters (i.e., Fig. 3.60). Provisional identification of the oleanane epimers in this figure was accomplished using coinjection of authentic standards (Fig. 2.19).

these demethylated hopanes and can be useful in distinguishing samples from different sources.

The structure of 17α(H),18α(H),21β(H) 28,30-bisnorhopane was first proven after its isolation from Monterey shale by Seifert et al. (1978). This compound is probably the same as the C_{28}-triterpane reported by Petrov et al. (1976) in Siva oil, Russia (Seifert, 1980). Later it was shown that the 17α,21β(H)-isomer coelutes with the 17β,21α(H)-isomer on most capillary columns and that a special Apiezon-coated column is necessary to separate them (Moldowan et al., 1984). Thus, these two compounds are usually quantified together. Unlike the 17β,21β(H)-isomers of the hopanes, the 17β,21β(H)-isomers of BNH and TNH are stable at thermal maturities within the oil window and occur in small amounts relative to the 17α,21β(H)- and 17β,21α(H)-isomers.

BNH and TNH do not appear to be generated from the kerogen, but are derived from the original, free bitumen in the rock [(Sec. 3.1.3.1; *BNH/*

(BNH + TNH)]. For example, note the large BNH peak (28,30-bisnorhopane) for immature bitumen from the Kimmeridge shale sample and its lower concentration in the related, but more mature Piper oil in Figure 3.45. As the source rock generates oil during thermal maturation, bitumen is diluted by products from the kerogen, and there is a systematic decrease in the (BNH+TNH)/hopanes ratio for oils or source rocks which contain these compounds (Moldowan et al., 1984). For example, the Bazhenov shale source rock and related oils from Western Siberia contain relatively high concentrations of TNH when immature, but the TNH/hopanes ratio decreases to zero in the late oil window (Fig. 3.53, bottom) (Peters et al., 1992a). The same figure (top) shows a similar trend for the BNH/hopanes ratio for oils from the Monterey Formation in California.

Grantham et al. (1980) indicate that the bisnorhopane/hopane ratio in North Sea oils appears to decrease with increasing thermal maturity, but other factors such as organic matter input and depositional environment may be important. Thus, a mature oil showing a BNH/hopane ratio of zero is not necessarily unrelated to a less mature potential source rock sample showing a nonzero value for the ratio. The presence of BNH and TNH in Kimmeridge source rock and related oils was used to differentiate these samples from other bitumens and oils in the North Sea (Peters et al., 1989).

Unlike Grantham et al. (1980), Hughes et al. (1985) note an **increase** in bisnorhopane/hopane with increasing maturity. They explain these conflicting observations as follows. For diverse samples (e.g., Hughes et al., 1985), source input is the major control on BNH/hopane ratios. However, among samples from a similar source (e.g., Grantham et al., 1980), maturity is the controlling factor.

We use the bisnorhopane/hopane ratio for comparing the maturity of *related* petroleums. It is our experience that this ratio decreases with thermal maturity. However, *unrelated* samples at the same level of maturity can show different bisnorhopane/hopane ratios because of differing organic matter input and/or depositional environment. Use of epimer ratios of BNH and TNH for determining the maturity of heavily biodegraded oils is discussed in Sec. 3.3.3 (C_{28}-C_{34} 30-nor-17α(H)-hopanes).

3.2.4.2 Steranes

20S/(20S + 20R); Sterane isomerization [%20S or 5α(H),14α(H),17α(H) 20S/($\alpha\alpha\alpha$20S + $\alpha\alpha\alpha$20R)]

Often applied; highly specific for immature to mature range (Fig. 3.55); measured on m/z 217 (Fig. 2.11) or preferably by GCMSMS analysis of C_{29} steranes.

Isomerization at C-20 in the C_{29} 5α(H),14α(H),17α(H)-steranes (Fig. 3.54) causes the 20S/(20S + 20R) ratio to rise from 0 to about 0.5 (0.52 to 0.55 = equilibrium; Seifert and Moldowan, 1986) (Fig. 3.46) with increasing maturity.

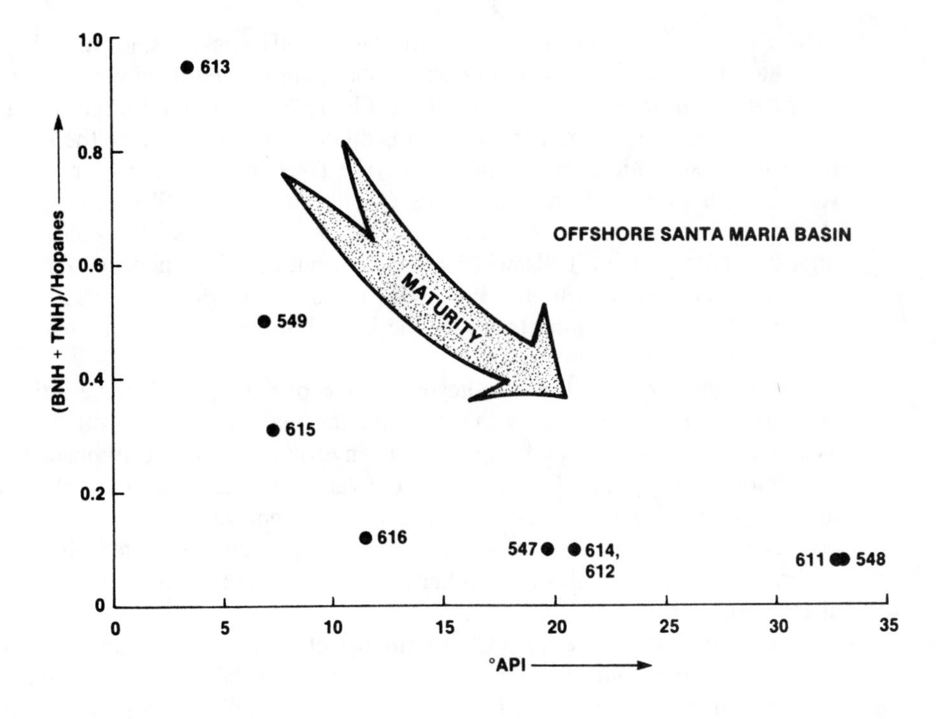

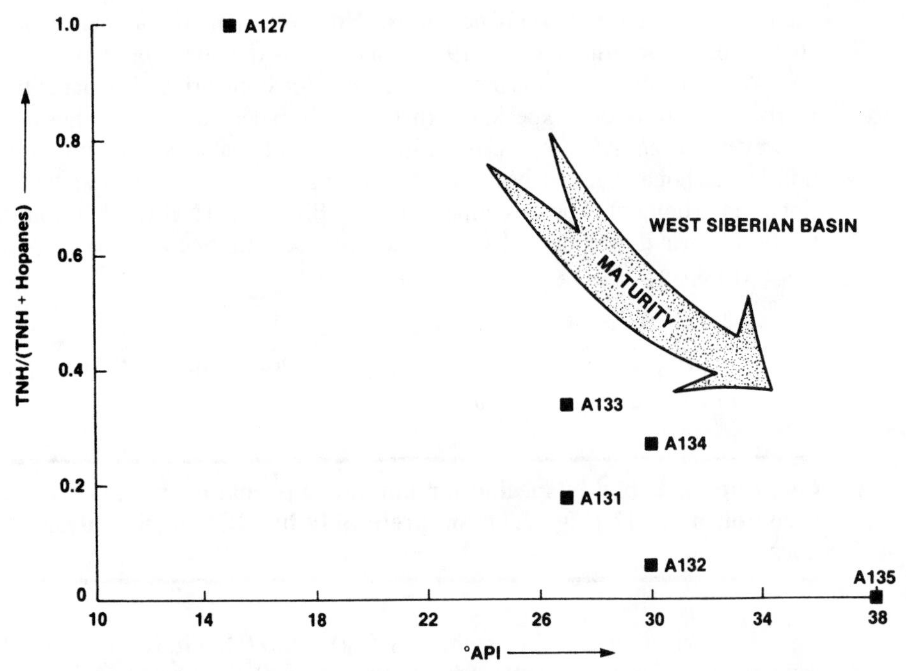

Figure 3.53 Decrease in (BNH+TNH)/hopane ratio with thermal maturity for Monterey oils from the offshore Santa Maria Basin, California (top).

Decrease in the TNH/(TNH+hopanes) ratio with thermal maturity (API gravity) for oils from Western Siberia (bottom).

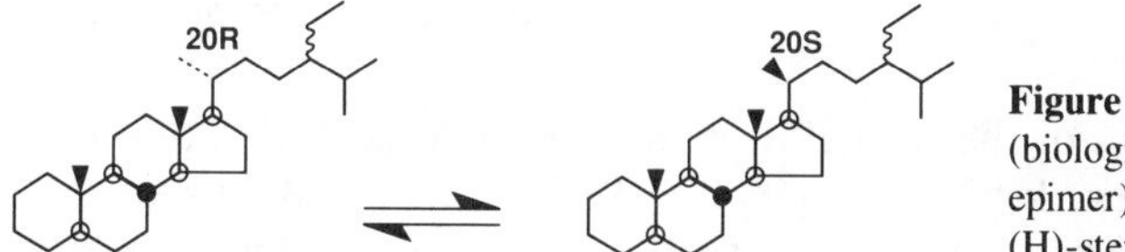

Figure 3.54 Equilibration between 20R (biological epimer) and 20S (geological epimer) for the C_{29} 5α(H),14α(H),17α(H)-steranes. See text.

Note: *However, a recent advance in GC methodology (Gallegos and Moldowan, 1992) shows that the chromatographic peak corresponding to the C_{29} $\alpha\alpha\alpha$20S isomer is normally contaminated by other C_{29} sterane isomers (probably $\alpha\beta\alpha$). Thus, all previously reported 20S/(20S + 20R) ratios are probably too high and the maximum (equilibrium?) value in oils and source rocks is 0.5 or less. Also, the 20S and 20R peaks each contain a mixture of 24S and 24R epimers that are usually ignored and measured together due to difficulties in their separation.*

Other isomerization reactions for the steranes are shown in Fig. 1.22. Only the R configuration at C-20 is found in steroid precursors in living organisms, and this is gradually converted during burial maturation to a mixture of the R and S sterane configurations. This ratio has been used in kinetic calculations for input to basin geochemical models (Mackenzie and McKenzie, 1983; Beaumont et al., 1985; Marzi and Rullkötter, 1992).

The sterane isomerization ratios are most often reported for the C_{29} compounds (24-ethylcholestanes or "stigmastanes," Fig. 1.10) because of the ease of analysis using the m/z 217 mass chromatogram. Isomerization ratios based on the C_{27} and C_{28} steranes commonly show interference by coeluting peaks. However, GCMSMS measurements show equivalent accuracy for C_{27}, C_{28}, and C_{29} 20S/(20S + 20R) ratios, all of which have equivalent potential as maturity parameters when measured by this method.

We do not recommend use of the 20S/(20S + 20R) ethylcholestane (C_{29}) ratio as an indicator of the onset of petroleum generation unless it is calibrated for each basin and source rock by comparison with other maturity and generation parameters. For example, Huang Difan et al. (1990) show correlations between vitrinite reflectance and two C_{29} sterane isomerization ratios, 20S/(20S + 20R) and $\beta\beta/(\beta\beta + \alpha\alpha)$, specific for the southern part of the Dagong oil field, China. Waples and Machihara (1991) used this approach to compare rocks from the Akita and Niigata basins, Japan. Mackenzie et al. (1980) indicate that petroleum generation begins at %20S values of about 40 percent. However, Seifert and Moldowan (1981) have observed numerous low-maturity oils with 20S/(20S + 20R) ratios in the range 0.23 to 0.29. The onset of petroleum generation is currently best estimated using hopane epimer ratios (see previous text) or the porphyrin maturity parameter (see following text).

Factors other than thermal maturity can affect sterane isomerization ratios. Facies effects on C_{29}-sterane 20S/(20S + 20R) ratios have been observed in a sequence of Lower Toarcian rocks from southwestern Germany (Moldowan et al., 1986) and in organic-rich lacustrine rocks from offshore West Africa (Hwang et al., 1989). Weathering in Phosphoria shale outcrops in Utah was reported to lead to preferential loss of the $\alpha\alpha\alpha$20S C_{29}-sterane diastereomer (Clayton and King, 1987). On the other hand, partial sterane biodegradation of an oil can result in an increase in $\alpha\alpha\alpha$20S/(20S + 20R) sterane ratios (C_{27}, C_{28}, and C_{29}) to above 0.55, presumably by selective removal of the $\alpha\alpha\alpha$ 20R epimer by bacteria (Rullkötter and Wendisch, 1982; McKirdy et al., 1983; Seifert et al., 1984). Peters et al. (1990) and Marzi and Rullkötter (1992) show that this ratio partly depends on the source rock and can decrease at high maturity. The results suggest that differential stability of epimers, generation of additional material from the kerogen, and other factors may affect this ratio. Based on measurements of absolute concentrations of 20S and 20R $\alpha\alpha\alpha$ ethylcholestanes in rock extracts and oils, Requejo (1992) concluded that greater stability of the 20S epimer compared to 20R is the main reason for increasing 20S/(20S + 20R) ratios with thermal maturity, and that there is no evidence for equilibration of 20S and 20R epimers.

Immature bitumens from hypersaline rocks can appear mature due to sterane (and hopane, see previous text) diagenetic pathways proposed by ten Haven et al. (1986). These immature hypersaline samples contain abundant 5α(H),14β(H),17β(H),20R and 20S and 5α(H),14α(H),17α(H),20R steranes, but virtually no 5α(H),14α(H),17α(H),20S steranes. This contradicts the relationship between these compounds proposed by Mackenzie et al. (1982a). Rullkötter and Marzi (1988) and Peakman et al. (1989) indicate that unusually high concentrations of $\alpha\beta\beta$ steranes in immature, hypersaline sediments may result from diagenetic reduction of Δ7 and Δ8(14),5α(H)-sterols via the Δ14,5α(H),17β(H)-sterenes. Care should be used when applying 20S/(20S + 20R) and especially $\beta\beta/(\beta\beta + \alpha\alpha)$ sterane ratios for maturity determinations of samples from hypersaline sources. Bitumens from hypersaline or evaporitic environments are characterized by many of the following characteristics: even to odd *n*-paraffin predominance, low pristane/phytane ratio, high gammacerane, low diasteranes/steranes ratio, and preferential preservation of C_{34} and/or C_{35} homohopanes.

$\beta\beta/(\beta\beta + \alpha\alpha)$; Sterane isomerization [%$\beta\beta$, or 14β(H),17β(H)/($\beta\beta + \alpha\alpha$) isomerization]

Often applied; highly specific for immature to mature range (Fig. 3.46); measured on m/z 217 (Fig. 2.11) or preferably by GCMSMS of C_{29} steranes

Isomerization at the C-14 and C-17 positions in the 20S and 20R C_{29} regular steranes causes an increase in the ratio $\beta\beta/(\beta\beta + \alpha\alpha)$ from nonzero values to about 0.7 (0.67 to 0.71 = equilibrium; Seifert and Moldowan, 1986). This ratio appears to be independent of source organic matter input (however, see the following) and somewhat slower to reach equilibrium than the 20S/(20S + 20R) ratio, thus making it effective at higher levels of maturity.

Plots of $\beta\beta/(\beta\beta + \alpha\alpha)$ versus 20S/(20S + 20R) for the C_{29} steranes are particularly effective in describing the thermal maturity of source rocks or oils (Seifert and Moldowan, 1986). The plots can be used to cross-check one maturity parameter versus another (Fig. 3.55). For example, data for any oils that plot far off the maturity trend line in the figure would be immediately reexamined in light of the disagreement between the two sterane maturity parameters. When such disagree-

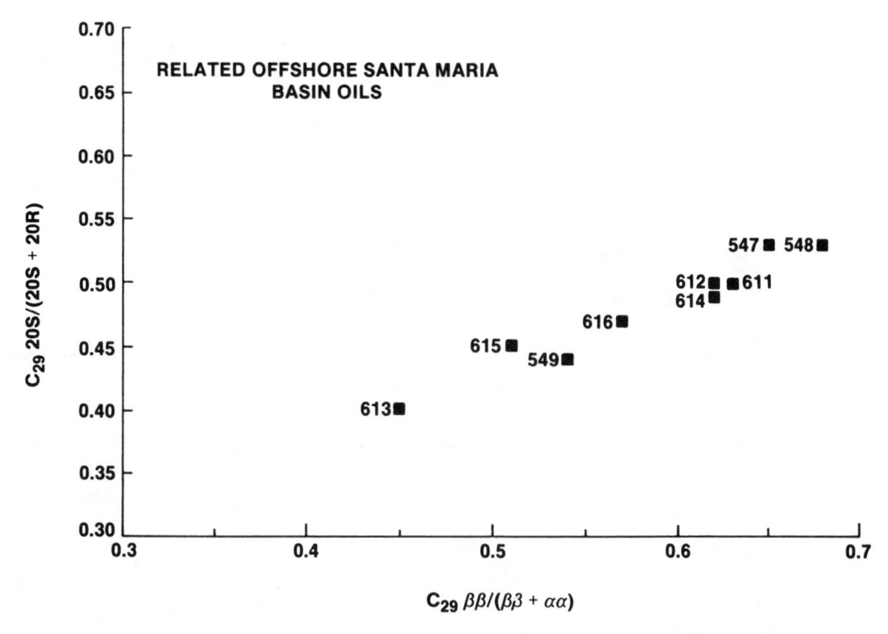

Figure 3.55 Correlation of thermal maturity parameters based on isomerization of asymmetric centers in the C_{29} steranes for offshore Santa Maria Basin oils, California. Same samples as in top half of Figure 3.53.

ments occur, they can sometimes be explained as resulting from some type of analytical error or poor sample quality. Alternatively, such variations can indicate samples that have experienced different heating rates in the subsurface (Mackenzie and McKenzie, 1983) or different levels of clay catalysis (Huang Difan et al., 1990).

Unlike older oils, the sterane isomerization ratios for Tertiary oils are generally not at equilibrium (Grantham, 1986b). This appears to result from insufficient time for complete sterane isomerization in Tertiary rocks, even though the generated oils are thermally mature based on bulk geochemical characteristics, including API gravity, sulfur content, and gross composition.

As for all sterane parameters, GCMSMS improves the accuracy of the C_{29}-sterane isomerization ratio measurements. In addition, GCMSMS facilitates accurate measurement of C_{27} and C_{28} $\beta\beta/(\beta\beta + \alpha\alpha)$ sterane ratios, which are not normally derived from routine GCMS because of coeluting compounds on the m/z 217 chromatograms.

Ten Haven et al. (1986) suggest early diagenetic formation of steranes in hypersaline environments. Abundant $\beta\beta$ steranes in these sediments could be due to reaction with sulfur. Schmid (1986) for example notes that heating $\alpha\alpha\alpha$ cholestane with sulfur yields little isomerization at C-20, but substantial amounts of $\alpha\beta\beta$ 20R and 20S isomers. Thus, $\beta\beta$ steranes in these environments may not originate during catagenesis. Despite this finding, the $\beta\beta/(\beta\beta + \alpha\alpha)$ ratio is very useful in maturity determinations for most oils and bitumens. McKirdy et al. (1983) suggest that both C_{29}-sterane maturation parameters may be affected by the source rock mineral matrix. For example, they observed that oils from a probable carbonate source plotted to the left of the Seifert and Moldowan (1981) empirical trend on the C_{29}-sterane diagram (Fig. 3.56). Similar observations were made by Huang Difan et al. (1990) for oils derived from gypsum-salt and carbonate-enriched rocks. Rullkötter and Marzi (1988) noted higher $\beta\beta/(\beta\beta + \alpha\alpha)$ ratios for bitumens from hypersaline rocks compared to adjacent shales. Laboratory heating experiments indicate that the $\beta\beta/(\beta\beta + \alpha\alpha)$ ratio is affected by different source rock types (Peters et al., 1990).

Biomarker maturation index (BMAI)

Rarely applied, highly specific for immature to mature range, measured on m/z 217 chromatogram or preferably by GCMSMS of C_{29} steranes.

In immature rocks, 24-ethyl-5α,14α,17α(H)-cholestane (20R) is the dominant C_{29} sterane. With increasing maturity, bitumens and oils show greater proportions of 24-ethyl-5α,14α,17α(H)-cholestane (20S) and 24-ethyl-5α,14β,17β-cholestane (20R + 20S) plus other epimers. As discussed, $\alpha\alpha\alpha$(20S) shows approximately the same thermal stability as $\alpha\alpha\alpha$(20R), while $\alpha\beta\beta$(20R) and $\alpha\beta\beta$(20S) are significantly more stable. These thermal stability relationships among epimers are supported by molecular mechanics calculations (Van Graas et al., 1982).

Seifert and Moldowan (1981) empirically derived a theoretical maturity curve (Fig. 3.56) which plots the [5α,14β,17β(20R)]/[5α,14α,17α(20R)] ratio versus [5α,14α,17α(20S)]/[5α,14α,17α(20R)] ratio for the C_{29} steranes. The distance from the origin along this curve yields a third maturation parameter called the Biomarker Maturation Index (BMAI). This plot is similar to that of the $\beta\beta/(\beta\beta + \alpha\alpha)$ versus 20S/(20S + 20R) ratios described (Fig. 3.55). The latter plot has been utilized in our most recent work because it gives a straight-line rather than curved-line relationship.

Controversy exists regarding the geochromatographic explanation for data plotting to the right of the BMAI curve in Fig. 3.56. Although some laboratory simulations suggest that expulsion and migration account for significant changes in biomarker epimer ratios (Fan Zhao-an and Philp, 1987; Jiang Zhusheng et al., 1988), other results for structurally similar biomarkers do not (Hoffman et al., 1984; Peters et al., 1990; Brothers et al., 1991). For example, Brothers et al. (1991) showed that no consistent redistribution of 5α(H)- versus 5β(H)-cholestane epimers occurred during laboratory migration experiments, although the column length used in the study may have been too small to adequately simulate the natural setting.

Diasteranes/regular steranes

Rarely applied for maturity, low specificity for early mature to mature range due to dependence on depositional environment; steranes measured on m/z 217 mass chromatograms, or preferably by GCMSMS ($M^+ \rightarrow$ m/z 217; where M^+ = m/z 400, 386, and 372); diasteranes measured on m/z 259 (Fig. 2.11), or preferably by GCMSMS.

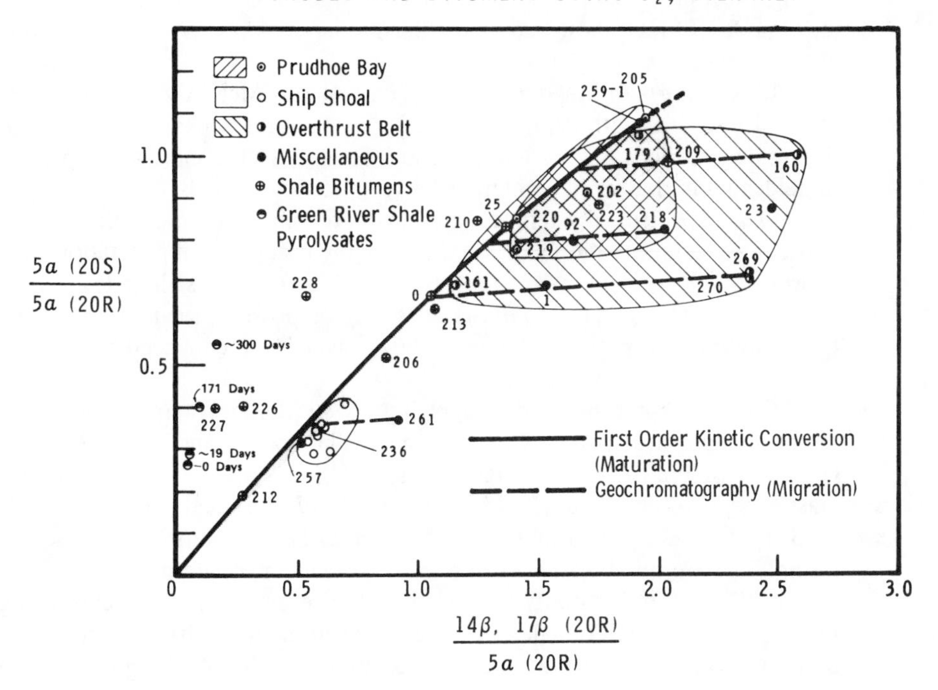

Figure 3.56 BMAI plot for various oils (Seifert and Moldowan, 1981). This curve was also used to calculate a Biomarker Migration Index (BMAI). Based on geological inferences, oils with sterane epimer ratios plotting to the right of the BMAI curve (e.g., those on the dashed lines) were explained by geochromatography during migration of the oils through rocks containing clays (Seifert and Moldowan, 1981; 1986). Differences in response to the two ratios to different heating rates of the source rocks were proposed as an alternative explanation (Mackenzie, 1984). No direct evidence for either hypothesis is available, however; Peters et al. (1990) show no significant change in 14β,17β(20R)/14α,17α (20R) C_{29}-sterane ratios related to expulsion in heating experiments. Thus, direct application of the BMAI method to assess relative migration of oils is not advised.

Diasteranes/steranes ratios are affected by both thermal maturity and inorganic characteristics of the source rock. As a result, these ratios are useful for maturity determination only when all oils or bitumens being compared are from the same source rock organic facies. Catalysis by acidic sites on clays has been proposed as the mechanism by which diasterenes are produced in sediments (Rubinstein et al., 1975). Acidic catalysis is necessary for the conversion of sterenes to diasterenes, which are the precursors of diasteranes (Fig. 3.35) (Kirk and Shaw, 1975). Thus, diasteranes/steranes ratios are typically low in carbonate source rocks and oils (Sec. 3.1.4). However, because diasteranes are found in certain highly calcareous rocks from the Adriatic that are very low in clays (Moldowan et al., 1992), other acid mechanisms may be effective. The Adriatic carbonate rocks that show high diasteranes are not petroleum source rocks. They contain only small amounts of oxidized organic matter. The high Eh of these sediments during deposition may account for the diasteranes.

Most reports of low diasteranes in carbonates (McKirdy et al. 1983; Rullkötter et al., 1985) deal with organic-rich, carbonate source rocks, where the original Eh during deposition was very low (anoxic). Thus, both Eh and clays appear important in determining diasterane content relative to steranes.

Once formed, diasteranes appear more stable than regular steranes. The diasteranes/steranes ratio increases dramatically past peak oil generation as shown by hydrous pyrolysis experiments (Peters et al., 1990). At these high levels of maturity, rearrangement of steranes to diasteranes may be possible even without clays, probably because of hydrogen exchange reactions which are enhanced by the presence of water (Rullkötter et al., 1984a).

Caution should be used in applying the diasteranes/steranes ratio because different versions are applied depending on the laboratory and the GCMS method. At Chevron, we typically use the ratio of [total C_{27} to C_{29} $13\beta(H),17\alpha(H)$(20S + 20R) diasteranes]/[total C_{27} to C_{29} $5\alpha(H),14\beta(H),17\beta(H)$ and $5\alpha(H),14\alpha(H),17\alpha(H)$ (20S + 20R)] steranes from GCMSMS data. Other laboratories may use this ratio at a single carbon number, for example C_{27}, because measurement of C_{27} $\alpha\beta$-diasteranes (20S + 20R) is usually free of interference on the m/z 217 chromatogram using GCMS in the MID mode. However, measurement of the C_{27} steranes is difficult because of interference from the C_{29} diasteranes and some C_{28} epimers. Some laboratories resort to a C_{27}-diasteranes/C_{29}-steranes ratio, but this is a poor compromise because this ratio introduces complications from source input variability.

The same diasteranes/steranes ratio determined by MID or GCMSMS analysis can result in somewhat different absolute values because the MID method involves more interfering peaks. The $13\beta(H),17\alpha(H)$(20S + 20R) isomers are the major diasteranes in rocks and petroleum (Ensminger et al., 1978). They can be measured using m/z 259 mass fragmentograms (Fig. 2.11), which are more specific for diasteranes than m/z 217. Minor amounts of $13\alpha(H),17\beta(H)$(20S + 20R) diasterane isomers also occur when the $13\beta(H),17\alpha(H)$ isomers are present, but they are generally not quantified.

20S/(20S + 20R) 13β(H),17α(H)-diasteranes

Rarely applied, specificity unknown, early mature range, measured on m/z 259 chromatograms (Fig. 2.11).

The ratio of 20S/(20S + 20R) for C_{27} to C_{29} diasteranes increases with thermal maturity (Mackenzie et al., 1980) in a similar fashion to that for the 22S/(22S + 22R) $17\alpha(H)$-homohopane ratio. The steric environment of the C-20 asymmetric carbon in these diasteranes is more like that of the C-22 carbon in the

17α(H)-homohopanes than the C-20 carbon in the steranes. The 20-position in the 13β,17α(H)-diasteranes equilibrates [20S/(20S + 20R) ~ 0.6] at about the onset of oil generation (Mackenzie et al., 1980). However, the 20S epimer is formed during the rearrangement of sterenes (Akporiaye et al., 1981), leading to 20S/(20S + 20R) ratios of about 0.1, even in very immature rocks.

3.2.4.3 Aromatic steroids

TA/(MA + TA); Monoaromatic steroid aromatization

Often applied, highly specific for immature to mature range (Fig. 3.46); measured using m/z 253 (monoaromatic steroid, Fig. 2.13) and m/z 231 (triaromatic steroid, Fig. 2.14) (Figure 2.11 shows principal fragments).

The aromatization of C-ring monoaromatic (MA) steroids to ABC-ring triaromatic (TA) steroids involves the loss of a methyl group at the A/B ring juncture (Fig. 3.57). Note that the asymmetric center at C-5 is lost during conversion of the MA to the TA compounds. Thus, maturation of MA-steroids yields TA-steroids with one less carbon. A conversion mechanism of this type is supported by comparisons of absolute concentrations of MA and TA steroids in petroleums (Requejo, 1992). TA/(MA + TA) increases from 0 to 100 percent during thermal maturation. The ratio has been applied to calibrations of basin models (Mackenzie, 1984). Evidence suggests that this ratio can be affected by migration (Hoffmann et al., 1984; Peters et al., 1990; Marzi and Rullkötter, 1992). The more polar TA-steroids are preferentially retained in the bitumen compared to the expelled oil.

C_{28}-TA/(C_{29}-MA + C_{28}-TA)

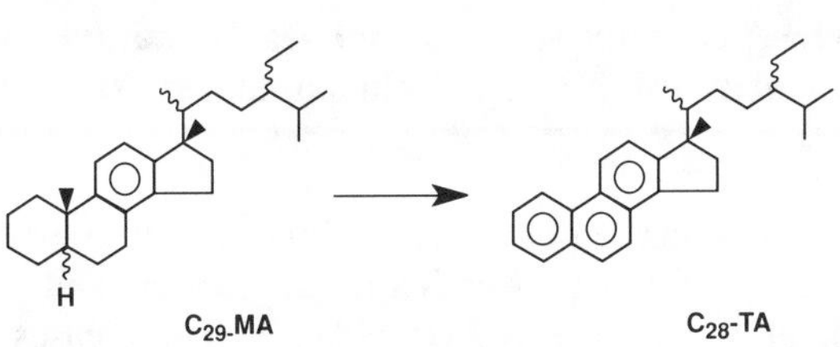

C_{29}-MA → C_{28}-TA

Figure 3.57 Conversion of C_{29}-monoaromatic to C_{28}-triaromatic steroids during thermal maturation. Note loss of the methyl group attached to C-10 (A/B ring juncture) and loss of the asymmetric center at C-5 in this reaction. See Fig. 1.12 for nomenclature related to numbering of the carbon atoms.

Several different TA/(MA + TA) maturity parameters are used in the literature. One of these ratios requires measurement of two C_{29} monoaromatic epimers [5α(H)20R, 5β(H)20R] and one C_{28} triaromatic (20R) (Mackenzie et al., 1981a). Palacas et al. (1986) modified the TA/(MA + TA) ratio used by Mackenzie et al. (1981a) to include C_{20} and C_{28} triaromatic steroids. The kinetics of the proposed aromatization reaction have been studied (Mackenzie and McKenzie, 1983). However, additional structure studies on C-ring monoaromatic steroids have shown that isomer distributions are much more complex than previously believed (Moldowan and Fago, 1986; Riolo and Albrecht, 1985; Riolo et al., 1986). For example, it is likely that MA-steroids with a methyl group rearranged from C-10 to C-5 also generate TA-steroids during thermal maturation. These rearranged or dia-MA-steroids are significant, sometimes dominant MA-steroids in oils and source rocks (Fig. 2.13). Most dia-MA-steroids and regular MA-steroids cannot be separately analyzed because of coelution. Thus, the C_{29}-MA in the ratio TA/(MA + TA) appears also to contain some C_{29} dia-MA-steroids.

Use of the TA/(MA + TA) ratio as a maturity parameter requires that the same version of the ratio be used for all samples. At Chevron, the TA/(MA + TA) parameter is measured as follows. The MA component includes the major known C_{29} MA-steroid peaks (four peaks labeled 12, 13, 15, and 16 from m/z 253 chromatogram, Fig. 2.13) representing twelve compounds: eight regular MA-steroids; 5α(H),10β(CH_3),20R,24R; 5α(H)20R,24S; 5α(H)20S,24R; 5α(H)20S,24S; 5β(H)20R,24R; 5β(H)20R,24S; 5β(H)20S,24R; and 5β(H)20S,24S; and four dia-5β(CH_3)MA-steroids; 5β(CH_3),10β(H)20R,24R; 20R,24S; 20S,24R; and 20S,24S. The TA component includes two C_{28} TA-steroid peaks (peaks 6 and 8 from m/z 231 chromatogram, Fig. 2.14) representing four epimeric compounds: 20R,24R; 20R,24S; 20S,24R; and 20S,24S. The 24R and 24S epimers for the MA- and TA-steroids are not resolved by the DB-1 chromatographic column we use under normal conditions.

Prior to 1985, we used the sum of all known C_{27} to C_{29} C-ring MA-steroid peaks (m/z 253) for MA and the sum of all C_{26} to C_{28} TA-steroid peaks (m/z 231) for TA in the expression TA/(MA + TA). However, this complex approach has been abandoned. Work from contract laboratories may include any of the above TA/(MA + TA) parameters or others.

MA(I)/MA(I + II)

Often applied, some interference from source input, early mature to late mature range (Fig. 3.46), measured using m/z 253 chromatogram (Fig. 2.11).

An increase in "side chain scission" (carbon-carbon cracking) with thermal maturity has been documented for aromatic steroids in oils (Seifert and Moldowan, 1978) and rocks (Mackenzie et al., 1981a). MA(I)/MA(I + II) increases from 0 to 100 percent during thermal maturation (Fig. 3.58). It is not known whether this increase is the result of: (1) conversion of long-chain to short-chain MA-steroids by carbon-carbon cracking, (2) preferential thermal degradation of the long- versus short-chain series, or (3) both. Moldowan et al. (1986) have shown that this parameter is influenced by diagenetic conditions, particularly Eh, in the source sediment. The origin of the short side chain aromatic steroids is unclear. They could be formed from the C_{27} to C_{29} MA-steroids and/or other precursors. Short side chain aromatic steroids could be derived from larger aromatic steroids by homolytic (free radical) scission of the bond between C-20 and C-22. However, the presence of substantial quantities of C_{21} MA-steroids suggests they are derived from sterols with

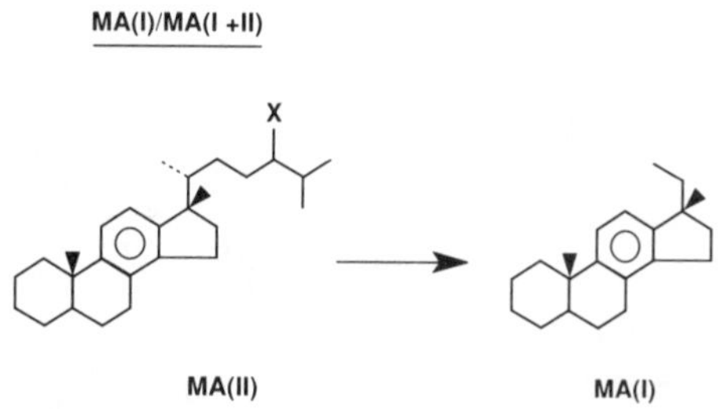

Figure 3.58 Conversion of Group II monoaromatic to Group I monoaromatic steroids by side chain cleavage during thermal maturation. Alternately, or in addition, short and long side chain steroids may have different natural product precursors and may undergo differential destruction during maturation. The exact position of the C-19 nuclear methyl group in the short chain compounds (Group I) is not well documented, and may reside at C-10 (shown), or at C-5, or as a mixture of the two.

functionalized side chains (i.e., oxidative cleavage of a C-22 double bond) during diagenesis. There are also C_{21} and C_{22} steroid natural products which could lead to short side chain MA-steroids.

Mackenzie et al. (1981a) use the C_{28} MA-steroid [mainly the 5β(H)20R isomer; Moldowan and Fago, 1986] as MA(II) and the C_{21} MA-steroid as MA(I) (Fig. 3.58). Our objection to this ratio is that the concentration of C_{28} relative to C_{27} through C_{29} MA-steroids is partly dependent on source input and that the C_{21} MA-steroid may be derived from any or all of the C_{27} to C_{29} MA-steroids. We use the sum of all major C_{27} to C_{29} MA-steroids as MA(II) and C_{21} plus C_{22} as MA(I) (peaks 4–16 and 1–2 in Fig. 2.13, respectively), in order to reduce the source input effect.

The aromatic steroid side chain scission reactions are based on carbon-carbon cracking and generally require more thermal energy to proceed than isomerizations. They appear to be most useful in the late oil-generative window when most other biomarker parameters are no longer effective as maturity indicators.

TA(I)/TA(I + II)

Often applied, some interference from source input, peak mature to late mature range (Fig. 3.46); measured using m/z 231 mass chromatogram (Fig. 2.14).

Most of the discussion on short- and long-chain MA-steroids applies to short- and long-chain TA-steroids (Fig. 3.59). In addition, the TA-steroids are probably derived from the MA-steroids [see previous discussion on "TA/(MA + TA)"]. Thus, the TA(I)/TA(I + II) ratio has the added advantage of being more sensitive at higher maturity than the same ratio for the MA-steroids. Heating experiments indicate that the TA(I)/TA(I + II) ratio increases due to preferential degradation of the long-chain triaromatic homologs rather than conversion of long- to short-chain homologs (Beach et al., 1989).

TA(I)/TA(I + II)

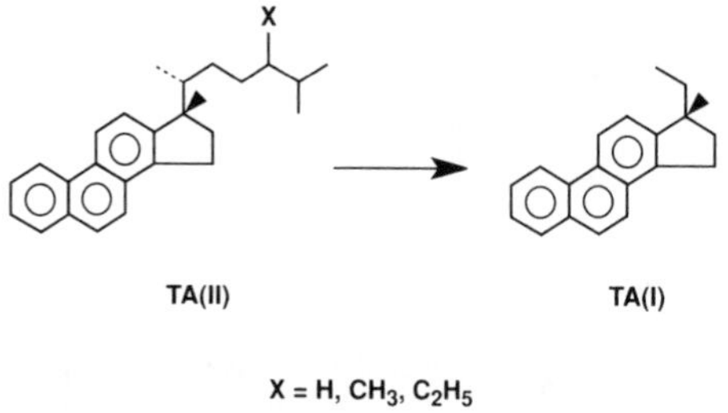

X

TA(II)

TA(I)

X = H, CH_3, C_2H_5

Figure 3.59 Proposed conversion of Group II triaromatic to Group I triaromatic steroids by side chain cleavage during thermal maturation. Beach et al. (1989) show that the TA(I)/TA(I + II) ratio increases due to preferential degradation of the long-chain triaromatics rather than conversion of long- to short-chain homologs.

The TA(I)/TA(I + II) parameter has been measured on two combinations of TA-steroids. Mackenzie et al. (1981a) use the C_{26} TA-steroid (20S) as TA(II) and the C_{20} TA-steroid as TA(I) (Fig. 3.46). We prefer to use the sum of C_{26}-C_{28} (20S + 20R) TA-steroids as TA(II) and the C_{20} and C_{21} TA-steroids as TA(I) (peaks 4–8 and peaks 1–2, respectively, in Fig. 2.14). As with the MA-steroid scission parameter, we use a summation of carbon numbers for the TA-scission parameter to reduce complications caused by the effects of source organic matter input. For example, preferential input of C_{27}, C_{28}, or C_{29} sterols in the source sediments could lead to a predominance of C_{26}, C_{27}, or C_{28} TA-steroids, respectively, unrelated to thermal maturity.

C_{26}-triaromatic 20S/(20S + 20R)

Rarely applied, highly specific (?) for mature to highly mature range; GCMSMS analyses of $M^+ \rightarrow 231$.

The first application of this maturity ratio shows good agreement with the two more commonly used maturity parameters based on isomerization in the C_{29} steranes (Fig. 3.60). Based on the figure, C_{26}-triaromatic 20S/(20S + 20R) ratio appears more sensitive at higher levels of maturity than either of the C_{29} sterane ratios. No other observations of this apparent maturity parameter using triaromatic compounds have yet been published.

The enhanced sensitivity of the C_{26}-triaromatic 20S/(20S + 20R) ratio compared to the more common steroid maturity parameters will be useful for studies of more highly mature oils and condensates. Isomerization of the C-20 position in the triaromatic steroids may be hindered because of the presence of an additional methyl group attached to C-17 compared to the steranes. This interpretation is supported by Schmid (1986, p. 158-159) who observed anomalously low 20S/(20S + 20R) ratios for this type of compound in laboratory heating experiments.

Ratios of 20S/(20S + 20R) for the C_{27} and C_{28} TA-steroids show similar trends to that for C_{26}-TA-steroids with maturity, although they are more erratic (unpublished data), possibly due to interference. Parameters based on all three carbon numbers (C_{26}-C_{28}) have the same potential for application when measured using GCMSMS.

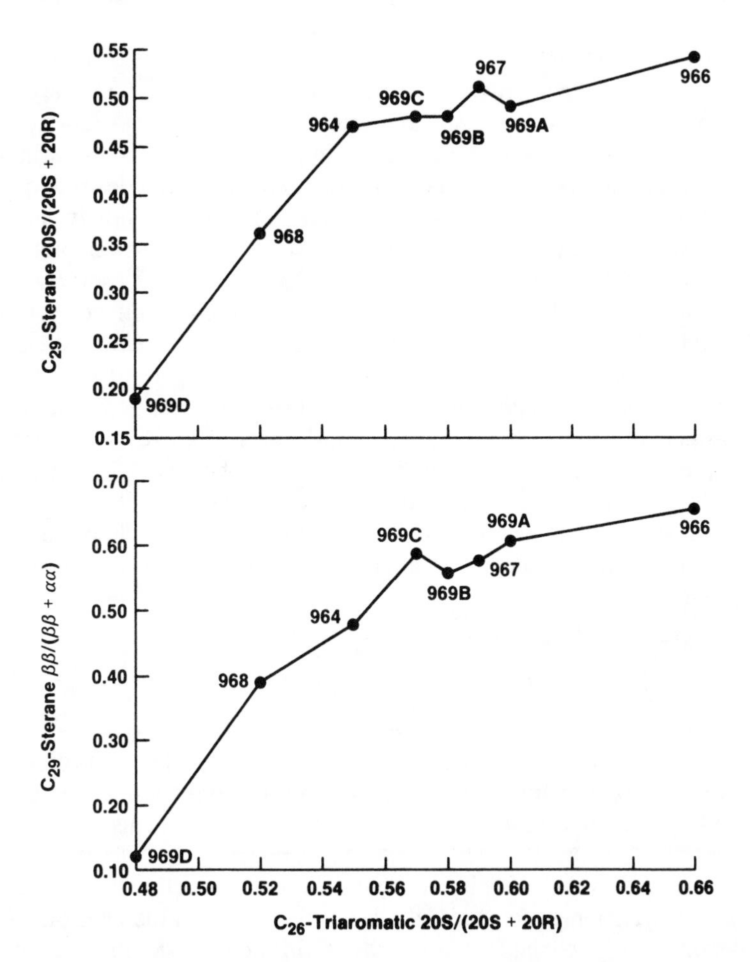

Figure 3.60 The C_{26} triaromatic steroid 20S/(20S + 20R) ratio for oils and bitumens from the Eel River Basin, California appears to be more sensitive at higher levels of maturity than the two C_{29} sterane biomarker maturity ratios shown. No other observations of this potential maturity parameter using triaromatic steroids have yet been published. These samples are the same as those shown in Fig. 3.52.

3.2.4.4 Monoaromatic Hopanoids

MAH parameter

Rarely applied, unknown specificity, early mature to late mature range (?), measured using m/z 191 for benzohopanes and m/z 365 for 8,14-secohopanoids (Fig. 2.11).

He Wei and Lu Songnian (1990) propose a new monoaromatic hopanoid (MAH) maturity parameter based on ring-D monoaromatized 8,14-secohopanoids and benzohopanes [8,14-secohopanoids/(8,14-secohopanoids + benzohopanes)] (Fig. 2.11). Benzohopanes are discussed in Sec. 3.1.3.3. They observed that aromatic fractions of crude oils and mudstone extracts from Liaohoe Basin, northeastern China show systematic increases in the MAH ratio with burial depth. Variations in the ratio are attributed to differential generation and destruction of monoaromatic secohopanoids and benzohopanes rather than conversion between the two groups of compounds. Based on a comparison of MAH with odd-even preference values for *n*-paraffins (OEP) and C_{31} hopane 22S/(22S + 22R) ratios for a limited number of samples, the authors conclude that the threshold of oil generation corresponds to MAH values of about 0.3. Because both monoaromatic 8,14-secohopanoids and benzohopanes are resistant to biodegradation compared to most saturated biomarkers, the authors suggest that the MAH maturity parameter can be used for maturity assessment of heavily biodegraded oils. Pairs of C_{28}-C_{31} demethylated counterparts of the regular D-ring monoaromatic 8,14-secohopanoids, most probably lacking the C-28 methyl group, have been observed in rock and oil samples from the offshore Korea Bay Basin in the Yellow Sea. Trends in the relative distributions of these demethylated compounds parallel those of the regular secohopanoids, suggesting a common origin.

3.2.4.5 Porphyrins

DPEP/etio ratio

Rarely applied, immature to mature range [as "etio/(etio + DPEP)" for vanadyl porphyrins, Fig. 3.46], measured from total averaged mass spectra or HPLC of demetallated porphyrin concentrates (not used at Chevron).

The decrease of this ratio with thermal maturity (Baker and Louda, 1986) is generally attributed to loss of the isocyclic ring from the DPEP molecules. However, the ratio may change as a result of differential stability of the two classes of molecules rather than conversion of DPEP directly to etio compounds. Furthermore, under oxidizing diagenetic conditions, the chlorin precursors to DPEP porphyrins can lose the isocyclic ring (Baker and Louda, 1986). Evidence for loss of the isocyclic ring during early diagenesis has been shown in the Messel shale (Ocampo et al., 1985a). Sundararaman et al. (1988a) and Sundararaman (1992) show that etio porphyrins are preferentially released during catagenesis or pyrolysis of kerogen, while DPEP porphyrins are dominant in immature bitumens. The DPEP/etio maturity parameter is rather crude because it involves many compounds and improvements have been minimal because of the complex procedures for extraction and isolation of metalloporphyrins.

Porphyrin maturity parameter (PMP)

Often applied at Chevron, highly specific for immature to early mature range (Fig. 3.46); separation and measurement by high performance liquid chromatography (HPLC).

The PMP ratio [$C_{28}E/(C_{28}E + C_{32}D)$ vanadyl porphyrins] increases with thermal maturity (Fig. 3.61) and is tied to generation from the kerogen (Sundararaman et al., 1988a and 1988b). Most other maturity parameters measure the extent of thermally induced chemical reactions not directly related to kerogen breakdown. PMP ratios of 0.2 in bitumens appear to signal the beginning of oil generation (Sundararaman et al., 1988a and 1988b). Comparable results were obtained for both natural and hydrous pyrolysis samples of cores from the Monterey Formation, Santa Maria Basin, California, where PMP increased sharply during the main phase of oil generation (Sundararaman, 1992). The ratio is determined using a rapid high performance liquid chromatographic separation of two vanadyl porphyrins; a C_{28} etioporphyrin and a C_{32} DPEP porphyrin. These two porphyrins may or may not be related as precursor and product. For example, the change in the $C_{28}E/(C_{28}E + C_{32}D)$ ratio with maturity could be caused by higher relative thermal stability of the C_{28} etio compared to the C_{32} DPEP compound. Detailed structures of these two porphyrins are unknown.

The mechanism by which porphyrins are generated from kerogen during maturation is not clear. Beato et al. (1991) pyrolyzed kerogen from the New Albany shale, Indiana, at 300°C and analyzed the generated nickel and vanadyl porphyrins using tandem mass spectrometry. They conclude that porphyrins are not released by carbon-carbon bond scission at 300°C, but are generated by a solubilization or desorption mechanism.

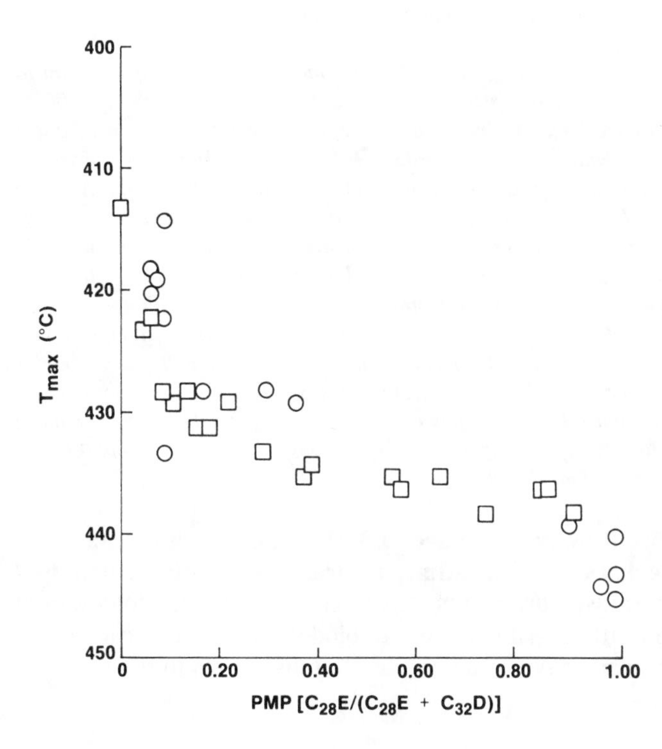

Figure 3.61 The porphyrin maturity parameter (PMP) shows a sharp increase at the onset of oil generation. The plot of Rock-Eval T_{max} versus PMP for source rock samples from the Williston Basin, North Dakota (circles) and San Joaquin Basin, California, (squares) shows that a strong shift in PMP from 0.3 to 1.0 occurs at T_{max} values near 435°C, consistent with the onset of oil generation (Sundararaman et al., 1988a).

3.3 Biodegradation

3.3.1 Concepts. Most workers consider aerobic bacteria to be the principal agents in subsurface degradation of petroleum (e.g., Milner et al., 1977; Palmer, 1984b). Anaerobic bacteria, such as sulfate reducers, can oxidize hydrocarbons, but probably do so much more slowly than aerobes. However, in terms of geologic time this difference in rate may not be critical. Anaerobes apparently require aerobes to initiate degradation of petroleum (Jobson et al., 1979).

Aerobic bacteria can catabolize oil only if several requirements are satisfied.

1. Access to surface recharge waters containing oxygen,
2. Temperatures no more than about 65 to 80°C,
3. The petroleum must be free of H_2S, which poisons the bacteria.

Note: *The discovery of bacteria near "black smoker" hydrothermal vents in the deep ocean (Baross and Deming, 1983) was interpreted to suggest that some organisms might actively degrade petroleum at much higher temperatures (~250°C) than previously suspected. However, these thermophilic (heat tolerant) archaebacteria are almost certainly not growing at 250°C. They probably were introduced into the samples by strong thermal convection of the water near the vents. The instability of vital macromolecules (e.g., proteins) and monomers (e.g., adenosine triphosphate) in living organisms above about 110°C suggests that this is a good estimate of the upper temperature limit for life. One anaerobic, thermophilic archaebacterium isolated from an undersea thermal vent, Pyrodictium brockii, shows optimal growth at 105°C (Brock and Madigan, 1991).*

Nonetheless, these above factors appear to limit most biodegraded petroleum to within a few hundred meters of the surface. For example, oils above about 1500 m (<~70°C) in the Shengli oilfield, China show biodegradative removal of n-alkanes, while those from greater depths are unaltered (Shi Ji-Yang et al., 1982). We observed an exceptional case of a biodegraded petroleum reservoired at 2500 m in the Adriatic Basin (Moldowan et al., 1992). The apparent activity of microbes at this depth was attributed to low reservoir temperatures resulting from a very low thermal gradient. In some cases, biodegraded oils in deep reservoirs may be the result of previous alteration at shallower depths, followed by burial to depths where bacteria are no longer active.

Because aerobic bacteria require molecular oxygen, active surface recharge waters are necessary for biodegradation to occur. Water washing typically accompanies biodegradation of petroleum and results in selective loss of lighter hydrocarbons, especially benzene, toluene, and other aromatics (Bailey et al., 1973). This loss results in an increase in sulfur, nitrogen, and oxygen (NSO-compounds), increased asphaltenes, and a decrease in API gravity of the oil.

The general sequence (however, see Sec. 3.3.3) of increasing resistance to biodegradation of biomarkers is: *n*-paraffins, isoprenoids, steranes, hopanes/diasteranes, aromatic steroids, porphyrins (Chosson et al., 1992; Moldowan et al., 1992). Because of their differential resistance to biodegradation, comparisons of the relative amounts of biomarker types can be used to rank oils as to the extent of biodegradation.

We have developed a scale for assessing the extent to which an oil has been biodegraded based on the relative abundances of the various hydrocarbon classes (Fig. 3.62) (modified from Moldowan et at., 1992). Others use similar scales (Volkman et al., 1983a and b). The figure shows the effects of various levels of biodegradation, ranked from 1 to 10, on the composition of a typical mature oil. This figure must be used cautiously because **biodegradation is a complex, "quasi-stepwise" process that cannot be described as a truly sequential alteration of compound classes** (Sec. 3.3.3). For example, some oils exposed to heavy biodegradation show significant alteration of the hopanes before all steranes are destroyed, although hopanes are generally considered more resistant to biodegradation than steranes. However, a sharp division occurs between ranks 5 and 6 (Fig. 3.62), after all isoprenoids are removed, but prior to degradation of the steranes.

Mixing of oil in the reservoir can result in characteristics which indicate heavy, plus light, or no biodegradation in the same oil. For example, Volkman et al. (1983a) observed demethylated hopanes (25-norhopanes) indicative of heavy biodegradation in otherwise nonbiodegraded paraffinic oils from Barrow Island, Western Australia. The demethylated hopanes are interpreted to have formed during severe biodegradation of an early accumulation of oil which occurred when the reservoir sands were very close to the surface. A second charge of nonbiodegraded oil subsequently mixed into the reservoir. Finally, microbe-resistant biomarkers also allow correlations between biodegraded oils that would be difficult or impossible using GC alone.

3.3.2 Nonbiomarker biodegradation parameters

3.3.2.1 n-Paraffin Envelope. *n*-Paraffins are typically the major peaks on gas chromatograms of mature oils and bitumens. The distribution of *n*-paraffins depends on organic source type, thermal maturity, expulsion, migration, and biodegradation. Biodegradation results in depletion of *n*-paraffins prior to significant alteration of any other compound class, including the acyclic isoprenoids pristane and phytane (Fig. 3.2) (Winters and Williams, 1969). Mildly biodegraded oils will show higher pristane/nC_{17} and phytane/nC_{18} ratios than related, nonbiodegraded oils. Decreased abundances of volatile paraffins ($<C_{15}$) can be related to low thermal maturity, migrational effects (phase separation), evaporative loss (weathering), or water washing.

More heavily degraded oils are characterized by a large, unresolved complex mixture (UCM or "hump") that rises above the baseline on gas chromatograms (Fig. 3.2). Biodegraded oils are best studied using gas chromatography-mass spectrometry of the biomarkers. Because of their relative resistance to degradation, the region where most biomarkers elute on the gas chromatogram of a biodegraded oil shows enhanced resolved and unresolved components. Using chromic acid oxidation followed by GCMS analysis of breakdown products, Gough and Rowland (1990) and Killops and Al-Juboori (1990) identified various bacterially resistant, branched alkanes in the UCM from petroleum.

3.3.2.2 Other "Nonbiomarkers". Connan (1984) states that alkylbenzenes, dialkylbenzenes, and trialkylbenzenes are highly susceptible to bacterial attack while

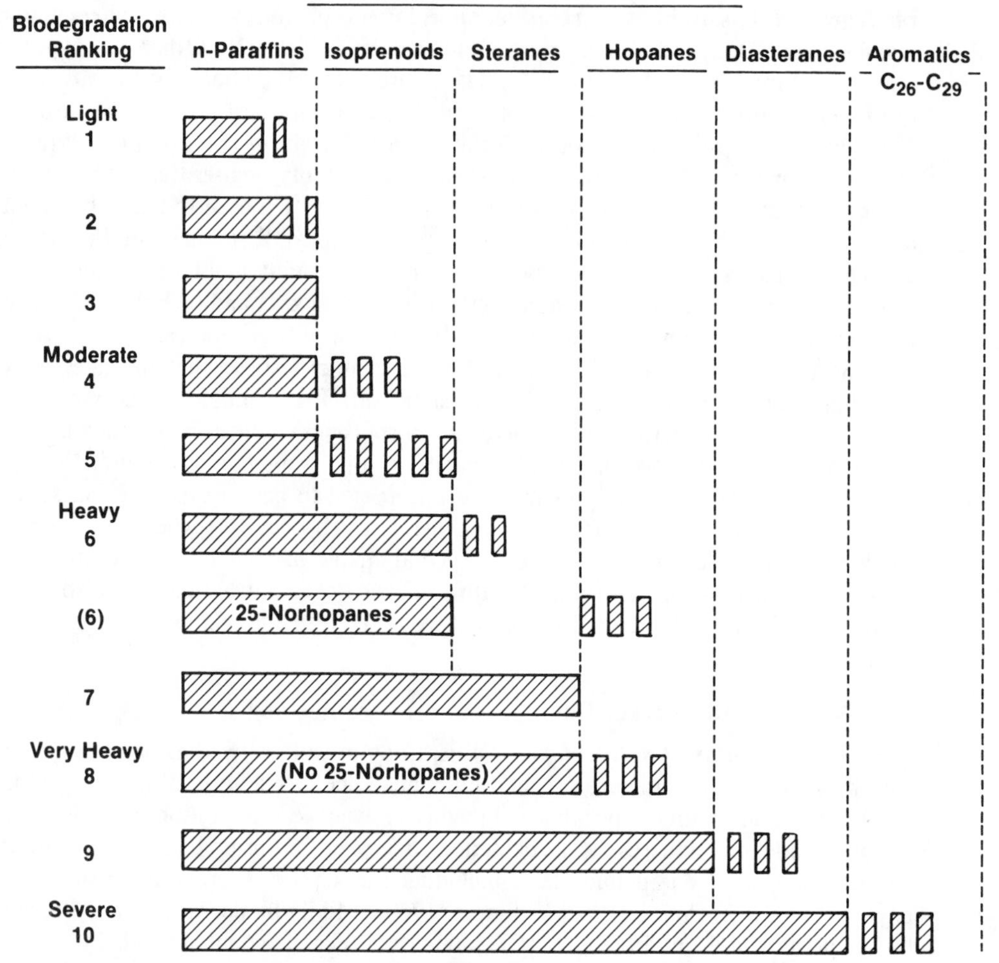

1 = Lower homologs of n-paraffins depleted.
2 = General depletion of n-paraffins.
3 = Only traces of n-paraffins remain.
4 = No n-paraffins, acyclic isoprenoids intact.
5 = Acyclic isoprenoids absent.
6 = Steranes partly degraded.
7 = Steranes degraded, diasteranes intact.
8 = Hopanes partly degraded.
9 = Hopanes absent, diasteranes attacked.
10 = C_{26}-C_{29} aromatic steroids attacked.

Figure 3.62 Effects of various levels of biodegradation on a typical mature oil. The figure can be used to rank the extent of biodegradation on a scale of 1 to 10. Biodegradation of petroleum is quasi-sequential in that "more resistant" compound classes can be attacked prior to complete destruction of a "less resistant" class.

di- and triaromatic fractions containing complex mixtures of alkylated naphthalenes, anthracenes, and phenanthrenes are less susceptible. He indicates that methylated naphthalenes appear to be fairly rapidly degraded but methylated phenanthrenes persist for longer times in oils degraded in the reservoir. Iso-, anteiso-, and cyclohexyl- or methylcyclopentylalkanes appear readily attacked by bacteria.

3.3.2.3 API Gravity and Heteroatom Content. API gravity is a "bulk parameter" which characterizes the whole oil and is affected by many variables (Sec. 3.2.2.2.3). Nonetheless, API gravity can sometimes be used to differentiate oils quickly from different sources. Biodegradation reduces oil API gravity and increases sulfur and other heteroatoms by selectively removing saturates and aromatics compared to NSO-compounds and asphaltenes. Thus, related oils can show different API gravities if some are biodegraded. Both API gravity and viscosity typically show a systematic relationship with the ratio of (NSO-compounds + asphaltenes)/(saturates + aromatics + NSO-compounds + asphaltenes) for related oils.

3.3.3 Biomarker biodegradation parameters. The following "quasi-stepwise" sequence describes the general order of susceptibility of various biomarker compound classes to biodegradation. This sequence is formalized in Fig. 3.62. The term "quasi-stepwise" is used to indicate that some components of more biodegradable compound classes may remain after biodegradation has already begun on the next most resistant class of compounds. Some of the conclusions described here are based on only one to a few reports and are thus tentative pending further studies.

n-paraffins (most susceptible)[1] > acyclic isoprenoids[2] > hopanes (25-norhopanes present)[3] ≥ steranes[4] > hopanes (no 25-norhopanes)[5] ~ diasteranes[6] > aromatic steroids[7] > porphyrins (least susceptible).[8]

1. Low molecular-weight *n*-paraffins are more susceptible than their high molecular-weight homologs.

2. No work has been published on biodegradation of extended isoprenoids ($>C_{20}$). Pristane (C_{19}), phytane (C_{20}), and other acyclic isoprenoids appear to be completely degraded prior to any change in polycyclic biomarkers (i.e., between ranks 5 and 6 in Fig. 3.62).

3–4. When 25-norhopanes are present, microbial attack appears to favor C_{27}-C_{32} > C_{33} > C_{34} > C_{35} 17α(H)-hopanes (based on one report, Peters and Moldowan, 1991). However, Rullkötter and Wendisch (1982) observed that the higher homologs in the series C_{27}-C_{32} were degraded faster than lower homologs, in conflict with the observations of Peters and Moldowan (1991). For the steranes degradation favors $\alpha\alpha\alpha$20R and $\alpha\beta\beta$20R > $\alpha\alpha\alpha$20S and $\alpha\beta\beta$20S and C_{27} > C_{28} > C_{29} > C_{30}.

4–5. When 25-norhopanes are absent, microbial attack favors C_{35} > C_{34} > C_{33} > C_{32} > C_{31} > C_{30} > C_{29} > C_{27} and 22R > 22S. For the steranes degradation favors $\alpha\alpha\alpha$20R (C_{27}-C_{29}) > $\alpha\alpha\alpha$20S(C_{27}) > $\alpha\alpha\alpha$20S (C_{28}) > $\alpha\alpha\alpha$20S(C_{29}) ≥ $\alpha\beta\beta$(20S + 20R)(C_{27}-C_{29}).

6. Degradation favors $C_{27} > C_{28} > C_{29}$.
7. Degradation favors C_{20}-C_{21} TA (water washing?) > C_{27}-C_{29} 20R MA ~ C_{26}-C_{28} 20R TA > C_{21}, C_{22} MA.
8. No evidence is available showing any biodegradation of porphyrins.

Connan (1984) and Connan and Dessort (1987) provide many examples of biodegraded oils and their biomarker compositions.

In the following section, we have arranged the compound classes and parameters in the approximate order of increasing resistance to biodegradation.

Isoprenoid and other organic acids. Although data are limited, the presence of 3(R),7(R),11(R)-tetramethylhexadecanoic acid (phytanic acid) may be a marker for the early stages of oil biodegradation (Mackenzie et al., 1983c). Other microbial degradation products, including carbon dioxide and organic acids, may react with minerals in carrier beds, resulting in increased secondary porosity (McMahon and Chapelle, 1991).

Isoprenoids. Pristane and phytane are resistant to biodegradation compared to the *n*-paraffins, but are less resistant than the polycyclic biomarkers (Seifert and Moldowan, 1979). Biodegraded oils commonly lack *n*-paraffins, but still show pristane, phytane, and other isoprenoids (e.g., Fig. 3.2, middle). Other biodegraded oils lack both *n*-paraffins and isoprenoids, but all steranes and hopanes appear intact. It is possible that some loss of isoprenoids can occur prior to complete biodegradative removal of the *n*-paraffins.

Steranes and diasteranes. Removal of steranes from petroleum by biodegradation occurs sometime after complete removal of C_{15} to C_{20} isoprenoids and before or after the hopanes, depending on circumstances that will be described. In general, the steranes are biodegraded from most to least effectively in the order: $\alpha\alpha\alpha$20R > $\alpha\alpha\alpha$20S > $\alpha\beta\beta$20R > $\alpha\beta\beta$20S > diasteranes (Seifert and Moldowan, 1979; Mackenzie et al., 1983c; Zhang Dajiang et al., 1988).

Where partial biodegradation of the steranes has occurred, the C_{27} to C_{29} $\alpha\alpha\alpha$20R steranes, which are in the "biological" configuration (Sec. 1.2.10.3), appear more susceptible to destruction than the $\alpha\beta\beta$(20R and 20S) epimers. Further, depletion of $\alpha\alpha\alpha$20S relative to the other epimers is observed in the order C_{27} > C_{28} > C_{29} (Seifert et al., 1984). For example, C_{27} $\alpha\alpha\alpha$20S can be nearly absent while C_{29} $\alpha\alpha\alpha$20S remains apparently unaltered. Preferential degradation of the $\alpha\alpha\alpha$20R sterane epimers, like that for the 22R-homohopanes (see following text), appears to reflect enzymatic specificity in the bacteria for the biological over the geological stereochemistry. Another example (Fig. 3.34) shows complete loss of C_{27} and C_{28} steranes, partial loss of C_{29} steranes with no apparent stereoisomer preference, and complete preservation of the C_{30}-steranes. This result is supported by Lin et al. (1989), who noted that C_{30} steranes are more resistant to biodegradation than the C_{27}-C_{29} homologs.

Rullkötter and Wendisch (1982) and McKirdy et al. (1983) found preferential removal of $\alpha\alpha\alpha$20R and $\alpha\beta\beta$20R steranes compared to their 20S epimers in oils that also show 25-norhopanes. However, Seifert et al. (1984) found complete removal of C_{27}, C_{28}, and C_{29} $\alpha\alpha\alpha$20R isomers, partial removal of $C_{27} > C_{28} > C_{29}$ $\alpha\alpha\alpha$20S sterane isomers, and no apparent loss of $\alpha\beta\beta$(20S + 20R) isomers from a seep oil from Greece. The oil showed no 25-norhopanes and no evidence of hopane destruction. Some of the differences in bacterial alteration between these and other oils appear to depend on bacterial populations and reservoir conditions (see Sterane versus Hopane Biodegradation in the text that follows).

Biodegradation can adversely affect some parameters. For example, both of the above oils show preferential removal of $\alpha\alpha\alpha$20R steranes during biodegradation. In the absence of other biodegradation indicators, this might be interpreted as increased maturation [i.e., higher 20S/(20S + 20R) ratio], although in the examples cited the 20S/(20S + 20R) ratios were near 1.0. Values of 20S/(20S + 20R) above 0.56 (equilibrium) are not possible without selective sterane biodegradation.

Several workers have tried (Connan et al., 1980; Rubinstein et al., 1977), but Goodwin et al. (1983) succeeded in causing sterane and hopane biodegradation in oil under laboratory conditions. Their work confirms the preferential removal of regular steranes compared to diasteranes and the preferred removal of $C_{27} > C_{28} > C_{29}$ steranes. Selective removal of the 5α20R C_{29}-sterane compared to the 20S epimer was also observed. In natural samples, Rullkötter and Wendisch (1982) also show a preferential removal of steranes in the order $C_{27} > C_{28} > C_{29}$. Biodegradation of the C_{27} and C_{28} diasteranes also appeared to be favored over C_{29}. Chosson et al. (1992) show that the degree of sterane and terpane biodegradation depends on the bacterial culture used. Significant amounts of steranes were removed from crude oil by only seven out of 73 strains of bacteria. These workers found the same sequence of sterane biodegradation based on stereochemistry and carbon number as that observed in nature.

Diasteranes are particularly resistant to biodegradation. Evidence shows complete destruction of C_{27}-C_{29} steranes prior to diasteranes (Seifert and Moldowan, 1979; McKirdy et al., 1983; Seifert et al., 1984; Connan, 1984). Requejo et al. (1989) show a mass chromatogram of an oil where nearly all steranes are degraded, but some diasteranes remain. Even in heavily biodegraded oils where steranes are totally removed, some diasteranes typically remain (Seifert and Moldowan, 1979). These authors observed that biodegradation of diasteranes appears to result in stereoselective destruction of 13β,17α(H)C_{27}(20S) over 13β,17α(H)C_{27}(20R).

Jiang Zhusheng et al. (1990) show a series of tricyclic compounds with a base peak at m/z 219 in a heavily biodegraded oil from the Kelamayi oilfield, Zhungeer Basin, northwest China, where steranes have been removed. These compounds show a similar carbon number distribution to that of steranes in a related, unaltered oil from the same field. The compounds might be useful for correlation of oils when steranes have been removed by biodegradation.

Hopanes. Like the 20R epimers in the steranes, the 22R epimers of the 17α,21β(H)-homohopanes are more susceptible to biodegradation than their counterparts in the S configuration. However, the 22R-homohopanes are more resistant to degradation than the 20R epimers of the steranes (Hoffman and Strausz, 1986; Requejo and Halpern, 1989; Lin et al., 1989).

The relative ease of biodegradation of the hopane homologs remains controversial. For example, laboratory experiments of Goodwin et al. (1983) indicate that higher rather than lower hopane homologs in the range C_{29} to C_{35} are more readily biodegraded. Conversely, others show evidence that the lower homologs are more readily biodegraded (Williams et al., 1986; Lin et al., 1989; Peters and Moldowan, 1991).

"Degraded" hopanes were first observed by Reed (1977) in asphalt deposits from the Uinta Basin. Seifert and Moldowan (1979) subsequently reported biodegraded California oils showing hopane degradation and postulated ring A/B demethylated hopane formation (25-norhopanes, discussion follows). However, many biodegraded oils, such as an example from the Gulf Coast noted by Seifert and Moldowan (1979) show degradative loss of hopanes without formation of demethylated hopanes. These observations initiated the controversial concept of two hopane biodegradation pathways (e.g., Brooks et al., 1988).

Hopane biodegradation in petroleum without formation of 25-norhopanes (10-desmethylhopanes) has been reported in a Texas Gulf Coast oil (Seifert and Moldowan, 1979), asphalts from Switzerland (Goodwin et al., 1983; Connan, 1984), and in seep oils from Western Greece (Seifert et al., 1984) and Yugoslavia (Moldowan et al., 1992). Fig. 3.63 shows another example of hopane biodegradation without 25-norhopane formation for an oil from Colombia. This example also demonstrates the resistance of oleanane to biodegradation compared to the C_{27}-C_{35} 17α(H)-hopanes. Goodwin et al. (1983) showed hopane degradation without loss of the methyl group at C-10 in laboratory simulations of crude oil biodegradation. Their study shows that 17α(H)-hopanes are more resistant to biodegradation than steranes, but more or less resistant than diasteranes, depending on experimental conditions. The 17α(H)-hopanes were removed more rapidly than moretanes with preferential removal of the higher molecular-weight homohopanes first. The 22R epimer of the homohopanes is removed faster than 22S. These observations agree with those for naturally occurring biodegraded oils and seeps. However, some of these observations conflict with results for biodegraded oils with 25-norhopanes (following), suggesting that at least two distinct biodegradation pathways exist for hopanes (see Sterane versus hopane biodegradation, following).

It has also been shown that 28,30-bisnorhopane (BNH) is demethylated during biodegradation to 25,28,30-trisnorhopane (TNH)(Seifert and Moldowan, 1978; Moldowan et al., 1984). However, TNH also occurs in nonbiodegraded petroleums, usually with some BNH (Section III.2.4.1). Thus, both appear to be products of early diagenetic bacterial reworking in sediments. A method for ranking oil maturities using percent $17\alpha,21\beta$(H)TNH/($\alpha\alpha + \beta\alpha + \beta\beta$)TNH is useful for heavily biodegraded oils where other biomarkers are degraded (Moldowan et al., 1984).

25-Norhopanes (10-desmethylhopanes). The 25-norhopanes are a series of compounds typical of many, but not all, heavily biodegraded oils (Reed, 1977; Seifert and Moldowan, 1979; Rullkötter and Wendisch, 1982; Goodwin et al., 1983; Seifert et al., 1984; Trendel et al., 1990). These compounds appear to result from the bacterial removal of the methyl group at C-10 from the regular hopanes. It has

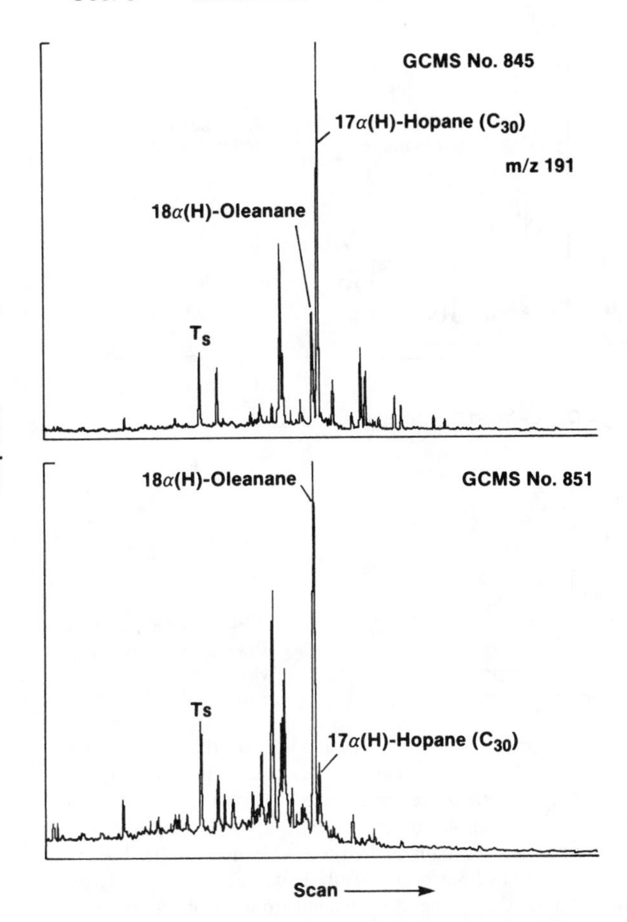

Figure 3.63 Saturate fraction of a very heavily biodegraded (rank 8, Table 3.3.1) seep oil (GCMS No. 851) (bottom) shows a different m/z 191 chromatogram compared to that for a related, moderately biodegraded (rank 5, Fig. 3.62) seep oil (GCMS No. 845) (top) from Colombia. Biodegradation has nearly completely removed the C_{27} to C_{35} 17α(H)-hopanes from seep No. 851, but 18α(H)-oleanane remains high because of its resistance to biodegradation compared to the hopanes. These seep oils do not contain 25-norhopanes.

also been proposed that 25-norhopanes are present in some oils in low concentrations, but are not observed until 17α(H)-hopanes are removed by heavy biodegradation (Chosson et al, 1992). In either case, the diagnostic m/z 191 ring A/B fragment of the hopanes is shifted to m/z 177 in the 25-norhopanes. For example, the m/z 177 fragmentogram for a biodegraded oil from the North Sea shows a series of 25-norhopanes (Fig. 3.64). Similar to the other compounds in the series, the C_{30} 17α(H)-hopane (m/z 191) loses a methyl group to form the C_{29} 25-nor-17α(H)-hopane (m/z 177) (Fig. 3.65). The mass spectrum of the latter is identical to that shown by Rullkötter and Wendisch (1982) for the isolated compound (Fig. 2.20).

The principle of quasi-stepwise biodegradation of biomarkers in petroleum discussed can be used to provide information on the migration history affecting a reservoir. In addition to 25-norhopanes, accepted by many as indicating heavy biodegradation, the biodegraded oil in Fig. 3.64 shows some characteristics of only mild to moderate biodegradation because, although most of the *n*-paraffins have been destroyed, the acyclic isoprenoids including pristane and phytane remain. Similar occurrences of demethylated hopanes in unaltered or slightly altered oils have

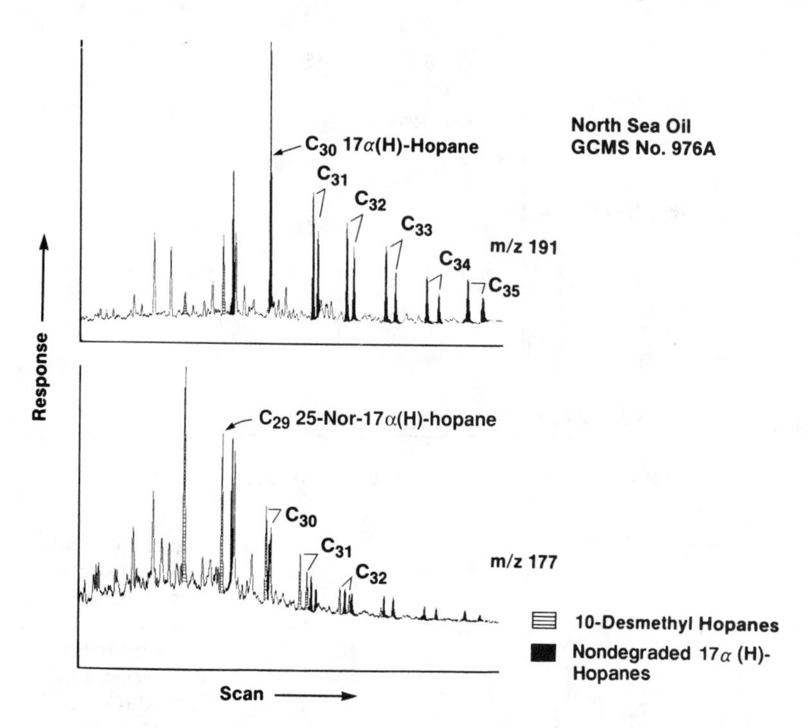

Figure 3.64 Mass chromatograms at m/z 191 (top) and m/z 177 (bottom) indicate that 25-norhopanes (also called 10-desmethylhopanes) are present in a biodegraded oil from the North Sea. The 25-norhopanes are believed to originate by loss of a methyl group from the C-10 position in hopanes. Thus, the single epimer of C_{30} $17\alpha,21\beta$(H)-hopane (top) has been partially altered to C_{29} 25-nor-17α(H)-hopane (bottom). The latter compound has been provisionally identified by the similarity of its mass spectrum to that of the synthetic compound from the literature (Fig. 2.20). The compound also shows the same relative retention time on the m/z 177 chromatogram as that of C_{29} 25-nor-17α(H)-hopane in the published work. The retention times of other 25-norhopanes on the m/z 177 trace are systematically reduced compared to those of their corresponding 17α(H)-hopane (22S + 22R) homologs. For example, the two C_{30} 25-nor-17α(H)-hopanes on the m/z 177 trace (bottom) correspond to the two C_{31} 17α(H)-hopane epimers (22S + 22R) on the m/z 191 trace (top).

also been explained by mixing a secondary pulse of oil to the reservoir with highly biodegraded residues from an earlier pulse (e.g., Alexander et al., 1983b; Philp, 1983; Volkman et al., 1983b; Talukdar et al., 1986, 1988; Sofer et al., 1986). Thus, the oil in the figure appears to represent a mixture of heavily biodegraded and mildly to moderately degraded oils. The 25-norhopanes are absent from the m/z 177 fragmentogram of the nonbiodegraded Piper oil, which is also from the North Sea and is related to the biodegraded oil.

Alternately, 25-norhopanes could form or be exposed at an earlier stage of biodegradation than previously suspected. In some cases 25-norhopanes have been reported in source rock extracts (Chosson et al., 1992). Noble et al. (1985c) ob-

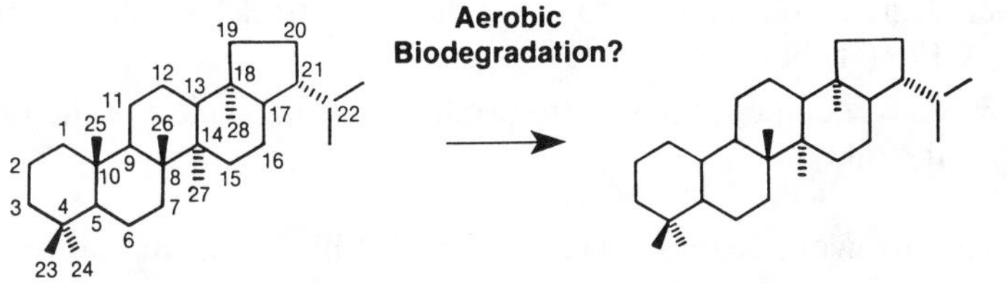

Figure 3.65 Proposed origin of 25-norhopanes by bacterial demethylation of 17α(H)-hopanes. The methyl group attached to the C-10 position in the C_{30} 17α(H)-hopane (left) is removed to produce the C_{29} 25-norhopane (right). The biodegraded Western Siberian oil discussed in the text shows a unique "reversed" distribution of C_{31} to C_{35} 17α(H)-hopanes, where the homolog abundance increases from C_{31} to C_{35}. The 25-norhopanes decrease from C_{30} to C_{34}, consistent with their origin from the 17α(H)-hopanes.

served 28,30-bisnorhopane, 25,28,30-trisnorhopane, 25,30-bisnorhopane, and other 25-norhopanes in some West Australian shales. These compounds are apparently not produced by thermal breakdown of kerogen in the extracted shales, indicating they are present in the shales as free hydrocarbons. This could indicate an unusual type of bacterial reworking of the organic matter or the inclusion of paleo-seepage oil into the source rock at the time of deposition.

Most reports of oils showing incomplete conversion of 17α(H)-hopanes to 17α(H)-25-norhopanes observe no particular resistance of any homohopane homolog to this type of biodegradation. However, Rullkötter and Wendisch (1982) note that among the C_{27}-C_{32} pseudohomologs biodegradation "proceeds from the higher molecular-weight end" in asphalts from Madagascar. Requejo and Halpern (1989) observed preferential preservation of the C_{35} homohopanes and Ts and Tm in biodegraded tars from the Monterey Formation.

However, Peters and Moldowan (1991) observed C_{26}-C_{34} 17α(H)-25-norhopanes in a biodegraded Western Siberian oil accompanied by nearly complete removal of the C_{27}-C_{32} 17α(H)-hopanes. The oil shows a unique, "reversed" C_{31} to C_{35} 17α(H)-homohopane distribution, where the homolog abundance increases from C_{31} to C_{35}. This observation contrasts with reports of the other type of hopane biodegradation without formation of 25-norhopanes where higher homohopanes are degraded faster (Goodwin et al., 1983; Seifert et al., 1984). Detailed examination of the m/z 177 mass chromatogram for this oil shows 25-norhopanes, which decrease from C_{30} to C_{34}. This novel C_{31} to C_{35} 17α(H)-hopane distribution and its inverse relationship with that of the C_{30} to C_{34} 25-norhopanes in the Western Siberian oil indicate that:

1. 17α(H)-hopanes and 25-norhopanes are related as precursor and product, respectively,

2. degradation proceeded by selective attack of the lower molecular-weight 17α(H)-hopanes,
3. conversion of precursor to product was incomplete at the time of production of the oil.

This work shows, at least for the biodegraded Western Siberian oil, that biodegradation of lower molecular-weight 17α(H)-hopanes occurs in preference to heavier homologs and that 25-norhopanes are formed. We conclude that the specificity of reactions that occur during biodegradation is highly dependent on the types of bacterial populations which alter oil. Similar conclusions were reached in studies by Brooks et al. (1988) and Requejo and Halpern (1989).

C_{28}-C_{34} 30-nor-17α(H)-hopanes. In heavily biodegraded oils where the hopanes have been altered, 30-nor-17α(H)-hopanes can provide supplementary information useful in correlation studies.

The 30-nor-17α(H)-hopanes are commonly found in petroleum from carbonate source rocks. Moldowan et al. (1992) found these compounds in related, non-biodegraded to severely biodegraded oils from the Adriatic Basin (Fig. 3.26). They used distributions of the C_{30}-C_{34} 30-nor-17α(H)-hopanes analyzed by GCMSMS to correlate the heavily biodegraded Adriatic oils. These compounds are also found in oils from Greece (Seifert et al., 1984).

The 30-nor-17α(H)-hopanes probably constitute a series of compounds ranging from C_{28} to C_{34}. Because the highest homolog in the series is C_{34}, precursors for the 30-nor-17α(H)-hopanes are probably about the same size as the hopanoids and serve similar functions in the cell membranes of prokaryotes (Sec. 1.2.10.2). Similar to the 30-nor-17α(H)-hopanes with which it appears related, the C_{28}-hopane pseudohomolog 29,30-bisnor-17α(H)-hopane is found mainly in carbonate-sourced oils. This compound is not favored as a thermal degradation product from the 17α(H)-hopanes containing a C-22 methyl branch because its formation would require cleavage of two carbon-carbon bonds at the C-22 position (Seifert et al., 1978).

Sterane versus hopane biodegradation. The relative extent of sterane and hopane biodegradation in oils appears to depend on various factors, including the type of biodegradation, environmental conditions, and the microbial population. A Malagasy asphalt, for example, contains only partially biodegraded steranes, but the 17α(H)-hopanes were partially converted to 25-norhopanes (Rullkötter and Wendisch, 1982). Volkman et al. (1983b) explain 17α(H)-25-norhopanes in non-biodegraded crudes as a result of the presence of biodegraded oil residues which have been dissolved by the nondegraded crude during accumulation in the reservoir. A biodegraded oil from West Siberia shows no evidence of sterane degradation but substantial conversion of 17α(H)-hopanes to 17α(H)-25-norhopanes (Peters and Moldowan, 1991). While the C_{27}-C_{32} 17α(H)-hopanes are severely depleted in this oil, the higher homologs remain and show resistance to biodegradation in the order $C_{35} > C_{34} > C_{33}$. This evidence supports early bacterial attack on 17α(H)-hopanes compared to steranes in cases where 25-norhopanes are formed. On the other hand, a biodegraded seep oil from Greece shows partial biodegradative loss of steranes without any apparent effect on the 17α(H)-hopanes (Seifert et al., 1984). Other seeps in the region show complete sterane degradation with only partial loss of 17α(H)-hopanes, but no formation of 17α(H)-25-norhopanes.

Thus, two heavy biodegradation pathways for oil appear to occur in nature:

1. a pathway where 25-norhopanes begin to form prior to sterane alteration,
2. a pathway where steranes are altered prior to hopanes and 25-norhopanes are not formed.

These conclusions were reached by Brooks et al. (1988) in a study of biodegraded oils from the Western Canada Basin. From our experience, it appears that pathway 1 can proceed to 25-norhopanes by preferential degradation of either lower or higher molecular weight 17α(H)-hopanes. Pathway 2 proceeds by preferential degradation of higher molecular-weight 17α(H)-hopanes without forming 25-norhopanes.

Tricyclic terpanes (cheilanthanes). The C_{19}-C_{45} tricyclic terpanes (Sec. 3.1.3.1) are highly resistant to biodegradation, surviving even when hopanes are removed (Reed, 1977; Seifert and Moldowan, 1979; Connan et al., 1980). Despite heavy biodegradation, seep oils from western Greece could still be correlated with subsurface oils using a combination of tricyclic terpane and aromatic steroid distributions, and stable carbon isotope ratios (Palacas et al., 1986).

Tricyclic terpanes can be altered at about the same level of biodegradation as the diasteranes (Connan, 1984; Lin et al., 1989). The tricyclics/17α(H)-hopanes ratio is likely to be altered in heavily biodegraded petroleums. Connan (1984) reports the eventual destruction of these tricyclic terpanes in a Swiss asphalt.

Other terpanes. Gammacerane is one of the last known saturated biomarkers to be biodegraded in petroleum (Seifert et al., 1984; Zhang Dajiang et al., 1988). Oleanane is also highly resistant (e.g., Fig. 3.63). Fowler et al. (1988) show mass chromatograms for biodegraded bitumens that indicate a general lack of steranes and triterpanes, although some of the more resistant compounds, like gammacerane, may not have been present prior to biodegradation.

Jiang Zhusheng et al. (1990) compared related unaltered and severely biodegraded oils from the Kelamayi oil field in northwestern China. Biodegradation resulted in concentration of C_{25} and C_{26} regular steranes, tricyclic, and tetracyclic terpanes, gammacerane, and 8,14-secohopanes. Several compounds were found in the altered, but not in the unaltered oil, including 25-norhopanes, and C-10 demethylated tricyclic and tetracyclic terpanes.

Aromatic steroids. The most thorough examination of possible biodegradation of aromatic steroids was reported by Wardroper et al. (1984). Until that time no biodegradative alteration of triaromatic (TA) or ring-C monoaromatic (MA) steroids had been reported (Connan, 1984), attesting to their bacterial resistance. Wardroper et al. (1984) indicate that C_{20}-C_{21} TA-steroids are among the first aromatic steroids to be depleted during degradation of petroleum, but it is not certain whether this is due to water washing or biodegradation. However, using biodegraded oils from the North Sea and the Molasse Basin showing severe degradation of steranes, diasteranes, and hopanes, they confirm that degradation of the C_{26}-C_{28} TA-steroids has occurred. The Molasse Basin oil also shows preferential degradation of the 20R versus 20S epimers of the C_{27}-C_{29} MA-steroids. The low molecular-weight (C_{21}, C_{22}) MA-steroids in these oils are more resistant to biodegradation than heavier homologs. These results are supported by laboratory biodegradation experiments (Wardroper et al., 1984).

Our work confirms that biodegradation of aromatic steroids is rare. Only two oils analyzed to date, seeps from Malagasy and Nigeria, have shown biodegradation of MA- and TA-steroids (unpublished data).

Porphyrins. Although further research is needed, it is generally believed that porphyrins are highly resistant to biodegradation. For example, Barwise and Park (1983) show that related, nonbiodegraded and extensively biodegraded oils of nearly equivalent maturity show similar porphyrin distributions (e.g., DPEP/etio ratios). Lin et al. (1989) observed no significant biodegradation of porphyrins in tar-sand bitumens from the Ardmore and Anadarko Basins, Oklahoma. A study of Colombian oils indicates that nickel porphyrins might be selectively destroyed by severe biodegradation, but that vanadyl porphyrin distributions remain as good correlation tools (S. E. Palmer, 1990, pers. comm.). Strong and Filby (1987) conclude that biodegradation has little effect on vanadyl porphyrin distributions in the highly biodegraded Alberta oil sands.

3.3.4 Effects of biodegradation on determination of maturity and correlation. Because biodegradation can alter commonly used maturity and correlation parameters, it is critical that the interpreter be aware of the extent of biodegradation achieved by every sample in each study. C_{29}-sterane 20S/(20S + 20R) ratios are among the first commonly used maturity parameters to be affected (level 6 in Fig. 3.62). Samples where biodegradation has affected this ratio typically show few, if any, *n*-paraffins, reduced isoprenoids, and anomalous 20S/(20S + 20R) ratios (i.e., > 0.55) because of preferential bacterial attack of the 20R epimer.

Maturity estimates based on triterpanes only begin to be affected after heavy to very heavy biodegradation (level 6 or greater in Fig. 3.62), where hopanes and moretanes are degraded.

The first detrimental effects of biodegradation on most correlations occur during degradation of the steranes (level 6 in Fig. 3.62). Preferential biodegradation of C_{27} over C_{28} over C_{29} steranes results in predictable changes in the locations of samples on ternary diagrams, like that in Fig. 3.31. If there is any evidence of sterane biodegradation on the ternary diagram or on m/z 217 mass chromatograms, the sterane ternary plot should not be used for correlation purposes. Alternate ternary plots can be prepared using more resistant biomarkers, such as diasteranes (Sec. 3.1.3.2) or monoaromatic steroids (Fig. 3.31).

Although only few oils of economic significance reach very high levels of biodegradation (e.g., levels 9 or 10 in Fig. 3.62), they can be volumetrically important. Examples of highly degraded oils include the Canadian and Venezuelan tar-sand deposits. Biomarker correlations for these oils are difficult because of the limited number of parameters that can be used to support interpretations. For example, one Colombian seep oil (GCMS No. 851) shows severe biodegradation of hopanes compared to another (GCMS No. 845) (Fig. 3.63), yet they are believed to be related because of similar porphyrin distributions, stable carbon isotope compositions, and abundant oleanane.

Sealed-tube pyrolysis of asphaltenes has been used to correlate heavily biodegraded seep oils from Venezuela (Cassani and Eglinton, 1988). Pyrolysis at 330°C or less releases biomarkers from the asphaltene structure whose distributions appear to be little altered from those in the original nonbiodegraded oil. Jones et al. (1988) used hydrous pyrolysis of asphaltenes and polar fractions from oils and heavily biodegraded asphaltic bitumens to show genetic relationships among these samples, even when their free hydrocarbon distributions contain little information.

4

Problem Areas and Further Work

This chapter describes areas requiring further research, including the application of biomarkers to migration, the kinetics of petroleum generation, correlation, depositional environments, and age determination. Readers who feel they have samples or ideas that might contribute to ongoing research efforts in these areas are encouraged to contact the authors.

4 PROBLEM AREAS

Continued advances in analytical instrumentation and molecular chemistry suggest that the use of biomarkers to solve geochemical problems will continue to grow. The following discussion outlines several areas where further biomarker research activity is likely.

4.1 Migration

Two processes appear to control biomarker distributions during oil migration: solubilization and geochromatography. Solubilization is poorly documented, but involves the incorporation of organic matter from rocks that are unrelated to the migrating petroleum. It is assumed that in most cases, the solubilized materials are of lower thermal maturity than those comprising the migrating petroleum. Solubilization has been observed in Mahakam Delta oils (Durand, 1983; Hoffmann et al., 1984; Jaffe et al., 1988a and b), in Australia (Philp and Gilbert, 1982, 1986), and in Angola (Fig. 4.1). Evidence of solubilization typically consists of "mixed" thermal maturity signals. For example, one parameter might indicate high maturity because the compounds used for the parameter are dominant components in the migrated fraction of the oil. Another parameter for the same oil might indicate lower thermal maturity because the compounds used for this parameter are dominant in the solubilized contaminants picked up by the migrating oil.

Although not well studied, oils showing significantly altered biomarker distributions because of solubilization of contaminants during migration are probably rare. In general, carrier beds and reservoir rocks are organic lean and the concentra-

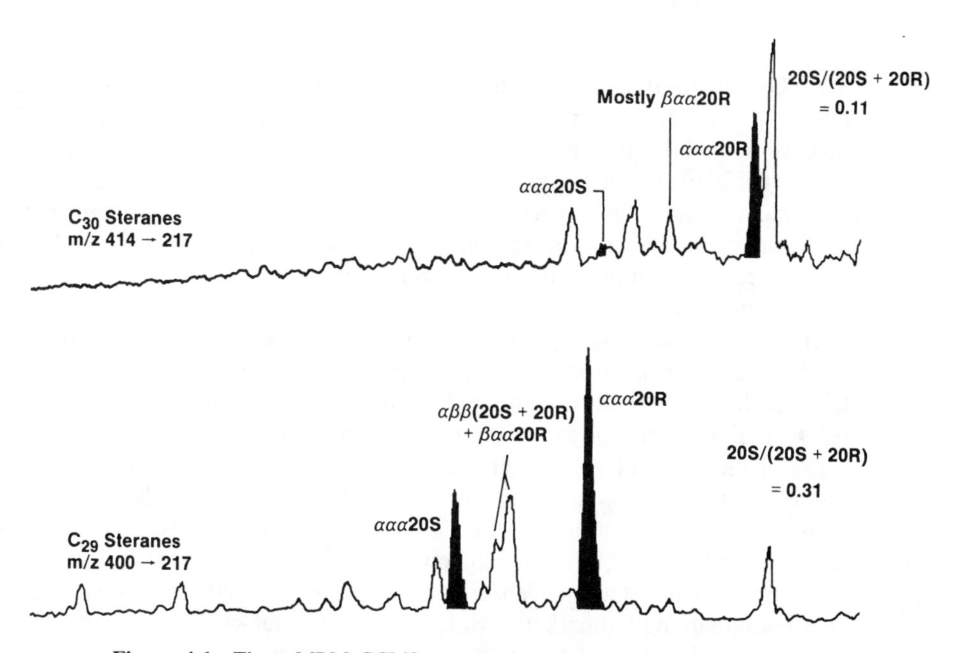

Figure 4.1 These MRM GCMS sterane chromatograms show leaching or solubilization of immature bitumen by oil. The oil (GCMS No. 929) was generated from Lower Cretaceous lacustrine source rocks in Angola. The oil migrated into an Upper Cretaceous reservoir containing sandstones with interbedded immature, organic-rich marine shales. The oil appears more mature based on the 20S/(20S + 20R) ratio for the C_{29} steranes (bottom) than for the C_{30} steranes (top). The more mature C_{29} sterane 20S/(20S + 20R) ratio (0.31) for the oil (bottom) results from combined inputs of mature oil (~ 0.5) and leached, immature bitumen. The less mature C_{30} sterane 20S/(20S + 20R) ratio (0.11) for the oil (top) reflects the maturity of the immature, leached bitumen, only because no C_{30} steranes (24-*n*-propylcholestanes) are found in lacustrine oils. Also note the prominent C_{30} 5β,20R sterane peak compared to the C_{29} 5β,20R sterane. The C_{30} 5β,20R sterane peak has been largely contributed from the immature bitumen.

tions of biomarkers solubilized from these sources are believed to be low compared to those in the migrating oil. However, in some cases such as migration of a biomarker-poor condensate through organic-rich coals, solubilization might be important.

> ***Note:*** *Petroleum associated with sulfide-rich sediments at active oceanic spreading centers represents an extreme example of solubilization and admixture of hydrocarbons generated over a wide range of thermal regimes (Kvenvolden and Simoneit, 1990). For example, hopanoid epimer ratios for extracts from the Guaymas Basin and Escanaba Trough indicate low maturity, while ratios of aromatic hydrocarbons (e.g., benzo(e)pyrene and benzo(a)pyrene) indicate high maturity. Unlike conventional petroleum, these oils were generated from organic matter in very young (Quaternary) sediment by hydrothermal activity.*

Geochromatography is the hypothesized process where biomarkers and other compounds migrate at different rates through the mineral matrix in a rock. Compounds with differing molecular weights, polarities, and stereochemistries should behave differently when exposed to various adsorptive/desorptive processes during movement out of the source rock (primary migration) or through carrier beds (secondary migration). For example, more polar molecules are expected to be more strongly retained on mineral surfaces. Both field (e.g., Seifert and Moldowan, 1978, 1981; Leythaeuser et al., 1984; Mackenzie et al., 1987) and laboratory studies (e.g., Carlson and Chamberlain, 1986; Kroos and Leythaeuser, 1988; Brothers et al., 1991) address geochromatographic effects on organic compounds. Carlson and Chamberlain (1986) address some of the principles controlling migration differences between biomarkers showing differential absorption on montmorillonite clay.

Applications of biomarker technology to studies of geochromatography are limited. Although it is not now possible to use biomarkers to describe absolute migration distances, an attempt has been made to use them to distinguish more- versus less-migrated oils (Seifert and Moldowan, 1981) (Fig. 3.56). Detailed studies of alternating sand and shale sequences in cores from the North Sea show little if any redistribution among biomarkers, while lower molecular-weight *n*-paraffins show large effects, probably related to differences in diffusivity and solubility (Leythaeuser et al., 1984; Mackenzie et al., 1988b). On the other hand, Peters et al. (1990) show evidence in hydrous pyrolysis experiments that biomarker differences, paralleling those reported for natural oils and bitumens, occur between expelled oils and the remaining unexpelled bitumen in the pyrolyzed rock chips. They observed increases in tricyclics/17α(H)-hopanes and diasteranes/steranes ratios in the expelled oils compared to the bitumens, but distributions of the hopane and sterane stereoisomers were unchanged. Seifert and Moldowan (1986) also propose that the ratio of tricyclic terpanes to pentacyclic terpanes is a migration parameter because tricyclics migrate faster than hopanes. More research is needed on geochromatography and its effects on biomarkers.

4.2 Kinetics

Geochemical basin models attempt to predict the thermal history of sedimentary basins, including the timing and quantities of petroleum generated and migrated (e.g., Welte and Yalcin, 1988; Ungerer, 1990). Accurate assessment of paleotemperatures is critical input for these modelling programs. Some of the more advanced basin models are calibrated using thermally-dependent biomarker ratios (Welte and Yalcin, 1987).

Mackenzie and McKenzie (1983) and Mackenzie et al. (1981b, 1984) first advocated the use of thermally dependent biomarker ratios to assess paleotemperatures for basin models. They monitored the progress of several biomarker reactions in samples at depth in basins with known temperature histories. Kinetic expressions, including apparent activation energies (E_a) and frequency factors (A), were derived for proposed steroid aromatization, and sterane and hopane isomerizations. They observed that the rate expressions for the monoaromatic to triaromatic steroid aromatization are consistent with laboratory heating experiments, while the hopane isomerization is not. Rullkötter and Marzi (1988) recommended adjustments to these apparent activation energies based on the need to account for both laboratory simulation and natural evolution kinetics.

Our understanding of the controls on kinetics of proposed biomarker reactions is limited. Rullkötter and Marzi (1988) derived kinetic constants based on the 20S/(20S + 20R) sterane isomerization during hydrous pyrolysis, which were used to reconstruct the geothermal history of the Michigan Basin (Marzi and Rullkötter, 1992). Abbott et al. (1990) attempted to derive a kinetic model to rationalize the use of these parameters. They found that direct chiral isomerization at C-20 in the regular steranes appears to be relatively unimportant under hydrous pyrolysis conditions. Marzi et al. (1990) observed wide variations in published kinetic expressions for the isomerization reaction at C-20 in the 5α(H),14α(H),17α(H)-steranes. They recommend that more precise kinetic expressions can be obtained if the results of both laboratory experiments (high temperature, short time) and natural series (low temperature, long time) can be combined. This approach assumes that the geologic and paleotemperature histories of the natural samples are known and that the chemical processes are the same in the laboratory and nature.

Until further research is completed, we recommend that published kinetic expressions for biomarker reactions be used only as rough constraints on paleotemperatures. At least three problems limit the use of biomarker thermal maturity ratios in basin modeling studies:

1. assumptions of simple precursor-product relationships may be oversimplified (e.g., Mackenzie and McKenzie, 1983),
2. some biomarker maturity ratios show reversals at high maturity (e.g., Lewan et al., 1986; Rullkötter and Marzi, 1988, Peters et al., 1990),
3. small analytical errors in biomarker maturity ratios can have a large effect on the kinetic expressions (Marzi et al., 1990; Gallegos and Moldowan, 1992).

Our results from hydrous pyrolysis of Monterey Phosphatic and Siliceous rock (Peters et al., 1990) show that steroid and hopanoid isomerizations are more complex than expected. During heating, kerogen-bound precursors generate steranes and hopanes showing lower levels of thermal maturity based on isomerization than those extracted from the unheated rock. Asymmetric centers in kerogen-bound steroids and hopanoids appear to be protected from isomerization compared to those of free steranes or hopanes in the bitumen. The Phosphatic and Siliceous rocks can show different levels of sterane or hopane isomerization when heated under the same time/temperature conditions. Further, maturity ratios based on these isomerizations unexpectedly decrease at high hydrous pyrolysis temperatures (>330°C) for the Phosphatic, but not for the Siliceous samples. Lewan et al. (1986) have observed a similar reversal in sterane isomerization for Phosphoria Retort shale at high hydrous pyrolysis temperature. Further, Strachan et al. (1989) have observed a reversal in the C_{29} $\alpha\alpha\alpha$20S/(20S + 20R) ratio at depth in a well in Australia.

Mineralogy clearly affects the kinetics of oil generation and the composition of the products (Tannenbaum et al., 1986). Soldan and Cerqueira (1986) observed variations in biomarker compositions for samples heated under hydrous pyrolysis conditions at the same temperatures using different water salinities.

4.3. Correlation

4.3.1. Test for indigenous bitumen. A critical question in any oil-source rock correlation is whether the bitumen in the rock is indigenous or represents a migrated oil. If unrecognized, migrated oil in a proposed "source rock" could result in a spurious oil-source rock correlation.

When used together, four principal tests can help establish whether a bitumen is indigenous to the fine-grained rock from which it has been extracted.

1. Consistency of bitumen/TOC and/or production index (PI) values with the level of thermal maturity of the kerogen (e.g., T_{max} or vitrinite reflectance),
2. Comparison of the thermal maturity of the bitumen (e.g., CPI or biomarker maturity ratios) with that of the kerogen,
3. Isotopic relationships between the bitumen and kerogen,
4. Comparison of biomarkers in the bitumen with those released from the kerogen by mild thermal or chemical degradation.

Seifert (1978) established relationships between source rocks and oils by pyrolyzing bitumen-free kerogen from the rocks and comparing the biomarkers in the pyrolyzates with those in oils. Early work was performed in an open pyrolysis system generating alkenes, while more recent studies use closed systems, such as hydrous pyrolysis, to produce a more petroleum-like product (Seifert and Moldowan, 1980; Seifert et al., 1979; Lewan, 1985).

Trapped and bound biomarkers in kerogen can sometimes be differentiated. For example, Summons et al. (1988a) used BBr_3 to release ether-linked compounds from kerogens as alkyl bromides. Using thin layer chromatography, these compounds were separated from trapped hydrocarbons and reduced using $LiAlD_4$ to deuterated alkanes (Chappe et al., 1980). The deuterated compounds are readily identified by GCMS because the mass of their parent ions (M^+) is increased by one amu (one dalton).

4.3.2 Hydrous pyrolysis of immature source rocks. Biomarker fingerprints of immature bitumens differ considerably from those of mature bitumens or oils from more deeply buried lateral equivalents of the same source rock. Mature bitumens consist largely of products generated from the kerogen. Prior to incorporation into the kerogen, these products were largely cell membrane lipids. Immature bitumen, on the other hand, may be largely derived from free lipids from the contributing organisms, which may show a different composition than those which are incorporated into the kerogen. Thus, a petroleum may correlate with bitumen from its mature source rock, while bearing little resemblance to bitumen from an immature equivalent of the same rock.

Hydrous pyrolyzates expelled from potential source rocks are compositionally similar to natural crude oils, except for elevated amounts of aromatic and polar materials. Noble et al. (1991) observed that the gas/oil ratios (GOR) for pyrolyzates from three types of rocks from Talang Akar, Indonesia, varied according to their petroleum generative potential and represented reasonable estimates of what might be expected from their subsurface maturation.

Figure 4.2 (from Moldowan et al., 1992) shows how to evaluate a possible correlation between immature source rock and petroleum using hydrous pyrolysis of the rock. The figure shows that the sterane distribution for Monte Prena bitumen is dominated by the C_{29} compounds. The Monte Prena hydrous pyrolyzate shows more C_{27} and C_{28} steranes, similar to the Katia oil. The immature monoaromatic steroids of Monte Prena bitumen also show a change in the hydrous pyrolyzate toward the composition of the Katia oil (Moldowan et al., 1992). Further details on the hydrous pyrolysis procedure are in Lewan (1985), Lewan et al. (1986), and Peters et al. (1990).

Eglinton and Douglas (1988) describe nonsystematic differences between source parameters in immature and mature (pyrolyzate) bitumens from Monterey, Kimmeridge, New Albany, and Green River shales. Attempts to correlate immature source rocks with oil (mature) are best approached by pyrolyzing whole or crushed samples of the immature rock rather than extracted rock. Like the oil, the resulting pyrolyzate will thus contain biomarkers from both the original bitumen and those cracked from the kerogen.

4.4 Isotopic Signatures of Individual Biomarkers

Bridging the gap between isotope and biomarker geochemistry is an important step in understanding paleoenvironments. In a study of oils from Silurian (Michigan

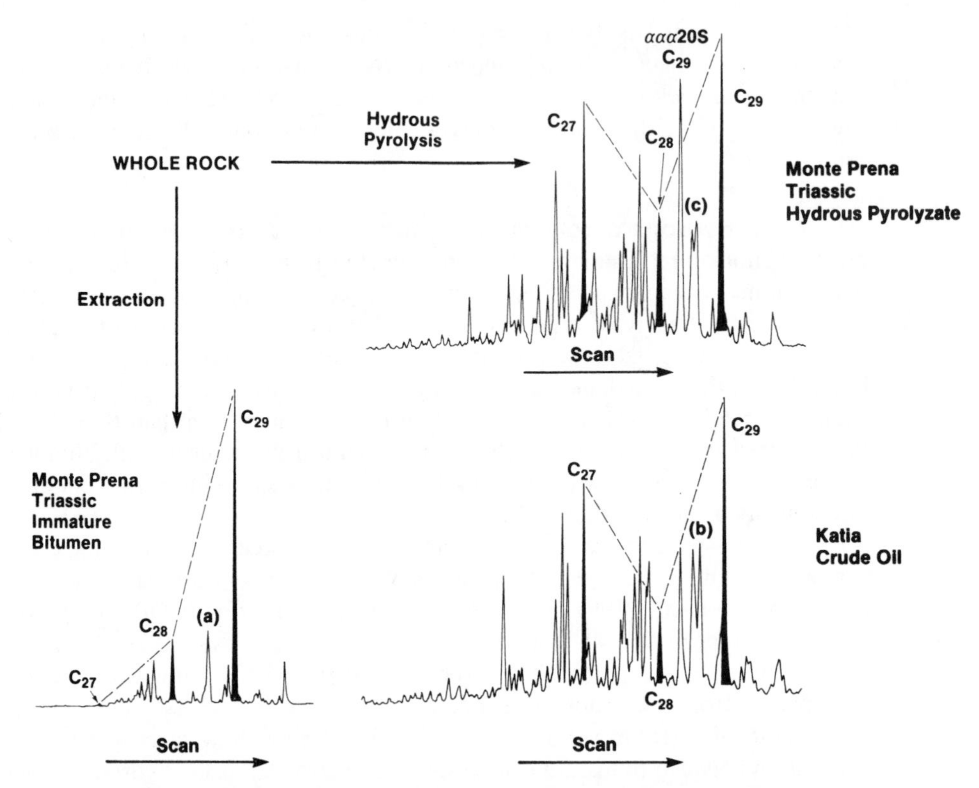

Figure 4.2 Hydrous pyrolysis enables correlation of an immature source rock (Monte Prena) with an oil (Katia) (Moldowan et al., 1992). The immature Monte Prena source rock was pyrolyzed under hydrous conditions (Peters et al., 1990; Lewan et al., 1986) in a pressure vessel at 320°C for 72 hours. The saturate fraction of oil generated from the heating experiment is compared by GCMS (m/z 217) sterane analysis with that of Katia oil. The results show a dramatic shift from the immature extract pattern, which contains regular sterane distributions heavily skewed toward C_{29}-steranes, to the mature oil pattern, which contains considerable C_{27}- and C_{28}-steranes. The pattern of the C_{27}-C_{29} $5\alpha,14\alpha,17\alpha$(H),20R-steranes ($\alpha\alpha\alpha$20R) indicated by the dotted line graph for the Monte Prena hydrous pyrolyzate is nearly identical to that for Katia oil.
(a) Mostly C_{29} $5\beta,14\alpha,17\alpha$(H),20R steranes.
(b) Mostly C_{29} $5\alpha,14\beta,17\beta$(H),20R steranes.
(c) Mixture C_{29} $5\beta,14\alpha,17\alpha$(H),20R + C_{29} $5\alpha,14\beta,17\beta$(H), 20S + 20R.

Basin) and Devonian (Western Canada) reefs, Summons and Powell (1986, 1987) identified a series of 1-alkyl-2,3,6-trimethylbenzenes. The structures of these compounds suggest they are diagenetic products of aromatic carotenoids of the green sulfur bacteria. Structural and isotopic evidence were combined to show that these compounds relate to a biosynthetic process (reductive tricarboxylic acid cycle) typical of *Chlorobiaceae*, presumed inhabitants of microbial communities in ancient restricted seas. Thus, a new class of biomarkers can be identified and its significance understood by combined use of biomarker and isotopic analysis.

A new technique called isotope-ratio-monitoring gas chromatography mass spectrometry (IRMGCMS) (Hayes et al., 1987; 1990) allows stable carbon isotopic analysis of biomarkers and other compounds as they elute from a gas chromatograph. The authors show that the stable carbon isotopic compositions of individual biomarkers are preserved during diagenesis. Biomarkers in the Eocene age Messel shale show a broad range in carbon isotopic compositions from −20.9 to −73.4 per mil. These biologically controlled isotopic compositions can be used to identify specific sources for some compounds and help reconstruct the carbon cycling within the lake and sediments which formed the Messel shale. For example, structures and isotopic compositions for several porphyrins allowed their differing origins from algal or bacterial sources to be determined.

By assuming that compounds with common biological origins show similar isotopic compositions, IRMGCMS becomes a powerful tool for reconstructing biogeochemical pathways for organic carbon. Using this hypothesis, Freeman et al. (1990) were able to reconstruct a variety of both primary and secondary pathways for carbon flow in the ancient Lake Messel. "Primary" refers to all products of photosynthesis, while "secondary" indicates derivation by subsequent processes, both biological and thermal. As another example, isotopic results (Hayes et al., 1990) show contrasting sources for acyclic isoprenoids and *n*-alkyl hydrocarbons in samples from the Cretaceous age, marine Greenhorn Formation. Pristane, phytane, and porphyrins comprise a group of compounds showing co-variant isotopic compositions, while *n*-alkanes and total organic carbon form another. The results are consistent with derivation of the pristane, phytane, and porphyrins from primary material (i.e. chlorophyll), while the *n*-alkanes are derived from secondary inputs.

IRMGCMS should prove to be a valuable method in oil : oil and oil : source rock correlation. The method might also prove useful for identifying co-mingled oils from different sources.

4.5 Age Determinations from Biomarkers

If the age distributions of more biomarkers were known, determination of the source rocks for oils would be greatly simplified. Because biomarkers represent a molecular record of life, a better understanding of their temporal evolution will necessarily improve our understanding of historical geology. Some major changes in earth history which might be reflected in biomarker distributions could include:

1. evolution of new organisms (e.g., land plants, angiosperms, diatoms, coccolithophorids, eukaryotes),
2. major extinctions or changes in biota (e.g., Cretaceous/Tertiary, Permian/Triassic),
3. major geologic events (e.g., oceanic anoxia, glaciation). Table 4.5 shows the limited data for certain biomarkers whose age distribution appears useful. Additional discussion of C_{27}-C_{28}-C_{29} steranes and monoaromatic steroids and their age significance is included in Sec. 3.1.3.2 and Sec. 3.1.3.3, respectively.

TABLE 4.5 Age Distributions for Some Biomarkers in Petroleum.

Biomarker	Organism	First geologic occurrence[1]
Terpanes		
Oleanane	Angiosperms	Cretaceous
Beyerane, kaurane, phyllocladane	Gymnosperms	Late Carboniferous
Gammacerane	Protozoa Bacteria	Late Proterozoic
28,30-Bisnorphane	Bacteria	Proterozoic
Steranes		
23,24-dimethylcholestane	Prymnesiophytes or Coccolithophores?	Triassic
4-methylsteranes	Dinoflagellates, Bacteria	Triassic
Dinosterane	Dinoflagellates	Triassic
24-*n*-propylcholestane	Marine Algae (e.g., *Sarcinochrysidales*)	Proterozoic
2- and 3-methylsteranes	Bacteria/Eukaryotes	Proterozoic
$C_{29}/(C_{27}$ to $C_{29})$ steranes	Eukaryotes	Varies Through Time
Isoprenoids		
Botryococcane	*Botryococcus braunii*	Jurassic
Biphytane	Archaebacteria	Proterozoic

[1] These suggested age boundaries are not rigorously documented for the related biomarkers and additional research may necessitate revisions.

4.6 Source Input and Depositional Environment

If an explorationist is able to determine the source rock for an oil, then probable migration pathways or types of petroleum to be expected can be used to develop new plays. Unfortunately, the source rock for a reservoired petroleum is commonly unknown and only oils are available for study. For this reason, the use of biomarker analyses on oils as a means of predicting the character of the source rock is a critical technology (e.g., Sec. 3.1.4).

Current understanding of the effects of depositional and diagenetic processes on geochemical parameters is limited, but useful. For example, as discussed earlier in the text, reducing (low Eh) depositional conditions for shale source rocks result in enchanced phytane vs. pristane, Ts versus Tm, and vanadyl versus nickel porphyrins while hindering formation of diasteranes versus steranes and short- versus long-side chain monoaromatic steroids. Alkaline (high pH) conditions appear to favor formation of Tm over Ts and moretanes over hopanes while hindering formation of diasteranes and phytane.

Although biomarkers are already powerful tools, further research is needed to establish more detailed, systematic understanding of source input and sedimentologic controls on biomarker compositions. The results of these studies will build on concepts of controls on organic facies developed by Demaison and Moore (1980) and Jones (1987). For example, the ratio of diasteranes/steranes in petroleum (Sec. 3.1.3.2) is commonly used to make qualitative inferences on the relative clay content in the source rock. The application of these and other parameters related to organic matter input and depositional environment is generally based on nonsystematic, empirical evidence (experience) and limited support from the literature or company memoranda. More systematic, statistical studies could improve the usefulness of these parameters. For example, Peters et al. (1986) used multivariate analysis to evaluate the significance of various biomarker and isotopic analyses for distinguishing oils derived from organic matter deposited in nonmarine shales, marine shales, and marine carbonates. Triangular plots of probabilities were developed that can be used to distinguish between the three source rock organic matter end-members.

Zumberge (1987a) used factor and stepwise discriminant analysis of tricyclic and pentacyclic terpane data for oils to predict source rock features or depositional environments. He was able to distinguish five categories of source rocks from the oil compositions: nearshore marine, deep-water marine, lacustrine, phosphatic-rich source beds, and Ordovician age source rocks. Telnaes and Dahl (1986) compared 45 oils of probable Kimmeridge source from the North Sea using multivariate analysis of biomarker data. They were able to separate oils from different fields and formations and make inferences about the significance of parameters in regard to depositional environment, maturity, or organic facies. Irwin and Meyer (1990) used multivariate analysis to distinguish organic and mineralogical compositions among four distinct lacustrine depositional settings in the Devonian Orcadian Basin of Scotland. Principal component analysis allowed identification of the biomarker parameters that were most useful in discriminating between the samples by environment or by relative thermal maturity.

Sample selection will be a critical factor in further studies of biomarkers for predicting source rock character. **We need well-defined source rock sequences where depositional and diagenetic processes are understood in order to relate these processes more accurately to measured biomarker parameters.** For example, detailed evaluation of the effects of various clay contents on biomarker ratios would require a sample suite where other variables, such as thermal maturity and organic matter input, were relatively uniform. Other sample suites that could be useful include: lacustrine rocks showing differences in paleosalinity, marine rocks showing differences in oxicity during deposition, and rocks containing organic matter specific to certain organisms. This work overlaps with the use of biomarkers to predict source rock age (Sec. 4.5).

*Glossary**

Acyclic

Refers to straight or branched carbon-carbon linkage in a compound without cyclic (ring) structures. May also refer to a noncyclic portion of a molecule.

Aerobic

Biofacies resulting from oxygenated water column or sediments containing more than 2.0 ml oxygen/l water (Tyson and Pearson, 1991). Maximum oxygen saturation in sea water is in the range 6 to 8.5 ml/l, depending on salinity and water temperature. Most organic matter is destroyed or severely altered when exposed to aerobic conditions for extended periods. Aerobic organisms require oxygen (Sec. 1.1).

Aliphatics

Hydrocarbons in petroleum that contain saturated and/or single unsaturated bonds and elute during liquid chromatography using nonpolar solvents. Includes alkanes and alkenes, but not aromatics (Sec. 1.2.3).

Alkane (Paraffin)

A saturated hydrocarbon, containing only hydrogen and carbon with no double bonds, which may be straight (normal), branched, or cyclic (Sec. 1.2.1). The simplest alkane is methane (CH_4), then ethane (C_2H_6), propane (C_3H_8), and so on.

* Supplementary definitions for many of the terms in this glossary can be found in Miles (1989).

Alkene (Olefin)

An unsaturated hydrocarbon, containing only hydrogen and carbon with one or more double bonds, but not aromatic. Alkenes do not normally occur in crude oils, but can occur as a result of rapid heating, as during turbo-drilling, steam treatment, or laboratory pyrolysis. Ethylene (C_2H_4) is the simplest alkene (Sec. 1.2.2).

Alkyl Group

A hydrocarbon substituent with the general formula C_nH_{2n+1} obtained by dropping one hydrogen from the fully saturated compound; e.g., methyl- ($—CH_3$), ethyl ($—CH_2CH_3$), propyl ($—CH_2CH_2CH_3$), or isopropyl [$(CH_3)_2CH—$]. In the literature alkyl groups attached to biomarkers are typically symbolized by the letter "R." In the text, we prefer to use the letter "X" so as not to confuse the reader with the "R" designation for stereochemistry (e.g., Fig. 1.20).

Amorphous

A term applied to kerogen lacking distinct form or shape as observed in microscopy (Sec. 3.1.2.5). Amorphous kerogen which fluoresces in ultraviolet light shows oil-generative potential. However, highly mature oil-prone kerogen and degraded gas-prone organic matter may appear amorphous, but do not fluoresce and do not act as sources for oil.

Amphipathic

Refers to organic compounds with polar (hydrophilic) and nonpolar (hydrophobic) ends; e.g., cholesterol, bacteriohopanetetrol, phospholipids (Sec. 1.3.1).

Anaerobic

Biofacies resulting from anoxic water column or sediments containing 0 ml oxygen/l water (Tyson and Pearson, 1991) or less than 0.1 ml oxygen/l water (Demaison and Moore, 1980).

Analyte

A compound to be analyzed.

Angiosperms

Flowering or "higher" land plants that originated during early Cretaceous time. Seeds are contained in a covered ovary. Oleanane is a biomarker believed to indicate angiosperm input to oil from source rocks of Cretaceous or younger age (Sec. 3.1.3.1).

Anoxic

Refers to water column or sediments containing less than 0.1 ml oxygen/l water (low Eh or redox potential) (Demaison and Moore, 1980), which is the threshold below which the activity of multicellular deposit feeders is significantly depressed. Between about 0.1 and 0.5 ml oxygen/l, nonburrowing epifauna can still survive above the sediment-water interface, but bioturbation by deposit feeders virtually ceases. Organic matter in sediments below anoxic water stands a better chance of preservation and is commonly more hydrogen-rich, more lipid-rich, and more abundant than under oxic water (Sec. 1.1). See Anaerobic.

API Gravity

An American Petroleum Institute measure of density for petroleum: API gravity = [141.5°/(specific gravity at 16°C) −131.5°]. The unusual form of the equation results in an easily recalled scale for classifying density. Fresh water has a gravity of 10° API. Heavy oils are $<$ 25° API; medium oils are 25° to 35° API; light oils are 35° to 45° API; condensates are $>$ 45° API (ranges vary depending on authors).

Archaebacteria

Primitive organisms containing ether-linked lipids built from phytanyl chains and characterized by their occurrence in extreme environments. They include extreme halophiles (hypersaline environments), thermoacidophiles (hot, acidic), thermophiles (hot), and methanogens (methane generated as a waste product of metabolism). Although generally considered prokaryotes due to their similarities to bacteria, archaebacteria show a number of eukaryotic features indicating they may represent a third distinct group (Woese et al., 1978).

Aromatics

Cyclic, planar organic compounds that are stabilized by pi-electron delocalization (e.g., Fig. 1.6). Includes aromatic hydrocarbons, such as benzene (C_nH_{2n-6}), cycloalkanoaromatics, such as monoaromatic steroids, some heterocyclic aromatic compounds, such as benzothiophenes, and porphyrins. Because of their high solubility, low molecular-weight aromatics like benzene and toluene are readily washed from petroleum by circulating groundwater. Aromatic compounds in petroleum are separated by liquid chromatography using polar solvents into aromatic and polar/porphyrin cuts (Fig. 2.1).

Aromatization Parameter (MA-steroid Aromatization)

A biomarker maturity parameter (Sec. 3.2.4.3) based on the hypothesis of irreversible thermal conversion of C_{29} monoaromatic to C_{28} triaromatic steroids [TA/(MA + TA)].

Asphaltenes

A complex mixture of heavy organic compounds precipitated from oils and bitumens by natural processes or in the laboratory by addition of excess *n*-pentane, *n*-hexane (Chevron), or *n*-heptane. After precipitation of asphaltenes, the remaining oil or bitumen consists of saturates, aromatics, and NSO-compounds.

Asymmetric Carbon (Center)

A carbon atom surrounded by four different substituents (Fig. 1.13), thus having no plane of symmetry.

Bacteriohopanetetrol

A C_{35}-hopanoid containing four hydroxyl groups found in the lipid membranes of prokaryotes (Fig. 1.20). This compound is presumed to be a major precursor for the hopanes in petroleum.

Base peak

The largest peak in the mass spectrum of a compound. Typically all other peaks in the spectrum are normalized to the base peak, which is assigned an intensity of 100 percent. For example, the mass spectrum of cholestane (a C_{27} sterane) shows a base peak at m/z 217 (Fig. 2.18), typical of many steranes.

Biodegradation

Alteration of organic matter or petroleum by the action of bacteria during migration, in the reservoir, or at surface seep locations (Sec. 3.3). Biodegradation of petroleum is limited to low temperatures (not more than about 65–80°C), shallow depths, and conditions where circulating groundwaters with dissolved oxygen are available to the (aerobic) bacteria. Water washing usually accompanies biodegradation. Microorganisms typically degrade petroleum by attacking the less complex, hydrogen-rich compounds first. For example, *n*-paraffins and acyclic isoprenoids are attacked prior to steranes and triterpanes.

Biogenic gas

See Dry gas.

Biological marker (Biomarker, Molecular Fossil)

Complex organic compounds composed of carbon, hydrogen, and other elements which are found in oil, bitumen, rocks, and sediments and show little or no change in structure from their parent organic molecules in living organisms. These compounds are typically analyzed using gas chromatography/mass spectrometry. Most, but not all, biomarkers are isopentenoids, composed of isoprene subunits. Biomarkers include pristane, phytane, steranes, triterpanes, porphyrins, and other compounds.

Biomarker

See Biological marker.

Bitumen

Organic matter extracted from fine-grained rocks using common organic solvents, such as methylene chloride. Unlike oil, bitumen is indigenous to the rock in which it is found (i.e., it has not migrated) (Sec. 1.1). If mistaken for bitumen, migrated oil impregnating a proposed source rock could result in an erroneous oil-source rock correlation.

Bitumen ratio

The ratio of extractable bitumen to total organic carbon (Bit/TOC) in fine-grained, nonreservoir rocks; ratio varies from near zero in shallow sediments to about 250 mg/g TOC at peak generation and decreases because of conversion of bitumen to gas at greater depths (Sec. 3.2.2.1.5). Anomalously high values for the bitumen ratio in immature sediments can be used to help show contamination by migrated oil or man-made products.

Boghead coal

A liptinite-rich coal dominated by algal remains.

Botryococcane

A saturated, irregular isoprenoid biomarker formed from precursors in *Botryococcus braunii*, a fresh/brackish water, colonial Chlorophycean alga found in ancient rocks and still living today (Sec. 3.1.3.1). Concentrated deposits of the alga may result in boghead coals or oil shales.

C_{15}+ Fraction

The fraction of oil or bitumen composed of compounds eluting after the C_{15} *n*-alkane from a gas chromatographic column (boiling point $nC_{15} \sim 271°C$). This fraction is not significantly altered by evaporation or sample preparation and is the most reliable for correlations involving oils and bitumens. For example, after extraction of the bitumen from a rock, removal of the solvent by rotoevaporation results in loss of the $<C_{15}+$ fraction. It is for this reason that oils are commonly "topped" to remove the $<C_{15}+$ fraction so that the remaining material ($C_{15}+$) can be more reliably compared to bitumens.

C_{30}-sterane index

Ratio of parts per thousand C_{30} steranes (24-*n*-propylcholestanes; proposed markers of marine algal input) to total C_{27} to C_{30} steranes (Sec. 3.1.3.1).

C_{35}-homohopane index

The percentage of C_{35} 17α(H)-homohopanes among the total C_{31} to C_{35} homohopanes (Sec. 3.1.3.1). High values indicate selective preservation of the C_{35} homolog caused by low Eh conditions or complexation of bacteriohopanoids with sulfur.

Carbon isotope ratio

See Stable carbon isotope ratio.

Carbon Preference Index (CPI)

Ratio of peak heights or peak areas for odd-to-even-numbered *n*-paraffins in the range nC_{24} to nC_{34} (Bray and Evans, 1961) (Sec. 3.2.2.2). If other ranges of *n*-paraffins are used, the results are usually referred to as odd-even predominance (OEP; Scalan and Smith, 1970).

Carbonate carbon

Carbon present in a rock as carbonate minerals (Sec. 3.1.1.1.1).

Carbonate rock

A sedimentary rock dominated by calcium carbonate ($CaCO_3$); in the text we define carbonate rocks as containing at least 50 wt % carbonate minerals; the consolidated equivalent of limy mud, shell fragments, or calcareous sand.

Carotane

(β-carotane, perhydro-β-carotene) A saturated tetraterpenoid biomarker ($C_{40}H_{78}$) typical of petroleum from saline, lacustrine environments (Sec. 3.1.3.1).

Carrier bed

A permeable rock which acts as a conduit for migrating petroleum.

Catagenesis

The thermal alteration of organic matter by burial and heating in the range of about 50 to 150°C, requiring millions of years; the level of maturity equivalent to the range 0.6 to 2.0 percent vitrinite reflectance including the "oil window" or "principal zone of oil formation" and "wet gas zone" (e.g., Fig. 1.2). The zone where kerogen decomposes to form oil and wet gas.

Chemotrophs

Organisms using redox reactions as an energy source; includes all higher animals, most microorganisms, and nonphotosynthetic plant cells.

Chiral

Many molecules containing asymmetric carbon atoms are chiral, that is, their mirror image structures differ in the spatial relationship between atoms in the same manner as right- and left-handed gloves. In solution, chiral molecules rotate plane polarized light either clockwise or counterclockwise. Certain chiral molecules are not asymmetric (Sec. 1.2.8.1).

Cholesterol

A sterol (steroid alcohol) containing 27 carbon atoms found in the lipid membranes of eukaryotes (Fig. 1.21). This compound is a precursor for the cholestanes in petroleum.

Chromatogram

A graphic representation of separated components as peaks versus time during chromatography (e.g., Figs. 3.1 to 3.4). Each peak on the chromatogram may represent more than one compound. Comparison of petroleum samples is typically accomplished using peak-height or peak-area ratios. However, internal standards allow direct comparisons of concentrations for specific peaks between samples.

Chromatography

Separation of mixtures of compounds based on their physiochemical properties. Liquid chromatography (LC; Sec. 2.4) and gas chromatography (GC; Sec. 2.5.1) are routinely used to separate petroleum components, based on their partitioning between a mobile and stationary phase in a chromatographic column. See also Thin-layer chromatography.

Coal

A rock containing greater than 50 wt % organic matter. Most, but not all coals are of higher plant origin, plot along the Type III (gas-prone) pathway on a van Krevelen diagram, and are dominated by vitrinite group macerals.

Coinjection

Chromatographic technique used to support identifications of unknown compounds (Sec. 2.5.4). A synthesized or isolated standard compound (some are commercially available) is mixed with the sample (a process called spiking) containing the compound to be identified. If the standard and unknown compounds coelute, the relative peak intensity of the unknown compound on chromatograms of the mixture will be higher than that for the neat (unspiked) sample (e.g., Fig. 2.19). Coelution supports, but does not prove that the compounds are identical. Proof of structure might include coelution of the unknown and standard on various chromatographic columns, identical mass spectra, and NMR or X-ray structural verification.

Collision-activated decomposition (CAD)

Dissociation of a projectile ion into fragment ions caused by collision with a target neutral species. For example, in a typical application of triple quadrupole mass spectrometry, parent ions separated in the first quadrupole collide with an inert collision gas (e.g., argon in the Finnigan MAT TSQ-70 instrument) and decompose into daughter ions in the second quadrupole (collision cell).

Condensate

A very light oil with an API gravity greater than 45°. The term is used by engineers to indicate petroleum which is gaseous under reservoir conditions but is liquid at surface pressures and temperatures.

Conformation

The three-dimensional arrangements that an organic molecule can assume by rotating carbon atoms or their substituents around single covalent carbon bonds. Although not fixed, one conformation may be more likely to occur than another. For example, cyclohexane shows the preferred "chair", in addition to "boat" (Fig. 1.9) and "twisted", conformations.

Covalent (Sigma) Bond

A chemical bond where valence electrons are shared equally between the bonding atoms (for example, C—C or C—H bonds). Because of equally shared electrons, covalent bonds are stronger than ionic bonds (as in NaCl), which are characterized by electrical asymmetry.

Cycloalkane (Naphthene, Cycloparaffin)

A saturated, cyclic compound containing only carbon and hydrogen. One of the simplest cycloalkanes is cyclohexane (C_6H_{12}). Steranes and triterpanes are branched naphthenes consisting of multiple condensed five- or six-carbon rings.

Dalton

One atomic mass unit (amu).

Daughter ion

An electrically charged product of a reaction of a particular parent ion.

Decarboxylation

Loss of one or more carboxyl (—COOH) groups from a compound. For example, decarboxylation of acetic acid (CH_3COOH) results in methane (CH_4) and carbon dioxide (CO_2). Decarboxylation of phytanic ($C_{20}H_{40}O_2$) acid results in pristane ($C_{19}H_{40}$).

Dehydration

Loss of one or more molecules of water from a compound. For example, dehydration of ethyl alcohol (CH_3CH_2OH) results in ethylene ($CH_2{=}CH_2$) and water.

Diagenesis

The biological and physiochemical alteration of organic matter and minerals which occurs in sediments after deposition, but prior to significant changes caused by heat; alteration processes occurring at maturity levels up to an equivalent of 0.6 percent vitrinite reflectance (e.g., Fig. 1.2).

Diahopane

A hopane where the methyl group attached to C-14 has been rearranged to the C-15 position.

Diasterane (Rearranged Sterane)

Rearrangement products produced from sterol precursors (e.g., Fig. 1.10).

The rearrangement involves migration of methyl groups and is favored by acidic conditions, clay catalysis, and/or high temperatures.

Diasterane index

The ratio of rearranged (diasteranes) to regular steranes (Sec. 3.1.3.2). Diasterane concentrations in petroleum depend on oxicity and pH of the depositional environment and clay content and thermal maturity in the source rock.

Diastereomer

Stereoisomers that are not related as nonsuperimposable mirror images. A stereoisomer that is not an enantiomer (mirror image). Diastereomers show different physical and chemical properties. An epimer is one type of diastereomer. See Epimer.

Disproportionation

Simultaneous oxidation-reduction reactions among the same molecules that result in both hydrogen-rich and hydrogen-poor products. For example, cracking of a reservoired petroleum results in formation of hydrogen-rich gases and a hydrogen-poor pyrobitumen residue.

Diterpanes

A class of biomarkers constructed from four isoprene subunits (e.g., Fig. 1.10).

Dry gas

Contains at least 98% methane/total hydrocarbons. Originates during diagenesis (microbial gas, sometimes called "biogenic" gas) by the action of methanogenic bacteria (Rice and Claypool, 1981), or during metagenesis.

Dwell time

The time spent by the detector at a given mass during scanning by the mass spectrometer. Longer dwell times result in more accurate quantitation of peaks by increasing signal-to-noise. Dwell times for MID GCMS and full scan GCMS are about 0.04 and < 0.0075 second/ion, respectively.

Dysaerobic

Biofacies resulting from water column or sediments containing between 0.2 and 2.0 ml oxygen/l water (Tyson and Pearson, 1991).

Dysoxic

Water column or sediments containing between 0.2 and 2.0 ml oxygen/l water (Tyson and Pearson, 1991). See Dysaerobic.

Eh

See Redox potential.

Electron impact

A GCMS ionization technique (Sec. 2.5.2.1). Compounds eluting from the GC are introduced by a transfer line into the ion source of the MS where they are bombarded with energetic electrons, typically at 70 eV (electron volts). This causes the compounds to ionize into a series of fragment ions characteristic of the structure of the particular compound. For example, $M + e^- \rightarrow M^{+\cdot} + 2e$-, where M is an eluting compound.

Eluent

A liquid used as the mobile phase to carry components to be separated through a chromatographic column. For example, in liquid chromatographic fractionation of oil or bitumen, hexane can be used as the eluent for the saturate fraction, while methylene chloride can be used for the polar fraction.

Enantiomers

A pair of stereoisomers of a molecule which have the same molecular formulae, but differ in the arrangement of substituents around every asymmetric carbon atom, thus, representing two mirror-image structures (e.g., Fig. 1.13). When only one asymmetric carbon atom is present, the mirror-image structure can only be obtained by transposing (inverting) any two of the four substituents on the asymmetric carbon atom (isomerization). In a molecule containing more than one asymmetric center, inversion of all the centers leads to the enantiomer. A pair of enantiomers represent isomers that are nonsuperimposable mirror images.

Epimer

In a molecule containing more than one asymmetric center, inversion of all the centers leads to the enantiomer (mirror image). Inversion of only one of the asymmetric centers yields an epimer (Sec. 1.2.8.1). Inversion of one or more than one of the asymmetric centers yields a diastereomer.

Eukaryote

A "higher" organism composed of cells or a single cell containing a distinct nucleus and nuclear membrane and other organelles; includes virtually all organisms except the prokaryotes (bacteria and cyanobacteria) and the archaebacteria.

Euphotic zone (Photic zone)

The zone in a water body where light penetrates sufficiently to support photosynthesis. Thickness of the zone varies with water clarity, but averages less than about 100 m.

Euxinic

Refers to anoxic, restricted depositional conditions with free hydrogen sulfide (H_2S). Present day examples include anoxic bottom water and sediments in

Lake Tanganyika, the Black Sea, the coast of Peru, and the northeastern Pacific. Anoxic sediments with organic matter and sulfate become euxinic through the activity of sulfate reducing bacteria.

Expulsion

The process of primary migration where oil or gas leaves the source rock; generally involves short distances (meters to tens of meters).

Extract

Refers to bitumen or oil removed from rocks using organic solvents.

FC-43

A standard compound (perfluorotributylamine) used to calibrate the mass scale of a mass spectrometer.

Fermentation

An anaerobic process by which some prokaryotes and eukaryotes (e.g., yeasts) perform oxidation-reduction reactions without an externally supplied electron acceptor, such as oxygen, to yield energy for growth. Some atoms of the energy source become more reduced while others become more oxidized.

"Fingerprint"

A chromatographic signature used in oil-oil or oil-source rock correlations. Mass chromatograms of steranes or terpanes are examples of fingerprints that can be used for qualitative or quantitative comparison of oils and bitumens.

Fragment ion

An electrically charged product of the fragmentation of a molecule or ion (Sec. 2.5.2.1). Fragment ions can dissociate further to generate other lower molecular weight fragments.

Fragmentogram

See Mass chromatogram.

Functional group

A site of chemical reactivity in a molecule that arises from differences in electronegativity or from a pi bond. For example, alcohol (—OH), thiol (—SH), carboxyl groups (—COOH), and double bonds (C=C) are common functional groups in organic compounds.

Gas

See Dry gas or Wet gas.

Gas Chromatography (GC)

An analytical technique designed to separate compounds (e.g., Fig. 2.6) where a mobile phase (inert carrier gas) passes through a column containing a stationary phase (high molecular-weight liquid). The stationary phase can be coated on the walls of the column or on a solid support packing material. When petroleum in solution with the mobile phase is injected into the GC, individual components move through the column at different rates as they partition between the mobile and stationary phases. The separated components are measured by a detector as they elute from the column.

Gas Chromatography-Mass Spectrometry

See GCMS.

Gasoline-range hydrocarbons

The fraction of crude oil boiling between about 15°C and 200°C, including low molecular-weight compounds, usually containing less than 12 carbon atoms.

Gas-prone

Organic matter that generates dominantly gases with only minor oil during thermal maturation (Sec. 3.1.2.4). Type III kerogen is gas-prone.

GCMS (Gas Chromatography-Mass Spectrometry)

The gas chromatograph separates organic compounds while the mass spectrometer is used as a detector to provide structural information (Fig. 2.5). Biomarker interpretations depend mainly on GCMS analysis.

GC-MSD (Gas Chromatograph-Mass Selective Detector)

A benchtop instrument which provides separation and detection of organic compounds, including biomarkers (Sec. 2.5.3.3).

GCMSMS (Gas chromatography-mass spectrometry-mass spectrometry)

Any method of GCMS analysis that uses the independent focusing capabilities of two or more sectors in a complex mass spectrometer. GCMSMS can be performed using double focusing magnetic, triple quadrupole, or hybrid mass spectrometer systems. See also Metastable reaction.

Gloeocapsamorpha prisca

A microorganism peculiar to Middle Ordovician rocks that appears responsible for the strong odd *n*-paraffin ($<C_{20}$) predominance in oils and bitumens of this age (Fig. 3.1).

Gymnosperms

Mostly conifers, including classes such as Coniferales, Ginkgoales, Bennettitales, and Cycadales. Dominant land-plants after their origin in the middle Paleozoic until the advent of angiosperms during the Cretaceous. Naked seeds are attached to cones.

Halophilic

Refers to organisms which prefer highly saline environments for growth, such as halophilic purple bacteria in certain modern lakes.

H/C Ratio

The atomic hydrogen to carbon ratio, typically used to describe kerogen type on van Krevelen diagrams (Sec. 3.1.2.4).

Heavy oil

Oil of low API gravity (less than about 25° API).

Heterocompounds (NSO-compounds)

A fraction (or individual compounds) separated from petroleum which contains various elements in addition to hydrogen and carbon, including nitrogen, sulfur, and/or oxygen.

Homohopane isomerization

A biomarker maturity ratio [22S/(22S + 22R)] describing the conversion of the biological 22R to the geological 22S configuration of homohopane molecules (Sec. 3.2.4.1). Typically this ratio is calculated for the C_{32} 17α(H)-homohopanes, but other carbon numbers in the range C_{31} to C_{35} are sometimes used.

Homologous Series

Compounds (homologs) showing similar structures, but differing by the number of methylene (-CH_2-) groups. For example, the homologous series of *n*-alkanes can be described by the formula C_nH_{2n+2} where n = 1,2,3,4, etc. (methane, ethane, propane, butane, etc.). Many biomarkers consist of homologous series.

Hopane

Pentacyclic hydrocarbons of the triterpane group believed to be derived primarily from bacteriohopanoids in bacterial membranes.

HPLC (High Performance Liquid Chromatography)

HPLC can be used to separate petroleum into saturate, aromatic, and polar fractions (Sec. 2.4.2). The method is typically automated. Larger amounts of materials can be separated by column chromatography.

Hydrocarbon

A compound containing only hydrogen and carbon. Sometimes this term is used to refer to petroleum.

Hydrogen index (HI)

A Rock-Eval parameter defined as (S2/TOC) × 100 and measured in units mg hydrocarbon/g TOC (Sec. 3.1.2.3). The HI is used in modified van Krevelen plots of HI versus OI to determine organic matter type (Peters, 1986).

Hydrogenation

Any reaction of hydrogen with an organic compound. Typically, hydrogen reacts with double bonds of unsaturated compounds, resulting in a saturated product. For example, hydrogen reacts with ethylene ($CH_2{=}CH_2$; where "=" is the double bond) to produce ethane ($CH_3{-}CH_3$).

Hydrous pyrolysis

A laboratory technique where potential source rocks are heated without air, under pressure, and with water to artificially increase the level of thermal maturity (Sec. 4.3). Oils generated during hydrous pyrolysis have been used for oil-source rock correlation where natural source rock extracts of suitable maturity are not available (Fig. 4.2).

Hypersaline

Describes brines showing salinities greater than about 35 parts per thousand.

Immature

Refers to organic matter subjected to insufficient temperature (or too short duration) for thermal generation of petroleum, that is, vitrinite reflectance less than about 0.6 percent. See Diagenesis.

Internal standard

A compound showing similar behavior to the unknown compounds in a mixture to be analyzed, but which is absent in the original mixture. Addition of the internal standard to the mixture allows more reliable quantitation of the unknown compounds. For example, 5β(H)-cholane is added to the saturate fractions of oils as an internal standard (Sec. 2.4.3). This compound is not found in significant amounts in natural oils, but its mass spectral characteristics are similar to steranes and other compounds in saturate fractions.

Inertinite

A maceral group composed of inert, hydrogen-poor organic matter with little or no petroleum generative potential. Type IV kerogen is dominated by inertinite.

Inspissation

"Drying up." For example, inspissation of seep oils can result in loss of gases and the lighter oil fractions due to reduced pressure (compared to that in the reservoir), evaporation and drying by exposure to sunlight, oxidation, and other related processes.

Isomers

Compounds with the same molecular formulae, but different arrangements of their structural groups (for example, *n*-butane and isobutane).

Isomerization (Configurational isomerization, Stereoisomerization)

Any rearrangement of the atoms in a molecule to form a different structure. During stereoisomerization, a hydride or hydrogen radical is believed to be removed from an asymmetric carbon atom, resulting in formation of a planar carbocation or radical intermediate, followed by reattachment of the hydride or hydrogen radical. Reattachment can occur on the same side of the planar intermediate as removal (thus, resulting in no change in configuration) or on the opposite side (resulting in an inverted or "mirror-image" configuration). Stereoisomerization only occurs when cleavage and renewed formation of the bonds results in an inverted configuration compared to the starting asymmetric center (Sec. 1.2.9).

Isopentenoids

Compounds composed of isoprene subunits (see Fig. 1.10). Most biomarkers are isopentenoids. Some biomarkers which are not isopentenoids include certain normal, iso- and anteisoalkanes.

Isoprene (Isopentadiene)

The basic structural unit composed of five carbon atoms found in many biomarkers, including mono-, sesqui-, di-, sester-, and triterpanes, steranes, and polyterpanes (Fig. 1.10).

Isoprenoid

A hydrocarbon composed of, or derived from, polymerized isoprene units. Typical acyclic isoprenoids include pristane (C_{19}) and phytane (C_{20}).

Isotopes

Atoms whose nuclei contain the same number of protons, but different numbers of neutrons (Sec. 3.1.2.2). For example, all carbon atoms have six protons, but there are isotopes containing 6, 7, and 8 neutrons resulting in atomic masses 12, 13, and 14. ^{12}C is the principal naturally-occuring isotope (98.89%). ^{13}C is the stable carbon isotope (1.11%), and ^{14}C is the unstable (radioactive) carbon isotope ($\sim 1 \times 10^{-11}$%).

Kerogen

Insoluble (in organic solvents) particulate organic matter comprised of various macerals. Kerogen originates from components of plants, animals, and bacteria that are preserved in sedimentary rocks. Kerogen can be isolated from rock by extracting bitumen with solvents and removing most of the rock matrix with hydrochloric and hydrofluoric acids.

Kerogen Type I, II, III, IV

See Type.

Lacustrine

Refers to lakes. For example, "lacustrine oils" were generated from organic matter deposited in a lake.

Light oil

Oil with a gravity of 35° to 45° API.

Limestone

See Carbonate rock.

Linked-scan mode

A GCMS mode in which two or more quadrupole, electrostatic, and/or magnetic fields are scanned simultaneously, thus, allowing detection of specific parent, daughter, or neutral loss relationships between ions.

Lipids

Oil-soluble, water-insoluble organic matter including fats, waxes, pigments, sterols, and hopanoids. Lipids are the precursors for petroleum (Silverman, 1971).

Liptinite

A maceral group composed of oil-prone, fluorescent, hydrogen-rich kerogen; includes: (1) structured liptinites, such as resinite, sporinite, or cutinite, and (2) unstructured amorphous liptinites, sometimes called amorphinite (Sec. 3.1.2.5).

Liquid chromatography ("LC")

A preparative column method used to separate oils and bitumens into saturate/aromatic and porphyrin/polar fractions prior to HPLC (Sec. 2.4.1).

MA steroid

See Monoaromatic steroid.

Maceral

Microscopically recognizable, particulate organic component of kerogen showing distinctive physiochemical properties which change with thermal maturity (Sec. 3.1.2.5). The three main maceral groups include liptinite, vitrinite, and inertinite.

Magnetic sector

A magnetic (versus electric or quadrupole) mass analyzer in mass spectrometers. Ions follow a curved path down the flight tube in the magnetic field (e.g., Fig. 2.8). The degree of curvature is related to the mass and velocity of the ion. Only those ions with a given m/z value reach the detector for a given magnetic field strength or accelerating voltage. All other ions collide with the walls of the flight tube.

Marl

A sedimentary rock containing calcareous clay formed under marine or especially freshwater conditions.

Mass chromatogram (Fragmentogram)

Intensity of a specific ion versus gas chromatographic retention time (e.g., Figs. 2.12, 2.13, and 2.14). Allows identification of carbon number and isomer distributions for selected compound types.

Mass spectrometry (MS)

A method used to supply information on the molecular structure of compounds, particularly biomarkers (Sec. 2.5.2). Molecules in the gaseous state (inserted directly into the mass spectrometer or eluting from a GC after separation) are ionized (usually by high-energy electrons). The resulting molecular and fragment ions are detected and displayed on the basis of increasing mass to charge ratio (m/z in a mass spectrum).

Mass spectrum

A plot of the relative intensities of ions, formed during bombardment of molecules by electrons, versus mass-to-charge (m/z) ratio. Mass spectra can be used for provisional compound identification because they represent a "fingerprint" which is often diagnostic of specific structures (Sec. 2.5.4).

Mature

Refers to organic matter in the oil window, that is, at a thermal maturity equivalent to vitrinite reflectance in the approximate range 0.6 to 1.4 percent (e.g., Fig. 3.46).

Maturity

The extent of heat driven reactions which convert sedimentary organic matter to petroleum and finally to gas and graphite. Different geochemical scales including vitrinite reflectance, pyrolysis T_{max}, and various biomarker maturity ratios, are used to indicate the level of thermal maturity of organic matter.

Metagenesis

The thermal destruction of organic molecules by cracking to gas which occurs after catagenesis, but prior to greenschist metamorphism (>200°C) in the range of about 150–200°C (e.g., Fig. 1.2); the level of maturity equivalent to the range 2.0–4.0 percent vitrinite reflectance.

Metastable ion

An ion that has been accelerated from the ion source and decomposes in one of the field free regions of the mass spectrometer, producing a broad, diffuse peak in the mass spectrum.

Metastable reaction (Metastable transition)

Decomposition of a metastable ion. In metastable reaction monitoring (MRM GCMS, also called MRM GCMSMS) parent mode, all or selected metastable ions (usually molecular ions) in the mass spectrometer decomposing to a single daughter ion are monitored using a linked-scan method (see "linked scan"). A daughter mode is also possible, that is, monitoring daughter ions from a given parent ion.

Methanogenesis

Generation of methane by bacterial fermentation; occurs only under anaerobic conditions where little sulfate is available in the water column or sediments (Sec. 3.1.3.1, discussion of acyclic isoprenoids).

Microscopic organic analysis (MOA)

Petrographic analysis of kerogen or rock in transmitted and/or reflected light. General use of the term includes both maceral (relative percentages of phytoclast types) and thermal maturity (vitrinite reflectance, TAI) analysis. MOA provides critical support for interpretations based on elemental analysis (Jones and Edison, 1978) and biomarkers.

MID (Multiple Ion Detection)

See Selected ion monitoring.

Mobile phase

See Gas chromatography.

Molecular fossil

See Biological marker.

Molecular ion (M^+ or $M^{\dot{+}}$)

An ion formed by the removal of one or more electrons from a molecule without fragmentation (Sec. 2.5.2.1). The mass of the molecular ion corresponds to the molecular weight of the parent compound (i.e., the sum of the masses of the most abundant isotopes of the atoms comprising the molecule, corrected for the gain or loss of electrons). For example, the mass of the positively charged molecular ion of methane is 1×12 plus 4×1.0078246 minus the mass of the electron. Most biomarker measurements are made at nominal or unit mass resolution which is reported as an integer representing the sum of the most abundant isotopes of the atoms in the molecule.

Monoaromatic (MA) Steroid

A class of biomarkers which contain one aromatic ring (in crude oils this is usually the C-ring; e.g., Fig. 3.39), probably derived from sterols.

Monoaromatic (MA) steroid triangle

Plot of C_{27}, C_{28}, and C_{29} MA-steroids on a ternary diagram (e.g., Fig. 3.38) used for correlation similar to the sterane triangle (e.g., Fig. 3.31). However, MA-steroids are probably derived from different sterol precursors than the steranes.

Monoterpane

A class of saturated biomarkers constructed from two isoprene subunits ($\sim C_{10}$) (Fig. 1.10).

Moretanes

$17\beta,21\alpha$(H)-pentacyclic triterpanes related to the hopanes (Sec. 3.2.4.1).

MRM

See Metastable reaction.

m/z

The mass-to-charge ratio of an ion in mass spectrometry measured in units of daltons per charge, with positive or negative values denoting cations or anions, respectively; that is, m/z 217 is a characteristic fragment of steranes (e.g., Fig. 2.11). Old papers may use the term "m/e" instead of "m/z." Incorrect use of "mass" and "daltons" as synonyms for m/z is especially confusing when applied to multiple-charged ions.

Multiple ion detection (MID)

See Selected ion monitoring.

Naphthenes

See Cycloalkanes.

NBS-22

A Pennsylvania No. 30 lubricating oil supplied by the National Bureau of Standards as one international standard for stable carbon isotope ratios (Sec. 3.1.2.2.1).

Neohopane

A hopane that has a methyl group at C-18 rearranged to the C-17 position (i.e., hopane II). Ts is the C_{27}-neohopane.

NMR

See Nuclear magnetic resonance.

NSO-compounds (resins)

Nitrogen-, sulfur-, and oxygen-containing compounds (nonhydrocarbons) which represent a pentane-soluble fraction separated from oils. Other fractions include saturates, aromatics, and asphaltenes.

Nuclear magnetic resonance (NMR)

"Proton NMR" (^{1}H NMR) is a spectroscopic method for determining molecular structure. The high resolution of the method commonly distinguishes stereoisomers that cannot be distinguished by mass spectroscopy. "^{13}C NMR" requires more sample than proton NMR and can identify the types of carbon atoms (e.g., aromatic versus saturate). New ^{13}C NMR methods allow determination of numbers of hydrogen atoms bound to individual carbon atoms (i.e., methyl, methylene, methine, or quaternary). New two-dimensional NMR techniques allow correlation of ^{13}C and proton NMR data and provide improved resolution and in some cases can rival X-ray diffraction crystallography in structural elucidation (Croasmun and Carlson, 1987).

O/C

The atomic oxygen to carbon ratio from elemental analysis of kerogen, typically used to describe kerogen type on van Krevelen diagrams. Because oxygen is not readily measured on kerogens, atomic O/C is not always used directly on van Krevelen plots (Jones and Edison, 1978) (Sec. 3.1.2.4).

Odd-even predominance

The ratio of odd- to even-numbered *n*-alkanes in a given range (Scalan and Smith, 1970) (Sec. 3.2.2.2).

Oil-prone

Organic matter that generates significant quantities of oil at optimal maturity. For example, Type I and II kerogens are highly oil-prone and oil-prone, respectively, (Sec. 3.1.2.4). Oil-prone kerogen is typically also more gas-prone than "gas-prone" kerogen. See Oil window.

Oil window

The maturity range in which oil is generated from oil-prone organic matter (~0.6–1.4% vitrinite reflectance), that is, within the catagenesis zone (~0.6–2.0% vitrinite reflectance) (e.g., Fig. 3.46).

Oleanane index

The ratio of oleanane (an angiosperm marker) to 17α(H)-hopane (a bacterial marker). Oleanane appears restricted to Cretaceous or younger rocks, but its absence cannot be used to prove age (Sec. 3.1.3.1).

Olefin

See Alkene.

Optical activity

The property of some compounds to rotate plane polarized light (Sec. 1.2.8.2). Biomarkers and other organic materials contain "asymmetric" carbon atoms resulting in left- or right-handed molecules (enantiomers). Enzymes in living organisms tend to produce one or the other of these optically active compounds. Solutions of these compounds rotate plane polarized light to varying degrees. Solutions of the same compounds produced without enzymes or living organisms are not optically active. See Chiral.

Organic facies

A mappable rock unit containing a distinctive assemblage of organic matter without regard to the mineralogy (after Jones, 1987).

Organic matter (Biomass)

Biogenic, carbonaceous materials. Organic matter preserved in rocks includes kerogen, bitumen, oil, and gas. Different types of organic matter can have different oil-generative potential.

Organic yield

A crude estimate (TOC is more accurate) of the amount of organic matter in a rock based on the volume of organic material which survives demineralization of the rock with acids during preparation of kerogen.

Overmature

See Postmature.

Oxic

Oxygenated water column or sediments containing more than 2.0 ml oxygen/l water (Tyson and Pearson, 1991). See Aerobic.

Oxidation

The process of (1) loss of one or more electrons by a compound, or (2) addition of oxygen to a compound, or (3) loss of hydrogen (dehydrogenation). Oxidation and reduction reactions are always coupled (redox reactions). For example, phytol (an alcohol) can be oxidized to phytanic acid.

Oxygen index

A Rock-Eval parameter defined as $(S3/TOC) \times 100$ and measured in mg carbon dioxide/g TOC. The OI is used on modified van Krevelen plots of HI versus OI to describe organic matter type (Peters, 1986).

Palynomorph

Organic-walled, acid-resistant microfossils useful in providing information on age, paleoenvironment, and thermal maturity (e.g., TAI).

Paraffin

See Alkane.

PDB

See Peedee belemnite.

Peedee belemnite (PDB)

An international primary stable isotope standard for carbon obtained from the carbonate fossil of a cephalopod, *Belemnitella americana* in the Cretaceous age Peedee Formation. Measurements are given in parts per thousand (per mil) relative to the standard using the standard "delta" notation. On the PDB scale a secondary standard, NBS-22 oil, measures about −29.81 per mil. Accurate conversion of PDB to NBS-22 values is not straightforward, especially for gases, and requires a correction (Sec. 3.1.2.2).

Petroleum

Mixture of organic compounds composed dominantly of hydrogen and carbon and found in the gaseous, liquid, or solid state in the earth; includes hydrocarbon gases, bitumen, migrated oil, and pyrobitumen, but not kerogen. In European usage, the term is sometimes restricted to refined products only.

pH

A measure of the tendency of an environment to supply protons (hydrogen ions) to a base or to take up protons from an acid (negative logarithm of the hydrogen ion concentration). A pH of 7.0 under standard conditions is "neutral" while lower and higher values are "acidic" and "basic," respectively. In nature, most pH values lie in the range 4 to 9.

Photosynthesis

The process where solar energy is captured and converted into chemical energy by phototrophic organisms (Sec. 1.1); requires chlorophyll and other light-trapping pigments, such as carotenoids or phycobilins; photosynthetic eukaryotes include higher green plants, multicellular green, brown, and red algae, and unicellular organisms, such as dinoflagellates and diatoms; photosynthetic prokaryotes include cyanobacteria, green bacteria, and purple bacteria.

Phototrophs

Organisms using light as an energy source (as opposed to chemotrophs, which use redox reactions for their energy); chlorophyll and other light-trapping pigments (such as carotenoids or phycobilins) are used by phototrophs in photosynthesis. Phototrophs include green cells of higher plants, cyanobacteria, photosynthetic bacteria, and nonsulfur purple bacteria.

Phytane (Ph)

A branched acyclic (no rings) isoprenoid hydrocarbon containing 20 carbon atoms which is a prominent peak eluting immediately after the C_{18} *n*-paraffin in petroleum on most GC columns (e.g., Fig. 2.7).

Phytoclast

An identifiable particle or maceral in kerogen, for example, phytoclasts of vitrinite are used for measurement of vitrinite reflectance.

Phytol

A branched acyclic (no rings) isoprenoid alcohol containing 20 carbon atoms (e.g., Fig. 1.14).

PMP

See Porphyrin maturity parameter.

PNA

See Polynuclear aromatics.

Polar compound

An organic compound with distinct regions of partial positive and negative charge. Polar compounds include alcohols, such as sterols, and aromatics, such as MA-steroids. Because of their polarity, these compounds are more soluble in polar solvents, including water, compared to nonpolar compounds of similar molecular weight (Sec. 2.4.1).

Polymer

A macromolecule composed of a large number of *repeating* subunits (monomers). Biopolymers include proteins (composed of amino acids) and polysaccharides (composed of sugars). Rubber is a polyunsaturated polymer composed of isoprene $[(C_5H_8)_n]$ subunits (Fig. 1.10). Kerogen is *not* a polymer.

Polynuclear aromatics

Molecules containing more than two aromatic rings.

Polyterpanes

A class of saturated biomarkers constructed of more than eight isoprene subunits ($\sim C_{40}+$).

Porphyrin maturity parameter (PMP)

A biomarker maturity parameter for petroleum based on generation (Sec. 3.2.4.5). PMP = C_{28} ETIO/(C_{28} ETIO + C_{32} DPEP). See Fig. 3.61.

Porphyrins

Complex biomarkers characterized by a tetrapyrrole ring, usually containing vanadium or nickel (e.g., Fig. 1.27), which are derived from various sources, including chlorophyll and heme.

Postmature

A high level of maturity where no further oil generation can occur, that is, vitrinite reflectance >1.4 percent.

Potential source rock

An oil-prone or gas-prone, organic-rich rock that has not yet generated oil or gas because of low thermal maturity.

Primary migration

See Expulsion.

Pristane (Pr)

A branched acyclic (no rings) isoprenoid hydrocarbon containing 19 carbon atoms which is a prominent peak eluting immediately after the C_{17} *n*-paraffin in petroleum on most GC columns (e.g., Fig. 2.7).

Production index (PI)

A Rock-Eval parameter (Sec. 3.2.2.1.1) useful in describing thermal maturity of source rocks or indicating contamination; PI = S1/(S1 + S2) (Peters, 1986). Also called "transformation ratio" (Tissot and Welte, 1984).

Prokaryotes

Primitive organisms composed of cells containing no distinct nucleus or other organelles; includes the bacteria and cyanobacteria.

Pseudohomologous series

Compounds showing similar structures, but differing in the length of a branched alkyl group. For example, the tricyclic terpanes (cheilanthanes) represent a pseudohomologous series.

Pyrobitumen

Thermally altered, solidified bitumen that is insoluble in common organic solvents.

Pyrogram

A plot showing detector response to products generated by pyrolysis (e.g., Fig. 3.43); for the Rock-Eval pyroanalyzer a pyrogram shows S1, S2, S3, T_{max}, the programmed temperature trace, and other information (Peters, 1986).

Pyrolysis

Breakdown of organic matter during heating in the absence of oxygen; as in Rock-Eval pyrolysis (for source rock evaluation), or hydrous pyrolysis (for simulating oil generation in source rocks). Rock-Eval employs "programmed temperature pyrolysis" because the temperature is programmed to increase at a selected rate during analysis. Hydrous pyrolysis typically employs a constant temperature for each experiment.

Quadrupole rods

An electrical (versus magnetic) mass analyzer in mass spectrometers (e.g., Fig. 2.9). Using a combination of DC (direct current) and RF (radio frequency) fields on the quadrupole rods serves as a "mass filter" for ions. Only ions of a given m/z value reach the detector for a given magnitude of the DC and RF fields. All other ions collide with the rods before reaching the detector.

Racemic mixture

A 50 : 50 mixture of enantiomers in solution resulting in no optical activity (Sec. 1.2.8.2). Racemic mixtures of enantiomers are typical of the nonbiologic (abiotic) synthesis of molecules containing asymmetric centers. Many biologically formed compounds are optically active.

Rearranged hopane

See Diahopane or Neohopane.

Rearranged sterane

See Diasterane.

Reconstructed ion chromatogram (RIC or Total Ion Chromatogram)

The magnitude of the total ion current during a GCMS analysis plotted versus scan or retention time. RIC and GC traces of petroleum samples are nearly identical.

Redox potential (Oxidoreduction potential)

A measure of the ability of an environment to supply electrons to an oxidizing agent, or to take up electrons from a reducing agent. In redox reactions there is transfer of electrons from an electron donor (the reducing agent or reductant) to an electron acceptor (the oxidizing agent or oxidant); expressed as the tendency of a reducing agent to lose electrons relative to the "standard reduction potential." Standard reduction potential is the electromotive force in volts given by a half-cell in which the reductant and oxidant are both present at 1.0 M concentration, 25°C, and pH 7.0, in equilibrium with an electrode which can reversibly accept electrons from the reductant species. The standard of reference is the reduction potential of the reaction: $H_2 = 2H^+ + 2e^-$, which is set at 0.0 volts under conditions in which the pressure of H_2 gas is 1.0 atm., $[H^+]$ is 1.0 M, pH is 7.0, and temperature is 25°C. Systems having a more negative standard reduction potential than the H_2-$2H^+$ couple have a greater tendency to lose electrons than hydrogen and vice versa. Most redox potentials of seawater, for example, lie in the range between +0.3 V for aerated water to −0.6 V for oxygen depleted (anoxic) bottom water.

Reduction

The process of (1) acceptance of one or more electrons by a compound, (2) removal of oxygen from a compound, or (3) addition of hydrogen (hydrogenation). Reduction and oxidation always occur as coupled reactions (redox reactions). For example, alkenes can be reduced to alkanes by hydrogenation.

Reflectance

See Vitrinite reflectance.

Reservoir rock

Any porous and permeable rock that contains petroleum; typically sandstones or carbonates.

Resinite

A maceral of the liptinite group derived from plant resins, for example, amber.

Resolution

(1) In gas chromatography the term refers to the efficiency with which adjacent peaks on the chromatogram are separated. "Baseline" separation indicates that two adjacent peaks are completely resolved (separated) and that the valley between the peaks reaches background (low) levels. (2) In mass spectrometry the term refers to the ratio of mass to the difference between two adjacent masses ($M/\Delta M$) that a mass spectrometer can just separate completely. Low and high resolution are about 1000 and >2000, respectively (no units).

Respiration

The process where stored chemical energy, such as that in carbohydrate, is released by oxidation within the living cell (opposite of photosynthesis).

Retention time

The time required for a compound to pass through a chromatographic column. An unknown compound can be provisionally identified if it shows the same retention time as a standard compound when injected into the same chromatographic column under the same conditions.

Rock-Eval

A commercially available pyrolysis instrument used as a rapid "screening" tool in evaluating the quantity, quality, and thermal maturity of rock samples (Sec. 3.1.2.3).

S1

A Rock-Eval pyrolysis parameter which is a measure of volatile organic compounds (mg HC/g TOC).

S2

A Rock-Eval pyrolysis parameter which is a measure of organic compounds generated by cracking of the kerogen (mg HC/g TOC).

S3

A Rock-Eval pyrolysis parameter which is a measure of organic carbon dioxide generated from the kerogen up to 390°C (mg CO_2/g TOC).

Sabkha

An evaporitic environment of sedimentation formed under arid to semiarid conditions, usually on restricted coastal plains; characterized by evaporite-salt, tidal-flood, and aeolian deposits along many modern coastlines; for example, Persian Gulf and Gulf of California.

Saturates (Saturate fraction)

Nonaromatic hydrocarbons in petroleum (Sec. 2.4.1). Includes normal and branched alkanes (paraffins), and cycloalkanes (naphthenes).

Scan mode

A GCMS operating mode where the detector records the entire mass range per unit of time (for example, 50-600 amu/3 seconds), resulting in a spectrum of masses for each analyzed peak (Sec. 2.5.3.2).

Screening

Rejection of inappropriate samples using rapid, inexpensive analyses to allow high-grading of other samples for more detailed analysis (Sec. 3.1.1.1.1). Large numbers of potential source rocks can be screened using Rock-Eval pyrolysis and total organic carbon analysis prior to further study. Similarly, benchtop GCMS in MID mode (Sec. 2.5.3.3) can be used to screen petroleum samples for their general biomarker composition prior to more detailed analysis, such as GCMSMS.

Secondary migration

Migration of petroleum along faults, unconformities, or through permeable rocks (carrier beds) after expulsion from the active source rock; unlike primary migration (expulsion) secondary migration generally involves long distances from tens of meters to hundreds of kilometers.

Selected ion monitoring (SIM; Multiple Ion Detection or MID)

Mass spectrometric monitoring of a specific mass/charge (m/z) ratio or limited number of ratios. For example, selected ion monitoring for m/z 217 results in a mass chromatogram dominated by steranes. Using the SIM method of monitoring one or a few masses results in better sensitivity than can be obtained using the full scan mode (Sec. 2.5.3).

Sesquiterpanes

A class of saturated biomarkers constructed from three isoprene subunits ($\sim C_{15}$) (Fig. 1.10).

Sesterterpanes

A class of saturated biomarkers constructed from five isoprene subunits ($\sim C_{25}$) (Fig. 1.10).

Shale

A sedimentary rock, commonly laminated, that is dominated by clay-size particles and shows fissility approximately parallel to the bedding. In this text, shales, calcareous shales, and limestones contain <25%, 25–50%, and >50 wt % carbonate, respectively.

SIM

See Selected ion monitoring.

Source rock

Fine-grained, organic-rich rock that could (potential source rock) or has already generated (effective or active source rock) trapped accumulations of petroleum. An effective source rock must satisfy requirements as to quantity, quality, and thermal maturity of organic matter (Peters and Cassa, 1992).

Spiking

See Coinjection.

Stable carbon isotope ratio

Relative amount of ^{13}C versus ^{12}C (nonradioactive isotopes) in organic matter; generally used to show relationships between oils or oils and source rocks (Sec. 3.1.2.2.1).

Stationary phase

See Gas chromatography.

Steranes

A class of tetracyclic, saturated biomarkers constructed from six isoprene subunits ($\sim C_{30}$) (Fig. 1.10). Steranes are derived from sterols, which are important membrane and hormone components in eukaryotic organisms (Sec. 1.2.10.3). Most commonly used steranes are in the range C_{26} to C_{30} and are detected using m/z 217 mass chromatograms.

Sterane isomerization

Refers to stereochemical conversions between the biological and geological configuration at several asymmetric centers which are used as indicators of thermal maturity; includes 20S/(20S + 20R) and $\beta\beta/(\beta\beta + \alpha\alpha)$ parameters (Secs. 1.2.9 and 3.2.4.2).

Sterane triangle

Plot of $\%C_{27}$, C_{28}, and C_{29} steranes on a ternary diagram used for correlation of oils and bitumens (Fig. 3.31). Relative location on plot can be used to infer organic matter input. For example, abundant $\%C_{29}$ usually, though not always, indicates major input from higher plant sterols.

Stereochemistry

The three-dimensional relationship between atoms in molecules (Sec. 1.2.8).

Stereoisomers

Molecules showing identical molecular formulae and the same linkage between atoms, but differing in the spatial arrangement of the atoms typically around an asymmetric carbon atom. Stereoisomers include enantiomers (mirror image structures) and diastereomers which differ at certain asymmetric centers, but are identical at others (Sec. 1.2.8).

Steroid

Same definition as "sterane" except that other elements in addition to carbon and hydrogen may be present, especially oxygen in alcohol groups. Certain structural elements may be rearranged or missing compared to sterols. For example, unlike sterols (Fig. 1.22) triaromatic steroid hydrocarbons lack an α-methyl group at C-10. Further, the methyl group at C-13 in sterols is rearranged to C-17 in triaromatic steroids (Fig. 3.41).

Stoichiometry

A branch of science that deals with the laws of definite proportions and conservation of matter during chemical reactions, that is, the predictable quantities of substances that enter into, and are produced by, chemical reactions. For example, carbon in saturated organic compounds is always bound to four substituents.

Structural isomer

Molecules showing the same molecular formulae, but differing in the linkage between atoms, for example, *n*-pentane and 2-methylbutane. Stereoisomers are a special form of structural isomer.

Suboxic

Water column or sediments containing 0 to 0.2 ml oxygen/l water. Biofacies resulting from these concentrations of oxygen are called quasi-anaerobic (Tyson and Pearson, 1991).

TAI

See Thermal alteration index.

Terpanes

A broad class of complex branched, cyclic alkane biomarkers including hopanes and tricyclic compounds (Sec. 1.2.10) commonly monitored using m/z 191 mass chromatograms.

Terpenoids (Isopentenoids)

A broad class of complex branched and cyclic biomarkers composed of isoprene (C_5) subunits, including mono- (two isoprene units), sesqui- (three), di- (four), sester- (five), tri- (six), tetraterpenoids (eight), and higher terpenoids. Terpenoids include the saturated terpanes (hydrocarbons) and compounds which may contain double bonds or other elements (in addition to carbon and hydrogen), such as oxygen. Certain structural units may be rearranged, missing, or added compared to terpenes and terpanes.

Tetrapyrrole pigments

Compounds required for photosynthesis that contain a macrocyclic nucleus composed of four linked pyrrole (nitrogen-containing) rings; that is, chlorophylls. Porphyrins contain a tetrapyrrole nucleus and are degradation products of these pigments.

Tetraterpanes

A class of saturated biomarkers constructed from eight isoprene subunits ($\sim C_{40}$) (Fig. 1.10).

Thermal alteration index

Various maturity scales based on changes in the color of spores and pollen from yellow to brown to black with thermal maturity. Although the TAI scale of Jones and Edison (1978) does not correspond to that of Staplin (1969), it was designed to correlate linearly with vitrinite reflectance. For example, 0.4 percent and 0.7 percent vitrinite reflectance are 2.4 and 2.7, respectively, on the Jones and Edison TAI scale (Sec. 3.2.2.1).

Thermal maturity

See Maturity.

Thermochemical sulfate reduction

A biological reduction of sulfate to sulfides in reservoirs below 200°C, where light hydrocarbons or other organic matter serve as reducing agents. This process may explain large accumulations of hydrogen sulfide (H_2S) in deep carbonate reservoirs and the apparent oxidation of light hydrocarbons to carbon dioxide (e.g., Krouse et al., 1988).

Thin-layer chromatography (TLC)

Separation of organic compounds using a thin layer of stationary phase coated on a glass plate. The sample is placed near the bottom edge of the plate which is placed vertically in a tray of solvent (mobile phase). Upward movement by capillary action results in partitioning of components between the mobile and stationary phases. Because it is time-intensive, TLC has been replaced by column chromatography and HPLC at most laboratories.

TLC

See Thin-Layer Chromatography.

T_{max}

A Rock-Eval pyrolysis thermal maturity parameter (Sec. 3.1.2.3) based on the temperature at which the maximum amount of pyrolyzate (S2) is generated from the kerogen in a rock sample. The beginning and end of the oil-generative window approximately correspond to T_{max} values of 435°C and 470°C, respectively (Peters, 1986).

Total ion chromatogram

See Reconstructed ion chromatogram.

Total Organic Carbon (TOC)

The quantity of organic carbon (excluding carbonate carbon) expressed as weight percent of the rock (Sec. 3.1.2.3). For rocks at a thermal maturity equivalent to vitrinite reflectance of 0.6% (beginning of oil window), Peters (1986) ranked TOC quantities as follows: Poor (TOC <0.5 wt %); Fair (0.5–1 wt %), Good (1–2 wt %), Very Good (>2 wt %). TOC decreases with maturity.

TOC

See Total organic carbon.

Triaromatic (TA-) steroids

A class of biomarkers containing three fused aromatic and one 5-membered naphthene rings (naphthenophenanthrenes); probably derived from monoaromatic steroids during maturation (e.g., Fig. 3.41, 3.57).

Triterpanes

A class of saturated biomarkers constructed from six isoprene subunits ($\sim C_{30}$) (Fig. 1.10).

Type I kerogen

Highly oil-prone organic matter showing Rock-Eval pyrolysis hydrogen indices over 600 mg hydrocarbon/g TOC when thermally immature; contains algal and bacterial input dominated by amorphous liptinite macerals, typically in lacustrine settings; Type I pathway on van Krevelen diagram (Sec. 3.1.2.4.1).

Type II kerogen

Oil-prone organic matter showing Rock-Eval pyrolysis hydrogen indices in the range 300–600 mg hydrocarbon/g TOC when thermally immature; contains algal, and bacterial organic matter dominated by liptinite macerals, such as exinite and sporinite; typically in marine settings; Type II pathway on van Krevelen diagram (Sec. 3.1.2.4.1).

Type III kerogen

Gas-prone organic matter showing Rock-Eval pyrolysis hydrogen indices in the range 50–200 mg hydrocarbon/g TOC when thermally immature; contains higher plant organic matter dominated by vitrinite macerals; typically in paralic marine settings; Type III pathway on van Krevelen diagram (Sec. 3.1.2.4.1).

Type IV kerogen

Inert organic matter showing Rock-Eval pyrolysis hydrogen indices below 50 mg hydrocarbon/g TOC in immature rocks; contains organic matter which has been recycled or extensively oxidized during deposition. Sometimes the Type IV pathway is not shown on van Krevelen diagrams, but it lies below the Type III pathway (Sec. 3.1.2.4.1).

Unsaturated

Refers to compounds containing one or more double or triple bonds, such as olefins (alkenes) (Sec. 1.2.2.2). Saturation of the double bond in the olefin called ethylene (C_2H_4) results in ethane (C_2H_6). Most unsaturated organic compounds are unstable in natural petroleum, except for aromatics like benzene (Sec. 1.2.3).

Urea Adduction

A procedure used to separate *n*-paraffins from other organic compounds in the saturate fraction of petroleum. Urea adduction is used to concentrate biomarkers for more reliable GCMS analysis. The method is typically applied to samples with low biomarkers, such as extremely waxy oils or condensates (e.g., Fig. 3.15).

Valence

The combining capacity of an element for other elements based on the number of electrons available for chemical bonding. For example, carbon typically shows a valence of four, indicating that it will form four covalent bonds with other elements.

Van Krevelen diagram

A plot of atomic H/C versus O/C (Section 3.1.2.4.1) originally used to characterize the compositions of coals, but now also used to describe different types of kerogen (see Type I, II, III, and IV) in rocks. Modified van Krevelen plots can be made using HI versus OI from Rock-Eval pyrolysis (Peters, 1986).

Vitrinite

A group of gas-prone macerals (Sec. 3.1.2.5) derived from land plant tissues. Particles of vitrinite are used for vitrinite reflectance (R_o) determinations of thermal maturity.

Vitrinite reflectance (R_o)

A maturation parameter for organic matter in fine-grained rocks (Sec. 3.2.2.1). Quoted reflectance values for individual rock samples are typically based on the average reflectance for at least 50–100 vitrinite phytoclasts (unless otherwise specified) in a polished kerogen slide. The value represents the percent of incident light (546 nm) reflected from the phytoclasts under an oil immersion microscope objective as measured by a photometer.

V/(V + Ni) Porphyrin ratio

This ratio of vanadyl- to nickel-porphyrins is used for correlation of oils and bitumens. The ratio is higher for oils and bitumens derived from marine source rocks deposited under anoxic compared to oxic or suboxic conditions (Sec. 3.1.3.4). Low ratios may indicate oxic to suboxic or lacustrine deposition. The ratio is sometimes expressed as Ni/(Ni + V).

Water washing

A process where formation or meteoric water removes light hydrocarbons, aromatics, and other soluble compounds from petroleum in the reservoir or during migration. Biodegradation commonly accompanies water washing of petroleum because bacteria can be introduced from the water.

Weathering

Processes related to the chemical action of air, water, and organisms. Weathering of seep oils or improperly sealed oil samples by subaerial exposure results in evaporative loss of light hydrocarbons. Weathering is commonly accompanied by biodegradation and water washing.

Wet gas

Contains ethane, propane, and heavier hydrocarbons and less than 98% methane/total hydrocarbons. Originates during catagenesis. The wet gas zone occurs below the bottom of the oil window and above the top of the gas window (1.4 to 2.0 percent vitrinite reflectance).

References

VAN AARSSEN, **B. G. K., and** DE **LEEUW, J. W.** (1989) On the identification and occurrence of oligomerized sesquiterpenoid compounds in oils and sediments of southeast Asia. 14th International Meeting on Organic Geochemistry, Paris, September 18–22, 1989, Abstract No. 63.

VAN AARSSEN, B. G. K., KRUK, C., HESSELS, J. K. C., and DE LEEUW, J. W. (1990a) *cis-cis-trans*-Bicadinane, a novel member of an uncommon triterpane family isolated from crude oils. *Tetrahedron Letters,* Vol. 31, 4645–4648.

VAN AARSSEN, B. G. K., COX, H. C., HOOGENDOORN, P., and DE LEEUW, J. W. (1990b) A cadinene biopolymer in fossil and extant dammar resins as a source for cadinanes and bicadinanes in crude oils from southeast Asia. *Geochimica et Cosmochimica Acta,* Vol. 54, p. 3021–3031.

ABBOTT, G. D., LEWIS, C. A., and MAXWELL, J. R. (1984) Laboratory simulation studies of steroid aromatization and alkane isomerization. *Organic Geochemistry,* Vol. 6, p. 31–38.

ABBOTT, G. D., WANG, G. Y., EGLINTON, T. I., HOME, A. K., and PETCH, G. S. (1990) The kinetics of sterane biological marker release and degradation processes during hydrous pyrolysis of vitrinite kerogen. *Geochimica et Cosmochimica Acta,* Vol. 54, p. 2451–2461.

ADAM, P., SCHMID, J. C., MYCKE, B., STRAZIELLE, C., CONNAN, J., HUC, A., RIVA, A., and ALBRECHT, P. (1992, in press) Structural investigation of non-polar sulfur cross-linked macromolecules in petroleum. *Geochimica et Cosmochimica Acta.*

AIZENSHTAT, Z. (1973) Perylene and its geochemical significance. *Geochimica et Cosmochimica Acta* Vol. 37, p. 559–567.

AKPORIAYE, E. E., FARRANT, R. D., and KIRK, D. N. (1981) Deuterium incorporation in the backbone rearrangement of cholest-5-ene. *Journal of Chemical Research* (S), p. 210–211.

ALAM, M., AND PEARSON, M. J. (1990) Bicadinanes in oils from the Surma Basin, Bangladesh. *Organic Geochemistry,* Vol. 15, p. 461–464.

ALBAIGES, J. (1980) Identification and geochemical significance of long chain acyclic isoprenoid hydrocarbons in crude oils. In: *Advances in Organic Geochemistry 1979* (A. G. Douglas and J. R. Maxwell, eds.) Pergamon, Oxford, p. 19–28.

ALBAIGES, J., BORBON, J., and WALKER, II., W. (1985) Petroleum isoprenoid hydrocarbons derived from catagenetic degradation of Archaebacterial lipids. *Organic Geochemistry,* Vol. 8, p. 293–297.

ALBRECHT, P., VANDENBROUCKE, M., and MANDENGUE, M. (1976) Geochemical studies on the organic matter from the Douala Basin (Cameroon). I. Evolution of the extractable organic matter and the formation of petroleum. *Geochimica et Cosmochimica Acta,* Vol. 40, p. 791–799.

ALEXANDER, R., KAGI, R., and WOODHOUSE, G. W. (1981) Geochemical correlation of Windalia oil and extracts of Winning Group (Cretaceous) potential source rocks, Barrow subbasin, Western Australia. *American Association of Petroleum Geologists Bulletin,* Vol. 65, p. 235–250.

ALEXANDER, R., KAGI, R., and NOBLE, R. (1983a) Identification of the bicyclic sesquiterpenes, drimane, and eudesmane in petroleum. *Journal of the Chemical Society, Chemical Communications,* p. 226–228.

ALEXANDER, R., KAGI, R. I., WOODHOUSE, G. W., and VOLKMAN, J. K. (1983b) The geochemistry of some biodegraded Australian oils. *Australian Petroleum Exploration Association Journal,* Vol. 23, p. 53–63.

ALEXANDER, R., LARCHER, A. V., KAGI, R. I., and PRICE, P. L. (1992) An oil-source correlation study using age specific plant-derived aromatic biomarkers. In: *Biological Markers in Sediments and Petroleum* (J. M. Moldowan, P. Albrecht, and R. P. Philp, eds.) Prentice Hall, Englewood Cliffs, N.J., p. 201–221.

AQUINO NETO, F. R., RESTLE, A., CONNAN, J., ALBRECHT, P., and OURISSON, G. (1982) Novel tricyclic terpanes (C_{19}, C_{20}) in sediments and petroleums. *Tetrahedron Letters,* Vol. 23, p. 2027–2030.

AQUINO NETO, F. R., TRENDEL, J. M., RESTLE, A., CONNAN, J., and ALBRECHT, P. A. (1983) Occurrence and formation of tricyclic and tetracyclic terpanes in sediments and petroleums. In: *Advances in Organic Geochemistry 1981* (M. Bjorøy et al., eds) J. Wiley and Sons, New York, p. 659–676.

AQUINO NETO, F. R., TRIGUIS, J., AZEVEDO, D. A., RODRIGUES, R., and SIMONEIT, B. R. T. (1989) Organic geochemistry of geographically unrelated *Tasmanites*. 14th International Meeting on Organic Geochemistry, Paris, September 18-22, 1989, Abstract No. 189.

ARTHUR, M. A., DEAN, W. E., and PRATT, L. M. (1988) Geochemical and climatic effects of increased marine organic carbon burial at the Cenomanian/Turonian boundary. *Nature,* Vol. 335, p. 714–717.

AYRES, M. G., BILAL, M., JONES, R. W., SLENTZ, L. W., TARTIR, M., and WILSON, A. O. (1982) Hydrocarbon habitat in main producing areas, Saudi Arabia. *American Association of Petroleum Geologists Bulletin,* Vol. 66, p. 1–9.

AZEVEDO, D. A., AQUINO NETO, F. R., SIMONEIT, B. R. T., and PINTO, A. C. (1992) Novel series of tricyclic aromatic terpanes characterized in Tasmanian tasmanite. *Organic Geochemistry,* Vol. 18, p. 9–16.

BAILEY, N. J. L., KROUSE, H. R., EVANS, C. R., and ROGERS, M. A. (1973) Alteration of crude oil by waters and bacteria—evidence from geochemical and isotopic studies. *American Association of Petroleum Geologists Bulletin,* Vol. 57, p. 1276–1290.

BAILEY, N. J. L., BURWOOD, R., and HARRIMAN, G. E. (1990) Application of pyrolyzate carbon isotope and biomarker technology to organofacies definition and oil correlation problems in North Sea basins. *Organic Geochemistry,* Vol. 16, p. 1157–1172.

BAKER, E. W., and LOUDA, J. W. (1983) Thermal aspects of chlorophyll geochemistry. In: *Advances in Organic Geochemistry 1981* (M. Bjorøy et al., eds.) J. Wiley and Sons, New York, p. 401–421.

BAKER, E. W., and LOUDA, J. W. (1986) Porphyrins in the geological record. In: *Biological Markers in the Sedimentary Record* (R. B. Johns, ed.) Elsevier, New York, p. 125–224.

BALOGH, B., WILSON, D. M., CHRISTIANSEN, P., and BURLINGAME, A. L. (1973) 17α(H)-hopane identified in oil shale of the Green River Formation (Eocene) by carbon-13 NMR. *Nature,* Vol. 242, p. 603–605.

BARNES, M. A., and BARNES, W. C. (1983) Oxic and anoxic diagenesis of diterpenes in lacustrine sediments. In: *Advances in Organic Geochemistry 1981* (M. Bjorøy et al., eds.) J. Wiley and Sons, New York, p. 289–298.

BAROSS, J. A., and DEMING, J. W. (1983) Growth of "black smoker" bacteria at temperatures of at least 250°C. *Nature,* Vol. 303, p. 423–426.

BARWISE, A. J. G. (1990) Role of nickel and vanadium in petroleum classification. *Energy and Fuels,* Vol. 4, p. 647–652

BARWISE, A. J. G., and PARK, P. J. D. (1983) Petroporphyrin fingerprinting as a geochemical parameter. In: *Advances in Organic Geochemistry 1981* (M. Bjorøy et al., eds.) J. Wiley and Sons, New York, p. 668–674.

BARWISE, A. J. G., and WHITEHEAD, E. V. (1980) Separation and structure of petroporphyrins. In: *Advances in Organic Geochemistry 1979* (A. G. Douglas and J. R. Maxwell, eds.) Pergamon, New York, p. 181–192.

BASKIN, D. K. (1979) A method of preparing phytoclasts for vitrinite reflectance analysis. *Journal of Sedimentary Petrology,* Vol. 49, p. 633–635.

BASKIN, D. K., and PETERS, K. E. (1992) Early generation characteristics of a sulfur-rich Monterey kerogen. *American Association of Petroleum Geologists Bulletin,* Vol. 76, p. 1–13.

BAUER, P. E., DUNLAP, N. K., ARSENIYADIS, S., WATT, D. S., SEIFERT, W. K., and MOLDOWAN, J. M. (1983) Synthesis of biological markers in fossil fuels. 1. 17α and 17β isomers of 30-norhopane and 30-normoretane. *Journal of Organic Chemistry,* Vol. 48, p. 4493–497.

BEACH, F., PEAKMAN, T. M., ABBOTT, G. D., SLEEMAN, R., and MAXWELL, J. R. (1989) Laboratory thermal alteration of triaromatic steroid hydrocarbons. *Organic Geochemistry,* Vol. 14, p. 109–111.

BEATO, B. D., YOST, R. A., VAN BERKEL, G. J., FILBY, R. H., and QUIRKE, M. E. (1991) The Henryville bed of the New Albany shale—III: Tandem mass spectrometric analyses of geoporphyrins from the bitumen and kerogen. *Organic Geochemistry,* Vol. 17, p. 93–105.

BEAUMONT, C., BOUTILIER, R., MACKENZIE, A. S., and RULLKÖTTER, J. (1985) Isomerization and aromatization of hydrocarbons and the paleothermometry and burial history of Alberta Foreland Basin. *American Association of Petroleum Geologists Bulletin,* Vol. 69, p. 546–566.

BENDORAITIS, J. G. (1974) Hydrocarbons of biogenic origin in petroleum—Aromatic triterpenes and bicyclic sesquiterpenes. In: *Advances in Organic Geochemistry 1973* (B. Tissot and F. Bienner, eds.) Éditions Technip, Paris, p. 209–224.

BERNER, R. A., SCOTT, M. R., and THOMLINSON, C. (1970) Carbonate alkalinity in the pore waters of anoxic marine sediments. *Limnology and Oceanography,* Vol. 14, p. 544–549.

BIDIGARE, R. R., KENNICUTT, M. C., ONDRUSEK, M. E., KELLER, M. D., and GUILLARD, R. R. L. (1990) Novel chlorophyll-related compounds in marine phytoplankton: Distributions and geochemical implications. *Energy and Fuels,* Vol. 4, p. 653–657.

BIRD, C. W., LYNCH, J. M., PIRT, F. J., REID, W. W., BROOKS, C. J. W., and MIDDLEDITCH, B. S. (1971) Steroids and squalene in *Methylococcus capsulatus* grown on methane. *Nature,* Vol. 230, p. 473.

BISSERET, P., ZUNDEL, M., and ROHMER, M. (1985) Prokaryotic triterpenoids. 2. 2β-Methylhopanoids from *Methylobacterium organophilum* and *Nostoc muscorum,* a new series of prokaryotic triterpenoids. *European Journal of Biochemistry,* Vol. 150, p. 29–34.

BLUNT, J. W., CZOCHANSKA, Z., SHEPPARD, C. M., WESTON, R. J., and WOOLHOUSE, A. D. (1988) Isolation and structural characterization of isopimarane in some New Zealand seep oils. *Organic Geochemistry,* Vol. 12, p. 479–486.

BODUSZYNSKI, M. M. (1987) Composition of heavy petroleums. I. Molecular weight, hydrogen deficiency, and heteroatom concentration as a function of atmospheric equivalent boiling point up to 1400°F (760°C). *Energy and Fuels,* Vol. 1, p. 2–11.

BOON, J. J., HINES, H., BURLINGAME, A. L., KLOK, J., RIJPSTRA, W. I. C., DE LEEUW J. W., EDMUNDS K. E., and EGLINTON G. (1983) Organic geochemical studies of Solar Lake laminated cyanobacterial mats. In: *Advances in Organic Geochemistry 1981* (M. Bjorøy et al., eds.) J. Wiley and Sons, New York, p. 207–227.

BOREHAM, C. J. (1992) Reversed-phase high performance liquid chromatography of metalloporphyrins. In: *Biological Markers in Sediments and Petroleum* (J. M. Moldowan, P. Albrecht, and R. P. Philp, eds.) Prentice Hall, Englewood Cliffs, N.J., p. 301–312.

BOREHAM, C. J., CRICK, I. H., and POWELL, T. G. (1988) Alternative calibration of the Methylphenanthrene Index against vitrinite reflectance: Application to maturity measurements on oils and sediments. *Organic Geochemistry,* Vol. 12, p. 289–294.

BOREHAM, C. J., FOOKES, C. J. R., POPP, B. N., and HAYES, J. M. (1989) Origins of etioporphyrins in sediments: Evidence from stable carbon isotopes. *Geochimica et Cosmochimica Acta,* Vol. 53, p. 2451–2455.

BOSTICK, N. H. (1979) Microscopic measurement of the level of catagenesis of solid organic matter in sedimentary rocks to aid exploration for petroleum and to determine former burial temperatures—A review. *Society of Economic Paleontologists and Mineralogists Special Publication,* No. 26, p. 17–43.

BOSTICK, N. H., and ALPERN, B. (1977) Principles of sampling, preparation and constituent selection for microphotometry in measurement of maturation of sedimentary organic matter. *Journal of Microscopy,* Vol. 109, p. 41–47.

BOUVIER, P., ROHMER, M., BENVENISTE, P., and OURISSON, G. (1976) Δ8,14-Steroids in the bacterium *Methylococcus capsulatus. Biochemistry Journal,* Vol. 159, p. 267–271.

BRASSELL, S. C., WARDROPER, A. M. K., THOMPSON, I. D., MAXWELL, J. R., and EGLINTON, G. (1981) Specific acyclic isoprenoids as biological markers of methanogenic bacteria in marine sediments. *Nature,* Vol. 290, p. 693–696.

BRASSELL, S. C., EGLINTON, G., and MO, F. J. (1986) Biological marker compounds as indicators of the depositional history of the Maoming oil shale. *Organic Geochemistry,* Vol. 10, p. 927–941.

BRASSELL, S. C., SHENG GUOYING, FU JIAMO, and EGLINTON, G. (1988) Biological markers in lacustrine Chinese oil shales. In: *Lacustrine Petroleum Source Rocks* (A. J. Fleet, K. Kelts, and M. R. Talbot, eds.) Blackwell, p. 299–308.

BRAY, E. E., and EVANS, E. D. (1961) Distribution of *n*-paraffins as a clue to recognition of source beds. *Geochimica et Cosmochimica Acta,* Vol. 22, p. 2–15.

BROCK, T. D., and MADIGAN, M. T. (1991) *Biology of Microorganisms.* Prentice Hall, Englewood Cliffs, N. J., 874 p.

BROOKS, J. D., GOULD, K., and SMITH, J. W. (1969) Isoprenoid hydrocarbons in coal and petroleum. *Nature,* Vol. 222, p. 257–259.

BROOKS, P. W. (1986) Unusual biological marker geochemistry of oils and possible source rocks, offshore Beaufort-Mackenzie Delta, Canada. In: *Advances in Organic Geochemistry 1985* (D. Leythaeuser and J. Rullkötter, eds.) Pergamon, p. 401–406.

BROOKS, P. W., FOWLER, M. G., and MACQUEEN, R. W. (1988) Biological marker and conventional organic geochemistry of oil sands/heavy oils, Western Canada Basin. *Organic Geochemistry,* Vol. 12, p. 519–538.

BROTHERS, L., ENGEL, M. H., and KROOS, B. M. (1991) The effects of fluid flow through porous media on the distribution of organic compounds in a synthetic crude oil. *Organic Geochemistry,* Vol. 17, p. 11–24.

BUCHARDT, B., CHRISTIANSEN, F. G., NOHR-HANSEN, H., LARSEN, N. H., and OSTFELDT, P. (1989) Composition of organic matter in source rocks. In: *Petroleum Geology of North Greenland* (F. G. Christiansen, ed.) Grønlands Geologiske Undersøgelse Bulletin 158, p. 32–39.

BURLINGAME, A. L., BAILLIE, T. A., DERRICK, P. G., and CHIZHOV, O. S. (1980) Mass spectrometry. *Analytical Chemistry,* Vol. 52, p. 214R–258R.

BURNHAM, A. K., CLARKSON, J. E., SINGLETON, M. F., WONG, C. M., and CRAWFORD, R. W. (1982) Biological markers from Green River kerogen decomposition. *Geochimica et Cosmochimica Acta,* Vol. 46, p. 1243–1251.

BURWOOD, R., CORNET, P. J., JACOBS, L., and PAULET (1990) Organofacies variation control on hydrocarbon generation: A Lower Congo Coastal Basin (Angola) case history. *Organic Geochemistry,* Vol. 16, p. 325–338.

CAHN, R. S., INGOLD, SIR C., and PRELOG, V. (1966) Specification of molecular chirality. *Angewandte Chemie International Ed.,* Vol. 5, p. 385–415.

CALLOT, H. J., OCAMPO, R., and ALBRECHT, P. (1990) Sedimentary porphyrins: Correlations with biological precursors. *Energy and Fuels,* Vol. 4, p. 635–639.

CALVERT, S. E. (1987) Oceanographic controls on the accumulation of organic matter in marine sediments. In: *Marine Petroleum Source Rocks* (J. Brooks and A. J. Fleet, eds.) London, Blackwell Scientific, p. 137–151.

CALVERT, S. E., KARLIN, R. E., TOOLIN, L. J., DONAHUE, D. J., SOUTHON, J. R., and VOGEL, J. S. (1991) Low organic carbon accumulation rates in Black Sea sediments. *Nature,* Vol. 350, p. 692–695.

CAMPBELL, R. M., DJORDJEVIC, N. M., MARKIDES, K. E., and LEE, M. L. (1988) Supercritical fluid chromatographic determination of hydrocarbon groups in gasolines and middle distillate fuels. *Analytical Chemistry,* Vol. 60, p. 356–362.

CARLSON, R. M. K., and CHAMBERLAIN, D. E. (1986) Steroid biomarker clay mineral adsorption free energies: Implications to petroleum migration indices. *Organic Geochemistry,* Vol. 10, p. 163–180.

CASAGRANDE, D. J. (1987) Sulfur in peat and coal. In: *Coal and Coal-bearing Strata: Recent Advances* (A. C. Scott, ed.) Geological Society Special Publication No. 32, p. 87–105.

CASPI, E., ZANDER, J. M., GREIG, J. B., MALLORY, F. B., CONNER, R. L., and LANDREY, J. R. (1968) Evidence for nonoxidative cyclization of squalene in the biosynthesis of tetrahymanol. *Journal of the American Chemical Society,* Vol. 90, p. 3563–3564.

CASSANI, F., and EGLINTON G. (1988) Organic geochemistry of Venezuelan extra-heavy oils. 1. Pyrolysis of asphaltenes: A technique for the correlation and maturity evaluation of crude oils. *Chemical Geology,* Vol. 56, p. 167–183.

CHAPPE, B., MICHAELIS W., and ALBRECHT, P. (1980) Molecular fossils of archaebacteria as selective degradation products of kerogen. In: *Advances in Organic Geochemistry 1979* (A. G. Douglas and J. R. Maxwell, eds.) Pergamon Press, Oxford, p. 265–274.

CHAPPE, B., ALBRECHT, P., and MICHAELIS, W. (1982) Polar lipids of archaebacteria in sediments and petroleum. *Science,* Vol. 217, p. 65–66.

CHICARELLI, M. I., KAUR, S., and MAXWELL, J. R. (1987) Sedimentary porphyrins: Unexpected structures, occurrence, and possible origins. In: *Metal Complexes in Fossil Fuels* (R. H. Filby and J. F. Branthaver, eds.) American Chemical Society Symposium Series 344, Washington, D. C., p. 41–67.

CHICARELLI, M. I., AQUINO NETO, F. R., and ALBRECHT, P. (1988) Occurrence of four stereoisomeric tricyclic terpane series in immature Brazilian shales. *Geochimica et Cosmochimica Acta,* Vol. 52, p. 1955–1959.

CHOSSON, P., CONNAN, J., DESSORT, D., and LANAU, C. (1992) In vitro biodegradation of steranes and terpanes: A clue to understanding geological situations. In: *Biological Markers in Sediments and Petroleum* (J. M. Moldowan, P. Albrecht, and R. P. Philp, eds.) Prentice Hall, Englewood Cliffs, N. J., p. 320–349.

CHUNG, H. M., BRAND, S. W., and GRIZZLE, P. L. (1981) Carbon isotope geochemistry of Paleozoic oils from Big Horn Basin. *Geochimica et Cosmochimica Acta,* Vol. 45, p. 1803–1815.

CLARK, J. P., and PHILP, R. P. (1989) Geochemical characterization of evaporite and carbonate depositional environments and correlation of associated crude oils in the Black Creek Basin, Alberta. *Canadian Petroleum Geologists Bulletin,* Vol. 37, p. 401–416.

CLAYPOOL, G. E., and KAPLAN, I. R. (1974) The origin and distribution of methane in marine sediments. In: *Natural Gases in Marine Sediments* (I. R. Kaplan, ed.) Plenum Press, New York, p. 99–140.

CLAYTON, J. L., and KING, J. D. (1987) Effects of weathering on biological marker and aromatic hydrocarbon composition of organic matter in Phosphoria shale outcrop. *Geochimica et Cosmochimica Acta,* Vol. 51, p. 2153–2157.

CLAYTON, J. L., KING, J. D., THRELKELD, C. N., and VULETICH, A. (1987) Geochemical correlation of Paleozoic oils, Northern Denver Basin—Implications for exploration. *American Association of Petroleum Geologists Bulletin,* Vol. 71, p. 103–109.

CONNAN, J. (1974) Diagenese naturelle et diagenese artificielle de la matiere organique a element vegetaux predominants. In: *Advances in Organic Geochemistry 1973* (B. P. Tissot and F. Bienner, eds.) Éditions Technip, Paris, p. 73–95.

CONNAN, J. (1981) Biological markers in crude oils. In: *Petroleum Geology in China* (J. F. Mason, ed.) Penn Well, Tulsa, p. 48–70.

CONNAN, J. (1984) Biodegradation of crude oils in reservoirs. In: *Advances in Petroleum Geochemistry,* Vol. 1 (J. Brooks and D. H. Welte, eds.) Academic Press, London. p. 299–335.

CONNAN, J., and DESSORT, D. (1987) Novel family of hexacyclic hopanoid alkanes (C_{32}-C_{35}) occurring in sediments and oils from anoxic paleoenvironments. *Organic Geochemistry,* Vol. 11, p. 103–113.

CONNAN, J., RESTLE, A., and ALBRECHT, P. (1980) Biodegradation of crude oil in the Aquitaine basin. *Physics and Chemistry of the Earth,* Vol. 12., p. 1–17.

CONNAN, J., BOUROULLEC, J., DESSORT, D., and ALBRECHT, P. (1986) The microbial input in carbonate-anhydrite facies of a sabkha palaeoenvironment from Guatemala: A molecular approach. *Organic Geochemistry,* Vol. 10, p. 29–50.

CORBETT, R. E., and SMITH, R. A. (1969) Lichens and fungi. Part VI. Dehydration rearrangements of 15-hydroxyhopanes. *Journal of the Chemical Society* (C), Vol. 1969, p. 44–47.

CORNFORD, C., NEEDHAM, C. E. J., and DE WALQUE, L. (1986) Geochemical habitat of North Sea oils. In: *Habitat of Hydrocarbons on the Norwegian Continental Shelf* (A. M. Spencer et al., eds.) Norwegian Petroleum Society, Graham and Trotman, Stavanger/London, p. 39–54.

CORNFORD, C., CHRISTIE, O., ENDRESEN, U., JENSEN, P., and MYHR, M. -B. (1988) Source rock and seep oil maturity in Dorset, Southern England. *Organic Geochemistry,* Vol. 13, p. 399–409.

COX, H. C., DE LEEUW, J. W., SCHENCK, P. A., VAN KONNIGSVELD, H., JANSEN, J. C., VAN DE GRAAF, B., VAN GEERESTEIN, V. J., KANTERS, J. A., KRUCK, C., and JANS, A. W. H. (1986) Bicadinane, a C_{30} pentacyclic isoprenoid hydrocarbon found in crude oil. *Nature,* Vol. 319, p. 316–318.

CROASMUN, W. R., and CARLSON, R. M. K. (1987) Two-dimensional NMR spectroscopy—Applications for chemists and biochemists. In: *Methods in Stereochemical Analysis,* (W. R. Croasmun and R. M. K. Carlson, eds.) VCH Publishers, New York, Vol. 9, p 534.

CURIALE, J. A. (1986) Origin of solid bitumens, with emphasis on biological marker results. *Organic Geochemistry,* Vol. 10, p. 559–580.

CURIALE, J. A. (1987) Steroidal hydrocarbons of the Kishenehn Formation, northwest Montana. *Organic Geochemistry,* Vol. 11, p. 233–244.

CURIALE, J. A. (1992) Molecular maturity parameters within a single oil family: A case study from the Sverdrup Basin, Arctic Canada. In: *Biological Markers in Sediments and Petroleum* (J. M. Moldowan, P. Albrecht, and R. P. Philp, eds.) Prentice Hall, Englewood Cliffs, N.J., p. 275–300.

CURIALE, J. A., and ODERMATT, J. R. (1989) Short-term biomarker variability in the Monterey Formation, Santa Maria Basin. *Organic Geochemistry,* Vol. 14, p. 1–13.

CZOCHANSKA, Z., GILBERT, T. D., PHILP, R. P., SHEPPARD, C. M., WESTON, R. J., WOOD, T. A., and WOOLHOUSE, A. D. (1988) Geochemical application of sterane and triterpane biomarkers to a description of oils from the Taranaki Basin in New Zealand. *Organic Geochemistry,* Vol. 12, p. 123–135.

Dahl, B., Speers, G. C., Steen A., Telnaes, N., and Johansen, J. E. (1985) Quantification of steranes and triterpanes by gas chromatographic-mass spectrometric analysis. In: *Petroleum Geochemistry in Exploration of the Norwegian Shelf* (B. M. Thomas et al., eds.) Graham and Trotman, London, p. 303–307.

Dahl, J., Moldowan J. M., McCaffrey M. A., and Lipton P. A. (1992) 3-Alkyl steranes in petroleum: Evidence for a new class of natural products. *Nature,* Vol. 355, p. 154–157.

Dastillung, M. and Albrecht, P. (1977) Δ2-Sterenes as diagenetic intermediates in sediments. *Nature,* Vol. 269, p. 678–679.

Deines, E. T. (1980) Biogeochemistry of stable carbon isotopes. In: *Organic Geochemistry* (G. Eglinton and M. T. J. Murphy, eds.) Springer Verlag, New York, p. 306–329.

Demaison, G. J. (1984) The generative basin concept. In: *Petroleum Geochemistry and Basin Evaluation* (G. J. Demaison and R. J. Murris, eds.) American Association of Petroleum Geologists Memoir 35, p. 1–14.

Demaison, G. J., and Moore, G. T. (1980) Anoxic environments and oil source bed genesis. *American Association of Petroleum Geologists Bulletin,* Vol. 64, p. 1179 –1209.

Demaison, G., Holck, A. J. J., Jones, R. W., and Moore, G. T. (1983) Predictive source bed stratigraphy; a guide to regional petroleum occurrence. In: *Proceedings of the 11th World Petroleum Congress 2,* PD1, J. Wiley and Sons, New York, p. 17–29.

Derenne, S., Largeau, C., Casadevall, E., and Connan, J. (1988) Comparison of torbanites of various origins and evolutionary stages. Bacterial contribution to their formation. Cause of lack of botryococcane in bitumens. *Organic Geochemistry,* Vol. 12, p. 43–59.

DeRosa, M., DeRosa, S., Gambacorta, A., Minale, L., and Bu'Lock, J. D. (1977a) Chemical structure of the ether lipids of thermophilic bacteria of the *Caldariella* group. *Phytochemistry,* Vol. 16, p. 1961–1965.

DeRosa, M., DeRosa, S., Gambacorta, A., and Bu'Lock, J. D. (1977b) Lipid structures in the *Caldariella* group of extreme thermoacidophile bacteria. *Journal of the Chemical Society, Chemical Communications,* p. 514–515.

Devon, T. K., and Scott, A. I. (1972) *Handbook of Naturally Occurring Compounds. Volume II Terpenes.* Academic Press, New York, p. 576.

Didyk, B. M., Simoneit, B. R. T., Brassell, S. C., and Eglinton, G. (1978) Organic geochemical indicators of palaeoenvironmental conditions of sedimentation. *Nature,* Vol. 272, p. 216–222.

Dimmler, A., Cyr, T. D., and Strausz, O. P. (1984) Identification of bicyclic terpenoid hydrocarbons in the saturate fraction of Athabasca oil sand bitumen. *Organic Geochemistry,* Vol. 7, p. 231–238.

Douglas, A. G., Sinninghe Damsté, J. S., Fowler, M. G., Eglinton, T. I., and de Leeuw, J. W. (1991) Unique distributions of hydrocarbons and sulphur compounds released by flash pyrolysis from the fossilized alga *Gloecapsomorpha prisca,* a major constituent in one of four Ordovician kerogens. *Geochimica et Cosmochimica Acta,* Vol. 55, p. 275–291.

Dow, W. G. (1977) Kerogen studies and geological interpretations. *Journal of Geochemical Exploration,* Vol. 7, p. 79–99.

Durand, B. (ed.) (1980) *Kerogen. Insoluble Organic Matter From Sedimentary Rocks.* Éditions Technip, Paris, p. 519.

Durand, B. (1983) Present trends in organic geochemistry in research on migration of hydrocarbons. In: *Advances in Organic Geochemistry 1981* (M. Bjorøy et al., eds.) J. Wiley and Sons, New York, p. 117–128.

DURAND, B., and MONIN, J. C. (1980) Elemental analysis of kerogens (C,H,O,N,S,Fe). In: *Kerogen, Insoluble Organic Matter from Sedimentary Rocks.* (B. Durand, ed.) Éditions Technip, Paris, p. 113–142.

EGLINTON, G., and CALVIN, M. (1967) Chemical fossils. *Scientific American,* Vol. 261, p. 32–43.

EGLINTON, G., SCOTT, P. M., BESKY, T., BURLINGAME, A. L., and CALVIN, M. (1964) Hydrocarbons of biological origin from a one-billion-year-old sediment. *Science,* Vol. 145, p. 263–264.

EGLINTON, T. I., and DOUGLAS, A. G. (1988) Quantitative study of biomarker hydrocarbons released from kerogens during hydrous pyrolysis. *Energy and Fuels,* Vol. 2, p. 81–88.

EKWEOZOR, C. M., and STRAUSZ, O. P. (1982) Tricyclic terpanes in the Athabasca oil sands: Their geochemistry. In: *Advances in Organic Geochemistry 1981* (M. Bjorøy et al., eds.) J. Wiley and Sons, New York, p. 746–766.

EKWEOZOR, C. M., and TELNAES, N. (1990) Oleanane parameter: Verification by quantitative study of the biomarker occurrence in sediments of the Niger delta. *Organic Geochemistry,* Vol. 16, p. 401–413.

EKWEOZOR, C. M., and UDO, O. T. (1988) The oleananes: Origin, maturation, and limits of occurrence in Southern Nigeria sedimentary basins. In: *Advances in Organic Geochemistry 1987* (L. Matavelli and L. Novelli, eds.) Pergamon Press, p. 131–140.

EKWEOZOR, C. M., OKOGUN, J. I., EKONG, D. E. U., and MAXWELL, J. M. (1979) Preliminary organic geochemical studies of samples from the Niger Delta (Nigeria). *Chemical Geology,* Vol. 27, p. 29–37.

EMERSON, S. (1985) Organic carbon preservation in marine sediments. In: *The Carbon Cycle and Atmospheric CO_2: Natural Variations from Archean to Present* (E. T. Sundquist and W. S. Broecker, eds.) American Geophysical Union, Geophysical Monograph 32, Washington D.C., p. 78–86.

ENGLAND, W. A. (1990) The organic geochemistry of petroleum reservoirs. *Organic Geochemistry,* Vol. 16, p. 415–425.

ENSMINGER, A. (1977) Evolution de composes polycycliques sedimentaires. (In French) These de Doctorat es-Sciences, University L. Pasteur, 149 p.

ENSMINGER, A. VAN DORSSELAER, A., SPYCKERELLE, C., ALBRECHT, P., and OURISSON, G. (1974) Pentacyclic triterpenes of the hopane type as ubiquitous geochemical markers: Origin and significance. In *Advances in Organic Geochemistry 1973* (B. Tissot and F. Bienner, eds.) Éditions Technip, Paris, p. 245–260.

ENSMINGER, A., ALBRECHT, P., OURISSON, G., and TISSOT, B. (1977) Evolution of polycyclic alkanes under the effect of burial (Early Toarcian shales, Paris Basin). In: *Advances in Organic Geochemistry 1975* (R. Campos and J. Goni, eds.) ENADIMSA, Madrid, p. 45–52.

ENSMINGER, A., JOLY, G., and ALBRECHT, P. (1978) Rearranged steranes in sediments and crude oils. *Tetrahedron Letters,* p. 1575–1578.

EPSTEIN, A. G., EPSTEIN, J. B., and HARRIS, L. D. (1977) Conodont color alteration—An index to organic metamorphism. *Geological Survey Professional Paper 995,* 27 p.

Espitalié, J., Madec, M., Tissot, B. and Leplat, P. (1977) Source rock characterization method for petroleum exploration. *Offshore Technology Conference,* OTC 2935, Houston, Texas, May 2-5, 1977, p. 439–444.

Espitalié, J., Marquis, F., and Sage, L. (1987) Organic geochemistry of the Paris Basin. In: *Petroleum Geology of Northwest Europe* (J. Brooks and K. Glennie, eds.) Graham and Trotman, London, p. 71–86.

Fan Pu, King, J. D., and Claypool, G. E. (1987) Characteristics of biomarker compounds in Chinese crude oils. In: *Petroleum Geochemistry and Exploration in the Afro-Asian Region* (R. K. Kumar et al., eds.) Balkema, Rotterdam, p. 197–202.

Fan Zhao-an, and Philp, R. P. (1987) Laboratory biomarker fractionations and implications for migration studies. *Organic Geochemistry,* Vol. 11, p. 169–175.

Farrimond, P., Eglinton, G., Brassell, S. C., and Jenkyns, H. C. (1989) Toarcian anoxic event in Europe: An organic geochemical study. *Marine and Petroleum Geology,* Vol. 6, p. 136–147.

Farrington, J. W., Davis, A. C., Tarafa, M. E., McCaffrey, M. A., Whelan J., and Hunt J.M. (1988) Bitumen molecular maturity parameters in the Ikpikpuk well, Alaska North Slope. *Organic Geochemistry,* Vol. 13, p. 303–310.

Filby, R. H., and Berkel, G. J. V. (1987) Geochemistry of metal complexes in petroleum, source rocks, and coals: An overview. In: *Metal Complexes in Fossil Fuels.* (R. H. Filby and J. F. Branthaver eds.) American Chemical Society Symposium Series 344, Washington D.C., p. 2–39.

Filby, R. H., and Branthaver, J. F. (eds.) (1987) *Metal Complexes in Fossil Fuels.* American Chemical Society Symposium Series 344, Washington D.C., 436 p.

Flesch, G., and Rohmer, M. (1988) Prokaryotic hopanoids: The biosynthesis of the bacteriohopane skeleton. Formation of isoprenic units from two different acetate pools and a novel type of carbon/carbon linkage between a triterpane and D-ribose. *European Journal of Biochemistry,* Vol. 175, p. 405–411.

Fowler, M. G., and Brooks, P. W. (1990) Organic geochemistry as an aid in the interpretation of the history of oil migration into different reservoirs at the Hibernia K-18 and Ben Nevis I-45 wells, Jeanne d'Arc Basin, offshore eastern Canada. *Organic Geochemistry,* Vol. 16, p. 461–475.

Fowler, M. G., and Douglas, A. G. (1987) Saturated hydrocarbon biomarkers in oils of Late Precambrian age from Eastern Siberia. *Organic Geochemistry,* Vol. 11, p. 201–213.

Fowler, M. G., Snowdon, L. R., Brooks, P. W., and Hamilton, T. S. (1988) Biomarker characterization and hydrous pyrolysis of bitumens from Tertiary volcanics, Queen Charlotte Islands, British Columbia, Canada. *Organic Geochemistry,* Vol. 13, p. 715–725.

Francois R. (1987) A study of sulphur enrichment in the humic fraction of marine sediments during early diagenesis. *Geochimica et Cosmochimica Acta,* Vol. 51, p. 17–27.

Freeman, K. H., Hayes, J. M., Trendel, J. M., and Albrecht, P. (1990) Evidence from carbon isotope measurements for diverse origins of sedimentary hydrocarbons. *Nature,* Vol. 343, p. 254–256.

Fu Jiamo, and Sheng Guoying (1989) Biological marker composition of typical source rocks and related oils of terrestrial origin in the People's Republic of China; a review. *Applied Geochemistry,* Vol. 4, p. 13–22.

Fu Jiamo, Sheng Guoying, Peng Pingan, Brassell, S. C., Eglinton, G., and Jigang, J. (1986) Peculiarities of salt lake sediments as potential source rocks in China. *Organic Geochemistry,* Vol. 10, p. 119–126.

Fu Jiamo, Sheng Guoying, and Liu Dehan (1988) Organic geochemical characteristics of major types of terrestrial source rocks in China. In: *Lacustrine Petroleum Source Rocks* (A. J. Fleet, K. Kelts, and M. R. Talbot, eds.) Blackwell, p. 279–289.

Fu Jiamo, Sheng Guoying, Xu Jiayou, Eglinton, G., Gowar, A. P., Jia Rongfen, Fan Shanfa, and Peng Pingan (1990) Application of biological markers in the assessment of paleoenvironments of Chinese non-marine sediments. *Organic Geochemistry,* Vol. 16, p. 769–779.

Fuex, A. N. (1977) The use of stable carbon isotopes in hydrocarbon exploration. *Journal of Geochemical Exploration,* Vol. 7, 155–188.

Gaffney, J. S., Premuzic, E. T., and Manowitz, B. (1980) On the usefulness of sulfur isotope ratios in crude oil correlations. *Geochimica et Cosmochimica Acta,* Vol. 44, p. 135–139.

Galimov, E. M. (1973) *Carbon Isotopes in Oil—Gas Geology*. Nedra, Moscow, 384 p.; National Aeronautics and Space Administration, Washington D.C. [Translation (1973) from Russian], 395 p.

Gallegos, E. J. (1971) Identification of new steranes, chromatography and mass spectrometry. *Analytical Chemistry,* Vol. 43, p. 1151–1160.

Gallegos, E. J. (1976) Analysis of organic mixtures using metastable transition spectra. *Analytical Chemistry,* Vol. 48, p. 1348–1351.

Gallegos, E. J. and Moldowan, J. M. (1992) The effect of hold time on GC resolution and the effect of collision gas on mass spectra in geochemical "biomarker" research. In: *Biological Markers in Sediments and Petroleum* (J. M. Moldowan, P. Albrecht, and R. P. Philp, eds.) Prentice Hall, Englewood Cliffs, N. J., p. 156–181.

Gallegos, E. J., Fetzer, J. C., Carlson, R. M., and Pena, M. M. (1991) High-temperature GC/MS characterization of porphyrins and high molecular weight hydrocarbons. *Energy and Fuels,* Vol. 5, p. 376–381.

Gardner, P. M., and Whitehead, E. V. (1972) The isolation of squalane from a Nigerian petroleum. *Geochimica et Cosmochimica Acta,* Vol. 36, p. 259–263.

Garrigues, P., Saptorahardjo, A., Gonzalez, C., Wehrung, P., Albrecht, P., Saliot, A., and Ewald, M. (1986) Biogeochemical aromatic markers in sediments from Mahakam Delta (Indonesia). *Organic Geochemistry,* Vol. 10, p. 959–964.

Gelpi, E., Schneider, H., Mann, J., and Oro, J. (1970) Hydrocarbons of geochemical significance in microscopic algae. *Phytochemistry,* Vol. 9, p. 603–612.

Goad, L. J., and Withers, N. (1982) Identification of 27-nor-(24R)24-methylcholesta-5,22-dien-3β-ol and brassicasterol as the major sterols of the marine dinoflagellate *Gymnodinium simplex*. *Lipids,* Vol. 17, p. 853–858.

Goodarzi, F., Brooks, P. W., and Embry, A. F. (1989) Regional maturity as determined by organic petrography and geochemistry of the Schei Point Group (Triassic) in the western Sverdrup Basin, Canadian Arctic Archipelago. *Marine and Petroleum Geology,* Vol. 6, p. 290–302.

GOODWIN, T. W. (1973) Comparative biochemistry of sterols in eukaryotic microorganisms. In: *Lipids and Biomembranes of Eukaryotic Microorganisms* (J. A. Erwin, ed.) Academic Press, New York, p. 1–40.

GOODWIN, N. S., PARK, P. J. D., and RAWLINSON, T. (1983) Crude oil biodegradation. In: *Advances in Organic Geochemistry 1981* (M. Bjorøy et al., eds.) J. Wiley and Sons, New York, p. 650–658.

GOODWIN, N. S., MANN, A. L., and PATIENCE, R. L. (1988) Structure and significance of C_{30} 4-methyl steranes in lacustrine shales and oils. *Organic Geochemistry,* Vol. 12, p. 495–506.

GOOSENS, H., DE LEEUW, J. W., SCHENCK, P. A., and BRASSELL, S. C. (1984) Tocopherols as likely precursors of pristane in ancient sediments and crude oils. *Nature,* Vol. 312, p. 440–442.

GOTH, K., DE LEEUW, J. W., PÜTTMANN, W., and TEGELAAR, E. W. (1988) Origin of Messel Oil Shale kerogen. *Nature,* Vol. 336, p. 759–761.

GOUGH, M. A., and ROWLAND, S. J. (1990) Characterization of unresolved complex mixtures of hydrocarbons in petroleum. *Nature,* Vol. 344, p. 648–650.

GRANSCH, J. A., and POSTHUMA, J. (1974) On the origin of sulfur in crudes. In: *Advances of Organic Geochemistry 1973* (B. Tissot and F. Bienner, eds.) Éditions Technip, Paris, p.727–739.

GRANTHAM, P. J. (1986a) The occurrence of unusual C_{27} and C_{29} sterane predominances in two types of Oman crude oil. *Organic Geochemistry,* Vol. 9, p. 1–10.

GRANTHAM, P. J. (1986b) Sterane isomerization and moretane/hopane ratios in crude oils derived from Tertiary source rocks. *Organic Geochemistry,* Vol. 9, p. 293–304.

GRANTHAM, P. J., and WAKEFIELD, L. L. (1988) Variations in the sterane carbon number distributions of marine source rock derived crude oils through geological time. *Organic Geochemistry,* Vol. 12, p. 61–73.

GRANTHAM, P. J., POSTHUMA, J., and DEGROOT, K. (1980) Variation and significance of the C_{27} and C_{28} triterpane content of a North Sea core and various North Sea crude oils. In: *Advances in Organic Geochemistry 1979* (A. G. Douglas and J. R. Maxwell, eds.) Pergamon Press, New York, p. 29–38.

GRANTHAM, P. J., POSTHUMA, J., and BAAK, A. (1983) Triterpanes in a number of Far-Eastern crude oils. In: *Advances in Organic Geochemistry 1981* (M. Bjorøy et al., eds.) J. Wiley and Sons, New York, p. 675–683.

GRANTHAM, P. J., LIJMBACH, G.W.M., POSTHUMA, J, HUGHES, CLARK M.W., and WILLINK, R. J. (1987) Origin of crude oils in Oman. *Journal of Petroleum Geology,* V. 11, p. 61–80.

GSCHWEND, P. M., CHEN, P. H., and HITES, R. A. (1983) On the formation of perylene in recent sediments: Kinetic models. *Geochimica et Cosmochimica Acta,* Vol. 47, p. 2115–2119.

GUADALUPE, M. F. M., CASTELLO BRANCO, V. A., and SCHMID, J. C. (1991) Isolation of sulfides in oils. *Organic Geochemistry,* Vol. 17, p. 355–361.

HADDON W. F. (1979) Computerized mass spectrometry linked scan system for recording metastable ions. *Analytical Chemistry,* Vol. 51, p. 983–988.

HALL, P. B., and DOUGLAS, A. G. (1983) The distribution of cyclic alkanes in two lacustrine deposits. In: *Advances in Organic Geochemistry 1981* (M. Bjorøy et al., eds.) J. Wiley and Sons, New York, p. 576–587.

HATCH, J. R., JACOBSON, S. R., WITZKE, B. J., RISATTI, J. B., ANDERS, D. E., WATNEY, W. L., NEWELL, K. D., and VULETICH, A. K. (1987) Possible Middle Ordovician organic carbon isotope excursion: Evidence from Ordovician oils and hydrocarbon source rocks, Mid-Continent, and East-Central United States. *American Association of Petroleum Geologists Bulletin,* Vol. 71, p. 1342–1354.

TEN HAVEN, H. L., and RULLKÖTTER, J. (1988) The diagenetic fate of taraxer-14-ene and oleanene isomers. *Geochimica et Cosmochimica Acta,* Vol. 52, p. 2543–2548.

TEN HAVEN, H. L., DE LEEUW, J. W., PEAKMAN, T. M., and MAXWELL, J. R. (1986) Anomalies in steroid and hopanoid maturity indices. *Geochimica et Cosmochimica Acta,* Vol. 50, p. 853–855.

TEN HAVEN, H. L., DE LEEUW, J. W., RULLKÖTTER, J., and SINNINGHE DAMSTÉ, J. S. (1987) Restricted utility of the pristane/phytane ratio as a palaeoenvironmental indicator. *Nature,* Vol. 330, p. 641–643.

TEN HAVEN, H. L., DE LEEUW, J. W., SINNINGHE DAMSTÉ, J. S., SCHENCK, P. A., PALMER, S. E., and ZUMBERGE, J. E. (1988) Application of biological markers in the recognition of palaeohypersaline environments. In: *Lacustrine Petroleum Source Rocks* (A. J. Fleet, K. Kelts, and M. R. Talbot, eds.) Geological Society Special Publication No. 40, p. 123–130.

TEN HAVEN, H. L., ROHMER, M., RULLKÖTTER, J., and BISSERET, P. (1989) Tetrahymanol, the most likely precursor of gammacerane occurs ubiquitously in marine sediments. *Geochimica et Cosmochimica Acta,* Vol. 53, p. 3073–3079.

TEN HAVEN, H. L., LITTKE, R., and RULLKÖTTER, J. (1992) Hydrocarbon biological markers in Carboniferous coals of different maturities. In: *Biological Markers in Sediments and Petroleum* (J. M. Moldowan, P. Albrecht, and R. P. Philp, eds.) Prentice Hall, Englewood Cliffs, N.J., p. 142–155.

HAYES, J. M., KAPLAN, I. R., and WEDEKING, K. M. (1983) Precambrian organic geochemistry, preservation of the record. In: *Earth's Earliest Biosphere, Its Origin and Evolution* (J. W. Schopf, ed.) Princeton University Press, New Jersey, p. 93–134.

HAYES, J. M., TAKIGIKU, R., OCAMPO, R., CALLOT, H. J., and ALBRECHT, P. (1987) Isotopic compositions and probable origins of organic molecules in the Eocene Messel shale. *Nature,* Vol. 329, p. 48–51.

HAYES, J. M., FREEMAN, K. H., POPP, B. N., and HOHAM, C. H. (1990) Compound-specific isotopic analyses: A novel tool for reconstruction of ancient biogeochemical processes. *Organic Geochemistry,* Vol. 16, p. 1115–1128.

HEISSLER, D., OCAMPO, R., ALBRECHT, P., RIEHL, J., and OURISSON, G. (1984) Identification of long-chain tricyclic terpene hydrocarbons (C_{21}-C_{30}) in geological samples. *Journal of the Chemical Society, Chemical Communications*, p. 496–498.

HE WEI, and LU SONGNIAN (1990) A new maturity parameter based on monoaromatic hopanoids. *Organic Geochemistry,* Vol. 16, p. 1007–1013.

HILLS, I. R., and WHITEHEAD, E. V. (1966) Triterpanes in optically active petroleum distillates. *Nature,* Vol. 209, p. 977–979.

HILLS, I. R., WHITEHEAD, E. V., ANDERS, D. E., CUMMINS, J. J., and ROBINSON, W. E. (1966) An optically active triterpane, gammacerane in Green River, Colorado, oil shale bitumen. *Journal of the Chemical Society, Chemical Communications*, Vol. 20, p. 752–754.

HO, E. S., MEYERS, P. A., and MAUK, J. L. (1990) Organic geochemical study of mineralization in the Keweenawan Nonesuch Formation at White Pine, Michigan. *Organic Geochemistry*, Vol. 16, p. 229–234.

HOEFS, J. (1980) *Stable Isotope Geochemistry*, Springer-Verlag, New York, p. 208.

HOERING, T. C. (1978) Molecular fossils from the Precambrian Nonesuch Shale. In: *Comparative Planetology* (C. Ponnamperuma, ed.) Academic Press, Chap. 15, p. 243–255.

HOFFMAN, C. G., and STRAUSZ, O. P. (1986) Bitumen accumulation in Grosmont Platform Complex, Upper Devonian, Alberta, Canada. *American Association of Petroleum Geologists Bulletin*, Vol. 70, p. 1113–1128.

HOFFMANN, C. F., MACKENZIE, A. S., LEWIS, C. A., MAXWELL, J. R., OUDIN, J. L., DURAND, B., and VANDENBROUCKE, M. (1984) A biological marker study of coals, shales, and oils from the Mahakam Delta, Kalimantan, Indonesia. *Chemical Geology*, Vol. 42, p. 1–23.

HONG, Z.-H., LI, H.-X., RULLKÖTTER, J., and MACKENZIE, A. S. (1986) Geochemical application of sterane and triterpane biological marker compounds in the Linyi Basin. *Organic Geochemistry*, Vol. 10, p. 433–439.

HORSTAD, I., LARTER, S. R., DYPVIK, H., AAGAARD, P., BJØRNVIK, A. M., JOHANSEN, P. E., and ERIKSEN, S. (1990) Degradation and maturity controls on oil field petroleum column heterogeneity in the Gullfaks field, Norwegian North Sea. *Organic Geochemistry*, Vol. 16, p. 497–510.

HUANG DIFAN, LI JINCHAO, and ZHANG DAJIANG (1990) Maturation sequence of continental crude oils in hydrocarbon basins in China and its significance. *Organic Geochemistry*, Vol. 16, p. 521–529.

HUANG, W.-Y., and MEINSCHEIN, W. G. (1979) Sterols as ecological indicators. *Geochimica et Cosmochimica Acta*. Vol. 43, p. 739–745.

HUGHES, W. B. (1984) Use of thiophenic organosulfur compounds in characterizing crude oils derived from carbonate versus siliciclastic sources. In: *Petroleum Geochemistry and Source Rock Potential of Carbonate Rocks*. (J. G. Palacas, ed.) American Association of the Petroleum Geologists, Studies in Geology 18, p. 181–196.

HUGHES, W. B., HOLBA, A. G., MILLER, D. E., and RICHARDSON, J. S. (1985) Geochemistry of greater Ekofisk crude oils. In: *Geochemistry in Exploration of the Norwegian Shelf* (B. M. Thomas, ed.), Graham and Trotman, p. 75–92.

HUNT, J. M. (1979) *Petroleum Geochemistry and Geology*. W. H. Freeman, San Francisco, 617 p.

HUSSLER, G., ALBRECHT, P., and OURISSON, G. (1984a) Benzohopanes, a novel family of hexacyclic geomarkers in sediments and petroleums. *Tetrahedron Letters*, Vol. 25, p. 1179–1182.

HUSSLER, G., CONNAN, J., and ALBRECHT, P. (1984b) Novel families of tetra- and hexacyclic aromatic hopanoids predominant in carbonate rocks and crude oils. *Organic Geochemistry*, Vol. 6, p. 39–49.

HUTTON, A. C., and COOK, A. C. (1980) Influence of alginite on the reflectance of vitrinite from Joadja, NSW, and some other coals and oils shales containing alginite. *Fuel,* Vol. 59, p. 711–714.

HUTTON, A. C., KANTSLER, A. J., COOK, A. C., and MCKIRDY, D. M. (1980) Organic matter in oil shales. *Journal of the Australian Petroleum Exploration Association,* Vol. 20, p. 44–67.

HWANG, R. J. (1990) Biomarker analysis using GC-MSD. *Journal of Chromatographic Science,* Vol. 28, p. 109–113.

HWANG, R. J., SUNDARARAMAN, P., TEERMAN, S. C., and SCHOELL, M. (1989) Effect of preservation on geochemical properties of organic matter in immature lacustrine sediments. 14th International Meeting on Organic Geochemistry, Paris, Sept. 18–22, 1989, Abstract No. 351.

ILLICH, H. A. (1983) Pristane, phytane, and lower molecular weight isoprenoid distributions in oils. *American Association of Petroleum Geologists Bulletin,* Vol. 67, p. 385–393.

IMBUS, S. W., ENGEL, M. H., ELMORE, R. D., and ZUMBERGE, J. E. (1988) The origin, distribution and hydrocarbon generation potential of the organic-rich facies in the Nonesuch Formation, Central North American Rift system: A regional study. *Organic Geochemistry,* Vol. 13, p. 207–219.

IRWIN, H., and MEYER, T. (1990) Lacustrine organic facies. A biomarker study using multivariate statistical analysis. *Organic Geochemistry,* Vol. 16, p. 176–210.

ISAACS, C. M., and PETERSEN, N. F. (1988) Petroleum in the Miocene Monterey Formation, California. In: *Siliceous Sedimentary Rock-Hosted Ores and Petroleum* (J. R. Hein, ed.), Van Nostrand Reinhold, New York, p. 83–116.

JACKSON, M. J., POWELL, T. G., SUMMONS, R. E., and SWEET, I. P. (1986) Hydrocarbon shows and petroleum source rocks in sediments as old as 1.7×10^9 years. *Nature,* Vol. 322, p. 727–729.

JACOBSON, S. R., HATCH, J. R., TEERMAN, S. C., and ASKIN, R. A. (1988) Middle Ordovician organic matter assemblages and their effect on Ordovician-derived oils. *American Association of Petroleum Geologists Bulletin,* Vol. 72, p. 1090–1100.

JAFFE, R., ALBRECHT, P., and OUDIN, J. L. (1988a) Carboxylic acids as indicators of oil migration. I. Occurrence and geochemical significance of C-22 diastereoisomers of the 17β(H),21β(H) C_{30} hopanoic acid in geological samples. *Organic Geochemistry,* Vol. 13, p. 483–488.

JAFFE, R., ALBRECHT, P., and OUDIN, J. L. (1988b) Carboxylic acids as indicators of oil migration. II. Case of the Mahakam Delta, Indonesia. *Geochimica et Cosmochimica Acta,* Vol. 52, p. 2599–2607.

JIANG, Z. S., and FOWLER, M. G. (1986) Carotenoid-derived alkanes in oils from northwestern China. *Organic Geochemistry,* Vol. 10, p. 831–839.

JIANG ZUSHENG, PHILP, R. P., and LEWIS, C. A. (1988) Fractionation of biological markers in crude oils during migration and the effects on correlation and maturation parameters. *Organic Geochemistry,* Vol. 13, p. 561–571.

JIANG ZHUSHENG, FOWLER, M. G., LEWIS, C. A., and PHILP, R. P. (1990) Polycyclic alkanes in a biodegraded oil from the Kelamayi oilfield, northwestern China. *Organic Geochemistry,* Vol. 15, p. 35–46.

JOBSON, A. M., COOK, F. D., and WESTLAKE, D. W. S. (1979) Interaction of aerobic and anaerobic bacteria in petroleum biodegradation. *Chemical Geology,* Vol. 24, p. 355–365.

JOHNS, R. B. (1986) *Biological Markers in the Sedimentary Record.* Methods in Geochemistry and Geophysics 24, Elsevier, New York, p. 364.

JONES, D. M., DOUGLAS, A. G., and CONNAN, J. (1988) Hydrous pyrolysis of asphaltenes and polar fractions of biodegraded oils. *Organic Geochemistry*, Vol. 13, p. 981–993.

JONES, R. W. (1984) Comparison of carbonate and shale source rocks. In: *Petroleum Geochemistry and Source Rock Potential of Carbonate Rocks.* (J. G. Palacas, ed.) American Association of Petroleum Geologists, Studies in Geology 18, p. 163–180.

JONES, R. W. (1987) Organic facies. In: *Advances in Petroleum Geochemistry*, Vol. 1 (J. Brooks and D. Welte, eds.) Academic Press, New York, p. 1–90.

JONES, R. W., and EDISON, T. A. (1978) Microscopic observations of kerogen related to geochemical parameters with emphasis on thermal maturation. In: *Low Temperature Metamorphism of Kerogen and Clay Minerals* (D. F. Oltz, ed.) Society of Economic Paleontologists and Mineralogists, Pacific Section, October 5, 1978, Los Angeles, p. 1–12.

KAPLAN, I. R. (1975) Stable isotopes as a guide to biogeochemical processes. *Proceedings of the Royal Society of London,* Vol. 189, p. 183–211.

KARLSEN, D. A., and LARTER, S. (1989) A rapid correlation method for petroleum population mapping within individual petroleum reservoirs: Applications to petroleum reservoir description. In: *Correlation in Hydrocarbon Exploration,* Norwegian Petroleum Society, Graham and Trotman, p. 77–85.

KATZ, B. J., and ELROD, L. W. (1983) Organic geochemistry of DSDP Site 467, offshore California, Middle Miocene to Lower Pliocene strata. *Geochimica et Cosmochimica Acta.,* Vol. 47, p. 389–396.

KAUFMAN, R. L., AHMED, A. S., and ELSINGER, R. J. (1990) Gas chromatography as a development and production tool for fingerprinting oils from individual reservoirs: Applications in the Gulf of Mexico. In: *Proceedings of the 9th Annual Research Conference of the Society of Economic Paleontologists and Mineralogists* (D. Schumacker and B. F. Perkins, eds.) New Orleans, p. 263–282.

KEELY, B. J., PROWSE, W. G., and MAXWELL, J. R. (1990) The Treibs hypothesis: An evaluation based on structural studies. *Energy and Fuels,* Vol. 4, p. 628–634.

KENIG, F., HUC, A. Y., PURSER, B. H., and OUDIN, J. L. (1990) Sedimentation, distribution and diagenesis of organic matter in recent carbonate environment, Abu Dhabi, U.A.E. *Organic Geochemistry,* Vol. 16, p. 735–747.

KILLOPS, S. D. (1991) Novel aromatic hydrocarbons of probable bacterial origin in a Jurassic lacustrine sequence. *Organic Geochemistry,* Vol. 17, p. 25–36.

KILLOPS, S. D., and AL-JUBOORI, M. A. H. A. (1990) Characterization of the unresolved complex mixture (UCM) in the gas chromatograms of biodegraded petroleums. *Organic Geochemistry,* Vol. 15, p. 147–160.

KIRK, D. N., and SHAW, P. M. (1975) Backbone rearrangements of steroidal 5-enes. *Journal of Chemical Society, Perkin I,* p. 2284–2294.

KLEEMANN, G., PORALLA, K., ENGLERT, G., KJØSEN, H., LIAAEN-JENSEN, S., NEUNLIST, S., and ROHMER, M. (1990) Tetrahymanol from the phototrophic bacterium *Rhodopseudomonas palustris*: first report of a gammacerane triterpene from a prokaryote. *Journal of General Microbiology*, Vol. 136, p. 2551–2553.

KNIGHTS, B. A., BROWN, A. C., CONWAY, E., and MIDDLEDITCH, B. S. (1970) Hydrocarbons from the green form of the freshwater alga *Botryococcus braunii. Phytochemistry,* Vol. 9, p. 1317–1324.

KOHL, W., GLOE, A., and REICHENBACH, H. (1983) Steroids from the myxobacterium *Nannocystis exedens. Journal of General Microbiology,* Vol. 129, p. 1629–1635.

KOHNEN, M. E. L., SINNINGHE DAMSTÉ, J. S., KOCK-VAN DALEN, A. C., and DE LEEUW, J. W. (1991) Di- or polysulfide-bound biomarkers in sulfur-rich geomacromolecules as revealed by selective chemolysis. *Geochimica et Cosmochimica Acta,* Vol. 55, p. 1375–1394.

KOLACZKOWSKA, E., SLOUGUI, N.-E., WATT, D. S., MARCURA, R. E., and MOLDOWAN, J. M. (1990) Thermodynamic stability of various alkylated, dealkylated, and rearranged 17α- and 17β-hopane isomers using molecular mechanics calculations. *Organic Geochemistry,* Vol. 16, p. 1033–1038.

VAN KREVELEN, D. W. (1961) *Coal.* Elsevier, New York, 514 p.

KROOS, B. M., and LEYTHAEUSER, D. (1988) Experimental measurements of the diffusion parameters of light hydrocarbons in water-saturated sedimentary rocks-II. Results and geochemical significance. *Organic Geochemistry,* Vol. 12, p. 91–108.

KROUSE, H. R., VIAU, C. A., ELIUK, L. S., UEDA, A., and HALAS, S. (1988) Chemical and isotopic evidence of thermochemical sulphate reduction by light hydrocarbon gases in deep carbonate reservoirs. *Nature,* Vol. 333, p. 415–419.

KRUGE, M. A., HUBERT, J. F., BENSLEY, D. F., CRELLING, J. C., and AKES, R. J. (1990) Organic geochemistry of a Lower Jurassic synrift lacustrine sequence, Hartford Basin, Connecticut, U.S.A. *Organic Geochemistry,* Vol. 16, p. 689–701.

KVENVOLDEN, K. A., and SIMONEIT, B. R. T. (1990) Hydrothermally derived petroleum: Examples from Guaymas Basin, Gulf of California, and Escanaba Trough, Northeast Pacific Ocean. *American Association of Petroleum Geologists Bulletin,* Vol. 74, p. 223–237.

LARCHER, A. V., ALEXANDER, R., and KAGI, R. I. (1987) Changes in configuration of extended moretanes with increasing sediment maturity. *Organic Geochemistry,* Vol. 11, p. 59–63.

DE LEEUW, J. W., and SINNINGHE DAMSTÉ, J. S. (1990) Organic sulfur compounds and other biomarkers as indicators of paleosalinity. In: *Geochemistry of Sulfur in Fossil Fuels* (W. L. Orr and C. M. White, eds.) American Chemical Society Symposium Series 429, Washington D.C., p. 417–443.

DE LEEUW, J. W., COX, H. C., VAN GRAAS, G., VAN DE MEER, F. W., PEAKMAN, T. M., BAAS, J. M. A., and VAN DE GRAAF, V. (1989) Limited double bond isomerization and selective hydrogenation of sterenes during early diagenesis. *Geochimica et Cosmochimica Acta,* Vol. 53, p. 903–909.

LEWAN, M. D. (1984) Factors controlling the proportionality of vanadium to nickel in crude oils. *Geochimica et Cosmochimica Acta,* Vol. 48, p. 2231–2238.

LEWAN, M. D. (1985) Evaluation of petroleum generation by hydrous pyrolysis experimentation. *Philosophical Transactions of the Royal Society of London,* Vol. A 315, p. 123–134.

LEWAN, M. D., BJORØY, M., and DOLCATER, D. L. (1986) Effects of thermal maturation on steroid hydrocarbons as determined by hydrous pyrolysis of Phosphoria Retort Shale. *Geochimica et Cosmochimica Acta,* Vol. 50, p. 1977–1987.

LEYTHAEUSER, D., MACKENZIE, A., SCHAEFER, R. G., and BJORØY, M. (1984) A novel approach for recognition and quantification of hydrocarbon migration effects in shale-sandstone sequences. *American Association of Petroleum Geologists Bulletin,* Vol. 68, p. 196–219.

LIJMBACH, G. W. M. (1975) On the origin of petroleum. *Proceedings 9th World Petroleum Congress,* Vol. 2, Applied Science Publishers, London, p. 357–369.

LIN, L. H., MICHAEL, G. H., KOVACHEV, G., ZHU, H., PHILP, R. P., and LEWIS, C. A. (1989) Biodegradation of tar-sands bitumens from the Ardmore and Anadarko Basins, Carter County, Oklahoma. *Organic Geochemistry,* Vol. 14, p. 511–523.

LONGMAN, M. W., and PALMER, S. E. (1987) Organic geochemistry of mid-continent Middle and Late Ordovician oils. *American Association of Petroleum Geologists Bulletin,* Vol. 71, p. 938–950.

LOUDA, J. W., and BAKER, E. W. (1986) The biogeochemistry of chlorophyll. In: *Organic Marine Chemistry* (M. L. Sohn, ed.) American Chemical Society Symposium Series, Vol. 305, p. 107–141.

LOUREIRO, M. R., and CARDOSO, J. N. (1990) Aromatic hydrocarbons in the Paraiba Valley oil shale. *Organic Geochemistry,* Vol. 15, p. 351–359.

LUDWIG, B., HUSSLER, G., WEHRUNG, P., and ALBRECHT, P. (1981) C_{26}-C_{29} triaromatic steroid derivatives in sediments and petroleums. *Tetrahedron Letters,* Vol. 22, p. 3313–3316.

MACKENZIE, A. S. (1984) Application of biological markers in petroleum geochemistry. In: *Advances in Petroleum Geochemistry,* Vol. 1, (J. Brooks and D. H. Welte, eds.) Academic Press, London, p. 115–214.

MACKENZIE, A. S., and MCKENZIE, D. (1983) Isomerization and aromatization of hydrocarbons in sedimentary basins formed by extension. *Geology Magazine,* Vol. 120, p. 417–470.

MACKENZIE, A. S., PATIENCE, R. L., MAXWELL, J. R., VANDENBROUCKE, M., and DURAND, B. (1980) Molecular parameters of maturation in the Toarcian shales, Paris Basin, France-I. Changes in the configuration of acyclic isoprenoid alkanes, steranes, and triterpanes. *Geochimica et Cosmochimica Acta,* Vol. 44, p. 1709–1721.

MACKENZIE, A. S., HOFFMANN, C. F., and MAXWELL, J. R. (1981a) Molecular parameters of maturation in the Toarcian shales, Paris Basin, France-III. Changes in aromatic steroid hydrocarbons. *Geochimica et Cosmochimica Acta,* Vol. 45, p. 1345–1355.

MACKENZIE, A. S., LEWIS, C. A., and MAXWELL, J. R. (1981b) Molecular parameters of maturation in the Toarcian shales, Paris Basin France—IV. Laboratory thermal alteration studies. *Geochimica et Cosmochimica Acta,* Vol. 45, p. 2369–2376.

MACKENZIE, A. S., BRASSELL, S. C., EGLINTON, G., and MAXWELL, J. R. (1982a) Chemical fossils: the geological fate of steroids. *Science,* Vol. 217, p. 491–504.

MACKENZIE, A. S., LAMB, N. A. and MAXWELL, J. R. (1982b) Steroid hydrocarbons and the thermal history of sediments. *Nature,* Vol. 295, p. 223–226.

MACKENZIE, A. S., DISKO, U., and RULLKÖTTER, J. (1983a) Determination of hydrocarbon distributions in oils and sediment extracts by gas chromatography-high resolution mass spectrometry. *Organic Geochemistry,* Vol. 5, p. 57–63.

MACKENZIE, A. S., LI REN-WEI, MAXWELL, J. R., MOLDOWAN, J. M., and SEIFERT, W. K. (1983b) Molecular measurements of thermal maturation of Cretaceous shales from the Overthrust Belt, Wyoming, USA. In: *Advances in Organic Geochemistry 1981* (M. Bjorøy et al., eds.) J. Wiley and Sons, New York, p. 496–503.

MACKENZIE, A. S., WOLFF, G. A., and MAXWELL, J. R. (1983c) Fatty acids in some biodegraded petroleums. Possible origins and significance. In: *Advances in Organic Geochemistry 1981* (M. Bjorøy et al., eds.) J. Wiley and Sons, New York, p. 637–649.

MACKENZIE, A. S., MAXWELL, J. R., COLEMAN, M. L., and DEEGAN, C. E. (1984) Biological marker and isotope studies of North Sea crude oils and sediments. In: *Proceedings of the Eleventh World Petroleum Congress,* J. Wiley and Sons, New York, Vol. 2, p. 45–56.

MACKENZIE, A. S., RULLKÖTTER, J., WELTE, D. H., and MANKIEWICZ, P. (1985) Reconstruction of oil formation and accumulation in North Slope, Alaska, using quantitative gas chromatography-mass spectrometry. In: *Alaska North Slope Oil/Source Rock Correlation Study* (L. B. Magoon and G. E. Claypool, eds.) American Association of Petroleum Geologists Studies in Geology No. 20, p. 319–377.

MACKENZIE, A. S., PRICE, I., LEYTHAEUSER, D., MÜLLER, P., RADKE, M., and SCHAEFER, R. G. (1987) The expulsion of petroleum from Kimmeridge clay source-rocks in the area of the Brae Oilfield, UK continental shelf. In: *Petroleum Geology of North West Europe* (J. Brooks and K. Glennie, eds.) Graham and Trotman, London, p. 865–877.

MACKENZIE, A. S., LEYTHAEUSER, D., ALTEBAUMER, F.-J, DISKO, U., and RULLKÖTTER, J. (1988a) Molecular measurements of maturity for Lias delta shales in N.W. Germany. *Geochimica et Cosmochimica Acta,* Vol. 52, p. 1145–1154.

MACKENZIE, A. S., LEYTHAEUSER, D., MÜLLER, P., QUIGLEY, T. M., and RADKE, M. (1988b) The movement of hydrocarbons in shales. *Nature,* Vol. 331, p. 63–65.

MAGOON, L. B., and CLAYPOOL, G. E. (1981) Two oil types on the North Slope of Alaska - Implications for future exploration. *American Association of Petroleum Geologists Bulletin,* Vol. 65, p. 644–652.

MAGOON, L. B., and CLAYPOOL, G. E. (1983) Petroleum geochemistry of the North Slope of Alaska: Time and degree of thermal maturity. In: *Advances in Organic Geochemistry 1981* (M. Bjorøy et al., eds.) J. Wiley and Sons, New York, p. 28–38.

MAGOON, L. B., and CLAYPOOL, G. E. (1984) The Kingak shale of north Alaska - Regional variations in organic geochemical properties and petroleum source rock quality. *Organic Geochemistry,* Vol. 6, p. 533–542.

MAIR, B. J., RONEN, Z., EISENBRAUN, E. J., and HORODYSKY, A. G. (1966) Terpenoid precursors of hydrocarbons from the gasoline range of petroleum. *Science,* Vol. 154, p. 1339–1341.

MANGO, F. D. (1991) The stability of hydrocarbons under the time-temperature conditions of petroleum genesis. *Nature,* Vol. 352, p. 146–148.

MANN, A. L., GOODWIN, N. S., and LOWE, S. (1987) Geochemical characteristics of lacustrine source rocks: A combined palynological/molecular study of a Tertiary sequence from offshore China. In: *Proceedings of the Indonesian Petroleum Association, Sixteenth Annual Convention.* Jakarta, Indonesian Petroleum Association, Vol. 1, p. 241–258.

MARZI, R., and RULLKÖTTER, J. (1992) Qualitative and quantitative evolution and kinetics of biological marker transformations—Laboratory experiments and application to the Michigan Basin. In: *Biological Markers in Sediments and Petroleum* (J. M. Moldowan, P. Albrecht, and R. P. Philp, eds.) Prentice Hall, Englewood Cliffs, N.J., p. 18–41.

MARZI, R., RULLKÖTTER, J., and PERRIMAN, W. S. (1990) Application of the change of sterane isomer ratios to the reconstruction of geothermal histories: Implications of the results of hydrous pyrolysis experiments. *Organic Geochemistry,* Vol. 16, p. 91–102.

MASON, G. M., TRUDELL, L. G., and BRANTHAVER, J. F. (1990) Review of the stratigraphic distribution and diagenetic history of abelsonite. *Organic Geochemistry,* Vol. 14, p. 585–594.

MAXWELL, J. R., DOUGLAS, A. G., EGLINTON, G., and MCCORMICK, A. (1968) The botryococcenes - hydrocarbons of novel structure from the alga *Botryococcus braunii Kutzing. Phytochemistry,* Vol. 7, p. 2157–2171.

MAXWELL, J. R., COX, R. E., EGLINTON, G., PILLINGER, C. T., ACKMAN, R. G., and HOOPER, S. N. (1973) Stereochemical studies of acyclic isoprenoid compounds—II. The role of chlorophyll in the derivation of isoprenoid-type acids in a lacustrine sediment. *Geochimica et Cosmochimica Acta,* Vol. 37, p. 297–313.

MCCAFFREY, M. A., FARRINGTON, J. W., and REPETA, D. J. (1989) Geochemical implications of the lipid composition of *Thioploca spp.* from the Peru upwelling region—15 S. *Organic Geochemistry,* Vol. 14, p. 61–68.

MCFADDEN, W. H. (1973) *Techniques of Combined Gas Chromatography Mass Spectrometry.* Wiley-Interscience, New York, 463 p.

MCKIRDY, D. M., ALDRIDGE, A. K., and YPMA, P. J. M. (1983) A geochemical comparison of some crude oils from Pre-Ordovician carbonate rocks. In: *Advances in Organic Geochemistry 1981* (M. Bjorøy et al., eds.) J. Wiley and Sons, New York, p. 99–107.

MCKIRDY, D. M., KANTSLER, A. J., EMMETT, J. K., and ALDRIDGE, A. K. (1984) Hydrocarbon genesis and organic facies in Cambrian carbonates of the Eastern Officer Basin, South Australia. In: *Petroleum Geochemistry and Source Rock Potential of Carbonate Rocks* (J. G. Palacas, ed.) American Association of Petroleum Geologists, Studies in Geology No. 18, p. 13–31.

MCKIRDY, D. M., COX, R. E., VOLKMAN, J. K., and HOWELL, V. J. (1986) Botryococcane in a new class of Australian nonmarine crude oils. *Nature,* Vol. 320, p. 57–59.

MCLAFFERTY, F. W. (1980) *Interpretation of Mass Spectra.* 3rd Ed., University Science Books, Mill Valley, California, p. 303.

MCMAHON, P. B., and CHAPELLE, F. H. (1991) Microbial production of organic acids in aquitard sediments and its role in aquifer geochemistry. *Nature,* Vol. 349, p. 233–235.

MEISSNER, F. F., WOODWARD, J., and CLAYTON, J. L. (1984) Stratigraphic relationships and distribution of source rocks in the Greater Rocky Mountain Region. In: *Hydrocarbon*

Source Rocks of the Greater Rocky Mountain Region (J. Woodward, F. F. Meissner, and J. L. Clayton, eds.) Rocky Mountain Association of Geologists, p. 1–34.

MELLO, M. R., GAGLIANONE, P. C., BRASSELL, S. C., and MAXWELL, J. R. (1988a) Geochemical and biological marker assessment of depositional environments using Brazilian offshore oils. *Marine and Petroleum Geology,* Vol. 5, p. 205–223.

MELLO, M. R., TELNAES, N., GAGLIANONE, P. C., CHICARELLI, M. I., BRASSELL, S. C., and MAXWELL, J. R. (1988b) Organic geochemical characterization of depositional paleoenvironments in Brazilian marginal basins. *Organic Geochemistry,* Vol. 13, p. 31–46.

MELLO, M. R., KOUTSOUKOS, E. A. M., HART, M. B., BRASSELLL, S. C., and MAXWELL, J. R. (1990) Late Cretaceous anoxic events in the Brazilian continental margin. *Organic Geochemistry,* Vol. 14, p. 529–542.

METZGER, P., CASADEVALL, E., POUET, M. J., and POUET, Y. (1985a) Structures of some botryococcenes: Branched hydrocarbons from the B-race of the green alga *Botryococcus braunii. Phytochemistry,* Vol. 14, p. 2995–3002.

METZGER, P., BERKALOFF, C., CASADEVALL, E., and COUTE, A. (1985b) Alkadiene- and botryococcene producing races of wild strains of *Botroycoccus braunii. Phytochemistry,* Vol. 24, p. 2305–2312.

MICHAEL, G. E., LIN, L. H., PHILP, R. P., LEWIS, C. A., and JONES, P. J. (1990) Biodegradation of tar-sand bitumens from the Ardmore/Anadarko Basins, Oklahoma-II. Correlation of oils, tar-sands, and source rocks. *Organic Geochemistry,* Vol. 14, p. 619–633.

MICHALCZYK, G. (1985) Determination of *n*- and *iso*-paraffins in hydrocarbon waxes—comparative results of analyses by gas chromatography, urea adduction, and molecular sieve adsorption (in German) *Fette-Seifen-Anstrichmittel,* Vol. 87, p. 481–486.

MILES, J. A. (1989) *Illustrated Glossary of Petroleum Geochemistry*. Oxford University Press, Oxford, 137 p.

MILNER, C. W. D., ROGERS, M. A., and EVANS, C. R. (1977) Petroleum transformations in reservoirs. *Journal of Geochemical Exploration,* Vol. 7, p. 101–153.

MISLOW, K. (1965) *Introduction to Stereochemistry,* W. A. Benjamin, New York, p. 193.

MOLDOWAN, J. M., and FAGO, F. J. (1986) Structure and significance of a novel rearranged monoaromatic steroid hydrocarbon in petroleum. *Geochimica et Cosmochimica Acta,* Vol. 50, p. 343–351.

MOLDOWAN, J. M., and SEIFERT, W. K. (1979) Head-to-head linked isoprenoid hydrocarbons in petroleum. *Science,* Vol. 204, p. 169–171.

MOLDOWAN, J. M., and SEIFERT, W. K. (1980) First discovery of botryococcane in petroleum. *Journal of the Chemical Society, Chemical Communications,* p. 912–914.

MOLDOWAN, J. M., SEIFERT, W. K., and GALLEGOS, E. J. (1983) Identification of an extended series of tricyclic terpanes in petroleum. *Geochimica et Cosmochimica Acta.,* Vol. 47, p. 1531–1534.

MOLDOWAN, J. M., SEIFERT, W. K., ARNOLD, E., and CLARDY, J. (1984) Structure proof and significance of stereoisomeric 28,30-bisnorhopanes in petroleum and petroleum source rocks. *Geochimica et Cosmochimica Acta,* Vol. 48, p. 1651–1661.

MOLDOWAN, J. M., SEIFERT, W. K., and GALLEGOS, E. J. (1985) Relationship between petroleum composition and depositional environment of petroleum source rocks. *American Association of Petroleum Geologists Bulletin,* Vol. 69, p. 1255–1268.

MOLDOWAN, J. M., SUNDARARAMAN, P., and SCHOELL, M. (1986) Sensitivity of biomarker properties to depositional environment and/or source input in the Lower Toarcian of S.W. Germany. *Organic Geochemistry,* Vol. 10, p. 915–926.

MOLDOWAN, J. M., FAGO, F. J., LEE, C. Y., JACOBSON, S. R., WATT, D. S., SLOUGUI, N. E., JEGANATHAN, A., and YOUNG, D. C. (1990) Sedimentary 24-*n*-propylcholestanes, molecular fossils diagnostic of marine algae. *Science,* Vol. 247, p. 309–312.

MOLDOWAN, J. M., LEE, C. Y., WATT, D. S., JEGANATHAN, A., SLOUGUI, N. E., and GALLEGOS, E. J. (1991a) Analysis and occurrence of C_{26}-steranes in petroleum and source rocks. *Geochimica et Cosmochimica Acta,* Vol. 55, p. 1065–1081.

MOLDOWAN, J. M., FAGO, F. J., CARLSON, R. M. K., YOUNG, D. C., DUYNE, G. V., CLARDY, J., SCHOELL, M., PILLINGER, C. T., and WATT, D. S. (1991b) Rearranged hopanes in sediments and petroleum. *Geochimica et Cosmochimica Acta,* Vol. 55, p. 3333–3353.

MOLDOWAN, J. M., FAGO, F. J., HUIZINGA, B. J., and JACOBSON, S. R. (1991c) Analysis of oleanane and its occurrence in Upper Cretaceous rocks. In: *Organic Geochemistry. Advances and Applications in the Natural Environment* (D. A. C. Manning, ed.) Manchester University Press, Manchester U.K., p. 195–197.

MOLDOWAN, J. M., SUNDARARAMAN, P., SALVATORI, T., ALAJBEG, A., GJUKIC, B., LEE, C. Y., and DEMAISON, G. J. (1992) Source correlation and maturity assessment of select oils and rocks from the Central Adriatic Basin (Italy and Yugoslavia). In: *Biological Markers in Sediments and Petroleum* (J. M. Moldowan, P. Albrecht, and R. P. Philp, eds.) Prentice Hall, Englewood Cliffs, N. J., p. 370–401.

MOOK, W. G. (1984, unpubl.) *Principles of Isotope Hydrology*. Course notes, Department of Hydrogeology and Geographical Hydrology, University of Groningen, Netherlands.

MORRISON, R. T., and BOYD, R. N. (1966) Organic Geochemistry. Allyn and Bacon, Inc., Boston, 1204 p.

MURPHY, M. T. J., MCCORMICK, A., and EGLINTON, G. (1967) Perhydro-β-carotene in Green River shale. *Science,* Vol. 157, p. 1040–1042.

MYCKE, B., NARJES, F., and MICHAELIS, W. (1987) Bacteriohopanetetrol from chemical degradation of an oil shale kerogen. *Nature,* Vol. 326, p. 179–181.

NES, W. R., and MCKEAN, M. L. (1977) *Biochemistry of Steroids and Other Isopentenoids*. University Park Press, Baltimore, 690 p.

NOBLE, R. A. (1986) A geochemical study of bicyclic alkanes and diterpenoid hydrocarbons in crude oils, sediments, and coals. Ph.D. Thesis, Department of Organic Chemistry, University of Western Australia, 365 p.

NOBLE, R. A., and ALEXANDER, R. (1989), Origin and significance of drimanes and related bicyclic alkanes in crude oils and ancient sediments. In: *Biomarkers in Petroleum,* Memorial Symposium for W. K. Seifert, Division of Petroleum Chemistry, Inc., American Chemical Society, Dallas Meeting, April 9–14, 1989, Preprints, p. 138.

NOBLE, R. A., KNOX, J., ALEXANDER, R., and KAGI, R. I. (1985a) Identification of tetracyclic diterpane hydrocarbons in Australian crude oils and sediments. *Journal of the Chemical Society, Chemical Communications*, p. 32–33.

NOBLE, R. A., ALEXANDER, R., KAGI, R. I., and KNOX, J. (1985b) Tetracyclic diterpenoid hydrocarbons in some Australian coals, sediments and crude oils. *Geochimica et Cosmochimica Acta,* Vol. 49, p. 2141–2147.

NOBLE, R., ALEXANDER, R., and KAGI, R. I. (1985c) The occurrence of bisnorhopane, trisnorhopane, and 25-norhopanes as free hydrocarbons in some Australian shales. *Organic Geochemistry,* Vol. 8, p. 171–176.

NOBLE, R. A., ALEXANDER, R., KAGI, R. I., and KNOX, J. (1986) Identification of some diterpenoid hydrocarbons in petroleum. *Organic Geochemistry,* Vol. 10, p. 825–829.

NOBLE, R. A., WU, C. H., and ATKINSON, C. D. (1991) Petroleum generation and migration from Talang Akar coals and shales offshore N. W. Java, Indonesia. *Organic Geochemistry,* Vol. 17, p. 363–374.

OCAMPO, R., CALLOT, H. J., ALBRECHT, P., and KINTZINGER, J. P. (1984) A novel chlorophyll c related petroporphyrin in oil shale. *Tetrahedron Letters,* Vol. 25, p. 2589–2592.

OCAMPO, R., CALLOT, H. J., and ALBRECHT, P. (1985a) Identification of polar porphyrins in oil shale. *Journal of the Chemical Society, Chemical Communications,* p. 198–200.

OCAMPO, R., CALLOT, H. J., and ALBRECHT, P. (1985b) Occurrence of bacteriopetroporphyrins in oil shale. *Journal of the Chemical Society, Chemical Communications,* p. 200–201.

OCAMPO, R., CALLOT, H. J., and ALBRECHT, P. (1989) Different isotope compositions of C_{32} DPEP and C_{32} etioporphyrin III in oil shale. *Naturwissenschaften,* Vol. 76, p. 419–421.

ORR, W. L. (1974) Changes in sulfur content and isotopic ratios of sulfur during petroleum maturation. Study of Big Horn Basin Paleozoic oils. *American Association of Petroleum Geologists Bulletin,* Vol. 50, p. 2295–2318.

ORR, W. L. (1986) Kerogen/asphaltene/sulfur relationships in sulfur-rich Monterey oils. *Organic Geochemistry,* Vol. 10, p. 499–516.

ORR, W. L., and GAINES, A. G. (1974) Observations on the rate of sulfate reduction and organic matter oxidation in the bottom waters of an estuarine basin: the upper basin of the Pettaquamscutt River (Rhode Island) In: *Advances in Organic Geochemistry 1973* (B. Tissot and F. Bienner, eds.) Éditions Technip, Paris, p. 791–812.

OURISSON, G., ALBRECHT, P., and ROHMER, M. (1979) The hopanoids. Palaeochemistry and biochemistry of a group of natural products. *Pure and Applied Chemistry,* Vol. 51, p. 709–729.

OURISSON, G., ALBRECHT, P., and ROHMER, M. (1982) Predictive microbial biochemistry, from molecular fossils to procaryotic membranes. *Trends in Biochemical Sciences,* Vol. 7, p. 236–239.

OURISSON, G., ALBRECHT, P., and ROHMER, M. (1984) The microbial origin of fossil fuels. *Scientific American,* Vol. 251, p. 44–51.

OURISSON, G., ROHMER, M., and PORALLA, K. (1987) Prokaryotic hopanoids and other polyterpenoid sterol surrogates. *Annual Review of Microbiology,* Vol. 41, p. 301–333.

PALACAS, J. G. (1984) Carbonate rocks as sources of petroleum: Geological and chemical characteristics and oil-source correlations. *Proceedings of the Eleventh World Petroleum Congress 1983,* London, Vol. 2, p. 31–43.

PALACAS, J. G., ANDERS, D. E., and KING, J. D. (1984) South Florida Basin—A prime example of carbonate source rocks in petroleum. In: *Petroleum Geochemistry and Source Rock Potential of Carbonate Rocks* (J. G. Palacas, ed.) American Association of Petroleum Geologists, Studies in Geology No. 18, p. 71–96.

PALACAS, J. G., MONOPOLIS, D., NICOLAOU, C. A., and ANDERS, D. E. (1986) Geochemical correlation of surface and subsurface oils, western Greece. *Organic Geochemistry,* Vol. 10, p. 417–423.

PALMER, S. E. (1984a) Hydrocarbon source potential of organic facies of the lacustrine Elko Formation (Eocene/Oligocene), Northeast Nevada. In: *Hydrocarbon Source Rocks of the Greater Rocky Mountain Region* (J. Woodward, F. F. Meissner, and J. L. Clayton, eds.) Rocky Mountain Association of Geologists, Denver, p. 491–511.

PALMER, S. E. (1984b) Effect of water washing on $C_{15}+$ hydrocarbon fraction of crude oils from northwest Palawan, Philippines. *American Association of Petroleum Geologists Bulletin,* Vol. 68, p. 137–149.

PATIENCE, R. L., YON, D. A., RYBACK, G., and MAXWELL, J. R. (1980) Acyclic isoprenoid alkanes and geochemical maturation. In: *Advances in Organic Geochemistry 1979* (A. G. Douglas and J. R. Maxwell, eds.) Pergamon, New York, p. 287–294.

PAYZANT, J. D., MONTGOMERY, D. S., and STRAUSZ, O. P. (1986) Sulfides in petroleum. *Organic Geochemistry,* Vol. 9, p. 357–369.

PEAKMAN, T. M., TEN HAVEN, H. L., RECHKA, J. R., DE LEEUW, J. W., and MAXWELL, J. R. (1989) Occurrence of (20R)- and (20S)-$\Delta 8(^{14})$ and Δ^{14} 5α(H)-sterenes and the origin of 5α(H),14β(H),17β(H)-steranes in an immature sediment. *Geochimica et Cosmochimica Acta,* Vol. 53, p. 2001–2009.

PEDERSEN, T. F., and CALVERT, S. E. (1990) Anoxia vs. productivity: What controls the formation of organic-carbon-rich sediments and sedimentary rocks? *American Association of Petroleum Geologists Bulletin,* Vol. 74, p. 454–466.

PETERS, K. E. (1986) Guidelines for evaluating petroleum source rock using programmed pyrolysis. *American Association of Petroleum Geologists Bulletin,* Vol. 70, p. 318–329.

PETERS, K. E., and CASSA M. R. (1992) Applied source-rock geochemistry. In: *The Petroleum System—From Source to Trap* (L. B. Magoon and W. G. Dow, eds.) American Association of Petroleum Geologists Memoir, submitted.

PETERS, K. E., and MOLDOWAN, J. M. (1991) Effects of source, thermal maturity, and biodegradation on the distribution and isomerization of homohopanes in petroleum. *Organic Geochemistry,* Vol. 17, p. 47–61.

PETERS, K. E., and SIMONEIT, B. R. T. (1982) Rock-Eval pyrolysis of quaternary sediments from Leg 64, Sites 479 and 480, Gulf of California. *Initial Reports Deep Sea Drilling Project,* Vol. 64, p. 925–931.

PETERS, K. E., ROHRBACK, B. G., and KAPLAN, I. R. (1981) Carbon and hydrogen stable isotope variations in kerogen during laboratory simulated thermal maturation. *American Association of Petroleum Geologists Bulletin,* Vol. 65, p. 501–508.

PETERS, K. E., WHELAN, J. K., HUNT, J. M., and TARAFA, M. E. (1983) Programmed pyrolysis of organic matter from thermally altered Cretaceous black shales. *American Association of Petroleum Geologists Bulletin,* Vol. 67, p. 2137–2146.

PETERS, K. E., MOLDOWAN, J. M., SCHOELL, M., and HEMPKINS, W. B. (1986) Petroleum isotopic and biomarker composition related to source rock organic matter and depositional environment. *Organic Geochemistry,* Vol. 10, p. 17–27.

PETERS, K. E., MOLDOWAN, J. M., DRISCOLE, A. R., and DEMAISON, G. J. (1989) Origin of Beatrice oil by co-sourcing from Devonian and Middle Jurassic source rocks, Inner Moray Firth, U.K., *American Association of Petroleum Geologists Bulletin,* Vol. 73, p. 454–471.

PETERS, K. E., MOLDOWAN, J. M., and SUNDARARAMAN, P. (1990) Effects of hydrous pyrolysis on biomarker thermal maturity parameters: Monterey Phosphatic and Siliceous Members. *Organic Geochemistry,* Vol. 15, p. 249–265.

PETERS, K. E., KONTOROVICH, A. EH., ANDRUSEVICH, V. E., MOLDOWAN, J. M., HUIZINGA, B. J., DEMAISON, G. J., and STASOVA, O. F. (1992a) Biomarker hydrocarbons in oils and rocks from the Middle Ob region of West Siberia, Russia. *American Association of Petroleum Geologists Bulletin,* submitted.

PETERS, K. E., ELAM, T. D., PYTTE, M. H., and SUNDARARAMAN, P. (1992b) Eocene and Miocene petroleum systems in the San Joaquin basin, California. In: *The Petroleum System—From Source to Trap* (L. B. Magoon and W. G. Dow, eds.) American Association of Petroleum Geologists Memoir, submitted.

PETROV, AL. A. (1987) *Petroleum Hydrocarbons.* Springer-Verlag, New York, p. 255.

PETROV, AL. A., PUSTIL'NIKOVA, S. D., ABRIUTINA, N. N., and KAGRAMONOVA, G. R. (1976) Petroleum steranes and triterpanes. *Neftekhimiia,* Vol. 16, p. 411–427.

PETROV, A. A., VOROBYOVA, N. S., and ZEMSKOVA, Z. K. (1990) Isoprenoid alkanes with irregular "head-to-head" linkages. *Organic Geochemistry,* Vol. 16, p. 1001–1005.

PHILP, R. P. (1982) Application of computerized gas chromatography/mass spectrometry to fossil fuel research. *Spectra* (Finnigan MAT), Vol. 8, p. 6–31.

PHILP, R. P. (1983) Correlation of crude oils from the San Jorges Basin, Argentina. *Geochimica et Cosmochimica Acta.,* Vol. 47, p. 267–275.

PHILP, R. P. (1985) *Fossil Fuel Biomarkers.* Methods in Geochemistry and Geophysics, 23. Elsevier, New York, 294 p.

PHILP, R. P., and GILBERT, T. D. (1982) Unusual distribution of biological markers in an Australian crude oil. *Nature,* Vol. 299, p. 245–247.

PHILP, R. P., and GILBERT, T. D. (1986) Biomarker distributions in oils predominantly derived from terrigenous source material. In: *Advances in Organic Geochemistry 1985* (D. Leythaeuser and J. Rullkötter, eds.) Pergamon Press, p. 73–84.

PHILP, R. P., and OUNG, J.-N. (1992) Biomarker distributions in crude oils as determined by tandem mass spectrometry. In: *Biological Markers in Sediments and Petroleum* (J. M. Moldowan, P. Albrecht, and R. P. Philp, eds.) Prentice Hall, Englewood Cliffs, N.J., p. 106–123.

PHILP, R. P., GILBERT, T. D., and FRIEDRICH, J. (1981) Bicyclic sesquiterpenoids and diterpenoids in Australian crude oils. *Geochimica et Cosmochimica Acta,* Vol. 45, p. 1173–1180.

PHILP, R. P., SIMONEIT, B. R. T., and GILBERT, T. D. (1983) Diterpenoids in crude oils and coals of South Eastern Australia. In: *Advances in Organic Geochemistry 1981* (M. Bjorøy et al., eds.) J. Wiley and Sons, New York, p. 698–704.

PHILP, R. P., OUNG, J.-N., and LEWIS, C. A. (1988) Biomarker determinations in crude oils using a triple-stage quadrupole mass spectrometer. *Journal of Chromatography,* Vol. 446, p. 3–16.

PHILP, R. P., LI JINGGUI, and LEWIS, C. A. (1989) An organic geochemical investigation of crude oils from Shanganning, Jianghan, Chaidamu and Zhungeer Basins, People's Republic of China. *Organic Geochemistry,* Vol. 14, p. 447–460.

POOLE, F. G., and CLAYPOOL, G. E. (1984) Petroleum source-rock potential and crude-oil correlation in the Great Basin. In: *Hydrocarbon Source Rocks of the Greater Rocky Mountain Region* (J. Woodward, F. F. Meissner, and J. L. Clayton, eds.) Rocky Mountain Association of Geologists, Denver, p. 491–511.

POOLE, C. F., and SCHUETTE, S. A. (1984) *Contemporary Practice of Chromatography*. Elsevier, New York, 708 p.

POWELL, T. G., and MCKIRDY, D. M. (1973) Relationship between ratio of pristane to phytane, crude oil composition and geological environment in Australia. *Nature,* Vol. 243, p. 37–39.

PRATT, L. M., SUMMONS, R.E., and HIESHIMA, G. B. (1991) Sterane and triterpane biomarkers in the Precambrian Nonesuch Formation, North American Midcontinent Rift, *Geochimica et Cosmochimica Acta,* Vol. 55, p. 911–916.

PREMUZIC, E. T., GAFFNEY, J. S., and MANOWITZ, B. (1986) The importance of sulfur isotope ratios in the differentiation of Prudhoe Bay crude oils. *Journal of Geochemical Exploration,* Vol. 26, p. 151–159.

PRICE, L. C., and BARKER, C. E. (1985) Suppression of vitrinite reflectance in amorphous rich kerogen — A major unrecognized problem. *Journal of Petroleum Geology,* Vol. 8, p. 59–84.

PRICE, P. L., O'SULLIVAN, T., and ALEXANDER, R. (1987) The nature and occurrence of oil in Seram, Indonesia. In: *Proceedings of the Indonesian Petroleum Association. Sixteenth Annual Convention,* Volume 1, Indonesian Petroleum Association, Jakarta, p. 141–173.

PROWSE, W. G., KEELY, B. J., and MAXWELL, J. R. (1990) A novel sedimentary metallochlorin. *Organic Geochemistry,* Vol. 16, p. 1059–1065.

PUSTIL'NIKOVA, S. D., ABRYUTINA, N. N., KAYUKOVA, G. P., and PETROV, AL. A. (1980) Equilibrium composition and properties of epimeric cholestanes. (In Russian, Chevron Research Company translation TCT-Tr. 242-80) *Neftekhimiia,* Vol. 20, p. 26–33.

RADKE, M. (1987) Organic geochemistry of aromatic hydrocarbons. In: *Advances in Petroleum Geochemistry* (J. Brooks and D. Welte, eds.) Academic Press, New York, p. 141–207.

RADKE, M. (1988) Application of aromatic compounds as maturity indicators in source rocks and crude oils. *Marine Petroleum Geology,* Vol. 5, p. 224–236.

RADKE, M., and WELTE, D. H. (1983) The methylphenanthrene index (MPI). A maturity parameter based on aromatic hydrocarbons. In: *Advances in Organic Geochemistry 1981* (M. Bjorøy et al., eds.) J. Wiley and Sons, New York, p. 504–512.

RADKE, M., SCHAEFER, R. G., LEYTHAEUSER, D., and TEICHMÜLLER, M. (1980) Composition of soluble organic matter in coals: relation to rank and liptinite fluorescence. *Geochimica et Cosmochimica Acta,* Vol. 44, p. 1787–1800.

RADKE, M., WELTE, D. H., and WILLSCH, H. (1982) Geochemical study on a well in the Western Canada Basin: Relation of the aromatic distribution pattern to maturity of organic matter. *Geochimica et Cosmochimica Acta,* Vol. 46, p. 1–10.

RADKE, M., WELTE, D. H., and WILLSCH, H. (1986) Maturity parameters based on aromatic hydrocarbons: Influence of the organic matter type. *Organic Geochemistry,* Vol. 10, p. 51–63.

RADKE, M., GARRIGUES, P., and WILLSCH, H. (1990) Methylated dicyclic and tricyclic aromatic hydrocarbons in crude oils from the Handil field, Indonesia. *Organic Geochemistry,* Vol. 15, p. 17–34.

RAEDERSTORFF, D., and ROHMER, M. (1984) Sterols of the unicellular algae *Nematochrysopsis roscoffensis* and *Chrysotila lamellosa:* Isolation of (24E)-24-*n*-propylidenecholesterol and 24-*n*-propylcholesterol. *Phytochemistry,* Vol. 23, p. 2835–2838.

RAISWELL, R., and BERNER, R. A. (1985) Pyrite formation in euxinic and semi-euxinic sediments. *American Journal of Science,* Vol. 285, p. 710–724.

REED, W. E. (1977) Molecular compositions of weathered petroleum and comparison with its possible source. *Geochimica et Cosmochimica Acta,* Vol. 41, p. 237–247.

REED, J. D., ILLICH, H. A., and HORSFIELD, B. (1986) Biochemical evolutionary significance of Ordovician oils and their sources. *Organic Geochemistry,* Vol. 10, p. 347–358.

REQUEJO, A. G. (1992) Quantitative analysis of triterpane and sterane biomarkers: Methodology and applications in molecular maturity studies. In: *Biological Markers in Sediments and Petroleum* (J. M. Moldowan, P. Albrecht, and R. P. Philp, eds.) Prentice Hall, Englewood Cliffs, N.J., p. 223–240.

REQUEJO, A. G., and HALPERN, H. I. (1989) An unusual hopane biodegradation sequence in tar-sands from the Point Arena (Monterey) Formation. *Nature,* Vol. 342, p. 670–673.

REQUEJO, A. G., HOLLYWOOD, J., and HALPERN, H. I. (1989) Recognition and source correlation of migrated hydrocarbons in Upper Jurassic Hareelv Formation, Jameson Land, East Greenland. *American Association of Petroleum Geologists Bulletin,* Vol. 73, p. 1065–1088.

RESTLE, A. (1983) Etude de nouveaux marquers biologiques dans des petroles biodegrades: cas naturels et simulations in vitro. D.Sc. Thesis, L'Universite Louis Pasteur de Strasbourg, 215 p.

RICHARDSON, J. S., and MIILLER, D. E. (1982) Identification of dicyclic and tricyclic hydrocarbons in the saturate fraction of a crude oil by gas chromatography-mass spectrometry. *Analytical Chemistry,* Vol. 54, p. 765–768.

RICHARDSON, J. S., and MIILLER, D. E. (1983) Biologically-derived compounds of significance in the saturate fraction of a crude oil having predominant terrestrial input. *Fuel,* Vol. 62, p. 524–528.

RIEDIGER, C. L., FOWLER, M. G., BROOKS, P. W., and SNOWDON, L. R. (1990) Triassic oils and potential Mesozoic source rocks, Peace River Arch area, Western Canada Basin. *Organic Geochemistry,* Vol. 16, p. 295–305.

RINALDI, G. G. L., LEOPOLD, V. M., and KOONS, C. B. (1988) Presence of benzohopanes, monoaromatic secohopanes, and saturate hexacyclic hydrocarbons in petroleum from carbonate environments. In: *Geochemical Biomarkers* (T. F. Yen and J. M. Moldowan, eds.), Harwood Academic Publishers, New York, p. 331–353.

RIOLO, J., and ALBRECHT, P. (1985) Novel rearranged ring C monoaromatic steroid hydrocarbons in sediments and petroleums. *Tetrahedron Letters,* Vol. 26, p. 2701–2704.

RIOLO, J., LUDWIG, B., and ALBRECHT, P. (1985) Synthesis of ring C monoaromatic steroid hydrocarbons occurring in geological samples. *Tetrahedron Letters,* Vol. 26, p. 2607–2700.

RIOLO, J., HUSSLER, G., ALBRECHT, P., and CONNAN, J. (1986) Distribution of aromatic steroids in geological samples: Their evaluation as geochemical parameters. *Organic Geochemistry*, Vol. 10, p. 981–990.

RISATTI, J. B., ROWLAND, S. J., YAN, D. A., and MAXWELL, J. R. (1984) Stereochemical studies of acyclic isoprenoids—XII. Lipids of methanogenic bacteria and possible contributors to sediments. *Organic Geochemistry*, Vol. 6, p. 93–104.

RIVA, A., CACCIALANZA, P. G., and QUAGLIAROLI, F. (1988) Recognition of 18β(H)-oleanane in several crudes and Tertiary-Upper Cretaceous sediments. Definition of a new maturity parameter. *Organic Geochemistry*, Vol. 13, p. 671–675.

ROHMER, M. (1987) The hopanoids, prokaryotic triterpenoids and sterol surrogates. In: *Surface Structures of Microorganisms and Their Interactions with the Mammalian Host* (E. Schriner et al., eds.) Proceedings of the Eighteenth Workshop Conference, Hocchst, Schloss Ringberg, October 20–23, 1987, VCH, p. 227–242.

ROHMER, M., and OURISSON, G. (1976) Methyl-hopanes *d'Acetobacter xylinum* et *d'Acetobacter rancens:* une nouvelle famille de composes triterpeniques. *Tetrahedron Letters*, Vol. 40, p. 3641–3644.

ROHMER, M., BOUVIER, P., and OURISSON, G. (1979) Molecular evolution of biomembranes: Structural equivalents and phylogenetic precursors of sterols. Proceedings of the National Academy of Sciences U.S.A. 76, p. 847–851.

ROHMER, M., BISSERET, P., and NEUNLIST, S. (1992) The hopanoids, prokaryotic triterpenoids and precursors of ubiquitous molecular fossils. In: *Biological Markers in Sediments and Petroleum* (J. M. Moldowan, P. Albrecht, and R. P. Philp, eds.) Prentice Hall, Englewood Cliffs, N.J., p. 1–17.

ROHRBACK, B. G. (1983) Crude oil geochemistry of the Gulf of Suez. In: *Advances in Organic Geochemistry 1981* (M. Bjorøy et al., eds.) J. Wiley and Sons, New York, p. 39–48.

ROWLAND, S. J. (1990) Production of acyclic isoprenoid hydrocarbons by laboratory maturation of methanogenic bacteria. *Organic Geochemistry*, Vol. 15, p. 9–16.

RUBINSTEIN, I., and ALBRECHT, P. (1975) The occurrence of nuclear methylated steranes in a shale. *Journal of the Chemical Society, Chemical Communications*, 1975, p. 957–958.

RUBINSTEIN, I., SIESKIND, O., and ALBRECHT P. (1975) Rearranged sterenes in a shale: Occurrence and simulated formation. *Journal of the Chemical Society, Perkin Transaction I*, p. 1833–1836.

RUBINSTEIN, I., STRAUSZ, O. P., SPYCKERELLE, C., CRAWFORD, R. J., and WESTLAKE, D. W. S. (1977) The origin of oil sand bitumens of Alberta. *Geochimica et Cosmochimica Acta*, Vol. 41, p. 1341–1353.

RULLKÖTTER, J., and MARZI, R. (1988) Natural and artificial maturation of biological markers in a Toarcian shale from northern Germany. *Organic Geochemistry*, Vol. 13, p. 639–645.

RULLKÖTTER, J., and NISSENBAUM, A. (1988) Dead Sea asphalt in Egyptian mummies: Molecular evidence. *Naturwissenschaften*, Vol. 75, p. 618–621.

RULLKÖTTER, J., and PHILP, P. (1981) Extended hopanes up to C_{40} in Thornton bitumen. *Nature*, Vol. 292, p. 616–618.

RULLKÖTTER, J., and WENDISCH, D. (1982) Microbial alteration of 17α(H)-hopane in Madagascar asphalts: Removal of C-10 methyl group and ring opening. *Geochimica et Cosmochimica Acta,* Vol. 46, p. 1543–1553.

RULLKÖTTER, J., LEYTHAEUSER, D., and WENDISCH, D. (1982) Novel 23,28-bisnorlupanes in Tertiary sediments. Widespread occurrence of nuclear demethylated triterpanes. *Geochimica et Cosmochimica Acta,* Vol. 46, p. 2501–2509.

RULLKÖTTER, J., AIZENSHTAT, Z., and SPIRO, B. (1984a) Biological markers in bitumens and pyrolyzates of Upper Cretaceous bituminous chalks from the Ghareb Formation (Israel). *Geochimica et Cosmochimica Acta,* Vol. 48, p. 151–157.

RULLKÖTTER, J., MACKENZIE, A. S. , WELTE, D. H., LEYTHAEUSER, D., and RADKE, M. (1984b) Quantitative gas chromatography-mass spectrometry analysis of geological samples. *Organic Geochemistry,* Vol. 6, p. 817–827.

RULLKÖTTER, J., SPIRO, B., and NISSENBAUM, A. (1985) Biological marker characteristics of oils and asphalts from carbonate source rocks in a rapidly subsiding graben, Dead Sea, Israel. *Geochimica et Cosmochimica Acta.* 49, p. 1357–1370.

RULLKÖTTER, J., MEYERS, P. A., SCHAEFER, R. G., and DUNHAM, K. W. (1986) Oil generation in the Michigan Basin: A biological marker carbon and isotope approach. *Organic Geochemistry,* Vol. 10, p. 359–375.

SAKATA, S., SUZUKI, N., and KANEKO, N. (1988) A biomarker study of petroleum from the Neogene Tertiary sedimentary basins in Northeast Japan. *Geochemistry Journal,* Vol. 22, p. 89–105.

SAMMAN, N., IGNASIAK, T., CHEN, C.-J., STRAUSZ, O. P., and MONTGOMERY, D. S. (1981) Squalene in petroleum asphaltenes. *Science,* Vol. 213, p. 1381–1383.

SCALAN, R. S., and SMITH, J. E. (1970) An improved measure of the odd-even predominance in the normal alkanes of sediment extracts and petroleum. *Geochimica et Cosmochimica Acta,* Vol. 34, p. 611–620.

SCHILDOWSKI, M., MATZIGKEIT, U., and KRUMBEIN, W. E. (1984) Superheavy organic carbon from hypersaline microbial mats. *Naturwissenschaften,* Vol. 71, p. 303–308.

SCHMID, J. C. (1986) Marqueurs biologiques soufres dans les petroles. Ph.D. Thesis, U. Strausbourg, France, 263 p.

SCHMID, J. C., CONNAN, J., and ALBRECHT, P. (1987) Occurrence and geochemical significance of long-chain dialkylthiocyclopentanes. *Nature,* Vol. 329, p. 54–56.

SCHOELL, M. (1984) Stable isotopes in petroleum research. In: *Advances in Petroleum Geochemistry,* Vol. 1 (J. Brooks and D. H. Welte, eds.) Academic Press, London, p. 215–245.

SCHOELL, M. (1988) Multiple origins of methane in the earth. *Chemical Geology,* Vol. 71, p. 1–10.

SCHOELL, M., and WELLMER, F.-W. (1981) Anomalous ^{13}C depletion in early Precambrian graphites from Superior Province, Canada. *Nature,* Vol. 290, p. 696–699.

SCHOELL, M., TESCHNER, M. WEHNER, H, DURAND, B., and OUDIN, J. L. (1983) Maturity related biomarker and stable isotope variations and their application to oil/source rock correlation in the Mahakam Delta, Kalimantan. In: *Advances in Organic Geochemistry 1981* (M. Bjorøy et al., eds.) J. Wiley and Sons, New York, p. 156–163.

SCHUBERT, K., ROSE, G., WACHTEL, H., HORHOLD, C., and IKEKAWA, N. (1968) Zum vorkommen von sterinen in bacterien. *European Journal of Biochemistry,* Vol. 5, p. 246.

SCHULZE, T., and MICHAELIS, W. (1990) Structure and origin of terpenoid hydrocarbons in some German coals. *Organic Geochemistry,* Vol. 16, p. 1051–1058.

SCHUMACKER, D., and KENNICUTT, M. C. (1989) *Geochemistry of Gulf Coast Oils and Gases.* Gulf Coast Section, Society of Economic Paleontologists and Mineralogists, 9th Annual Research Conference, New Orleans, Dec. 1988, p. 15.

SCHWARK, L., and PÜTTMAN, W. (1990) Aromatic hydrocarbon composition of the Permian Kuperschiefer in the Lower Rhine Basin, NW Germany. *Organic Geochemistry,* Vol. 16, p. 749–761.

SEIFERT, W. K. (1975) Carboxylic acids in petroleum in sediments. *Fortschritte der Chemie Organischer Naturstoffe,* Vol. 32, p. 1–49.

SEIFERT, W. K. (1977) Source rock/oil correlations by C_{27}-C_{30} biological marker hydrocarbons. In: *Advances in Organic Geochemistry 1974* (R. Campos and J. Goni, eds.), ENADIMSA, Madrid, p. 21–44.

SEIFERT, W. K. (1978) Steranes and terpanes in kerogen pyrolysis for correlation of oils and source rocks. *Geochimica et Cosmochimica Acta,* Vol. 42, p. 473–484.

SEIFERT, W. K. (1980) Impact of Treib's discovery of porphyrins on present day biological marker organic geochemistry. In: *Proceedings of the Treibs International Symposium* (A. Prashnowsky, ed.), Munich, July 1979, Halbigdruck Publ., Wurzburg, Germany, p. 13–35.

SEIFERT, W. K., and MOLDOWAN, J. M. (1978) Applications of steranes, terpanes, and monoaromatics to the maturation, migration, and source of crude oils. *Geochimica et Cosmochimica Acta,* Vol. 42, p. 77–95.

SEIFERT, W. K., and MOLDOWAN, J. M. (1979) The effect of biodegradation on steranes and terpanes in crude oils. *Geochimica et Cosmochimica Acta,* Vol. 43, p. 111–126.

SEIFERT, W. K., and MOLDOWAN, J. M. (1980) The effect of thermal stress on source-rock quality as measured by hopane stereochemistry. *Physics and Chemistry of the Earth,* Vol. 12, p. 229–237.

SEIFERT, W. K., and MOLDOWAN, J. M. (1981) Paleoreconstruction by biological markers. *Geochimica et Cosmochimica Acta,* Vol. 45, p. 783–794.

SEIFERT, W. K., and MOLDOWAN, J. M. (1986) Use of biological markers in petroleum exploration. In: *Methods in Geochemistry and Geophysics* (R. B. Johns, ed.) Vol. 24, p. 261–290.

SEIFERT, W. K., MOLDOWAN, J. M., SMITH, G. W., and WHITEHEAD, E. V. (1978) First proof of a C_{28}-pentacyclic triterpane in petroleum. *Nature,* Vol. 271, p. 436–437.

SEIFERT, W. K., MOLDOWAN, J. M., and JONES, R. W. (1980) Application of biological marker chemistry to petroleum exploration. *Proceedings of the Tenth World Petroleum Congress,* Bucharest, Romania. September 1979. Paper SP8, Heyden, p. 425–440.

SEIFERT, W. K., CARLSON, R. M. K., and MOLDOWAN, J. M. (1983) Geomimetic synthesis, structure assignment, and geochemical correlation application of monoaromatized petroleum steranes. In: *Advances in Organic Geochemistry 1981* (M. Bjorøy et al., eds.) J. Wiley and Sons, New York, p. 710–724.

SEIFERT, W. K., MOLDOWAN, J. M., and DEMAISON, G. J. (1984) Source correlation of biodegraded oils. *Organic Geochemistry,* Vol. 6, p. 633–643.

SHI JI-YANG, MACKENZIE, A. S., ALEXANDER, R., EGLINTON, G., GOWAR, A. P., WOLFF, G. A., and MAXWELL, J. R. (1982) A biological marker investigation of petroleums and shales from the Shengli oilfield, the People's Republic of China. *Chemical Geology,* Vol. 35, p. 1–31.

SHIEA, J., BRASSELL, S. C., and WARD, D. M. (1990) Mid-Chain branched mono- and dimethyl alkanes in hot spring cyanobacterial mats: A direct biogenic source for branched alkanes in ancient sediments? *Organic Geochemistry,* Vol. 15, P. 223–231.

SIERRA, M. G., CRAVERO, R. M., LABORDE, M. A., and RUVEDA, E. A. (1984) Stereoselective synthesis of (+/−)-18,19-Dinor-13β(H),14α(H)-cheilanthane: the most abundant tricyclic compound from petroleums and sediments. *Journal of the Chemical Society, Chemical Communications,* p. 417–418.

SIESKIND, O., JOLY, G., and ALBRECHT, P. (1979) Simulation of the geochemical transformation of sterols: Superacid effects of clay minerals. *Geochimica et Cosmochimica Acta,* Vol. 43, p. 1675–1679.

SILVERMAN, S. R., (1971) Influence of petroleum origin and transformation on its distribution and redistribution in sedimentary rocks. *Proceedings of the Eighth World Petroleum Congress,* Applied Science Publication, p. 47–54.

SILVERMAN, S. R., and EPSTEIN, S. (1958) Carbon isotopic compositions of petroleums and other sedimentary organic materials. *American Association of Petroleum Geologists Bulletin,* Vol. 42, p. 998–1012.

SIMONEIT, B. R. T. (1986) Cyclic terpenoids of the geosphere. In: *Biological Markers in the Sedimentary Record.* (R. B. Johns, ed.) Elsevier, New York, p. 43–99.

SINNINGHE DAMSTÉ, J. S., and DE LEEUW, J. W. (1990) Analysis, structure and geochemical significance of organically-bound sulphur in the geosphere: State of the art and future research. *Organic Geochemistry,* Vol. 16, p. 1077–1101.

SINNINGHE DAMSTÉ, J. S., DE LEEUW, J. W., DALEN, A. C. K., DE ZEEUW, M. A., DELANGE, F., RIJKSTRA, I. C., and SCHENCK, P. A. (1987) The occurrence and identification of series of organic sulfur compounds in oils and sediment extracts. I. A study of Rozel Point oil (U.S.A.) *Geochimica et Cosmochimica Acta,* Vol. 51, p. 2369–2391.

SINNINGHE DAMSTÉ, J. S., RIJKSTRA, I. C., DE LEEUW, J. W., and SCHENCK, P. A. (1988) Origin of organic sulfur compounds and sulfur-containing high molecular weight substances in sediments and immature crude oils. *Organic Geochemistry,* Vol. 13, p. 593–606.

SINNINGHE DAMSTÉ, J. S., VAN KOERT, E. R., KOCK-VAN DALEN, DE LEEUW, J. W., and SCHENCK, P. A. (1990) Characterization of highly branched isoprenoid thiophenes occurring in sediments and immature crude oils. *Organic Geochemistry,* Vol. 14, p. 555–567.

SMITH, H. M. (1968) Qualitative and quantitative aspects of crude oil composition. *U.S. Bureau of Mines Bulletin 642,* p. 136.

SMITH, G. W., FOWELL, D. T., and MELSOM, B. G. (1970) Crystal structure of 18α(H)-oleanane. *Nature,* Vol. 219, p. 355–356.

SNOWDON, L. R. (1980) Resinite—A potential petroleum source in the Upper Cretaceous/Tertiary of the Beaufort Mackenzie Basin. In: *Facts and Principles of World Petroleum Occurrence.* (A. D. Miall, ed.) Canadian Society of Petroleum Geologists Memoir 6, p. 509–521.

SOFER, Z. (1984) Stable carbon isotope compositions of crude oils: application to source depositional environments and petroleum alteration. *American Association of Petroleum Geologists Bulletin,* Vol. 68, p. 31–49.

SOFER, Z. (1988) Biomarkers and carbon isotopes of oils in the Jurassic Smackover Trend of the Gulf Coast States, U.S.A. *Organic Geochemistry,* Vol. 12, p. 421–432.

SOFER, Z., ZUMBERGE, J. E., and LAY, V. (1986) Stable carbon isotopes and biomarkers as tools in understanding genetic relationship, maturation, biodegradation, and migration in crude oils in the Northern Peruvian Oriente (Maranon) Basin. *Organic Geochemistry,* Vol. 10, p. 377–389.

SOLDAN, A. L., and CERQUEIRA, J. R. (1986) Effects of thermal maturation on geochemical parameters obtained by simulated generation of hydrocarbons. *Organic Geochemistry,* Vol. 10, p. 339–345.

STACH, E., MACKOWSKY, M.-TH., TEICHMÜLLER, M., TAYLOR, G. H., CHANDRA, D., and TEICHMÜLLER, R. (1982) *Coal Petrology*. Gebrüder Borntraeger, Berlin, 535 p.

STAHL, W. J. (1978) Source rock-crude oil correlation by isotopic type-curves. *Geochimica et Cosmochimica Acta,* Vol. 42, p. 1573–1577.

STAPLIN, F. L. (1969) Sedimentary organic matter, organic metamorphism, and oil and gas occurrence. *Canadian Petroleum Geologists Bulletin,* Vol. 17, p. 47–66.

STEEN, A. (1986) Gas chromatographic/mass spectrometric (GC/MS) analysis of C_{27}-C_{30}-steranes. *Organic Geochemistry,* Vol. 10, p. 1137–1142.

STRACHAN, M. G., ALEXANDER, R., SUBROTO, E. A., and KAGI, R. I. (1989) Constraints upon the use of 24-ethylcholestane diastereomer ratios as indicators of the maturity of petroleum. *Organic Geochemistry,* Vol. 14, p. 423–432.

STRONG, D., and FILBY, R. H. (1987) Vanadyl porphyrin distribution in the Alberta oil-sand bitumens. In: *Metal Complexes in Fossil Fuels* (R. H. Filby and J. F. Branthaver, eds.) American Chemical Society Symposium Series 344, Washington D.C., p. 154–172.

SUMMONS, R. E., and CAPON, R. J. (1988) Fossil steranes with unprecedented methylation in ring A. *Geochimica et Cosmochimica Acta,* Vol. 52, p. 2733–2736.

SUMMONS, R. E., and CAPON, R. J. (1991) Identification and significance of 3β-ethyl steranes in sediments and petroleum. *Geochimica et Cosmochimica Acta,* Vol. 55, p. 2391–2395.

SUMMONS, R. E., and JAHNKE, L. L. (1992) Hopenes and hopanes methylated in ring A: Correlation of the hopanoids from extant methylotrophic bacteria with their fossil analogues. In: *Biological Markers in Sediments and Petroleum* (J. M. Moldowan, P. Albrecht, and R. P. Philp, eds.) Prentice Hall, Englewood Cliffs, N.J., p. 182–194.

SUMMONS, R. E., and POWELL, T. G. (1986) *Chlorobiaceae* in Palaeozoic sea revealed by biological markers, isotopes, and geology. *Nature,* Vol. 319, p. 763–765.

SUMMONS, R. E., and POWELL, T. G. (1987) Identification of arylisoprenoids in a source rock and crude oils: Biological markers for the green sulfur bacteria. *Geochimica et Cosmochimica Acta,* Vol. 51, p. 557–566.

SUMMONS, R. E., and WALTER, M. R. (1990) Molecular fossils and microfossils of prokaryotes and protists from Proterozoic sediments. *American Journal of Science,* Vol. 290-A, p. 212–244.

SUMMONS, R. E., VOLKMAN, J. K., and BOREHAM, C. J. (1987) Dinosterane and other steroidal hydrocarbons of dinoflagellate origin in sediments and petroleum. *Geochimica et Cosmochimica Acta,* Vol. 51, p. 3075–3082.

SUMMONS, R. E., BRASSELL, S. C., EGLINTON, G., EVANS, E., HORODYSKI, R. J., ROBINSON, N., and WARD, D. M. (1988a) Distinctive hydrocarbon biomarkers from fossiliferous sediment of the Late Proterozoic Walcott Member, Chuar Group, Grand Canyon, Arizona. *Geochimica et Cosmochimica Acta,* Vol. 52, p. 2625–2637.

SUMMONS, R. E., POWELL, T. G., and BOREHAM, C. J. (1988b) Petroleum geology and geochemistry of the Middle Proterozoic McArthur Basin, Northern Australia: III. Composition of extractable hydrocarbons. *Geochimica et Cosmochimica Acta,* Vol. 52, p. 1747–1763.

SUNDARARAMAN, P. (1985) High-performance liquid chromatography of vanadyl porphyrins. *Analytical Chemistry,* Vol. 57, p. 2204–2206.

SUNDARARAMAN, P. (1992) Comparison of natural and laboratory simulated maturation of vanadylporphyrins. In: *Biological Markers in Sediments and Petroleum* (J. M. Moldowan, P. Albrecht, and R. P. Philp, eds.) Prentice Hall, Englewood Cliffs, N.J., p. 313–319.

SUNDARARAMAN, P., and BOREHAM, C. J. (1991) Vanadyl 3-nor C_{30}DPEP: Indicator of depositional environment of a lacustrine sediment. *Geochimica et Cosmochimica Acta,* Vol. 55, p. 389–395.

SUNDARARAMAN, P., BIGGS, W. R., REYNOLDS, J. G., and FETZER, J. C. (1988a) Vanadylporphyrins, indicators of kerogen breakdown and generation of petroleum. *Geochimica et Cosmochimica Acta,* Vol. 52, p. 2337–2341.

SUNDARARAMAN, P., MOLDOWAN, J. M., and SEIFERT, W. K. (1988b) Incorporation of petroporphyrins into biomarker correlation problems. In: *Geochemical Biomarkers* (T. F. Yen and J. M. Moldowan, eds.) Harwood Academic Publ., New York, p. 373–382.

TALUKDAR, S., GALLANGO, O., and CHIN-A-LIEN, M. (1986) Generation and migration of hydrocarbons in the Maracaibo Basin, Venezuela: An integrated basin study. *Organic Geochemistry,* Vol. 10, p. 261–279.

TALUKDAR, S., GALLANGO, O., and RUGGIERO, A. (1988) Generation and migration of oil in the Maturin Subbasin, Eastern Venezuelan Basin. *Organic Geochemistry,* Vol. 13, p. 537–547.

TANNENBAUM, E., RUTH, E., HUIZINGA, B. J., and KAPLAN, I. R. (1986) Biological marker distribution in coexisting kerogen, bitumen and asphaltenes in Monterey Formation diatomite, California. *Organic Geochemistry,* Vol. 10, p. 531–536.

TAPPAN, H. N. (1980) *The Paleobiology of Plant Protists.* W. H. Freeman and Company, San Francisco, 1028 p.

TEGELAAR, E. W., DE LEEUW, J. W., DERENNE, S., and LARGEAU, C. (1989) A reappraisal of kerogen formation. *Geochimica et Cosmochimica Acta,* Vol. 53, p. 3103–3106.

TELNAES, N., and DAHL, B. (1986) Oil-oil correlation using multivariate techniques. *Organic Geochemistry,* Vol. 10, p. 425–432.

THOMAS, J. B., MANN, A. L., BRASSELL, S. C., and MAXWELL, J. R. (1989) 4-Methyl steranes in Triassic sediments: Molecular evidence for the earliest dinoflagellates. 14th International Meeting on Organic Geochemistry, Paris, Sept. 18–22, 1989, Abstract No. 177.

THOMPSON, K. F. M. (1983) Classification and thermal history of petroleum based on light hydrocarbons. *Geochimica et Cosmochimica Acta,* Vol. 47, p. 303–316.

TISSOT, B. P., and WELTE, D. H. (1984) *Petroleum Formation and Occurrence.* Springer-Verlag, New York, 699 p.

TISSOT, B., CALIFET-DEBYSER, Y., DEROO, G., and OUDIN, J. L. (1971) Origin and evolution of hydrocarbons in early Toarcian shales, Paris Basin, France. *American Association of Petroleum Geologists Bulletin,* Vol. 55, p. 2177–2193.

TISSOT, B. P., DURAND, B., ESPITALIÉ, J., and COMBAZ, A. (1974) Influence of the nature and diagenesis of organic matter in formation of petroleum. *American Association of Petroleum Geologists Bulletin,* Vol. 58, p. 499–506.

TISSOT, B. P., DEROO, G., and HOOD, A. (1978) Geochemical study of the Uinta Basin: Formation of petroleum from the Green River formation. *Geochimica et Cosmochimica Acta,* Vol. 42, p. 1469–1485.

TORNABENE, T. G. (1978) Non-aerated cultivation of *Halobacteriam cutirubrum* and its effects on cellular squalenes. *Journal of Molecular Evolution,* Vol. 11, p. 253–257.

TREIBS, A. (1936) Chlorophyll and hemin derivatives in organic mineral substances. *Angewandte Chemie,* Vol. 49, p. 682–686.

TRENDEL, J. M., RESTLE, A., CONNAN, J., ALBRECHT, P. (1982) Identification of a novel series of tetracyclic terpene hydrocarbons (C_{24}-C_{27}) in sediments and petroleums. *Journal of the Chemical Society, Chemical Communications,* p. 304–306.

TRENDEL, J., BUILHEM, J., CRISP, P., REPETA, D., CONNAN, J., and ALBRECHT, P. (1990) Identification of two C-10 demethylated C_{28} hopanes in biodegraded petroleum. *Journal of the Chemical Society, Chemical Communications,* p. 424–425.

TRIFILIEFF, S. (1987) Etude de la structure des fractions polaires de pétroles (résines et asphalténes) par dégradations chimiques sélectives. D.Sc. Thesis, U. Strasbourg, France, 206 p.

TRIFILIEFF, S., SIESKIND, O., and ALBRECHT, P. (1992) Biological markers in petroleum asphaltenes: Possible mode of incorporation. In: *Biological Markers in Sediments and Petroleum* (J. M. Moldowan, P. Albrecht, and R. P. Philp, eds.) Prentice Hall, Englewood Cliffs, N. J., p. 350–369.

TSUDA, K., HAYATSU, R., KISHIDA, Y., and AKAGI, S. (1958) Steroid studies. VI. Studies of the constitution of sargasterol. *Journal of the American Chemical Society,* Vol. 80, p. 921–925.

TYSON, R. V., and PEARSON, T. H. (1991) Modern and ancient continental shelf anoxia: an overview. In: *Modern and Ancient Continental Shelf Anoxia* (R. V. Tyson and T. H. Pearson, eds.) Geological Society Special Publication No. 58, p. 1–24.

UDO, O. T., and EKWEOZOR, C. M. (1990) Significance of oleanane occurrence in shales of the Opuama Channel Complex, Niger delta. *Energy and Fuels,* Vol. 4, p. 248–254.

UNGERER, P. (1990) State of the art of research in kinetic modelling of oil formation and expulsion. *Organic Geochemistry,* Vol. 16, p. 1–25.

VALISOLALAO, J., PERAKIS, N., CHAPPE, B., and ALBRECHT, P. (1984) A novel sulfur containing C_{35} hopanoid in sediments. *Tetrahedron Letters,* Vol. 25, p. 1183–1186.

VAN DORSSELAER, A. (1974) Triterpenes de sediments. Ph.D. Thesis, U. Strasbourg, France, 113 p.

VAN DORSSELAER, A., ALBRECHT, P., and OURISSON, G. (1977) Identification of novel 17α(H)-hopanes in shales, coals, lignites, sediments, and petroleum. *Bulletin de la Societe Chimique de France,* No. 1–2, p. 165–170.

VAN GRAAS, G. W. (1990) Biomarker maturity parameters for high maturities: Calibration of the working range up to the oil/condensate threshold. *Organic Geochemistry,* Vol. 16, p. 1025–1032.

VAN GRAAS, G., BAAS, J. M. A., DE GRAAF, V., and DE LEEUW, J. W. (1982) Theoretical organic geochemistry. I. The thermodynamic stability of several cholestane isomers calculated by molecular mechanics. *Geochimica et Cosmochimica Acta,* Vol. 46, p. 2399–2402.

VASSOEVICH, N. B., AKRAMKHODZHAEV, A. M., and GEODEKYAN, A. A. (1974) Principal zone of oil formation. In: *Advances in Organic Geochemistry 1973,* (B. Tissot and F. Bienner, eds.) Technip, Paris, p. 309–314.

VENKATESAN, M. I. (1989) Tetrahymanol: Its widespread occurrence and geochemical significance. *Geochimica et Cosmochimica Acta,* Vol. 53, p. 3095–3101.

VILLAR, H. J., PUTTMAN, W., and WOLF, M. (1988) Organic geochemistry and petrography of Tertiary coals and carbonaceous shales from Argentina. *Organic Geochemistry,* Vol. 13, p. 1011–1021.

VLIERBLOOM, F. W., COLLINI, B., and ZUMBERGE, J. E. (1986) The occurrence of petroleum in sedimentary rocks of the meteor impact crater at Lake Siljan, Sweden. *Organic Geochemistry,* Vol. 10, p. 153–161.

VOLKMAN, J. K. (1986) A review of sterol markers for marine and terrigenous organic matter. *Organic Geochemistry,* Vol. 9, p. 84–99.

VOLKMAN, J. K. (1988) Biological marker compounds as indicators of the depositional environments of petroleum source rocks. In: *Lacustrine Petroleum Source Rocks* (A. J. Fleet, K. Kelts, and M. R. Talbot, eds.) Geological Society Special Publication No. 40, p. 103–122.

VOLKMAN, J. K., and MAXWELL, J. R. (1986) Acyclic isoprenoids as biological markers. In: *Biological Markers in the Sedimentary Record* (R. B. Johns, ed.) Elsevier, New York, p. 1–42.

VOLKMAN, J. K., GILLAN, F. T., JOHNS, R. B., and EGLINTON, G. (1981) Sources of neutral lipids in a temperate intertidal sediment. *Geochimica et Cosmochimica Acta,* Vol. 45, p. 1817–1828.

VOLKMAN, J. K., ALEXANDER, R., KAGI, R. I., NOBLE, R. A., and WOODHOUSE, G. W. (1983a) A geochemical reconstruction of oil generation in the Barrow Sub-basin of Western Australia. *Geochimica et Cosmochimica Acta,* Vol. 47, p. 2091–2106.

VOLKMAN, J. K., ALEXANDER, R., KAGI, R. I., and WOODHOUSE, G. W. (1983b) Demethylated hopanes in crude oils and their application in petroleum geochemistry. *Geochimica et Cosmochimica Acta,* Vol. 47, p. 785–794.

VOLKMAN, J. K., ALEXANDER, R., KAGI, R. I., and RULLKÖTTER, J. (1983c) GC-MS characterization of C_{27} and C_{28} triterpanes in sediments and petroleum. *Geochimica et Cosmochimica Acta*, Vol. 47, p. 1033–1040.

VOLKMAN, J. K., BANKS, M. R., DENWER, K., and AQUINO NETO, F. R. (1989) Biomarker composition and depositional setting Tasmanite oil shale from northern Tasmania, Australia. 14th International Meeting on Organic Geochemistry, Paris, September 18–22, 1989, Abstract No. 168.

VOLKMAN, J. K., KEARNEY, P., and JEFFREY, S. W. (1990) A new source of 4-methyl and 5α(H)-stanols in sediments: prymnesiophyte microalgae of the genus *Pavlova*. *Organic Geochemistry,* Vol. 15, p. 489–497.

WAKEHAM, S. G., SCHAFFNER, C., GIGER, W., BOON, J. J., and DE LEEUW, J. W. (1979) Perylene in sediments from the Namibian Shelf. *Geochimica et Cosmochimica Acta,* Vol. 43, p. 1141–1144.

WALDO, G. S., CARLSON, R. M. K., MOLDOWAN, J. M., PETERS, K. E., and PENNER-HAHN, J. E. (1991) Sulfur speciation in heavy petroleums: Information from X-ray absorption near-edge structure. *Geochimica et Cosmochimica Acta,* Vol. 55, p. 801–814.

WALTERS, C. C., and CASSA, M. R. (1985) Regional organic geochemistry of offshore Louisiana. *Transactions Gulf Coast Association Geological Society,* Vol. 35, p. 277–286.

WANG, T.G., and SIMONEIT, B. R. T. (1990) Organic geochemistry and coal petrology of Tertiary brown coal in the Zhoujing mine, Baise Basin, South China. 2. Biomarker assemblage and significance. *Fuel,* Vol. 69, p. 12–20.

WAPLES, D.W., and MACHIHARA, T. (1991) *Biomarkers for Geologists*. American Association of Petroleum Geologists, Methods in Exploration Series, No. 9, 91 p.

WARBURTON, G. A., and ZUMBERGE, J. E. (1982) Determination of petroleum sterane distributions by mass spectrometry with selective metastable ion monitoring. *Analytical Chemistry,* Vol. 55, p. 123–126.

WARDROPER, A. M. K., HOFFMANN, C. F., MAXWELL, J. R., BARWISE, A. J. G., GOODWIN, N. S., and PARK, P. J. D. (1984) Crude oil biodegradation under simulated and natural conditions-II. Aromatic steroid hydrocarbons. *Organic Geochemistry,* Vol. 6, p. 605–617.

WATSON, J. T. (1985) Gas chromatography/mass spectrometry. American Chemical Society Audio Course, p. 126.

WELTE, D. H., and YALCIN, M. N. (1987) Formation and occurrence of petroleum in sedimentary basins as deduced from computer aided modelling. In: *Petroleum Geochemistry and Exploration in the Afro-Asian Region*. (R. K. Kunar, P. Dwiwedi, V. Banerjie, and V. Gupta, eds.) Balkema, Rotterdam, p. 17–23.

WELTE, D. H., and YALCIN, M. N. (1988) Basin modelling—A new comprehensive method in petroleum geology. *Organic Geochemistry,* Vol. 13, p. 141–151.

WESTON, R. J., PHILP, R. P., SHEPPARD, C. M., and WOOLHOUSE, A. D. (1989) Sesquiterpanes, diterpanes and other higher terpanes in oils from the Taranaki Basin of New Zealand. *Organic Geochemistry,* Vol. 14, p. 405–421.

WHITEHEAD, E. V. (1973) Molecular evidence for the biogenesis of petroleum and natural gas. In: *Proceedings of Symposium on Hydrogeochemistry and Biogeochemistry* (E. Ingerson, ed.) Vol. II, The Clarke Co., p. 158–211.

WHITEHEAD, E. V. (1974) The structure of petroleum pentacyclanes. In: *Advances in Organic Geochemistry 1973* (B. Tissot and F. Bienner, eds.) Éditions Technip, p. 225–243.

WILLIAMS, L. A. (1984) Subtidal stromatolites in Monterey Formation and other organic-rich rocks as suggested contributors to petroleum formation. *American Association of Petroleum Geologists Bulletin,* Vol. 68, p. 1879–1893.

WILLIAMS, J. A., BJORØY, M., DOLCATER, D. L., and WINTERS, J. C. (1986) Biodegradation in South Texas Eocene oils—Effects on aromatics and biomarkers. *Organic Geochemistry,* Vol. 10, p. 451–461.

WINTERS, J. C., and WILLIAMS, J. A. (1969) Microbiological alteration of crude oil in the reservoir. American Chemical Society, Division of Petroleum Chemistry, New York Meeting Preprints, 14(4), E22–E31.

WITHERS, N. (1983) Dinoflagellate sterols. In: *Marine Natural Products 5* (P. J. Scheuer, ed.) Academic Press, p. 87–130.

WOESE, C. R., MAGRUM, L. J., and FOX, G. E. (1978) Archaebacteria. *Journal of Molecular Evolution,* Vol. 11, p. 245–252.

WOLFF, G. A., LAMB, N. A., and MAXWELL, J. R. (1986) The origin and fate of 4-methyl steroid hydrocarbons I. 4-methyl sterenes. *Geochimica et Cosmochimica Acta,* Vol. 50, p. 335–342.

YON, D. A., MAXWELL, J. R., and RYBACH, G. (1982) 2,6,10-trimethyl-7-(3-methylbutyl)-dodecane, a novel sedimentary biological marker compound. *Tetrahedron Letters,* Vol. 23, p. 2143–2146.

ZENG, X., LIU, S., and MA, S. (1988) Biomarkers as source input indicators in source rocks of several terrestrial basins of China. In: *Geochemical Biomarkers* (T. F. Yen and J. Moldowan, eds.) Harwood Academic, Chur, p. 25–49.

ZHANG DAJIANG, HUANG DIFAN, and LI JINCHAO (1988) Biodegraded sequence of Karamay oils and semi-quantitative estimation of their biodegraded degrees in Junggar Basin, China. *Organic Geochemistry,* Vol. 13, p. 295–302.

ZUMBERGE, J. E. (1983) Tricyclic diterpane distributions in the correlation of Paleozoic crude oils from the Williston Basin. In: *Advances in Organic Geochemistry 1981* (M. Bjorøy et al., eds.) J. Wiley and Sons, New York, p. 738–745.

ZUMBERGE, J. E. (1984) Source rocks of the La Luna (Upper Cretaceous) in the Middle Magdalena Valley, Colombia. In: *Geochemistry and Source Rock Potential of Carbonate Rocks* (J. G. Palacas, ed.) American Association of Petroleum Geologists, Studies in Geology No. 18, p. 127–133.

ZUMBERGE, J. E. (1987a) Prediction of source rock characteristics based on terpane biomarkers in crude oils: A multivariate statistical approach. *Geochimica et Cosmochimica Acta,* Vol. 51, p. 1625–1637.

ZUMBERGE, J. E. (1987b) Terpenoid biomarker distributions in low maturity crude oils. *Organic Geochemistry,* Vol. 11, p. 479–496.

ZUNDEL, M., and ROHMER, M. (1985) Prokaryotic triterpenoids. 1. 3β-methylhopanoids from *Acetobacter sp.* and *Methylococcus capsulatus. European Journal of Biochemistry,* Vol. 150, p. 23–27.

Index

Abelsonite, 46
Abietane, *illus*. 172, *table* 211
Abietic acid, 146
Accretionary wedge oils, 132
Acid hydrolysis, 133
Acidification of ground rock sample, 133
Acyclic alkanes, 15, *table* 16
Acyclic biomarkers as indicators of biological input or depositional environment, *table* 141
Acyclic diterpanes, *illus*. 18, 45-47
Acyclic isoprenoids, 24, 35, 119, 151-54
Acyclic sesquiterpanes, *illus*. 18
Acyclic triterpanes, *illus*. 20
Adriatic Basin, Italy and Yugoslavia, 147, 189, 192, 228
  homohopane distribution, *illus*. 147
Aerobic degradation, 7-8, 252, *illus*. 261
Aerobic metabolism, 7
Aerobic organisms (aerobes), *illus*. 3, 137
Africa:
  Carbonate platform oils, 132
Agathis, 173
Age determinations from biomarkers, 273, *table* 274
Age of petroleum, 131-32
Akata Formation, 155
Akita Basin, Japan, 239
Alkylbenzenes, *illus*. 74, 253
Alaska:
  Point Barrow, oils, 132
  Prudhoe Bay, oils, 132
  Swanson River Unit, Cook Inlet, Miocene oil, *illus*. 145
Albert Shale, Moncton Basin:
  oils, 164
Albertite, 114
Algal organic matter, *table* 141-2, 164
Algonkian Chuar Shale, 160, *illus*. 183, 189
Aliphatic hydrocarbon, 13
Alkanes, 13, 15
  iso- or branched, 15
  n-, 15, *table* 16, *table* 141, *table* 211
  sigma bond, 11-12
  See also Paraffins
Alkenes, 13
  pi bond, 12
Alkylbenzenes, *illus*. 74, 253
Alkylcyclohexanes, *illus*. 70
Alkyldibenzothiophenes, 138, *table* 142
3-Alkyl steranes, 192-95, *illus*. 193, *illus*. 195
Alumina, 53
Alumina column chromatography, 52-54
  short-column chromatography, *illus*. 54

Alum shale, *illus.* 183
Amazon, rubber, 17
Amphipathic lipids, 37, *illus.* 37, *illus.* 42
Amplifier saturation, 97
Amposta crude oil, 152
Anadarko Basin, 208
Anaerobic degradation, 7-8, 252
Anaerobic metabolism, 7
Anaerobic organisms (anaerobes), *illus.* 3
Analyte, 63
Angiosperm (flowering plant), *table* 142, 155, 170, *table* 274
Angiosperm dammar resins, 169
Angola:
  Lower Cretaceous lacustrine source rock, *illus.* 145, *illus.* 183
  Pre-Salt source rock, 127
  Upper Cretaceous shale, *illus.* 183
Angolan oils, 132, *illus.* 151, *illus.* 181, *illus.* 267
Anhydrite, 138, 201
Animal bone, *illus.* 124
Anoxia, 7, 8, *table* 141-2, *illus.* 149
Anteisoparaffins, 24
Anthraxolite, 114
API gravity, 127, 136, 140, 219-20, *illus.* 238, 255
Araucariaceae, 173
Archaebacteria, 23, 24, 43, 44-45, *table* 141, *illus.* 149, 151-2, 154, *table* 274
Ardmore Basin, 208
Argentina:
  Tertiary coals and shales, 174
Argillites, 151
Aromatic fraction, *illus.* 41, 52, *illus.* 53, 56, *illus.* 57
Aromatic hopanoids, 197, 249-50
Aromatic hydrocarbons, 12-13, *illus.* 74, *table* 142
Aromatic steroids, 197, 245-49, 263-64
Aromatization parameter, *illus.* 226, 245-46
Artificial (laboratory) maturation, 112, 265, 271, *illus.* 272
Asia:
  cadinanes, bicadinanes, tricadinanes in oils, 168, *illus.* 168, 169
  oil, *illus.* 194-95
  rubber from southeast, 17
Asphalt:
  use in mummification process, 9
  Switzerland, 258
Asphaltenes, 113, *illus.* 129, 130, *illus.* 130, 148, 265
Asphaltic-rich zones, 122
Asymmetric carbon, *illus.* 27, 27-28, *illus.* 31, 31-32, *illus.* 33-34, 35, 39
Athabasca oil, 112, 113
  tar sand, 170
Atomic H/C, 49, *illus.* 134, *table* 134, 212
  versus O/C, 135, 215
Atomic O/C, estimation of, 136
Atomic orbitals, *illus.* 10, *illus.* 11, *illus.* 12
Atomic S/C, 136
Anthrasteroids, 204
Australia:
  botryococcane in petroleum, 154
  coorongite, 136, 154
  diterpanes in oils, 170
  high-wax oils and condensates, 118
  McArthur Basin, 1
  northwestern shelf oil, *illus.* 144
  terrestrial origin oil, 165
  tetracyclic diterpanes, 171
  Yallourn lignite in, 41

Bacteria, *table* 141-42
Bacterial CH_4, *illus.* 124
  See also Methane
Bacterial photosynthesis, 4
Bacteriochlorophylls, 4, 45, 149, 206
Bacteriohopane, 37, 42-43, 148
Bacteriohopanetetrol, *illus.* 36, *illus.* 37, 38, 41, 42, 43, 146, 175
Baker alumina oxide acid powder, 53
Bakken shale, *illus.* 124
Bangladesh:
  Tertiary oils, Surma Basin, 169
Barrow Island, Western Australia, 253
Barrow Subbasin, Australia:
  oils and rock extracts, 161
Baseline threshold, 105
Base peak, 68
Bazhenov Formation, *illus.* 129
Beatrice oil, 104, 105, 122, 128, *illus.* 165, 186, *illus.* 187
Belemnitella americana (*see* PDB)
Benchtop quadrupole GCMS, *table* 69, 81-82, 111, 197
Benthic algae, 179
Benzene, 13, *illus.* 13
Benzohopanes, *illus.* 74, 202, 204
Benzopyrenes, *illus.* 74, *illus.* 104, 105
Benzothiophenes, 138, *table* 142
Betulins, 155
Beyerane, 171, *illus.* 171-72, *table* 274
Bicadinanes, 168, *illus.* 168, *illus.* 169
  See also Cadinanes
Bicarbonate (HCO_3^-), 123, *illus.* 124
Bicyclic diterpanes, *illus.* 19, 170
  See also Tricyclic diterpanes
Bicyclic sesquiterpanes, *illus.* 18, 165-67, *table* 211
Bicyclic terpanes, 146
Big Horn Basin oils, Wyoming, 126
Biodegradation, 113, 115, *illus.* 117, 119, 252, *illus.* 254
  biomarker parameters, 255-65
  biomarkers resistance to, 9, *illus.* 254
  concepts, 252-53

effects of maturity and correlation determination, 264-65
nonbiomarker parameters, 253-55
Biodegraded petroleums, 115, *illus.* 117, 119, 190, *illus.* 254, *illus.* 259-60
Biological markers (*see* Biomarkers)
Biomarker analysis, 48-109
Biomarker biodegradation parameters, *illus.* 254, 255-65
aromatic steroids, 263-64
hopanes, 257-58
isoprenoid and other organic acids, 256
isoprenoids, 256
C_{28}-C_{34} 30-nor-17α(H)-hopanes, 262
25-norhopanes (10-desmethylhopanes), 258-62
porphyrins, 264
sterane versus hopane biodegradation, 262-63
steranes and diasteranes, 256-57
terpanes, 263
tricyclic terpanes (cheilanthanes), 263
Biomarker laboratory organization, 49-50
Biomarker Maturation Index (BMAI), 242, *illus.* 243
Biomarker maturity parameters, 221-27, *illus.* 226
aromatic steroids, 245-49
criteria for ideal, 220-21
monoaromatic hopanoids, 249-50
porphyrins, 250-51
steranes, 237-45
terpanes, 227-37
Biomarker parameters, 49, 68
correlation/source/depositional environment, 140-208
aromatic steroids and hopanoids, 197-204
porphyrins, 205-8
steranes, 178-97
terpanes, 143-78
Biomarker quantitation, 96
Biomarkers, 1, 5, 9
in biodegraded oil, *illus.* 41
GCMS to evaluate, 57, *illus.* 58
characteristics of, 2, *illus.* 6
examples of, 40-47
mass chromatograms for, 68, *illus.* 75-81
mass spectra for, 89, *illus.* 90-93
mass spectrometric fragmentations to monitor, 68, *illus.* 70-74
maturation to oil-generative window ranges, *illus.* 226
modes of GCMS analysis for, *table* 69
modifiers/nomenclature related to, *table* 26
oldest indigenous, 1
in Proterozoic rocks/oils, 2
resistance to weathering/biodegradation, 9
sterane/triterpane, 9
stereochemistry, 25, 34-35
usefulness of, 2, 5, 113, 115
Biomarker separation/analysis:
gas chromatography/mass spectrometry, 57-109
geochemistry support, 48-49
organization, 49-50
sample availability/quality/selection, 50-52
sample cleanup/separation, 52-57
Biopolymers, 168, *illus.* 168
Bioturbation, 7, 8
25,30-Bisnorhopane, 132
28,30-Bisnorhopane, *illus.* 20, *illus.* 72, 132, *table* 142, 156, 158-159, *illus.* 224, *table* 274
29,30-Bisnorhopane, *illus.* 144, 235-37
Bisnorlupanes, *table* 142, 178, *table* 211
Bis-phytane [1,1-bis(phytane)], *illus.* 22, 23
Bitumen, 7, 112, *table* 133
Bitumen ratio, 218
Bitumen/total organic carbon (TOC) (*see* Bitumen ratio)
Black Sea:
anoxic sediments from, 8
Black smoker, 252
BMAI (*see* Biomarker Maturation Index (BMAI))
BNH/(BNH + TNH), 237
(BNH + TNH)/Hopanes, 235-37
BHN/TNH ratio, 158-159
Boghead coals, 136, 154
Bohai Basin, China oil, *illus.* 183
Boscan crude oil, 206
Botryococcane, *illus.* 20, 23, 24, *illus.* 73, *illus.* 116, *table* 141, *illus.* 153, 154-55, *table* 211, *table* 274
Botryococcene, 154, *illus.* 155
Botryococcus braunii, illus. 116, 119, 135, *table* 141, 154
Brazil:
β-Carotane, 164
lacustrine oil, 132
petroleums, 186
Bristol, U.K., 207
Brockman activity II, 53
Brownlee Spheri-10 silica guard column, *illus.* 55
Bucomazi Formation, Angola, 132, *illus.* 183, 207
Butane, 14, *illus.* 14, 15
n-, *illus.* 15

C_{29}Ts (*see* Norneohopane)
C_{30}* (*see* Diahopane)
C_{29}Ts/(C_{29}17α(H)-hopane + C_{29}Ts), 162, *illus.* 163, 234
See also Norneohopane
C_3-pathway, 124
C_4-pathway, 124
CAD (*see* Collision-activated decomposition)

Cadinanes (mono-, bi-, tricadinanes), 165, *illus*. 168, 168-69
Cadinene oligomers, 168
CAI (*see* Conodont alteration index (CAI))
Calcite, 192
Calibration (GCMS), 66-68
California:
  bitumen from, 94
  Miocene in, 136
  oil from, 56, *illus*. 97, *table* 108
  petroleum, 104-105
Calvin Pathway (*see* C_3-pathway)
CAM Pathway (*see* C_4-pathway)
Cameroon, 151
Canada:
  Athabasca, 112, 113, 170
  Jeanne d'Arc Basin, 234
  Mackenzie Delta, 178
  Rankin Formation oils/bitumens, 88
Canyon Diablo (CD) Troilite, *table* 125
Cape Verde Rise, 215
Capillary columns, 59, 115, 206
  methylsilicone, *illus*. 166
Carbocation, 32, 34, *illus*. 34
Carbohydrate, 2
Carbon, 1, *illus*. 3, 10, 11, 122, *illus*. 123, *table* 125
  atomic orbitals, *illus*. 10,
  graphite, 2
  hybridized sp^2 atomic orbitals of, *illus*. 12
  hybridized sp^3 atomic orbitals of, *illus*. 11
  isotopes of, 122, *illus*. 123
Carbonate-cemented horizons, 122
Carbonate Platform Oils, 132
Carbonate rocks, 208, *illus*. 209
Carbonate source rocks, 119, 137, *table* 142, 175
  gammacerane, 159
Carbonate versus clastic source rocks, 190
Carbonate versus shale source rocks, 208-9
  characteristics of petroleum, *table* 209
Carbon cycle, *illus*. 3, 5, 126
Carbon dioxide (CO_2), *illus*. 3, 4, *illus*. 124
Carbon-hydrogen isotopic crossplots, 132
Carbon isotopes (*see* Stable carbon isotopes)
Carbonium ion (*see* Carbocation)
Carbon preference index (CPI), *table* 209, 212, 219
Carbon-sulfur bonds, 148
Carneros, California:
  oil from, 56, *illus*. 78-79, *table* 108, *illus*. 200, *illus*. 203, 232
β-Carotane, *illus*. 22, *illus*. 73, *illus*. 93, *table* 141, 164, *illus*. 165
Carotenoids, 4, *illus*. 149, 164
Carrier gas, 59, 60
Catagenesis, *illus*. 6, 9, 40, 45, 113, 135, 211
Catalysis (*see* Clay-mediated acid catalysis)
Caving, *table* 216
CD (*see* Canyon Diablo (CD) Troilite)
Cell membrane, *illus*. 42
Chad, Miandoum oil, *illus*. 166-167
Cheilanthanes (*see* Tricyclic terpanes)
Chelation, 207
Chemical ionization (CI), 61, 63, 206
China:
  Liaohoe Basin oils, extracts, 250
  northwest oils, 159
  Maoming Shale, 154
  salt lake sediments, 138
  Tertiary oils, 159
Chinese oil shales, 136
Chiral, 27, 28
Chirality, 27-29, *illus*. 29
Chlorins, 45
Chlorobiaceae, 206, 272
Chlorobium, 137
Chlorophyll, 4, 41, 45, *illus*. 46, *illus*. 149
Chloroplasts, 45
5β(H)-cholane, *illus*. 57, *illus*. 75-77, 96
Cholestane, *illus*. 21, 23, *illus*. 30, 43-45, *illus*. 92, 192
  homologous series of, *illus*. 193
Cholesterols, *illus*. 37, 41, 43-45, *illus*. 191
Chromatography column, 53, *illus*. 54
  See also Column chromatography
Chromatography (gas) (*see* Gas chromatography)
Chromatography (HPLC) (*see* High performance liquid chromatography)
Chrysophyte algae, *table* 142, 186
Chuar Formation, Arizona, 160, *illus*. 183, 189
CI (*see* Chemical ionization)
Cis-cis-trans-bicadinane, 168, *illus*. 168
Clay-mediated acidic catalysis, 162, 243-4
Clay-poor:
  carbonate sediments, *illus*. 137
  source rocks, 190
Closed-tube pyrolysis, 112, 265
Coal, *illus*. 3, 4, 41, 46, *illus*. 124, 135, 178
Coalification, 135
Coccolithophores, 185
Cold trapping, 59
Coelution, *illus*. 94, 119
Coinjection, 94, 96
Cold spot in transfer line, 105, *illus*. 106
Collision-activated decomposition (CAD), 86, 88, 89, 225
Collision-induced fragmentation (*see* Collision-activated decomposition (CAD))
Colombia, oils and seep oils, *illus*. 189, *illus*. 259
Colorado organic-rich shale, 136
Column bleed, *illus*. 97, 98, *illus*. 100
Column chromatography, 52-53
  adsorption of sample on alumina, 53

alumina column, *illus*. 54
preparation/loading of, 53-54
separation of oils and bitumens, *illus*. 53
Column overload, 98, *illus*. 101
Compound class distributions, 212
Compound separation, 59-60, *illus*. 60
Concentrations (biomarkers), *illus*. 41, *illus*. 231
Condensate, *illus*. 97, 114, 118, 190, *illus*. 226
Conifers, *table* 141, 170, 173-74
Connecticut:
Hartford Basin, black shales, 232
Conodont alteration index (CAI), 217
Contamination, 51, 52, *table* 216
Cook Inlet, Alaska:
Miocene oil, *illus*. 145
Cooper Basin, Australia, 205
Coorongite, 136, 154
Core, 51
Correlation, 270-71
concept, 111
biomarker parameters, 140-208
immature source rocks hydrous pyrolysis, 271
indigenous bitumen test, 270-71
nonbiomarker parameters, 115-140
oil-oil, 113-14
oil-source rock, 112-13
oil and source rock screening, 111
organic input and depositional environment, 114
solid bitumens, 114
Covalent bonds, 12
CPI (*see* Carbon preference index)
Crassulacean Acid Metabolism (CAM) (*see* C_4-pathway)
Cretaceous, *illus*. 124, 131, 132, 136, *table* 142, 155, 185, 215
Cupressaceae, 173
Cuttings, 51
Cyanobacteria, *illus*. 3, *table* 141
Cyclic alkanes, 15-17
Cyclic biomarkers as indicators of biological input or depositional environment, *table* 141-42
Cyclohexane, 15, *illus*. 16, 17
Cyclohexylalkanes (*see* Alkylcyclohexanes)
Cyclopentane, 15, *illus*. 16

***Dacrydium*, 173**
Dagong oil field, 239
Damar oil, *illus*. 153
Dammar resin, 169
Data problems (GCMS), 97-107
baseline threshold, 105
cold spots, 105, *illus*. 106
column bleed, 98, *illus*. 100
column overload, 98, *illus*. 101
data system overload, 98, *illus*. 102
defocusing by *n*-paraffins, 105, *illus*. 107
dirty ion source, 105
incorrect mass range monitored, 98, 103, *illus*. 103
ion sampling frequency, 98, *illus*. 99
maximum column temperature, 105
poor GC resolution, 97
poor signal-to-noise, *illus*. 97, 97-98
Data processing (GCMS), *illus*. 58, 66-68
software, 68
Data system overload, 98, *illus*. 102
Daughter mode GCMSMS, 89
Dead carbon, 136
Dead Sea asphalt deposits, 9
Deasphalting, 130
Decorah Formation, Iowa, 139, *illus*. 183
Deep Sea Drilling Project (DSDP), 133
Defocusing by *n*-paraffins, 105, *illus*. 107
Degraded aromatic diterpanes, 205
Denver Basin:
oil, 113
Deoxophylloerythroetioporphyrins (*see* DPEP porphyrins)
Depositional environment, 114-15, 126, 274-5
10-desmethylhopanes (*see* 25-Norhopanes)
Desulfovibrio, 137
Deuterated standards, 57
Deuterium, 57, 125, 132
Devonian age flagstone, *illus*. 165
Devonian pinnacle reefs, Western Canada, 112
Devonian vitrinite, 215
Dia/(Dia + Regular) monoaromatic steroids hydrocarbons, 201
Dia/(Dia + Regular) steranes (*see* Diasteranes/regular steranes)
Diacholestanes, *illus*. 21, *illus*. 92
Diagenesis, 5, *illus*. 6, 7, *illus*. 38-39, 45, 146
Diahopane [17α(H)-diahopane], *illus*. 75-77, 160, *illus*. 161, 162-64, *illus*. 163
Dialkylbenzenes, 253
Diamond, *illus*. 124
Diasteranes, *illus*. 71, *illus*. 169, *illus*. 191, 257
lack of in Dead Sea asphalt, 9
13α,17β(H), *illus*. 71
C_{27}-C_{28}-C_{29} Diasteranes (*see* Diasterane triangle)
Diasteranes [20S/(20S + 20R], 244-45
Diasteranes/regular steranes, 190-92, *table* 209, *illus*. 226, 242-44
Diasterane triangle, 190, *table* 191
Diasterenes, *table* 142, *illus*. 191
Diastereomer, 27, 28, 32, 148
Diatoms, 185
Dihydrophytol, 149

Dillinger Ranch Field:
oil, *illus*. 118
2,6-dimethyloctane, *illus*. 18, *illus*. 25
Dimethylphenanthrene, *illus*. 173, 205
Dinoflagellates, 114, *table* 142, 185, 196-7
Dinosteranes, *illus*. 71, *table* 142, 196-7, *table* 274
Dinosterol, 114
Dipentene (*see* Limonene)
Diploptene, 175, *illus*. 176
Diplopterol, 175, *illus*. 176
Direct combustion TOC method, 133
Diterpanes, 17, *illus*. 18
acyclic, *illus*. 18, 45-47
bicyclic, *illus*. 19
tetracyclic, *illus*. 19
tricyclic, *illus*. 19
Diterpenoids, 146
Douala Basin:
argillities, 151
Double bond, 12
DPEP/etio (*see* Etio/(Etio + DPEP))
DPEP porphyrins, 45, *illus*. 46, 205, 250
Drill stem tests (DST), 121
Drilling fluids, 51, 121, 218
Drimane, *illus*. 70, 143, 146, 165, *illus*. 167
Dry Piney Field, *illus*. 118
DSDP (*see* Deep Sea Drilling Project)
DST (*see* Drill stem tests)
Dwell time, *table* 69, 81, 98
Dysoxic, 148

East Berlin Formation black shale, 232
Eel River Basin, California:
bitumen from, 94, *illus*. 236, *illus*. 249
oils from, *illus*. 97, *illus*. 236, *illus* 249
petroleum, 104-105
Effective source rock, 212
Egypt:
mummification process, 9
Eh, 137, 146-47, 148, 192, 207
Electrostatic mass analyzer (sector), *illus*. 64
Electronic zero level, 97
Electron impact (EI) ionization, 61-62, 206
Electron multiplier, *illus*. 58, 59, 62, 66, *illus*. 87
Elemental analysis, 135
sulfur content, 136-39
van Krevelen diagrams, types of organic matter, 135-36
Elko Formation, Nevada:
oils, 160
organic facies, 184-185
Enantiomer, 27, *illus*. 27, *illus*. 29, 29
Ent-isocopalane, *illus*. 20, 146
Eocene age Messel shale, 45, *illus*. 116, 206, 273
Eocene-Oligocene Elko Formation, Nevada, 184
Eocene rocks (California), 127 *illus*. 134
Epimer, 27, 197, *illus*. 227
Eromanga Basin, Australia, 205
Ergostane, *illus*. 21, 23, 192
Error analysis (GCMS), 107, *table* 108-9
Escanaba Trough, 268
Ethane, 15, *table* 16
Ether:hexane, 53
Ethylcholestane, *illus*. 41, *illus*. 193
Ethylene molecule, *illus*. 12
Etio/(Etio + DPEP), 250
nickel, *illus*. 226
vanadyl, *illus*. 226
Etioporphyrins, 45, *illus*. 46, 205
Eubacteria, 44-45
Eudesmane, *illus*. 18, *table* 142, 165-67, *table* 211
Eukaryotes, 24, *illus*. 42, 43-45, 206, 273
Euphotic zone, 2, 7
Euxinic sediments, 7, 137
Evaporite source rock, gammacerane, 159
Expulsion/migration of compounds, 112-13

Farnesane, 17, *illus*. 18
Fermentation, *illus*. 3, 4
Fichtelite, *illus*. 19, 170-71, *illus*. 172
FID (*see* Flame ionization detector)
Field ionization, 61, 63
Filletino oil, *illus*. 147
Fingerprint, 9, 68, *illus*. 97
Flame ionization detector (FID), *illus*. 61, 66, 213
Flight tube, *illus*. 64
Florida:
high diasteranes in clay-poor limestones, 192
Fossil fuel, 4
Fragmentogram (*see* Mass chromatogram)
France:
Toarcian, hydrogen index versus oxygen index, *illus*. 134
Fucosterol, 40
Full scan GCMS, 69, *table* 69, 81

Gallium, 46
Gamma-carotane, 164, *illus*. 165
Gammacerane, *illus*. 20, 28, *illus*. 29, *illus*. 71, *illus*. 75-77, 132, *table* 142, *table* 274
in asphalt, 9
Gammacerane index, 132, 150, *illus*. 151, 159-60
Gas (hydrocarbon), *illus*. 6, 211

Gas chromatogram, *illus*. 61, *illus*. 116-118, *illus*. 120, *illus*. 165
Gas chromatographic fingerprints, 68, 113, *illus*. 116-18, 119, *illus*. 120, 121-22
Gas chromatography (GC), 49, 51, 111, 115, *illus*. 116-18, 121
compound separation by, 59-60, *illus*. 60
solving production problems, 121
Gas chromatography-mass spectrometry (GCMS), 57
absolute biomarker quantitiation, 96
data problems, 97-109
functions of, 57, *illus*. 58, 59-61
gas chromatography in, 59-61
mass spectra/compound identification/quantitiation, 89-96
mass spectrometry in, 61-68
operating modes, 68-89
Gas chromatography/mass spectrometry/mass spectrometry (GCMSMS), 52, *table* 69, 82-83, *illus*. 83-85, 86-89
Gas/liquid chromatography (*see* Gas chromatography)
Gas-oil ratio (GOR), 219, 271
Gasoline-range hydrocarbons, 127
Gas-prone organic matter, 127, 128, *illus*. 134, *table* 134
GC (*see* Gas chromatography)
GC combustion-MS (*see* IRMGCMS)
GC-high resolution-MS, 52
GCMS (*see* Gas chromatography-mass spectrometry)
GC-MSD (*see* Benchtop quadrupole GCMS)
GCMSMS (*see* Gas chromatography/mass spectrometry/mass spectrometry)
Generation (*see* Oil-generative window)
Geochemistry:
biomarker interpretations, 49
support, 48-49
Geochromatography, 268
Germany:
Permian age Kuperschiefer sequence, 150
Toarcian shales, 148, 192, 201
Gilsonite, 114, 208
Gippsland Basin oils, 170, *illus*. 171-73, 228
Glamoč oil, *illus*. 147
Gloeocapsamorpha prisca, *illus*. 116, 119, 139, *table* 141
Grahamite, 114
Grand Canyon, Arizona, 160
Graphite (pure carbon), 2, 211
Great Basin, 160
Greece:
oils and seeps, 202
Prinos oil, *illus*. 144
Greenhorn Formation, 273
Greenland:
Tertiary, hydrogen index versus oxygen index, *illus*. 134
Green River Formation, Utah
abelsonite, 46
oil shale, 94, 136, 151, *illus*. 243
source rocks (Eocene), *illus*. 134
Green River marl and oils, 159, 164
Greenschist metamorphism, 9
Green sulfur bacteria, *table* 142
Guard column, *illus*. 55
Guatemala samples, 201
Guaymas Basin, 268
Gulf Coast, oils, 113, *illus*. 117, 119, 146, 150, 175, 220
Gulf of California:
anoxic versus oxic sediments from, 8
Gulf of Mexico (*see* Gulf Coast)
Gullfaks Field, Norwegian North Sea, 121
Guttenberg oil rock, 139
Gymnosperms, 170, *table* 274
Gypsum, 138

Halophilic bacteria, 24, 114, 154
Hamilton Dome, Wyoming:
oil from, 41, *illus*. 41, 56-57, 59, *illus*. 67, *illus*. 75-77, *table* 108, 147, *illus*. 193
Hanifa-Hadriya, Saudi Arabia:
oil-prone source rocks, 215
Hartford Basin, Connecticut, black shale, 232
Hatch-Slack Pathway (*see* C_4-Pathway)
Helium, 59
Hemiterpanes, 17, *illus*. 18
Heteroatom content, 255
Hevea brasiliensis, 17
Hexacyclic hopanoids (hexahydrobenzohopanes) *illus*. 72, *table* 142, 176-77, *illus*. 177
Hexane, 15, 53-55
boat and chair conformations of, *illus*. 16
HI (*see* Hydrogen index)
High performance liquid chromatography (HPLC), 53, 54-56, 205, 206
High-resolution capillary columns, 59
High signal-to-noise output, 59
High-sulfur crude oils, 137
High-temperature gas chromatography/electron impact mass spectrometry (HTGC/EIMS), 206
High-temperature gas chromatography/field ionization mass spectrometry (HTGC/FIMS), 206
Histogram, *illus*. 216
Homodrimane, *illus*. 166-7
Homohopanes, 35, *illus*. 75-77, 95, *illus*. 144-45, *illus*. 147, 227
index, 8, 146-49
isomerization, *illus*. 36, *illus*. 226, *illus*. 227, 227-28
Homologous series, *illus*. 15-16

Homopregnane, *table* 142
Hopane, 15, *illus*. 16, *illus*. 20, 35, *illus*. 36, *illus*. 67, 68, *illus*. 72, 95, *illus*. 177, *illus*. 226, 257-58, *illus*. 260-261
Hopane epimer ratios (*see* Homohopane isomerization)
Hopanoids, 41, 146, 176, 197, 249-250
  in prokaryotes, *illus*. 44
Hopenepolyols, *illus*. 179
HPLC (*see* High performance liquid chromatography)
HTGC/EIMS (*see* High-temperature gas chromatography/electron impact mass spectrometry)
HTGC/FIMS (*see* High-temperature gas chromatography/field ionization mass spectrometry)
Hump (*see* Unresolved complex mixture (UCM))
Hybrid mass spectrometers, 87
Hydrocarbon, 2, 7, 24, 211-212
  aliphatic, 13
  aromatic, 13, 24
  benzene, 13
  generative potential, *table* 133
Hydrogen, 1, *illus*. 30, 31, 59, *table* 125
Hydrogen atoms, 3-4, 57
Hydrogen isotopes, 57, 88, 122, *table* 125, 132
Hydrogenation, 40, *illus*. 93
Hydrogen index (HI), 133, *table* 134, 214
  versus oxygen index, *illus*. 134
Hydrogen sulfide, 4,7, 137, 207, 252
Hydrotroilite, 137
Hydrous pyrolysis, 112, *illus*. 272
  immature source rock, 271
Hydroxyl, 148
Hypersaline, 24, 132, *table* 141-42, 159-60, 230, 240-42

Iatroscan, 122
Immature organic matter, 212
INCOS computer system, 66, 96, 98
Indiana, 251
Indigenous bitumen, test for, 270-71
Indirect TOC method, 133
Indonesia:
  Mahakam oil, 152
  rubber, 17
Inertinite, 135, 136, 140
Internal standards, *illus*. 53, 56, *illus*. 57, *illus*. 75-77, *illus*. 78-79, *illus*. 80-81, *illus*. 84-85, 96
Ion beam, 63-5, *illus*. 64-65
Ion sampling frequency, 98, *illus*. 99
Ionization, *illus*. 58, 59, 60, *illus*. 61, 61, 63, *illus*. 87, 206
Ion source, dirty, 105
  See also Ionization
Iosene (*see* Phyllocladane)
Iowa, oil, *illus*. 116, *illus*. 183
IRMGCMS. Isotope-ratio-monitoring gas chromatography mass spectrometry, 273
Iron, 46, 137
Isobutane, *illus*. 15
ISCO Foxy programmable fraction collector, *illus*. 55
Isocyclic ring, 250
Isomerization, 32, 34, 40
Isomers, 15, 35, 185
Isopentenoids, 17
Isoparaffins, 24
Isopimarane, 170, *illus*. 171-72, *table* 211
Isoprene, 17, *illus*. 18, *illus*. 25
Isoprene rule, 17-25
Isoprenoid acids, 256
Isoprenoid alkanes, 23
Isoprenoid/*n*-paraffin ratios, 118-19
Isoprenoids, 17, *illus*. 25, 149-154, 256
  [1,1′-Bis(phytane)], *illus*. 73
  irregular, *illus*. 22, *illus*. 73, *table* 142
  regular, *illus*. 22, *illus*. 73
Isoprenoid side chains, 143
Isotope-ratio-monitoring gas chromatography mass spectrometry (IRMGCMS), 273
Isotopic signatures of biomarkers, 271-73
Isotopic standards, commonly used, *table* 125
Italy:
  bituminous shale, 139

Japan, 239
Jordanian oil, 162, *illus*. 163
Jurassic source rock, *illus*. 104, 104-5, 128, 136, 185, 186, 232

Katia oil, *illus*. 147, *illus*. 272
Kaurane, *illus*. 19, *illus*. 171-2, 173, *table* 211, *table* 274
Kelamayi oils, 164, 257
Kenai reservoir, Alaska:
  Miocene oil, *illus*. 145
Kerogen, *illus*. 3, 5, 7, 43, 126, 127, *illus*. 129, 135, 210-211
  definition, 5, 139
  Type I (very oil prone, *illus*. 134, 135-36, 218
  Type II (oil prone), *illus*. 134, 135, 136, 218
  Type III (gas prone), *illus*. 134, 135, 136, 218
  Type IV (inert), 135, 136, 218
Kerogen atomic H/C, 49, *illus*. 134, 212

Kimmeridge shale, *illus*. 124, 159, 186, *illus*. 223-24
Kinetics, 269-70
Kingak, 232, *illus*. 232
Kogolym, Middle Ob region, West Siberia:
oil, *illus*. 129
Korea Bay, Yellow Sea, 250
Kuperschiefer, Germany, 150

Labdane, *illus*. 19, 170, *illus*. 171-72
Lacustrine algae, 135, *illus*. 155
Lacustrine source rocks, 1, 97, 119, 138, *table* 141-2, 150, *illus*. 151, 152, *illus*. 153, *illus*. 183, *table* 211
Devonian, 104, 128
La Luna Formation, *illus*. 189
Land plants, *illus*. 124
LCMS (*see* Liquid chromatography-mass spectrometry)
Leaching, *illus*. 267
Liaohoe Basin, China, 250
Liassic-Triassic-sourced oils, 189
Lichens, 181
Light oils, 114
Lignite, 170, 171
Lignitic siltstone facies, 160
Limnic coals, 173
Limonene, *illus*. 25
Linked-scanning, 83, 86, 225
Linyi Basin, China, 233
Lipid bilayer, 42
Liptinite, 135, 136, 140
Liquid chromatography-mass spectrometry (LCMS), 206
particle-beam, 206
Lithology, 52, 192, 218
Long-chain dialkylthiacyclopentanes, 138
Low-sulfur crude oils, 137
Lupane, 178, 197, *table* 211
Lycopane, *illus*. 22, 152, 164
Lycopene, 164

C_{27}-C_{28}-C_{29} Monoaromatic steroids (*see* Monoaromatic steroid triangle)
m/z (*see* Mass/charge ratio (m/z))
m/z 239 fingerprint, 204
m/z 191 fingerprint (*see* Terpane fingerprint)
m/z 267 fingerprint, 204
MA(I)/MA(I + II), *illus*. 226, 246-47
MA-steroids (*see* Monoaromatic steroids)
Maceral, 5, 49, 130, 135, 139
analysis, 139-140
MacKenzie Delta, Canada:
petroleums, 178
Magnesium, 45
Magnetic mass spectrometers, 64
Magnets, 63, *illus*. 64
Mahakam Delta rocks, 227
Mahakam oil, Indonesia, 152
Malagasy (Madagascar), 261, 264
Manganese, 46
Maoming shale, 154
Maracaibo Basin, Venezuela:
oils and seep oils, *illus*. 189, 210
Marine calcareous, 208
Marine carbonate, *illus*. 124, *illus*. 184, 208, 275
Marine deltaic, 208
Marine evaporitic, 208
Marine Iabe-Landana source rock, 132
Marine petroleum, 119, 186-189, 196, *table* 211
Marine plants, *illus*. 124
Marine rocks, 1
Marine shale, 138, *illus*. 184, 275
Marine siliceous lithology, 208
Marine versus terrestrial input, 131, *table* 141-2, 156, 209-10, *table* 211
Marsh gas (*see* Methane (CH_4))
Mass analysis, *illus*. 58, 59, 63-65
Mass/charge ratio (m/z), 56, 57, 63, *illus*. 67, 68, *illus*. 70-74
Mass chromatogram, *illus*. 61, *illus*. 64, 66, *illus*. 67, 68, *illus*. 75-81, *illus*. 231, *illus*. 260
Mass fragmentogram (*see* Mass chromatogram)
Mass range monitoring, incorrect, 98-103, *illus*. 103
Mass spectrometric fragmentation, 68, *illus*. 70-74
Mass spectrometry, 61-68
chemical ionization, 63
data processing and calibration, 66-68
detection and scan analysis, 65-66
electron impact ionization, 61-62
field ionization, 63
ionization, *illus*. 58, 59, 61
mass analysis, 63-65, *illus*. 65
scan analysis, *illus*. 67
Mass spectra and compound identification, 89-95
Mass spectrum, 62, 64, *illus*. 64, 66, *illus*. 67, 89, *illus*. 90-93, *illus*. 95
cholestane, *illus*. 92, 122
Maturation, *illus*. 118, 119
biomarker maturity parameters, 221-51, *illus*. 226
concepts, 210-12
ideal biomarker maturity parameters, 220-21
nonbiomarker maturity parameters, 213-20, *table* 214
Mature organic matter, 212
Maximum column temperature too low, 105

Maui, New Zealand, oil, 170, 173
McArthur Basin, Australia, 1
McKittrick, California, 232
Meso-pristane, 27, *illus*. 28
Messel oil shale (Germany), 45, *illus*. 116, 206, 273
Metagenesis, 5, *illus*. 6, 9
Metalloporphyrin, 45
See also Porphyrins
Metastable reaction monitoring (MRM), 82-86, *table* 69, 156, *illus*. 177, *illus*. 181
Methane (CH_4), 2, *illus*. 3, 4, 11, *illus*. 11, 15, *table* 16, *illus*. 124, 126
Methanogenic bacteria (methanogens), *illus*. 3, 4, 24, *table* 142
Methanol (CH_3OH), 52
Methylalkylcyclohexanes, *illus*. 70
Methylbenzenes, *illus*. 74
Methylbutadiene (*see* Isoprene)
2-methylbutane, *illus*. 18
2-methyl,3-ethylheptane, *illus*. 18, *illus*. 25, *illus*. 31, *illus*. 34
Methylene chloride, 53
Methylhopanes, *table* 142, 178, *illus*. 179
Methylococcus capsulatus, 43, 196
Methylotrophic bacteria (methylotrophs) *illus*. 3, 178, *illus*. 179
Methylphenanthrenes, *illus*. 173, 205, 220
Methylpropane, *illus*. 15
Methyl radical, *illus*. 12
Methylsteranes, *illus*. 71, *illus*. 194-195, *table* 274
4-Methyl steranes, *table* 142, 196-97, *table* 211, *table* 274
Miandoum oil, *illus*. 166-67
Michigan:
Nonesuch Shale, White Pine copper mine, 222
Microscopy, 139, 140
MID (*see* Multiple ion detection)
Mid-chain monomethylakanes, *table* 141
Middle Ordovician rock, *illus*. 116, 119
organic matter, 126
Migration, 119, 220, *illus*. 243, 266-68
Migration/expulsion of compounds, 112-13
Minas Field, Sumatra:
oil, *illus*. 116, *illus*. 153
Miocene oil, *illus*. 145
Miocene rocks (California), 127, 136, 159
Mississippian Albert Shale:
β-carotane in oils, Moncton Basin, 164
Mitochondria, 45
M. Jurassic source rock, *illus*. 104, 104-5, 128
Mobile phase, 59, *illus*. 60
Modified direct TOC method, 133
Molasse Basin oils, 264
Molecular formula, 15
Molecular fossils, 1
Molecular ions, 62
Molecular sieves, 54, 56, *illus*. 107
Moncton Basin oils, 164
Monoaromatic hopanoids (MAH), 249-50
Monoaromatic steroid
aromatization, 245-46, *illus*. 245
Monoaromatic steroid
side chain scission, 246-7, *illus*. 247
Monoaromatic steroids, *illus*. 41, *illus*. 57
C-ring, *illus*. 74, *illus*. 78-79, 197-201, *illus*. 199-200
Monoaromatic steroid triangle, 197-201, *illus*. 198
Monoterpanes, 17, *illus*. 18, *illus*. 25
Monte Prena pyrolyzate, *illus*. 272
Monterey Formation, California, 112, 136, *illus*. 147, 148, *illus*. 183, 206
Monterey oils, *illus*. 124, *illus*. 147, 159
Montmorillonite, *illus*. 191, 268
Moray Firth, United Kingdom:
Beatrice oil from, *illus*. 104, 104, 128, 164, *illus*. 165
Moretane (C-22) isomerization, 228
Moretanes, 35, *illus*. 36, 228-29
Moretanes/hopanes
([βα/(αβ + ββ)-hopanes], 228-30
MRM GCMS (*see* Metastable reaction monitoring)
Multiple ion detection (MID), 68-81, *table* 69, 222
mass chromatogram for monoaromatic (MA-) steroids, *illus*. 78-79
mass chromatogram for saturate fraction biomarkers, *illus*. 75-77
mass chromatogram for triaromatic (TA-) steroids, *illus*. 80-81
Mummies:
use of asphalt, 9

***n*-alkane homologs, 15, *table* 16, *table* 141**
Nannocystis exedens, 43
Naphthenes, 15-17, 146, *table* 209
National Bureau of Standards (NBS) Oil Sample, *table* 125
Natural gas, *illus*. 124
NBS-22, *table* 125
Negative correlation, 111
Negative picking, 51
Neat sample, *illus*. 94
Neiber Dome, Wyoming, monoaromatic steroid distribution, *illus*. 200
Neutral loss mode GCMSMS, 89
Neutrons, 122, *illus*. 123
New Albany shale, Indiana, 251
New Zealand oils, 170
Nickel, 45
and vanadium content, 140, 205, 206-7

Niger Delta, 155, 235
Nigeria:
  crude oil, 89
Nitrogen, 52, 53, *table* 125, 181
Niigata Basin, Japan, 239
NMR (*see* Nuclear magnetic resonance)
Nonbiomarker biodegradation parameters, 253-55
  API gravity and heteroatom content, 255
  n-paraffin envelope, 253
Nonbiomarker maturity parameters, 213
  oils, 218-20
  rocks, 213-18
Nonbiomarker parameters for correlation/source/depositional environment 115-40
  elemental analysis, 135
    sulfur content, 136-39
    van Krevelen diagrams, organic matter type, 135-36
  gas chromatography, 115-18
    fingerprints, 119-21
    isoprenoid/*n*-paraffin ratios, 118-19
    pristane/phytane ratio, 118
    TLC-FID, 122
  maceral analysis, 139-40
  nickel and vanadium content, 140
  stable isotope ratios, 122-32
    stable carbon isotopes, 126
    sulfur and hydrogen isotopes, 132
  total organic carbon and pyrolysis, 132-4
Nonesuch Shale, White Pine copper mine, Michigan, 222
Nonmarine carbonate minerals, *illus*. 124
Nonmarine petroleum, 196, 210, *illus*. 211
Nonmarine shale, *illus*. 184, 275
Norcholestanes, 180-82, *illus*. 180, *illus*. 181, *illus*. 183
Norhopane, *table* 142
25-Norhopanes, *illus*. 72, 95, *illus*. 95, 258-62, *illus*. 260, *illus*. 261
30-Norhopanes, *illus*. 90-91, *table* 142, *illus*. 177
Norian/Rhaetian (Upper Triassic), 196
Norisopimarane, 170, *illus*. 171-72
Normal alkanes, 15, *table* 16
Normal paraffins (*see* Normal alkanes)
C_{28}-C_{34} 30-nor-17α(H)-hopanes, 262
30-Normoretane, *illus*. 75-77, *illus*. 90
Norneohopane [18α(H)-Norneohopanes (C_{29}Ts)], 160, *illus*. 161, 162-64, *illus*. 163, *illus*. 231
Nortetracyclane, *illus*. 171-72
North Sea:
  bitumens, *illus*. 187-88
  oil from, 95, *illus*. 95, *illus*. 124, 127, 164, *illus*. 187-88, *illus*. 260
  Jurassic source rock, 136, 275
North Somerset, U.K.:
  black shales, 196-97
n-paraffins, 24, 35, 105, *illus*. 107, 119, *illus*. 120
  See also Normal alkanes
n-paraffin envelope, 253
NSO-compounds (nitrogen/sulfur/oxygen), 113, *illus*. 129, *illus*. 130, 220
Nuclear magnetic resonance (NMR), 50, 89, 95, 160, 168
Nuclear membrane, *illus*. 42
Nugget Sandstone, *illus*. 107

Oceanic, 232
Odd:even preference (OEP), 219
OEP (*see* Carbon preference index)
OI (*see* Oxygen index (OI))
Oil, 7, *illus*. 41, *illus*. 124, 218-19
  API, CPI, OEP, 219
  elemental analysis, 135
Oil-generative window (oil window), *illus*. 6, *illus*. 120, 198, 212, 214, *illus*. 214, *illus*. 226
Oil-oil correlation, 113-14, 132, 270-73
Oil-prone organic matter, 127, 128, 132, 139
Oil samples, 52
  separation of bitumens and, *illus*. 53
Oil-source rock correlation, 112-13, 270-73
Oil and source rock screening, 111
  geochemical parameters:
    Rock-Eval, source rock thermal maturity, *table* 214
    source rock generative potential (quantity) *table* 133
    type of hydrocarbon generated (quality), *table* 134
Oklahoma:
  Pennsylvania age coal from, 173
Oleanane, *illus*. 71, *illus*. 75-77, 235, *table* 211, *table* 274
  18α(H), 89, *illus*. 94, 89, 94, *illus*. 94, 132, *table* 142, 155-6, *illus*. 226, *illus*. 236, *illus*. 259
  18β(H), 94, *illus*. 94, 155-56, *illus*. 157, *illus*. 226, *illus*. 236
Oleanane/C_{30} hopane, 155-56, *illus*. 184
Oleanane index, 155-56, 235
Oman oil, 131, 132, 150, *illus*. 231
Optical activity, 29-30
Opus computer system, 66
Orcadian Basin, Scotland, 275
Ordovician
  oils, *illus*. 124
  organic matter, 126, 131, *table* 141
  rock, *illus*. 116, 119
Organic acids, 256
Organic carbon, 2, *illus*. 3, 5
Organic chemistry, 10-40
Organic compounds, structures of, 14-15

Organic facies, 9, 111, 126
Organic input, 114-115
Organic-lean rocks, 51, 192
  Toarcian age, 51
Organic matter, *illus.* 3, 5, 135, *illus.* 137, 212
  in source rocks, *table* 133-34, *illus.* 134
Organic-rich coals, 268
Outcrop, 51
Overthrust Belt, Wyoming, 233
Oxic, 8, 137, 148, 192, 207
Oxicity, 146
Oxic versus anoxic, *illus.* 6, 7, 8
Oxic water column, 7
Oxidation-reduction (redox) reaction, 3, 4
Oxygen, 3, 7, 43, 52
Oxygen index (OI), 133, 214
  versus hydrogen index, *illus.* 134

Paleoenvironment, 150
Paleogene age oil, 132
Paleo-oceanographic controls, 9
Paleozoic oils, 113, *illus.* 183, 185
Palynomorph thickness variations, 217
Papua New Guinea:
  oils, *illus.* 161, 162, 190, *illus.* 191
Paraffins, 15, 24, 35
  See also Alkanes
Parent/daughter ions relationship, 82, *illus.* 83, *illus.* 87, *table* 88
Parent mode, *table* 69, 82-83, 87-89
  GCMSMS using C_{26} to C_{30} steranes, *table* 88
  metastable reaction monitoring (MRM), *table* 69
Pavlova, 196
PDB (*see* Peedee belemnite standard)
Peak broadening, *illus.* 101
Peak interference, 104, *illus.* 104
Peat, *illus.* 124
Peedee belemnite standard (PDB), *table* 125
Peedee Formation, *table* 125
Pentacyclic aromatics, 197
Pentacyclic triterpanes, *illus.* 20, 146, 175, *illus.* 177
Perhydro-β-carotane (*see* Carotane)
Perkin Elmer LC 55 UV detector, 55
Permian Phosphoria Formation, Wyoming:
  oils, *illus.* 118, 119, 136
Permian source rock, 205
Perylene, *illus.* 74, 204
Petapahan oil, *illus.* 153
Petrographic analysis of maceral composition (MOA), 49, 139-140, 173
Petroleum, *illus.* 3, 5, 7
  alumina column for sample separation, *illus.* 54
  alkanes in, 11-12, *illus.* 16
  aromatic hydrocarbons in, 13
  biological lipids, 30
  cyclopentyl or cyclohexyl carbon in, 15
  hopanes in, *illus.* 36
  monoaromatic steroid structures, *illus.* 199
  optical activity, 30
  organic origin of, 45
  porphyrins in, 45-47
  steranes in, *illus.* 38
  triaromatic steroids, *illus.* 202
Petroleum generation, 217, *illus.* 226
Petroleum geochemistry:
  biodegradation concept, 252-53
    biomarker biodegradation parameters, 255-65
    nonbiomarker biodegradation parameters, 253-55
  carbonate versus shale source rock, 208-9
  correlation concept, 111-15
    biomarker parameters for correlation/source/depositional environment, 140-210
    biomarker parameters for correlation/source/depositional environment, 115-40
  maturation concept, 210-212
    biomarker maturity parameters, 221-51
    ideal biomarker maturity parameters, criteria for, 220-21
    nonbiomarker maturity parameters, 213-20
Petroleums derived from organic matter input, *table* 211
Petroporphyrins, 205
pH, 192
Phacoides, 232
Phosphatic Member of Monterey Formation 136, 270
Phosphoria Formation, *illus.* 124
Photosynthesis, 2, *illus.* 3, 4, 45, 124
Photosynthetic pathways, 124
Phototrophs, 2, *table* 141
Phototrophic organisms, 2, 149, 160
Phthalates, 52
Phyllocladane, *table* 141, 170, *illus.* 171-72, *table* 274
Phytane, *illus.* 18, 23, 24, 35, 45, *illus.* 46, *illus.* 61, 149
Phytoclasts, 139
Phytol, 23, *illus.* 28, 35, *illus.* 149, 150
Phytoplankton (*see* Plankton)
Phytyl side chain, *illus.* 46, 149
PI (*see* Production index)
Pi bond, *illus.* 12, *illus.* 13
Pimarane, *illus.* 19, 170, *illus.* 172, *table* 211
Pinda Formation, 132
Pineview Field, Wyoming:
  paraffinic oil, Nugget Sandstone, *illus.* 107

Piper oil, 159, 186, *illus*. 187, *illus*. 223-24
Plankton, 24, 124, 136, 185
Plankton-fixed carbon, 124
PMP (*see* Porphyrin maturity parameter)
Podocarpaceae, 173
Podocarpane, *illus*. 172
Point Barrow, Alaska:
  oils, 132
Pokachev rock, *illus*. 129
Polar coordinate (star) diagrams, 122
Polycadinene, *illus*. 168
Polycyclic aromatic hydrocarbons, 52
Polynuclear aromatics, 105
Polysaccharides, 2, 17
Polyterpanes, 17, *illus*. 22
Ponca City crude, 24-25
Porphyrins, 8, 45-47, 148, 205-208, 250-51, 264
  distributions, 207-8
  maturity parameter (PMP), *illus*. 226, 250-51, *illus*. 251
Porphyrin-polar fraction, 54, *illus*. 54
Positive correlation, 111
Positive picking, 51-52
Postmature organic matter, 212
Potential source rock, 212
Pregnane, *table* 142
Precambrian organic matter, 131, 185, 189
Precision calculation from ten consecutive analyses, 107, *table* 108-09
Pre-Ordovician carbonate-sourced oils, 132
Pre-Salt source rocks (Angola), 127
Primary productivity, *illus*. 6, 8, 273
Prinos Field, Greece:
  oil, *illus*. 116, *illus*. 144
Pristane, *illus*. 18, 23, 24, *illus*. 28, 35, 45, *illus*. 46, *illus*. 61, *illus*. 149, 150
Pristane/phytane ratios, 8, 113, 118, 138, *table* 141, 149-51, *table* 211
Pristene, 150
Production index (PI), 214, 270
Prokaryote, 24, *illus*. 42, 43, *illus*. 44
Propane, 15
Propylcholestanes, *illus*. 21, *table* 142, *illus*. 193
Proterozoic rocks, 1-2, 24, *table* 274
Protons, 122, *illus*. 123
Protozoa, *table* 142
Prudhoe Bay, Alaska:
  oils, 132, 160, 162, *illus*. 243
Prymnesiophyte algae, 196
Pseudohomologous series, *illus*. 163
Pteridosperm, 205
Pyrite, 137
Pyrobitumen, 211
Pyrodictium brockii, 252
Pyrolysis, 213-15, 265
  total organic carbon (TOC) and, 132-34
  pyrogram, *illus*. 213
Pyrolyzates, *illus*. 116, 127, 271, *illus*. 272

Quadrupole mass analyzer, 65
Quadrupole mass spectrometers, *illus*. 65
Quadrupole rods, 63, 64, *illus*. 65, *illus*. 87
Qualitative fluorescence, 139
Quantitation (GCMS), 96
Quantitative estimate of oil co-sources, 127-28

Racemic mixture, 29
Raney nickel, 138, 139, 148, 193
Rankin Formation, Canada, 88
Ravni Kotari, Yugoslavia:
  oil, *illus*. 144-145, *illus*. 147
Reconstructed ion chromatogram (RIC), 66
Redox cycle, *illus*. 3
Redox potential, 146
Regular steranes/17α(H)-hopanes, 178-80, *table* 211
  See also Steranes, 17α(H)-hopanes
Repeat formation tests (RFT), 121
Reproducibility (*see* Error analysis)
Resolution (GC), 97
Resolution (MS), 65
Respiration, 2, *illus*. 3, 4
Response factor, 96
Retene, 170, *illus*. 173, 205
Retention time (GC), 96
Reversed circulation, 121
RFT (*see* Repeat formation tests)
Rhaetogonyaulax rhoetica, 197
Rhizobium, 181
Rhodo-DPEP, 46
Rhodo-etio, 46
Rhodo-porphyrins, 46
Rhuthenium tetroxide, 148
RIC (*see* Reconstructed ion chromatogram)
Rimuane, 170, *illus*. 171-72
R_0 (*see* Vitrinite reflectance)
Rock-Eval pyrolysis, 8, 49, 51, 111, *table* 133-34, 212, 213, *illus*. 213, *table* 214
Rock-Eval II plus TOC, 133
Rock extract (bitumen), 52
  South America, *illus*. 157
  Upper Hilliard shales, Wyoming, *illus*. 158
Rocks, 213
  bitumen (transformation) ratios, 218
  pyrolysis, 213-15
  thermal alteration index (TAI), 217-18
  van Krevelen diagrams, 218
  vitrinite reflectance, 215-17
Rock samples, 51-52
Rotoevaporation, 127
Rovesti oil, *illus*. 147
Rozel Point, Utah:
  lacustrine-sourced oils, 138
Rubber, 17, *illus*. 22
Ruhr area, Germany:
  paralic marine coals, 173

Russia:
Siva oil, 236
Timan-Pechora basin oil, 130, *illus.* 130
West Siberia, *illus.* 129, 136

Sabkha, *illus.* 199, 201
Sag River oil, *illus.* 232
Salym oil, *illus.* 129
Sample cleanup, separation, 52
column chromatography (cleanup), 53-54
high-performance liquid chromatography (separation) 54-56
internal standards and preliminary analyses, 56-57, 96
Sample compound quantitation formula, 96
Sample selection, 50-52
San Joaquin Basin, *illus.* 251
Santa Barbara, California:
Monterey Formation oil, *illus.* 183
Type II-S kerogens, 136
Santa Maria Basin, California:
Type II-S kerogens, 136
oils, *illus.* 138
Santa Maria-3 oil, Italy, *illus.* 177
Sarcinochrysidales, 186, *table* 274
Saturate-aromatic fraction, 53
alumina column, *illus.* 54
Saturated carbon atoms, 11
Saturated hydrocarbons, 12, *illus.* 70-73
methane, 15
Saturate fraction, *illus.* 41, 52
Saudi Arabia:
hydrogen index versus oxygen index, *illus.* 134
Jurassic age source rock, 136
Scan analysis, 65-66, *illus.* 67
Scanning, 64
Scan rate, 97-98, *illus.* 99
Scission (*see* Side chain cleavage)
Scotland:
Orcadian Basin, 275, torbanites, 136
Screening analysis, 48-49
oil-source rock screening, 111, *table* 133-34, *table* 214
17,21-Secohopane, *illus.* 20, 146, 175-76
See also Tetracyclic terpanes
Secohopanoids:
aromatic, *illus.* 74, 249-250
Secondary porosity, 256
Sedimentary porphyrins, 45
Sedimentary rocks, 135
Selected ion mode (SIM) (*see* Multiple ion detection)
Selected ion monitoring (SIM), *table* 69
See also Multiple ion detection (MID)
Selected metastable ion monitoring (SMIM), 82-83
Senonian limestone, 9
Sesquiterpanes, 17, 170
acyclic, *illus.* 18
bicyclic, *illus.* 18, 165-67, *table* 211
Sesterterpanes, 17, *illus.* 19, 146
SFC MS (*see* Supercritical fluid chromatography (SFC) mass spectrometry (MS))
Shengli oilfield, China, 222
Ship Shoal oil, Louisiana, *illus.* 243
Short-column chromatography, *illus.* 54, 87
See also Alumina column chromatography
Shublik shale, 232, *illus.* 232
Side chain cleavage reactions, 222, 225, *illus.* 226, 246-8, *illus.* 247-8
Sidewall core, 51
Sigma bond, 11-12, *illus.* 12, *illus.* 13, *illus.* 34
Signal-to-noise ratio, *illus.* 97, 97-98
Silicon, 11
Silurian age, 196
SIM (Selected ion mode) (*see* Multiple ion detection)
Single bond, 12
SMOW (*see* Standard Mean Ocean Water)
Solid bitumens, 114
Solubilization, *illus.* 267
Source rock, 2, *illus.* 6, 8-9, 50, 115
geochemical parameters:
generative potential (quantity), *table* 133
type of hydrocarbon generated (quality), *table* 134
thermal maturity, *table* 214
screening, 111
See also Oil and source rock screening
South America:
lacustrine oils, 132
oleananes, *illus.* 157
South Florida Basin, 174
Spanish oils, 152
Spiking, 94
Spirotriterpane, 89
Squalane, *illus.* 20, 23, *illus.* 73, *table* 142, 154
Squalene, *illus.* 44, 146, 154
Stable carbon isotopes, 49, 111, 113, 123-5, *illus.* 124, *table* 125, 126-32, 175
age determination and depositional environment, 131-32
marine versus terrestrial input, 131
oil and bitumen correlations, 126-27
quantitative estimates of oil cosources, 127-28
Stable carbon isotope type-curves, 128-31, *illus.* 129-30
Stable isotope ratios, 122-25
stable carbon isotopes, 126-32
standards and notation, 125-26
sulfur and hydrogen isotopes, 132

Standard Mean Ocean Water (SMOW), *table* 125
Stationary phase, 59, *illus*. 60, *illus*. 100
C_{30}-Sterane index, 132, 186-89
Sterane isomerization [20S/(20S + 20R); ββ/(ββ + αα], *illus*. 38-9, *illus*. 99, *illus*. 226, 237, 239-42, *illus*. 239, *illus*. 241, *illus*. 267
Steranes, 17, *illus*. 21, 25, *illus*. 26, 38-40, *illus*. 41, *illus*. 57, *illus*. 61, 68, *illus*. 70, *illus*. 84-85, *table* 108, *illus*. 169, 178-97, 237-45, 256, *illus*. 272
  carbon labeling system,, *illus*. 26
  C_{29}- and C_{30}- distribution, *illus*. 187-88
  fingerprinting, 68
  mass chromatograms for, *illus*. 61
  in sediment, *illus*. 38
  from sterols, *illus*. 39
C_{26} Steranes, 180-82, *illus*. 180
C_{27}-C_{28}-C_{29} Steranes (*see* Sterane triangle)
C_{30}/C_{27} to C_{30}) Steranes (*see* C_{30}-sterane index)
C_{30}-Steranes, 17, *illus*. 21, 156, 186-189, *illus*. 187-89, *illus*. 191, *illus*. 194-95, *table* 211
Sterane and terpane mass chromatograms for immature bitumen, *illus*. 223-24
Sterane triangle, *table* 114, 182-6, *illus*. 184
Steranes/17α(H)-hopanes, 178-180, *table* 211
Sterane versus hopane biodegradation, 262-63
Sterenes, 34, *illus*. 191, 193, 240, 243
Stereochemistry, 25, *illus*. 30, *illus*. 31,
  nomenclature (R, S, α, β), 30-32, *illus*. 33
Steroid internal standards, 56, *illus*. 57
Stereoisomer, 27, 32
Stereoisomerization, 32-40, *illus*. 34
  See also Isomerization
Steroids, 41
  in eukaryotes, *illus*. 44
Sterols, 38-39
  conversion to diasterenes, 190, *illus*. 191
  in eukaryotes, *illus*. 37, *illus*. 44
  in eukaryotic organism, *illus*. 38, *illus*. 42
Stigmastane, *illus*. 21, 23, *illus*. 41, 192
Strasbourg, France, 207
Structural nomenclature, notation, *illus*. 14, *illus*. 18-22, *illus*. 26
Suboxic, 148, 207
Sudan, waxy oils, 131
Sugar, glucose, 2
Sulfate reduction, sulfite oxidation (*see* Sulfur)
Sulfur, 4, 52, *table* 125, 136, *illus*. 137, *table* 209, *table* 211, 212, 252
Sulfur bacteria, 4
Sulfur content, 136-39
Sulfur isotope, 113, *table* 125, 132
  hydrogen and, 132
Supercritical fluid chromatography (SFC) mass spectrometry (MS), 206
Surma Basin oils, 169
Swab runs, 121
Swanson River Unit, Alaska:
  Miocene oil, Kenai reservoir, *illus*. 145
Sweden:
  L. Ord. alum shale, *illus*. 183
Symbionts, 181

TAI (*see* Thermal alteration index)
TA(I)/TA(I + II), 225, 247-48
TA/(MA + TA) (*see* Aromatization parameter)
Talang Akar, Indonesia, 271
Tandem magnetic instruments, 86
Tandem mass spectrometry (*see* Triple quadrupole)
Taranaki Basin, New Zealand:
  oil fields, 170, 173
Tarraco crude oil, 152
Tasmanites, 135, 140, *table* 142, 146, 170, 175
TA-steroids (*see* Triaromatic steroids)
TCD (*see* Thermal conductivity detector)
Ternary diagram (*see* Monoaromatic steroid triangle; Sterane triangle)
Terpane fingerprint, *illus*. 61, 68, *illus*. 97, 143-146, *illus*. 144-45
Terpanes, 35-38, *illus*. 61, 68, *illus*. 70, *illus*. 107, 143-78, 227-37, 263
  mass chromatograms for Oman-area oils, *illus*. 231
Terpenoids, 17, 23
Terrestrial organic matter, *illus*. 116, *table* 141-2, 150, *table* 211
  marine versus terrestrial input, 131, 156, *table* 211
Terrigenous higher plants, 155, 179
Tertiary Akata Formation marine shales, 155
Tertiary Niger delta crude oil, 155
Tertiary oils, 169, 185
Tertiary rock, 136, 159
Tetracyclic aromatics, 197
Tetracyclic diterpanes, *illus*. 19, 170, 171, 173-74, *table* 211
C_{24} Tetracyclic terpane, *illus*. 20, *table* 142
Tetracyclic terpanes, *illus*. 20, *illus*. 70, *illus*. 75-77, 143, 146, 175-76
Tetraedron, illus. 116
Tetrahymanol, 159, 160
Tetrapyrrole, 41, 45, *illus*. 46
  See also Porphyrins
Tetraterpanes, *illus*. 22
Thermal alteration index (TAI), 139, 212, 214, 217-18
Thermal conductivity detector (TCD), 214
Thermal maturity (*see* Maturation)
Thermal stress parameters, 212-214, 225
Thermoacidophilic bacteria, 24, 152

Thermochemical sulfate reduction, 212
Thermospray, 206
Thin-layer chromatography/flame ionization detection (TLC-FID), 122
Thiobacillus, 137
Thiophene-rich XANES, 139
Thioploca, 43, 159
TIC (Total ion chromatogram) (*see* Reconstructed ion chromatogram)
Timan-Pechora Basin, Russia:
  asphaltene-poor oils, 130-131, *illus*. 130
TLC-FID (*see* Thin-layer chromatography/flame ionization detection)
Tm (*see* Ts/(Ts + Tm))
T_{max}, 214, *illus*. 251
TNH/(TNH + hopanes), *illus*. 238
Toarcian Formation, Paris basin, 197
Toarcian shales, 148, 192, 201
TOC (*see* Total organic carbon)
Tocopherols, 24, 149
Torbanite, 136, 154
Total organic carbon (TOC), 49, 51, 111, 214
  and pyrolysis, 132-34, *table* 133
Transfer line, 60, 105
Trans-rubber, *illus*. 22
Transformation ratio (*see* Bitumen ratio)
Trans-trans-trans-bicadinane, 168, *illus*. 168
Trialkylbenzenes, 253
C_{26} Triaromatic steroid 20S/(20S + 20R), 248, *illus*. 249
C_{26}-C_{27}-C_{28} Triaromatic steroids, 201-2, *illus*. 203
Triaromatic (TA) steroids, *illus*. 41, *illus*. 57, *illus*. 74, *illus*. 80-81, *illus*. 102, 201, *illus*. 202, *illus*. 203, 225, 247-8
Triassic age, *table* 142, 196
Tricadinane, 168, *illus*. 168
  See also Cadinanes
Tricyclic diterpanes, *illus*. 19, 146, 170, *illus*. 171, *table* 211
  See also Bicyclic diterpanes
Tricyclics/17α(H)-hopanes, 143, 174-75, *illus*. 226, 230-32, *illus*. 231
C_{30} Tricyclic terpane, *illus*. 75-77, *illus*. 93
Tricyclic terpanes, *illus*. 20, *illus*. 72, 132, *table* 142, *illus*. 145, 146, 174-75, 263
  terrigenous indicators, 146
Tricyclohexaprenane, *illus*. 20
Tricyclohexaprenol, 146, 175
Trimethyldecylchromans, *table* 142
Trimethylnaphthalene, *illus*. 173, 205
Trinidad:
  oils, 119, *illus*. 120
Triple quadrupole, 83, 86-87, *illus*. 87, *illus*. 183
  GCMSMS results for diasteranes, 190
Triple sector quadrupole (TSQ) (*see* Triple quadrupole)
25,28,30-Trisnorhopane, *illus*. 20, *table* 142, 156, 158-159, 235-37
Trisnorhopane (Tm) (*see* Ts/(Ts + Tm))
Trisnorhopane II (Ts) (*see* Ts/(Ts + Tm))
Triterpanes, 17, *illus*. 20, 25
  acyclic, *illus*. 20
  carbon labeling system, *illus*. 26
  pentacyclic, *illus*. 20
  from prokaryote, *illus*. 44
  tetracyclic, *illus*. 20
  tricyclic, *illus*. 20, 146
Troilite, 137
Ts/C_{30} 17α(H)-hopane, 234-35
Ts/(Ts + Tm), *illus*. 72, 114, 162, 222, 225, *illus*. 226, 233-34, *illus*. 233
Turkey:
  sterane-rich seep oil from, *illus*. 101-102
Type of organic matter, 135-36
Type-curves, 128-31
Tyumen Formation, *illus*. 129

UCM (*see* Unresolved complex mixture)
Ultraviolet detector, *illus*. 55
Ultraviolet spectrophotometry, 206
United Kingdom:
  oil from, 104, *illus*. 104
Urea adduction, 56, *illus*. 107, *illus*. 153, 154
Unresolved complex mixture (UCM), 115, *illus*. 117, 253
Ursane skeletons, 197
Utah:
  abelsonite from Green River Formation, 46
  Green River oil shale, 94, *illus*. 134, 136
UV detector (*see* Ultraviolet detector)
UV-visible spectrophotometer, 54

Vanadium, 140, 205
Vanadyl ion, 45, 206-207
Van-Egan, W. Siberia:
  oil, *illus*. 183
Van Krevelen diagram, *illus*. 134, 135-36, 218
Venezuela:
  Cretaceous age source rock, 136
  oils and seep oils, *illus*. 189
Vitrinite, 135, 136, 140
Vitrinite reflectance (R_0), 49, 133, 139, 192, 212, 215-217, *illus*. 226
  histogram, *illus*. 216
  problems affecting interpretation of vitrinite reflectance, *table* 216
  relationship between Chevron TAI and, *table* 217
Voltziales, 173
V/Ni, 113
V/(V + Ni) porphyrins, 148, 206-7, *table* 211

Walvis Bay sediment, 204
Watchet, U.K., black shale, 196-97
Waters model 590 HPLC pump, 54, *illus*. 55
Water washing, 220
Weathering, 51, 217, 239
biomarkers resistance to, 9
Wellhead, 121
Wet-gas zone, 214, *illus*. 226
West Siberia:
Jurassic age source rock, 136
Kogolym oil, *illus*. 129
oils, *illus*. 238
Whatman Partisil 10 Silica column, 54
White Pine copper mine, Michigan, 222
Whole core, 51
Williston Basin, *illus*. 124, *illus*. 251
Wood, *illus*. 124
Wyoming:
biomarker classes in oil from, 41, *illus*. 41
organic-rich shale, 136
rock extract, *illus*. 157-58
standard oil from, 41, *illus*. 41, 56-57, 59, *illus*. 67, *illus*. 75-77, *illus*. 80-81, *table* 108, 147, *illus*. 193

XANES (*see* X-ray absorption near edge spectroscopy)
X-ray absorption near-edge spectroscopy (XANES), 139
X-ray diffraction crystallography, 50, 89

Yallourn lignite, 41
Yugoslavia:
Ravni Kotari-3 oil, *illus*. 145
seep oils, 258

Zhungeer Basin, 164, 257
Zooplankton, 136
Zostera, 210